Gallium Nitride (GaN) II

SEMICONDUCTORS
AND SEMIMETALS
Volume 57

Semiconductors and Semimetals

A Treatise

Edited by R. K. Willardson
    CONSULTING PHYSICIST
    SPOKANE, WASHINGTON

Eicke R. Weber
DEPARTMENT OF MATERIALS SCIENCE
AND MINERAL ENGINEERING
UNIVERSITY OF CALIFORNIA AT
BERKELEY

Gallium Nitride (GaN) II

SEMICONDUCTORS
AND SEMIMETALS

Volume 57

Volume Editors

JACQUES I. PANKOVE
DEPARTMENT OF ELECTRICAL ENGINEERING
UNIVERSITY OF COLORADO, BOULDER

THEODORE D. MOUSTAKAS
DEPARTMENT OF ELECTRICAL ENGINEERING
BOSTON UNIVERSITY

ACADEMIC PRESS
San Diego   London   Boston
New York   Sydney   Tokyo   Toronto

This book is printed on acid-free paper.

COPYRIGHT © 1999 BY ACADEMIC PRESS

ALL RIGHTS RESERVED.
NO PART OF THIS PUBLICATION MAY BE REPRODUCED OR TRANSMITTED IN ANY FORM OR BY ANY MEANS, ELECTRONIC OR MECHANICAL, INCLUDING PHOTOCOPY, RECORDING, OR ANY INFORMATION STORAGE AND RETRIEVAL SYSTEM, WITHOUT PERMISSION IN WRITING FROM THE PUBLISHER.

The appearance of the code at the bottom of the first page of a chapter in this book indicates the Publisher's consent that copies of the chapter may be made for personal or internal use of specific clients. This consent is given on the condition, however, that the copier pay the stated per-copy fee through the Copyright Clearance Center, Inc. (222 Rosewood Drive, Danvers, Massachusetts 01923), for copying beyond that permitted by Sections 107 or 108 of the U.S. Copyright Law. This consent does not extend to other kinds of copying, such as copying for general distribution, for advertising or promotional purposes, for creating new collective works, or for resale. Copy fees for pre-1999 chapters are as shown on the title pages; if no fee code appears on the title page, the copy fee is the same as for current chapters. 0080-8784/99 $30.00

ACADEMIC PRESS
525 B Street, Suite 1900, San Diego, CA 92101-4495, USA
1300 Boylston Street, Chestnut Hill, Massachusetts 02167, USA
http://www.apnet.com

ACADEMIC PRESS LIMITED
24–28 Oval Road, London NW1 7DX, UK
http://www.hbuk.co.uk/ap/

International Standard Book Number: 0-12-752166-6
International Standard Serial Number: 0080-8784

PRINTED IN THE UNITED STATES OF AMERICA
98 99 00 01 02 QW 9 8 7 6 5 4 3 2 1

Contents

PREFACE xi
LIST OF CONTRIBUTORS xv

Chapter 1 Hydride Vapor Phase Epitaxial Growth of III-V Nitrides
Richard J. Molnar

    LIST OF ACRONYMS AND ABBREVIATIONS 1
  I. INTRODUCTION 2
 II. NITRIDE HVPE GROWTH 4
    1. Reactor Design 5
    2. GaN Growth Kinetics and Thermochemistry 7
    3. Heteronucleation and Film Growth 10
    4. Doping of HVPE GaN 11
    5. Growth of Other III-V Nitrides by HVPE 13
 III. GaN FILM CHARACTERIZATION 15
    1. Film Morphology/Structure 15
    2. Electrical Properties 18
    3. Optical Properties 20
 IV. LIGHT-EMITTING DIODES 22
  V. HVPE FOR NITRIDE SUBSTRATES 24
    1. Defect Reduction 25
    2. Epitaxial Overgrowths 25
    3. Film/Substrate Separation 26
 VI. CONCLUSIONS 27
    REFERENCES 28

Chapter 2 Growth of III-V Nitrides by Molecular Beam Epitaxy
T. D. Moustakas

  I. INTRODUCTION 33
 II. BACKGROUND OF MOLECULAR BEAM EPITAXY TECHNIQUES 34
 III. NITROGEN SOURCES USED FOR THE GROWTH OF III-V NITRIDES
    BY MOLECULAR BEAM EPITAXY 40
    1. *Ammonia as a Nitrogen Source* 40

    2. *Nitrogen Plasma Sources* 41
    3. *Nitrogen Ion Sources* 50
IV. GaN FILMS 51
    1. *Film Growth* 51
    2. *Film Structure and Microstructure* 60
    3. *Doping Studies* 70
    4. *Electronic Structure and Optoelectronic Properties* 80
V. InGaAlN ALLOYS 89
    1. *Growth* 89
    2. *Phase Separation and Long-range Atomic Ordering* 89
VI. MULTIQUANTUM WELLS 101
    1. $In_xGa_{1-x}N/Al_yGa_{1-y}N$ *Multiquantum Wells* 101
    2. *GaN/AlGaN Multiquantum Wells* 105
VII. DEVICE APPLICATIONS 110
    1. *Device Processing* 110
    2. *Devices* 111
VIII. CONCLUSIONS 119
REFERENCES 121

Chapter 3   Defects in Bulk GaN and Homoepitaxial Layers

Zuzanna Liliental-Weber

I. INTRODUCTION 129
II. POLARITY OF THE CRYSTALS 130
III. DEFECT DISTRIBUTION 135
IV. NANOTUBES 140
V. PL AND POINT DEFECTS 142
VI. INFLUENCE OF ANNEALING 145
VII. LARGER–DIMENSION BULK GaN CRYSTALS 146
VIII. HOMOEPITAXIAL LAYERS 147
    1. *Influence of Polarity* 147
    2. *Pinholes and Growth Rate* 152
IX. SUMMARY 153
REFERENCES 154

Chapter 4   Hydrogen in III-V Nitrides

Chris G. Van de Walle and Noble M. Johnson

I. INTRODUCTION 157
    1. *General Features of Hydrogen in Semiconductors* 158
    2. *Presence of Hydrogen During Nitride Growth and Processing* 160
II. THEORETICAL FRAMEWORK 161
    1. *Theoretical Approaches Used for Hydrogen in Nitrides* 161
    2. *Isolated Interstitial Hydrogen in GaN* 164
    3. *Hydrogen Molecules* 168
    4. *Interaction of Hydrogen with Shallow Impurities* 168
III. EXPERIMENTAL OBSERVATIONS 175
    1. *Intentional Hydrogenation and Diffusion in GaN* 175
    2. *Effects of Hydrogen During Growth/Hydrogen on GaN Surfaces* . . . 176

3. *Passivation and Activation of Mg Acceptors*	. . . .	177
4. *Passivation of Other Acceptor Dopants*	. . . .	179
5. *Local Vibrational Modes of the Mg—H Complex in GaN.*	. . . .	180
IV. CONCLUSIONS AND OUTLOOK	. . . .	181
REFERENCES	. . . .	183

Chapter 5 Characterization of Dopants and Deep Level Defects in Gallium Nitride
W. Götz and N. M. Johnson

I. INTRODUCTION	185
II. MATERIALS PREPARATION	187
III. SHALLOW DOPANTS	187
1. *Hall-Effect Measurement*	187
2. *n-type, Si-doped GaN*	188
3. *p-type, Mg-doped GaN*	190
4. *Thickness Dependence of Electronic Properties for "Thick" GaN Films* . . .	193
IV. DEEP LEVEL DEFECTS	195
1. *Rectifying Devices and Capacitance Transient Methods*	195
2. *Conventional (Thermal) Deep Level Transient Spectroscopy*	197
3. *Optical Deep Level Transient Spectroscopy*	200
V. CONCLUSIONS	203
REFERENCES	205

Chapter 6 Stress Effects on Optical Properties
Bernard Gil

I. INTRODUCTION	209
II. THE CRYSTALLINE STRUCTURES OF III-NITRIDES	213
III. EFFECTS OF STRAIN FIELDS ON THE ELECTRONIC STRUCTURE OF WURTZITE III NITRIDES	213
1. *The Problem of the Determination of the Bandgap of GaN*	213
2. *Influence of Strain Fields on the Band Structure of III Nitrides*	217
3. *Deformation Potentials of Wurtzite GaN*	227
IV. EXCITON OSCILLATOR STRENGTHS AND LONGITUDINAL-TRANSVERSE SPLITTINGS	234
1. *Exchange Interaction and Strain-Induced Variations of the Oscillator Strengths for Γ_5 and Γ_1 Excitons*	234
2. *Strain-Induced Variations of the Longitudinal-Transverse Splittings* . . .	239
V. ORIGIN OF THE STRAIN	244
VI. PHONONS UNDER STRAIN FIELDS	254
VII. SHALLOW VS DEEP-LEVEL BEHAVIOR UNDER HYDROSTATIC PRESSURE	257
VIII. INFLUENCE OF STRAIN FIELDS ON OPTICAL PROPERTIES OF GaN-AlGaN QUANTUM WELLS	259
1. *Introduction*	259
2. *Dispersion Relations in the Valence Band of GaN*	260
3. *Valence Band Levels and Excitons in GaN-AlGaN Quantum Wells*	262
IX. SELF-ORGANIZED QUANTUM BOXES	267
X. CONCLUSION	268
REFERENCES	270

Chapter 7 Strain in GaN Thin Films and Heterostructures
Christian Kisielowski

I.	Thin-Film Growth at Low Temperatures	275
	1. Introduction	275
	2. Origin of Stresses	277
	3. Growth Modes	283
II.	Stress/Strain Relations	286
	1. Elastic Constants of GaN	286
	2. Experimental Stress/Strain Calibrations	289
III.	Control of Hydrostatic and Biaxial Stress and Strain Components	292
	1. Stress-Controlling Parameters	292
	2. Lattice Constants	297
	3. Engineering of Materials Properties by Active Utilization of Stress and Strain	299
IV.	Strained AlN/InN/GaN Heterostructures	304
	1. Al and In Distribution in $Al_xGa_{1-x}N/In_xGa_{1-x}N/GaN$ Quantum Well Structures	304
	2. Strain-Induced Transition from a Two-Dimensional to a Three-Dimensional Growth Mode	308
V.	Perspectives	313
	References	315

Chapter 8 Nonlinear Optical Properties of Gallium Nitride
Joseph A. Miragliotta and Dennis K. Wickenden

I.	Introduction	319
II.	Background	322
III.	Second-Order Nonlinear Optical Phenomena	325
	1. Second-Harmonic Generation	326
	2. The Electro-Optic Effect (LEO)	338
IV.	Third-Order Nonlinear Optical Phenomena	341
	1. Third-Harmonic Generation (THG)	342
	2. Electric Field-Induced Second-Harmonic Generation (EFISH)	346
	3. Two-Photon Absorption (TPA)	351
	4. Degenerate Four-Wave Mixing	355
V.	Potential Devices	361
VI.	Conclusions	366
	References	366

Chapter 9 Magnetic Resonance Investigations on Group III-Nitrides
B. K. Meyer

I.	Introduction	371
II.	Magnetic Resonance — The Basis of Identification	373
	1. Electron Paramagnetic Resonance (EPR)	373
	2. Optically Detected Magnetic Resonance (ODMR)	375
	3. Electrically Detected Magnetic Resonance (EDMR)	377
III.	Shallow Donors in Cubic and Hexagonal GaN (EPR Results)	377
	1. The Conduction Band/Shallow Donor Spin Resonance	377
	2. The Conduction Band/Shallow Donor Resonance in AlGaN	382

3. *Overhauser Shift and Nuclear Double Resonance*	383
IV. SHALLOW AND DEEP DONORS IN GaN (ODMR RESULTS)	385
1. *ODMR and Nuclear Double Resonance Investigations*	385
2. *Magneto-Optical and Infrared Absorption Experiments*	390
V. SHALLOW AND DEEP ACCEPTORS IN GaN	391
VI. DEFECTS INDUCED BY PARTICLE IRRADIATION IN GaN AND AlN	394
VII. DEVICE-RELATED MAGNETIC RESONANCE STUDIES	396
VIII. TRANSITION METAL IMPURITIES	399
IX. OUTLOOK	403
REFERENCES	404

Chapter 10 GaN and AlGaN Ultraviolet Detectors
M. S. Shur and M. Asif Khan

I. INTRODUCTION	407
II. PRINCIPLE OF OPERATION	409
1. *Photovoltaic Detectors*	411
2. *Photoconductive Detectors*	412
III. DETECTIVITY AND NOISE EQUIVALENT POWER (NEP)	415
IV. GaN PHOTODETECTOR FABRICATION	417
V. GaN-BASED PHOTOCONDUCTIVE DETECTORS	419
VI. GaN-BASED PHOTOVOLTAIC DETECTORS	424
1. *GaN p-n Junction Photodetectors*	424
2. *GaN Schottky Barrier Photodetectors*	425
3. *GaN Metal-Semiconductor-Metal Photodetectors*	430
VII. OPTOELECTRONIC AlGaN/GaN FIELD EFFECT TRANSISTORS	430
VIII. CONCLUSIONS AND FUTURE CHALLENGES	436
REFERENCES	437

Chapter 11 III–V Nitride-Based X-ray Detectors
C. H. Qiu, J. I. Pankove, and C. Rossington

I. INTRODUCTION	441
II. MATERIALS REQUIREMENTS AND CURRENT STATUS	443
III. THE PHOTOCONDUCTIVITY RESPONSE OF NITRIDES	448
1. $\eta\mu\tau$ *Product*	448
2. *Photocurrent Decay*	453
3. *Mg-Doped GaN*	457
IV. EFFECTS OF IRRADIATION ON GaN	458
V. X-RAY RESPONSE OF PROTOTYPE DIODES	460
VI. SUMMARY AND FUTURE WORK	461
1. *X-ray and γ-ray Astronomy*	462
2. *Radiography*	463
3. *Crystallography*	463
4. *Environmental Monitoring*	463
REFERENCES	463

INDEX	467
CONTENTS OF VOLUMES IN THIS SERIES	473

Preface

This book addresses issues related to the growth, structure, properties, and applications of III-V nitrides, using the same tutorial approach that was adopted in the first volume of the series (Volume 50). Our objective is for the book to be useful to both newcomers and experts and to have a lasting influence in the field. The chapters first present an historical overview of the topic, then describe the state of the art, anticipate future developments, and present a complete list of references.

Chapter 1 discusses the growth of GaN and other III-V nitrides by the Hydride-Vapor-Phase-Epitaxy (HVPE) method. It addresses issues related to the kinetics of growth, doping, properties, and applications. The significant progress made over the past few years in controlling the impurities and the heteroepitaxy together with the inherently high growth rate makes the method an attractive one for the formation of nitride substrates.

Chapter 2 discusses the growth of III-V nitrides by the Molecular-Beam-Epitaxy (MBE) method. It addresses issues related to the various nitrogen sources used, the heteroepitaxial growth and structure of GaN in their wurtzite and zinc blende polymorphs, n- and p-type doping, and optical and transport properties of such films. Furthermore, it addresses the growth and structure of InGaAlN alloys with special emphasis on the recently discovered phase separation and long-range atomic ordering phenomena. This method produces multiple-quantum wells (MQWs) with excellent structural and optical properties. The applications of MBE-grown III-V nitrides to various optical and electronic devices are also discussed.

Chapter 3 discusses the crystal structure of bulk GaN grown from liquid Ga under high hydrostatic pressure of nitrogen and homoepitaxial GaN layers grown on these crystals. In particular it addresses issues related to crystal polarity, defect distribution (stacking faults, dislocation loops, and Ga-precipitates) and formations of nanotubes. The influence of crystal polarity and pinholes in the homoepitaxial growth of GaN layers is also addressed.

Chapter 4 discusses the theory and experimental observations of hydrogen. The theoretical work is based on pseudopotential-density-functional calculations. The work includes isolated interstitial hydrogen, hydrogen molecules, and interaction of hydrogen with shallow impurities. The experimental work includes hydrogenation and diffusion studies, passivation of acceptors, and formation of Mg—H complexes.

Chapter 5 addresses the characterization of dopants and deep-level defects using Hall effect and capacitance transient methods. The thermal activation energies of Si- and Mg-dopants were determined to be in the ranges of 12–16 meV and 160–185 meV, respectively. Deep levels in n- and p-type GaN were found to have concentrations below 10^{16} per cubic centimeter.

Chapter 6 discusses the effect of residual strain fields in III-V nitride films and heterostructures. More specifically it addresses the origin of strain and the effects of strain field on the electronic structure and phonon spectra in bulk films. It also addresses the behavior of shallow and deep levels under hydrostatic pressure. Furthermore, the optical properties of GaN/AlGaN quantum wells, as well as how strain can be used for the formation of self-organized quantum boxes during film growth, are discussed.

Chapter 7 addresses the effects of strains in thin layers of semiconductors. There are the uniaxial or biaxial strains due to lattice mismatch to the substrate and the hydrostatic strains due to point defects and impurities. It is possible to decompose a biaxial stress into a hydrostatic stress and two shear stresses. A knowledge of the nature of strains and stresses can lead to stress engineering to control certain material properties, for example, to achieve a smooth surface or to avoid film cracking.

Chapter 8 discusses the nonlinear optical properties of GaN. More specifically it addresses second-order and third-order nonlinearities in this material and shows that the observed optical nonlinearities are of sufficient magnitude for realistic device applications such as frequency conversion and phase modulation.

Chapter 9 discusses the study of shallow and deep donors and acceptors in cubic and wurtzite structures in GaN and AlGaN films using magnetic resonance spectroscopies. Such probes were used also to study defects induced by particle irradiation, electroluminescence processes, and transition metal impurities.

Chapters 10 and 11 discuss the use of GaN and AlGaN for ultraviolet and X-ray detectors. Both photoconductive and photovoltaic detectors are discussed. Photoconductive detectors with variable $\eta\mu\tau$ products were fabricated by varying the resistivity of the GaN films. These detectors were also characterized by studying their time response. The responsivity of these detectors to both UV light and X-rays is discussed. Photovoltaic detectors having the p-n junction or Schottky barrier configuration are addressed.

We hope that the adapted tutorial approach of these chapters by seasoned experts will have a positive influence in the training of a new generation of researchers in this field.

JACQUES I. PANKOVE
THEODORE D. MOUSTAKAS

List of Contributors

Numbers in parenthesis indicate the pages on which the authors' contribution begins.

BERNARD GIL, (209), *Centre National de la Recherche Scientifique, Group d'Etude des Semiconducteurs, Université de Montpellier II, Case courrier 074, 34095 Montpellier cedex 5, France*

WERNER GÖTZ, (185), *Hewlett-Packard Company, Optoelectronics Division, 370 West Trimble Road, San Jose, California*

NOBLE M. JOHNSON, (157, 185), *Xerox Palo Alto Research Center, Palo Alto, California*

M. ASIF KHAN, (407), *Department of Electrical Engineering, University of South Carolina, Columbia, South Carolina*

CHRISTIAN KISIELOWSKI, (275), *National Center for Electron Microscopy, National Laboratory, Berkeley, California*

ZUZANNA LILENTAL-WEBER, (129), *Center for Advanced Materials, Materials Science Division, Lawrence Berkeley National Laboratory, Berkeley, California*

B. K. MEYER, (371), *I Physics Institute, Justus Liebig University Giessen, Heinrich-Buff Ring 16 35392, Giessen, Germany*

JOSEPH A. MIRAGLIOTTA, (319), *Applied Physics Laboratory, Johns Hopkins University, 1100 J. H. Road, Laurel, Maryland*

RICHARD J. MOLNAR, (1), *Massachusetts Institute of Technology, Lincoln Laboratory, Lexington, Massachusetts*

THEODORE D. MOUSTAKAS, (33), *Photonics Center, Department of Electrical Engineering, Boston University, Boston, Massachusetts*

JACQUES I. PANKOVE, (441), *Department of Electrical and Computer Engineering, University of Colorado, Boulder, Colorado*

C. H. QIU, (441), *Astralux, Inc., 2500 Central Avenue, Boulder, Colorado*

C. ROSSINGTON, (441), *Lawrence Berkeley National Laboratory, 1 Cyclotron Road, Berkeley, California*

M. S. SHUR, (407), *Center for Integrated Electronics and Electronics Manufacturing and Department of Electrical, Computer and Systems Engineering, Rensselaer Polytechnic Institute, Troy, New York*

CHRIS VAN DE WALLE, (157), *Xerox Palo Alto Research Center, 3333 Coyote Hill Road, Palo Alto, California*

DENNIS K. WICKENDEN, (319), *Applied Physics Lab, Johns Hopkins University, 1100 J. H. Road, Laurel, Maryland*

CHAPTER 1

Hydride Vapor Phase Epitaxial Growth of III-V Nitrides

Richard J. Molnar

MASSACHUSETTS INSTITUTE OF TECHNOLOGY
LINCOLN LABORATORY
LEXINGTON, MASSACHUSETTS

LIST OF ACRONYMS AND ABBREVIATIONS	1
I. INTRODUCTION	2
II. NITRIDE HVPE GROWTH	4
1. Reactor Design	5
2. GaN Growth Kinetics and Thermochemistry	7
3. Heteronucleation and Film Growth	10
4. Doping of HVPE GaN	11
5. Growth of Other III-V Nitrides by HVPE	13
III. GaN FILM CHARACTERIZATION	15
1. Film Morphology/Structure	15
2. Electrical Properties	18
3. Optical Properties	20
IV. LIGHT-EMITTING DIODES	22
V. HVPE FOR NITRIDE SUBSTRATES	24
1. Defect Reduction	25
2. Epitaxial Overgrowths	25
3. Film/Substrate Separation	26
VI. CONCLUSIONS	27
REFERENCES	28

LIST OF ACRONYMS AND ABBREVIATIONS

AFM	atomic force microscopy
C-V	capacitance-voltage
CVD	chemical vapor deposition

DH	double heterostructure
DLTS	deep-level transient spectroscopy
ECR-MBE	electron-cyclotron-resonance-plasma-assisted molecular beam epitaxy
ELO	epitaxial lateral overgrowth
FWHM	full width at half-maximum
HVPE	hydride vapor phase epitaxy
LED	light-emitting diode
MBE	molecular beam epitaxy
MIS	metal-insulator-semiconductor
OMVPE	organometallic vapor phase epitaxy
PL	photoluminescence
RF	radio frequency
SEM	scanning electron microscopy
TEM	transmission electron microscopy
VPE	vapor phase epitaxy

I. Introduction

There has been a recent resurgence of interest in wide-bandgap semiconductors, such as SiC, II-VI compounds and III-V nitrides. This is due to the realization of the theoretical limits of most well-developed semiconductors (e.g., Si, GaAs) for both electrical and optical devices, as well as the need for short-wavelength light emitters for performance enhancements in emerging technologies such as optical storage and full-color displays. The slow but steady development of growth techniques to synthesize these difficult-to-grow wide-bandgap materials has enabled demonstration of a variety of devices with performance advantages over existing technologies. Producibility and reproducibility of the epitaxial materials used to fabricate these devices remain the key issues that need to be addressed for the commercialization of these devices.

The III-V nitrides (InN, GaN, AlN and their solid solutions), which are the focus of this book, are unique in that they all have large, direct bandgaps and are extremely structurally stable materials. This is to be contrasted with the indirect-bandgap polytypes of SiC and direct-bandgap II-VI materials. With the latter, researchers have demonstrated injection lasers that emit in the green; however, these devices suffer from device degradation believed to be due to their "softness." If the longevity of the nitride-based commercially available high-brightness light-emitting diodes (LEDs) and injection lasers is compared with similar devices fabricated from II-VI materials, the advantages of the III-V nitrides become apparent. The nitrides are also being pursued for high-temperature, high-power electronics due to GaN's high electron-saturation velocity, wide bandgap, high thermal conductivity and the possibility of heterojunction-based devices.

Historically, hydride vapor phase epitaxy (HVPE), and its closely related technique, halide vapor phase epitaxy, have played an important role in the development of semiconductor material systems, such as the arsenides, phosphides, and nitrides. Their ability to produce high-quality thin and thick films has enabled the study of fundamental parameters of these semiconductors, as well as the commercialization of devices, such as LEDs, which can be inexpensively mass produced by this economical growth method. While the HVPE growth of arsenides and phosphides has been studied extensively and is well developed, the HVPE growth of the nitrides is still poorly understood. This results from complications such as the lack of a native substrate and fundamental differences in the growth chemistry, which makes nitride HVPE growth more complicated. Regardless, there have been several reports of the growth of GaN films, which exhibit electrical and optical properties comparable to the best reported for GaN grown by organometallic vapor phase epitaxy (OMVPE) or molecular beam epitaxy (MBE). In fact, it is remarkable that GaN films that would be considered to be of state-of-the-art quality, even by present standards, were grown by HVPE in the 1970s (Ilegems, 1972; Crouch *et al.*, 1978). The high growth rates ($>1\ \mu m/min$) one can obtain with HVPE have also enabled its usage as a quasi-bulk growth technique for GaP (Huang *et al.*, 1992) and GaN (Detchprohm *et al.*, 1994).

The HVPE process has played an important historical role in the development of the nitrides in that it was the first (Maruska and Tietjen, 1969) and, until the early 1980s, the most popular method of growing epitaxial layers of gallium nitride (Ban, 1972; Ilegems, 1972). This technique was largely abandoned in the early 1980s because of apparent difficulties in reducing the native shallow-donor concentration to nondegenerate levels and thus enabling *p*-type doping. This was presumed to be due to a nitrogen vacancy defect, which would be thermodynamically favored at the high growth temperatures typically used in HVPE GaN growth. This interpretation predominated in spite of reports of nondegenerate films grown by HVPE (Ilegems, 1972; Crouch *et al.*, 1978) and careful growth studies, which suggested that the incorporation behavior of this donor was inconsistent with a nitrogen vacancy defect (Seifert *et al.*, 1983). With the advent of high-purity source materials and improved heteronucleation schemes, the growth of nondegenerate material has been reported by several groups.

II. Nitride HVPE Growth

The HVPE process is a chemical vapor deposition method, which is usually carried out in a hot wall reactor, at atmospheric pressure. A novel aspect of this method is that the group III precursors are usually synthesized within the reactor vessel, upstream from the substrate, by either the reaction of a halide containing gas, such as hydrogen chloride (HCl), with a group III metal at high temperature, or, as in the case with halide vapor phase epitaxy, by the reaction of a group V chloride (e.g., $AsCl_3$) with either a group III metal or III-V bulk compound. In the case of GaN, gallium monochloride is usually synthesized upstream in the reactor by reacting HCl gas with liquid Ga metal at 800–900 °C. The GaCl is transported to the substrate where it is reacted with NH_3 at 900–1100 °C to form GaN, via the reaction

$$GaCl + NH_3 \leftrightarrow GaN + HCl + H_2$$

Because the group III element is transported to the substrate as a volatile halogen compound (usually a chloride), this technique is often referred to as chloride-transport vapor phase epitaxy (VPE). Due to the relatively low vapor pressure of the metal chlorides at room temperature, these molecules will tend to condense on unheated surfaces. This is the primary motivation for the use of a hot wall reactor and the *in situ* synthesis of the metal chlorides. This avoids the complicated vapor delivery systems and gas inlet heating necessary if the chlorides are synthesized/stored externally. Additionally, the metal halides' hydroscopic and corrosive nature would make maintaining source integrity during storage and processing somewhat difficult if the metal-halogen was presynthesized. However, as difficulties can result from the use of highly corrosive HCl gas, which can quickly destroy reactor equipment if care is not taken to avoid air leaks, several groups have reported using presynthesized $GaCl_3$ instead of *in situ* synthesis with HCl gas (Nickl et al., 1974; Lee et al., 1996; Tsuchiya, 1996). Using this precursor, they have been able to produce single-crystalline GaN and with improvements in metal-halogen source purity, this technique may eventually prove to be viable for nondegenerate film growth.

For the case of *in situ* synthesis, Ga metal can be readily obtained with a purity of 99.999999%, and with a properly designed gas delivery system, HCl purity approaching 99.9999% can be attained. Chlorine has been the most common halogen used for transport due to the higher vapor pressures of the metal chlorides vs bromides and iodides. Hydrogen chloride is usually preferred over chlorine gas for both the higher purities available and handling considerations. There have been, however, reports on using bromine or even iodine for metal transport. These compounds have lower dissociation temperatures than the chlorides and were usually utilized in studies investigating reduced growth temperatures (Morimoto et al., 1973; Born and Robertson, 1980).

The HVPE process tends to create copious amounts of NH_3Cl, $GaCl_3$, and $GaCl_3 \cdot NH_3$, which can condense on and eventually clog exhaust lines unless they are heated to sufficiently high temperatures ($>150\,°C$) and/or operated at reduced pressure.

1. REACTOR DESIGN

The majority of reports on GaN films grown by HVPE have been based on horizontal reactor designs. One such design, developed at RCA laboratories, is schematically illustrated in Fig. 1. This design has been popular for the growth of a variety of III-V and II-VI materials. However, it is important to recognize that the nature of the chemistry involved in GaN growth by HVPE differs substantially from that of other III-V semiconductors. For instance, in GaAs growth by HVPE or halide vapor phase epitaxy, thermal disassociation of the arsenic compounds results in the formation of As_4 and As_2 molecules, which typically remain volatile and chemically reactive and thus participate in the film growth. In GaN HVPE growth, the thermal disassociation of NH_3 results in the formation of N_2 molecules, which are extremely stable and essentially unreactive at the temperatures of interest. In fact, the viability of HVPE GaN growth lies in the relatively sluggish disassociation of NH_3, which enables the effective transport of reactive nitrogen to the growth surface (Ban, 1972). Therefore, the growth

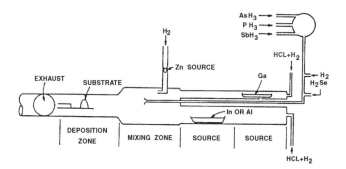

FIG. 1. Typical horizontal HVPE reactor design.

of stoichiometric, homogeneous films over large areas requires the efficient and uniform transport of NH_3 to the growth surface to prevent the growth of dark, metal-rich films (Wickenden et al., 1971).

Another difficulty in the growth of GaN by HVPE is the strong propensity for GaN and gas-phase adducts to form, leading to undesirable gas phase reactions, particulation, and wall deposition problems (which can induce reactor vessel cracking upon cooldown). Also, owing to this strong propensity for GaN formation, horizontal reactor designs have typically suffered from severe gas-phase depletion effects (Ilegems, 1972), although recently there have been reports of improved uniformity in horizontal coaxial flow designs (Safvi et al., 1996). Additionally, as NCl_3 is a highly explosive compound, there have not been reports of the halide vapor phase growth of nitride materials. For this reason, throughout this chapter, HVPE will refer to the hydride vapor phase epitaxial process, unless noted otherwise.

The films we have deposited at Lincoln Laboratory were grown in a vertical reactor schematically represented in Fig. 2 (Molnar et al., 1995; Molnar et al., 1996a). The vertical reactor design facilitates substrate rotation during growth to improve the film uniformity. The substrate holder can be raised and lowered isothermally into a counterflow tube, in which a mixture of NH_3 and carrier gas is passed. The substrate can then be slowly lowered and cooled in an NH_3 environment to minimize film decomposi-

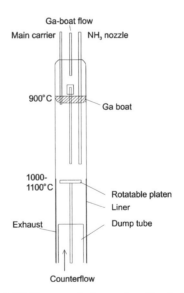

FIG. 2. Vertical HVPE reactor design used at Lincoln Laboratory.

tion. All of the growths reported in this chapter were on the (0001) plane of sapphire. In order to improve the heteronucleation density of the GaN layer on (0001) sapphire, two different procedures have been employed: a GaCl pretreatment (Naniwae *et al.*, 1990; Molnar *et al.*, 1995); or an RF-sputtered ZnO buffer layer technique (Detchprohm *et al.*, 1992; Molnar *et al.*, 1996b).

2. GaN GROWTH KINETICS AND THERMOCHEMISTRY

This chapter will not provide a rigorous presentation of the thermodynamics and kinetics of HVPE GaN growth. Interested readers are referred to previous publications directed towards such issues (Liu and Stevenson, 1978; Shintani and Minagawa, 1974; Seifert *et al.*, 1981) and Newman's discussion in volume I of this series. Rather, I hope to give an intuitive look at several key issues in HVPE GaN growth and relate these to other material systems to emphasize the challenges of GaN HVPE growth.

Although the chemistry involved in the growth of GaN by HVPE presents difficulties in equipment design, it also presents distinct advantages. For example, unlike OMVPE, the HVPE process is inherently carbon-free, making the growth of higher-purity films somewhat easier. Additionally, the presence of efficient etchant halogen species helps remove excess metallic species from the growth surface, thus inhibiting the formation of metal-rich material or phase-separated Ga droplets. This self-stabilizing effect, as well as higher adatom mobility for chemisorbed GaCl, may account for the higher growth rates attainable by HVPE as compared with other epitaxial growth methods. Self-stabilization also allows for the growth under Ga-rich conditions, which have been empirically found to result in higher material quality with other growth techniques, such as molecular beam epitaxy. It also means that the ammonia flow required for stoichiometric growth is lower. For example, we have found an optimized ammonia flow rate of ~ 800 sccm for growth on a 2-in.-diameter substrate, several times lower than that reported for comparably sized OMVPE reactors, even though the HVPE growth rate is about an order of magnitude higher (~ 40 μm/h). This is significant, as ammonia has been identified as a major source of impurities in VPE nitride growth, as well as a significant contributor to operation costs.

Additional HCl can be intentionally introduced downstream of the Ga boat to increase effectively the Cl/Ga ratio at the growth surface. This has been found to have a beneficial role in improving film properties (Jacob *et al.*, 1978). We have observed similar behavior (Molnar *et al.*, 1997) and

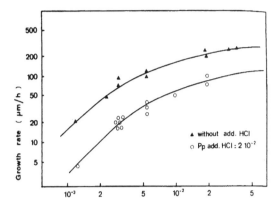

FIG. 3. Variation of the growth rate vs GaCl partial pressure, with and without additional HCl. (Reprinted with permission from Jacob et al. 1978.)

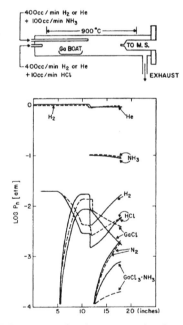

FIG. 4. Variation of partial pressures of various gas species along growth tube determined by mass spectroscopy. The apparatus and experimental conditions are schematically represented in upper drawing. (-----: H_2 carrier gas; ——: He carrier gas.) (Reprinted with permission from Ban 1972.)

attribute it to enhanced lateral growth rates, which result in larger grain size as well as possible enhanced impurity desorption. As can be seen in Fig. 3, introducing free HCl has a moderate reducing effect on growth rate over the entire range of GaCl partial pressures investigated.

As mentioned, the formation of gas phase adducts occurs rapidly upon the mixing of ammonia and GaCl in the reactor tube. Evidence for this was presented by Ban, who performed mass spectroscopic investigations into the growth chemistry of HVPE GaN. The variation in partial pressure of several gaseous species with position in a hot-walled reactor is shown in Fig. 4. As the GaCl and ammonia mix, an adduct is rapidly formed. Also, at 900 °C the rate of ammonia decomposition is relatively low, but increases rapidly with temperature (Ban, 1972).

In discussing the thermochemistry of HVPE GaN growth, a discussion of nitrogen vacancies is appropriate, as they have widely been held responsible for the degenerate doping levels observed in GaN films. This model was adopted based largely upon the known thermal instability of GaN as well as impurity analysis, which seemed to indicate Si and O levels too low to account for the residual donor levels. It is, however, difficult to obtain very accurate determinations of impurity densities without normalization standards. Additionally, other impurities, such as Se and Ge, are known to act as donors in GaN, and the effect of one or more impurities appears now to be a plausible explanation of the residual donors common in GaN films. Evidence for this was offered in 1983 by Seifert *et al.*, who presented systematic data on the role of residual water vapor in the ammonia gas and the residual donor concentration. As can be seen in Fig. 5, water impurities in the ppm range can easily account for the 10^{19} cm^{-3} donor levels in GaN films grown over a wide range of temperatures. As shown in Fig. 6, by

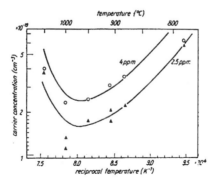

FIG. 5. Free carrier concentration vs growth temperature with (♦) and without (○) ammonia purification. (Reprinted with permission from Seifert *et al.* 1983.)

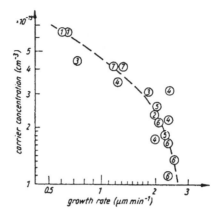

FIG. 6. Free carrier concentration vs growth rate. ① 785 °C, ② 880 °C, ③ 900 °C, ④ 920 °C, ⑤ 950 °C, ⑥ 1000 °C, and ⑦ 1050 °C. (Reprinted with permission from Seifert et al. 1983.)

examining samples grown over a wide range of temperatures, the authors also found that the residual donor concentration was not well correlated with the growth temperature, which is inconsistent with nitrogen vacancies being the predominant donor in these films. As accurate secondary ion mass spectroscopy results normalized by ion-implanted standards are now available, the nitrogen vacancy model has become doubtful as the most plausible explanation for the residual donor in GaN.

3. HETERONUCLEATION AND FILM GROWTH

One of the difficulties in growing high-quality GaN by HVPE, as with other epitaxial growth techniques, is achieving high-quality nucleation on the substrate material. A variety of substrates have been utilized, including Al_2O_3 (sapphire), $MgAlO_4$ (spinel), SiC, Si, YAG, GGG, GaAs, and ZnO (sputter-deposited). Sapphire has been by far the most popular choice because of its comparatively low cost, high quality, large diameter, and chemical compatibility. Both c-plane (0001) and r-plane ($1\bar{1}02$) substrates have been studied extensively. The r-plane has been found to give a higher growth rate. However, the GaN film deposits with its a-plane ($11\bar{2}0$) parallel to the sapphire's r-plane surface, leading to rough, ridged surfaces bounded by ($1\bar{1}00$) planes. GaN films grown on c-plane (0001) substrates grow (0001)-oriented. However, the low surface energy of this plane results in poor heteronucleation, resulting in rough film morphology usually dominated by the sapphire's surface features/contaminants. With improvements to wafer polishing as well as improved heteronucleation schemes, this effect can be well suppressed. Growth on (0001) SiC substrates has been reported by several groups (Wickenden et al., 1971; Melnik et al., 1996). Interestingly, high-quality growth can be achieved without first depositing an AlN nucleation layer, which has been found necessary for OMVPE growth on SiC. This may afford the potential of GaN/SiC heterojunction-based devices as well as vertical conduction in LEDs and lasers.

4. DOPING OF HVPE GaN

As already mentioned, HVPE-grown GaN has traditionally been characterized by high (10^{18} to 10^{20} cm^{-3}) residual shallow donor levels. Therefore, the vast majority of doping studies focused on finding a suitable acceptor dopant to compensate these donors and to try to achieve p-type material and p-n junctions. However, the HVPE process, being a hot-wall process, can potentially suffer from pronounced chemical interaction between the hot quartz reactor components and some gas phase dopant species. This potential exists when a particular dopant oxide is more thermodynamically stable than SiO_2 (quartz). In this case, potential exchange reactions can result in degradation of quartz reactor components, the formation of wall deposits of the oxidized dopant species, and the leaching of Si contaminants into the growth process. This presents difficulties in introducing acceptors, which are thermally ionized at room temperature in HVPE-grown GaN, as the two most shallow acceptors in GaN, Mg, and Ca, both form very stable oxides and thus are prone to reacting with the quartz reactor, as shown in Table I.

Mg-doped HVPE GaN films have been reported (Sano and Aoki, 1976), although p-type conductivity was not attained. Mg-doped p- on n-type structures grow by HVPE have been reported recently (Nikolaev et al., 1998). Photoluminescence (PL) studies of ion-implanted acceptors in GaN indicate that Zn may be a third choice as a dopant for achieving a shallow acceptor, as shown in Fig. 7. Thus, most acceptor doping studies have focused on introducing Zn acceptors (Jacob et al., 1977a). As an interesting historical note, Saparin et al., reported in 1984 that the blue luminescence in Zn-doped, HVPE-grown GaN could be substantially enhanced by irradiating the sample with an electron beam (Saparin et al., 1984). In light of later studies on Mg-doped OMVPE-grown GaN films (Amano et al., 1989), this is likely the result of the electron beam disassociation of Zn-H complexes.

The large ionization energy of Zn acceptors in GaN, the formation of hydrogen complexes and the typically high residual donor levels make the

TABLE I

PARTIAL PRESSURE OF OXYGEN OVER VARIOUS OXIDES AT 1000 K
($-\log p_{O_2}$ for oxygen in equilibrium)

Oxide	pO*	Oxide	pO*
Au_2O_3	−5.5	SiO	36.3
Ag_2O_3	−3.3	RaO	46.3
PtO	−1.3	Al_2O_3	47.2
PdO	−1.1	LiO	48.5
IrO_2	0.9	BaO	48.6
Rh_2O	4.2	Sc_2O_3	50.3
CdO	16.6	SrO	51.2
In_2O_3	24.1	BeO	52.0
ZnO	25.8	MgO	52.0
Ga_2O_3	26.2	CaO	55.5

*Reed (1971).

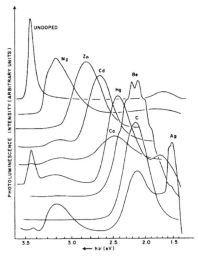

FIG. 7. Characteristic PL spectra of undoped and ion-implanted GaN with indicated impurities, at 78 K. (Reprinted with permission from Pankove and Hutchby 1976.)

achievement of p-type conductivity in Zn-doped GaN difficult. As shown in Fig. 8, Jacob et al. (1978) were able to obtain high-resistivity films that exhibit a p-type Seebeck coefficient. They refer to these films as π-type and were able to make efficient m-i-n LEDs utilizing such layers.

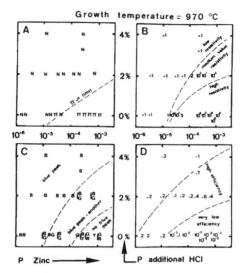

FIG. 8. Electrical and optical properties of GaN vs Zn and HCl partial pressures (in atm). (A) Conductivity type; (B) resistivity (Ω-cm); (C) CL emission bands (B, blue; G, green; Y, yellow; R, red); (D) CL external efficiency (%). (Reprinted with permission from Jacob et al. 1978.)

5. GROWTH OF OTHER III-V NITRIDES BY HVPE

a. Cubic GaN

Cubic GaN has attracted much interest owing to potential doping and cleaving advantages. As this phase is not thermodynamically favored over the wurtzite phase, predominantly cubic films have only been observed in situations where the cubic phase was stabilized by the symmetry of the substrate. While such films have been grown by several growth techniques, the structural quality of these films has been poor because of the lack of a lattice-matched substrate. To this end, several researchers have reported on growing cubic GaN films by HVPE to achieve thick films of improved structural quality (Tsuchiya et al., 1997; Yamaguchi et al., 1996). The most popular choice for substrate has been (100) GaAs due to the potential of first depositing a homoepitaxial GaAs buffer layer to heal surface polishing damage and to form a wetted buffering surface nitridation layer by conversion with ammonia. It has been reported that low growth rates (<4 μm/h) were required to suppress the introduction of hexagonal phases into the films (Tsuchiya et al., 1994; Tsuchiya et al., 1996). Further work is required to develop material of device quality.

b. AlN and AlGaN

The growth of aluminum-bearing nitrides by HVPE, as with magnesium-doped films, is complicated by strong exchange reactions between the AlCl and the hot quartz reactor, which leads to excessive impurity incorporation in the films and to quartz reactor degradation. There have been only a few reports of HVPE growth of AlN (Yim *et al.*, 1973), AlGaN (Hagen *et al.*, 1978; Baranov *et al.*, 1978). Often the use of Grafoil or tungsten liners was employed to reduce exposure of the quartz to the aluminum chloride.

c. InN and InGaN

As the bandgap of GaN is in the UV and GaN LEDs tend to have reduced apparent brightness owing to the relatively short wavelength of emission, there has been interest in growing InN and InGaN compounds by HVPE for tailoring the bandgap into the visible region of the spectrum. The thermal instability of InN imposes severe limitations on the growth temperature such that In-bearing growing compounds must be deposited at temperatures less than 800 °C for InGaN and less than 500 °C for InN. One problem associated with this is the very low cracking efficiency of NH_3 at these temperatures, making it extremely difficult to grow stoichiometric material. Even more important is the apparent inability to grow InN with InCl (monochloride). Successful attempts at growing In-bearing nitrides have been with $InCl_3$, which is the predominant species at lower temperatures. It has been speculated that the formation of a gas-phase adduct with ammonia ($InCl_3 \cdot NH_3$) is essential for the subsequent deposition of InN at the reduced growth temperature. Even under such conditions, the thermal instability and poor ammonia cracking efficiency make it difficult to avoid In inclusions in the resultant films. It is clear that further work is necessary to resolve these issues.

d. Scandium, Yttrium, and Rare-Earth Nitrides

The relative simplicity and flexibility of HVPE has made it a powerful technique for growing films of a variety of different compounds. This has allowed researchers to evaluate the materials' characteristics of the compounds and determine their suitability for device applications. GaN is a poignant example. Other examples are some rare-earth nitrides, which were grown at RCA laboratories in the early 1970s (Dismukes *et al.*, 1970; Dismukes *et al.*, 1972). The authors found that these compounds crystallize in the NaCl-structure, and ScN in particular exhibited excellent stability and a sizable optical bandgap (~2.2 eV). Large residual donor levels ($10^{21}\,cm^{-3}$), probably the result of exchange reactions with the quartz reactor, coupled with reasonably high electron mobilities suggest that this semiconductor may find applications if high-purity material becomes available.

III. GaN Film Characterization

In this section, I review growth behavior and material properties observed for films grown in the vertical reactor described in Section II.1.

1. FILM MORPHOLOGY/STRUCTURE

Samples deposited directly on sapphire without any surface pretreatment varied from being highly transparent to brown. As shown in Fig. 9, these consist of clearly defined hexagonal islands, typical for low-nucleation-density growth. These islands tend to grow more laterally with increasing growth temperature and/or reduced growth rates.

These surface features are similar to those reported for one-step OMVPE-grown GaN films on sapphire (Nakamura, 1991), although the size of the features is somewhat larger. This suggests that with the development of techniques to improve the heteronucleation density of the GaN film on the sapphire substrate and to increase the film thickness, HVPE may offer a potential advantage in growing large-cell, low-defect GaN films. The high lateral rate can also be seen on samples, which contain large densities of screw defects due to surface contamination, as shown in Fig. 10. In these samples, terrace widths of several microns and cell diameters of hundreds of microns are commonly observed, suggestive of high lateral/vertical growth ratios.

The forementioned results clearly point out that, as with other GaN growth processes, the film heteronucleation is critical for the growth of

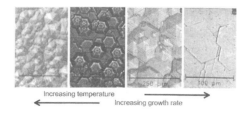

FIG. 9. Surface morphologies for GaN films deposited directly on sapphire.

FIG. 10. Optical micrograph of screw defect in HVPE GaN film. (Scale bar represents 100 μm.)

high-quality material. Attempts to deposit a low-temperature buffer layer followed by a high-temperature-growth step resulted in polycrystalline material. However, by utilizing either a GaCl or ZnO pretreatment, highly transparent GaN films are obtained, which show few morphological features, as shown in Fig. 11. The atomic force microscopy (AFM) images, one of which is shown in Fig. 12 for a 40-μm-thick film growth with ZnO pretreatment, reveal the presence of ∼5-Å terraces on the growth surface that are ∼75 nm wide. This is indicative of a step-flow mode of growth, and the terrace height corresponds to what one would expect for monoatomic steps ($c = 5.19$ Å). Thicknesses of up to 74 μm have been grown without thermally induced cracking observable by optical microscopy. Above this value or for cases where surface preparation is nonoptimal, growth striae and/or thermal cracking (usually originating in the sapphire) can be observed. X-ray diffraction studies yield a θ rocking curve full width at half

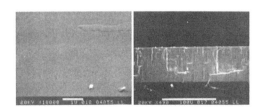

FIG. 11. Surface and cross-sectional scanning electron micographs for sample grown with GaCl pretreatment.

1 HYDRIDE VAPOR-PHASE EPITAXIAL GROWTH OF III-V NITRIDES   17

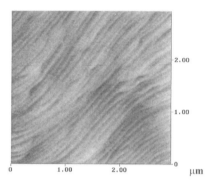

FIG. 12. AFM micrograph of GaN growth surface for 40-μm-thick GaN film.

maximum (FWHM) of 4.8–7.2 arc min (Romano et al., 1996) for the samples measured. These values are comparable to those reported for high-quality OMVPE material.

Cross-sectional transmission electron microscopy (TEM) of the GaN/sapphire interface in samples grown by the two nucleation schemes reveals a substantially different interface, as indicated in Fig. 13. For the sample

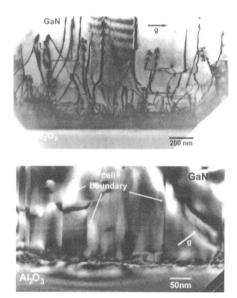

FIG. 13. Cross-sectional TEM images of GaN/Al$_2$O$_3$ interfaces for films grown with GaCl pretreatment (top) and ZnO pretreatment (bottom).

grown with a GaCl pretreatment, a ~200-nm-thick layer with a high density of stacking faults is formed at the interface. Above this region, high densities of dislocations thread upwards, while undergoing significant interaction/annihilation, resulting in dislocation densities of ~10^8 cm^{-3} after several tens of microns of growth. For the samples grown utilizing the ZnO pretreatment, the highly faulted region is absent. However, dislocation interaction reduces the defect densities to levels similar to those for the GaCl pretreatment. It should also be noted that for most of the samples grown on the ZnO-coated sapphire wafer, there is no evidence of any ZnO at the interface, suggesting that most of the film is thermochemically desorbed early in the GaN film growth. This is consistent with an apparent insensitivity to the ZnO film thickness within the range investigated, consistent with previous reports (Detchprohm *et al.*, 1993). Even though there is no discernible ZnO at the GaN/sapphire interface, this ZnO pretreatment does appear to improve the quality of the GaN film growth markedly.

2. ELECTRICAL PROPERTIES

Electrical transport measurements as a function of temperature taken in optimized films grown by both pretreatment methods and with varying film thicknesses are shown in Fig. 14. The temperature dependence of the electron concentrations is well explained by the presence of Si donors and an additional deeper donor, the nature of which is unclear at the moment.

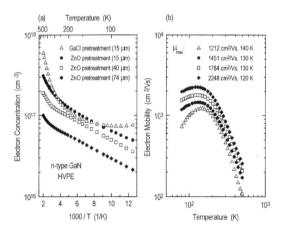

FIG. 14. (a) Electron concentration vs reciprocal temperature; and (b) electron mobility vs temperature for unintentionally doped n-type HVPE-grown GaN.

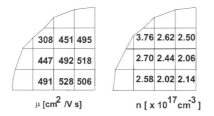

FIG. 15. Wafer map of room-temperature Hall measurements of one-fourth (one quadrant) of a 2-in.-diameter GaN film. The other three quadrants exhibit similar characteristics due to the high rotational rate (38 rpm).

For the 74-μm-thick sample, the carrier concentration does not deviate from this Arrhenius behavior at low temperatures, as is typically observed for GaN films doped at this level (Molnar et al., 1993). This may be the result of low concentrations of compensating defects in these films. The deeper level is found to vary in a range of 100–200 meV from sample to sample. As can be seen these samples exhibit high electron mobility, particularly in consideration of the concentration of residual donors ($\mu = 880$ cm^2/V·s, n = 7×10^{16} cm^{-3} at 293 K and $\mu = 2248$ cm^2/V·s, n = 4×10^{16} cm^{-3} at 120 K, for a 74-μm-thick ZnO pretreated sample), which is another indication of minimal compensation in these films. The data of Fig. 14 also indicate that the average transport properties of these films improve with film thickness. This is consistent with defect annihilation observed in the TEM results. Additionally, the electrical homogeneity of these films is excellent as evidenced by the small variation in mobility and carrier concentration over a 2-in. substrate as shown in Fig. 15 (Molnar et al., 1996a). The films' electrical properties are typically uniform to within 30% with the exception of a 1- to 2-mm edge band, which is common in chemical vapor deposited films.

Deep-level defects were investigated by deep-level transient spectroscopy (DLTS) measurements for the sample in Fig. 14 grown using the GaCl pretreatment. Schottky barriers were formed on the GaN surface by evaporating Au contacts. The spectrum shown in Fig. 16 exhibits the signature of three deep levels, which are labeled DLN_1, DLN_2, and DLN_3. DLN_2 appears as a shoulder of the DLN_3 peak and, therefore, no analysis was attempted. The deep levels DLN_1 and DLN_3 are characterized by activation energies for electron emission to the conduction band of (0.21 ± 0.02) and (0.65 ± 0.03) eV, respectively (inset). The concentration of these levels is $(1.2 \pm 0.5) \times 10^{14}$ and $(4.5 \pm 0.5) \times 10^{15}$ cm^{-3}. Due to the surface-sensitive nature of DLTS measurements, one can conclude that the density of deep-level defects is comparatively low after ~ 15 μm of growth.

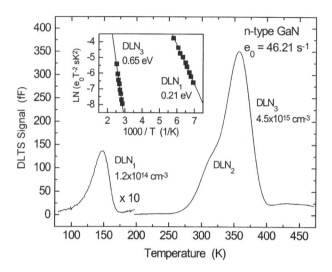

FIG. 16. DLTS spectrum for GaN, grown by HVPE with GaCl pretreatment.

3. OPTICAL PROPERTIES

The 2-K PL spectrum of these films is shown in Fig. 17. The low-energy portion of the spectrum is magnified for both spectra by a factor of 25. The inset shows the near-bandedge region of the spectra with a portion of the spectra magnified. PL lines are labeled BX and L1 to L5. An arrow indicates the first phonon replica of the BX line. All the samples investigated exhibit intense near-bandedge emission associated with donor-bound excitons at 3.468 eV and FWHM = 2.42–5 meV. A high-energy shoulder is believed to be due to the free exciton. The FWHM of the donor-bound excitonic peak does not appear to be correlated with the films' electrical and structural quality, as the highest-mobility, lowest-defect-density sample also showed the broadest exciton peak (~ 5 meV). We are not able to detect any yellow emission centered at 2.2 eV, although weak red emissions centered at 2.0 and 1.8 eV are detected in some of the samples.

Optical pumping of the HVPE GaN films at 77 K and room temperature was also investigated, for which the spectra are shown in Fig. 18. Stimulated emission was observed with a threshold of 0.7 MW/cm^2 at 77 K and 1.8 MW/cm^2 at room temperature. These thresholds are low for bulk GaN materials and are consistent with the high quality of the material.

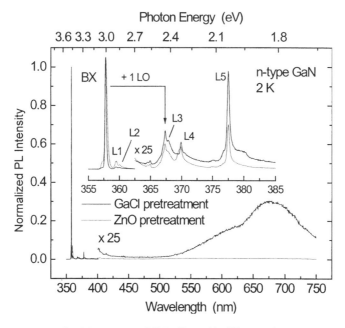

FIG. 17. Normalized PL spectra of GaN films with different substrate pretreatments.

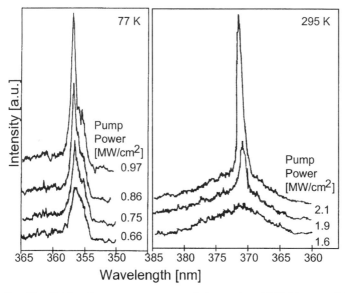

FIG. 18. 77-K and room-temperature optical pumping of HVPE-grown GaN.

IV. Light-Emitting Diodes

As already mentioned, HVPE has played an important historic role in the development of nitride semiconductors. This is illustrated by the first demonstration of metal-insulator-semiconductor (MIS) LEDs based upon GaN, shown in Fig. 19 (Pankove et al., 1971). These diodes were shown to emit light with peak wavelengths spanning the UV to red, as indicated in Fig. 20. In fact, the first commercial nitride LED was a GaN MIS homojunction device with a reported quantum efficiency of 0.03% (Ohki et al., 1982). However, their high series resistance, due to the high resistivity of the top i-layer, meant these devices tended to have high (>10 V) and widely varying operating voltages as indicated in Fig. 19. While these devices did not necessarily afford performance improvements over SiC devices, their historical significance cannot be overstated.

As already mentioned, difficulty in achieving p-type GaN by HVPE was likely due to the lack of a shallow dopant species that is compatible with the quartz-wall reactors used for HVPE growth. Elements such as Mg and Ca are relatively shallow acceptors ($\Delta E = 150$ meV), but react with the quartz reactor surfaces degrading the growth tube and producing significant Si and O contaminants, which incorporate as donors. Zn was the most commonly used acceptor species but its relatively deep nature makes achieving p-type conductivity difficult. Mg-doped LEDs, emitting violet light, were reported by Maruska et al. (1972). They were able to minimize wall reactions by using graphite foil liners.

Jacob et al. (1977a) and Jacob and Bois (1977b) reported on highly efficient LED structures grown by HVPE with low turn-on voltages. In

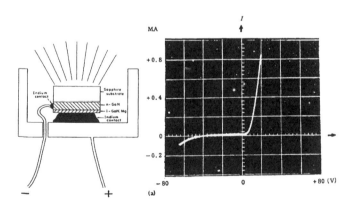

FIG. 19. Typical device structure and current-voltage characteristics of GaN LED. (Reprinted with permission from Maruska et al. 1973.)

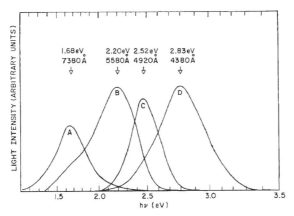

FIG. 20. Emission bands observed in GaN LEDs. (Reprinted with permission from Pankove 1973.)

order to explain the high efficiency and low turn-on voltage of these devices the authors propose a field-assisted tunneling process in the thin ($\sim 1000\,\text{Å}$) insulating region, where the field is largest, as shown in Fig. 21. As the thickness of this region is somewhat difficult to control, operating voltages were difficult to control.

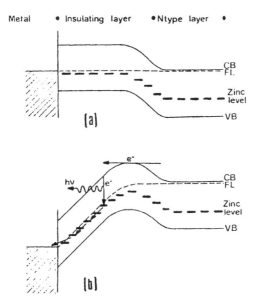

FIG. 21. Energy band diagram for m-i-n LED under (a) zero bias and (b) forward bias. (Reprinted with permission from Jacob and Bois 1977b.)

V. HVPE for Nitride Substrates

While extended structural defects, such as dislocations, do not appear to quench luminescent activity in the nitrides as aggressively as in other semiconductors (Lester et al., 1995), they may serve as optical scattering centers in coherent light emitters (Liau et al., 1996) as well as limiting their lifetimes (Nakamura et al., 1998). They may also increase leakage currents in electronic devices operating at high temperatures, through the introduction of deep levels (Khan et al., 1995).

In addition, current-crowding effects are expected to be predominant in lateral-conduction injection lasers deposited on insulating substrates (such as sapphire) by OMVPE or MBE because the devices' underlying conductive layers are generally <10 μm thick (Liau et al., 1996). While SiC substrates offer the possibility of vertical conduction, lasers and LEDs grown on SiC substrates currently suffer from cracking problems related to tensile thermal stress in the nitride epilayer as well as an interfacial conduction barrier, which increases the generation of heat in the device. Finally, device stability may be limited at high injection currents by diffusion of impurities (such as contact metals) along extended structural defects, which usually exist at high densities (10^9–10^{10} cm^{-2}) in samples grown on both sapphire and SiC substrates (Lester et al., 1995). Bulk growth techniques at very high pressure (~ 15 kbar) (Karpinski et al., 1982) have succeeded in growing low dislocation ($<10^7$ cm^{-2}) material. This material is generally composed of comparatively small crystallites (~ 1 cm^2 or less).

As such limited area will likely inhibit commercialization of this technology, there has recently been a renewed interest in the growth of GaN by HVPE. It is hoped that this high-growth-rate technique will facilitate the growth of low-defect, thick (≥ 20 μm), large-area substrates or buffers (Detchprohm et al., 1992; Molnar et al., 1995; Perkins et al., 1996). Alternatives to the use of a low-temperature nucleation layer or so-called "two-step" growth, such as GaCl pretreatments (Naniwae et al., 1990), ZnO (Detchprohm et al., 1992) or AlN (Lee et al., 1996) sputtered-deposited buffer layers or epitaxial GaN or AlN starting layers (Akasaki et al., 1991) have greatly improved film morphology. HVPE GaN layers grown using such techniques have been shown to have comparatively low defect densities ($<10^8$ cm^{-2}). Epitaxial overgrowths on these HVPE-grown GaN buffers, grown by OMVPE and MBE, have been shown to replicate the structure of the underlying HVPE layer, resulting in substantially reduced defect densities in the overgrown device layer.

In order for this approach to afford yet lower defect densities, a clearer understanding of the defect evolution in the layers needs to be developed.

An apparent bottleneck at $\sim 10^8$ dislocations/cm^2 results from the fact that HVPE is essentially an epitaxial technique that requires a large-area seed crystal on which to nucleate. Because GaN substrates do not exist, foreign substrates, such as sapphire or SiC, are usually employed. Heteroepitaxially associated misfit in lattice constant and isomorphism manifest themselves as a high density of structural defects, such as dislocations and inversion domains. Threading dislocations in these wurtzitic materials show only weak mutual annihilation due to their propagation generally parallel to the growth direction and, hence, weak interaction, particularly after several microns of growth. The presence of mixed dislocations, which are prevalent in the HVPE GaN films, has been recently shown to promote this interaction (Romano et al., 1997). This may account for the somewhat lower defect densities achieved in HVPE epilayers. In fact, it has recently been demonstrated that dislocation densities as low as 3×10^6 cm^{-2} can be achieved in very thick (300 μm) layers (Vaudo, private communication). However, unless schemes are developed to separate the GaN film from its underlying substrate, thermal strain and associated cracking and/or bowing will limit the maximum useful film thickness and ease of device processing.

1. DEFECT REDUCTION

The predominant difficulties with the employment of HVPE as a bulk growth technique are the structural defects and strain that result from the use of a foreign substrate. As dislocations do tend to interact, growing thicker layers does afford a net reduction of defects. However, techniques for substantial and efficient reduction of dislocations are being developed to address the defect problem. An approach of particular significance is the use of epitaxial lateral overgrowth for suppressing/bending a significant fraction of the threading dislocations in the material (Usui et al., 1997). By laterally overgrowing oxide stripes deposited on OMVPE-grown GaN films, they were able to achieve about two orders of magnitude reduction in defect density to the mid-10^7 cm^{-2} level, as compared with growth directly on sapphire. The authors also reported an apparent suppression of thermal cracking in the films on cooldown, allowing them to grow 100-μm-thick samples without cracks.

2. EPITAXIAL OVERGROWTHS

In order to demonstrate the utility of using such films as substrates for subsequent device growth, we investigated films overgrown on $\sim$13-μm-

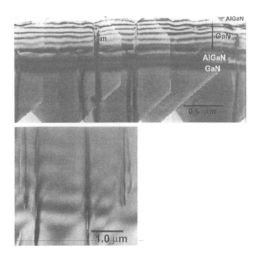

FIG. 22. Cross-sectional TEM micrographs of MBE-grown AlGaN/GaN double heterostructure (top) and 2-μm-thick OMVPE-grown GaN (bottom) deposited on ∼13-μm-thick HVPE buffers.

thick HVPE buffers by electron-cyclotron-resonance-plasma-enhanced MBE (ECR-MBE) and OMVPE. Cross-sectional TEM of the interface shows that by both techniques the overlayer replicates the defect structure of the HVPE film, resulting in dislocation densities of $\sim 10^8\,\mathrm{cm}^{-2}$ for both types of films. While this is comparable to state-of-the-art OMVPE material grown on sapphire, the dislocation density in the film grown by ECR-MBE is far lower than is observed for MBE films grown on sapphire and comparable to the best reported for growth on SiC. This suggests that by further optimization of the HVPE film, substantial improvements in the quality of OMVPE- and MBE-grown material can be achieved by use of HVPE substrates. From the AlGaN/GaN double heterostructure grown by ECR-MBE and shown in Fig. 22, optically pumped lasing was observed with a threshold of 0.5 $\mathrm{MW/cm}^{-2}$ and clearly defined cavity modes (Aggarwal et al., 1996). To our knowledge, this is the first observation of lasing in MBE-grown material.

3. FILM/SUBSTRATE SEPARATION

In order to facilitate current injection, heat sinking and device processing, it would be very desirable to be able to separate the HVPE-grown GaN layer from the underlying substrate. Detchprohm et al. reported on sepa-

1 HYDRIDE VAPOR-PHASE EPITAXIAL GROWTH OF III-V NITRIDES

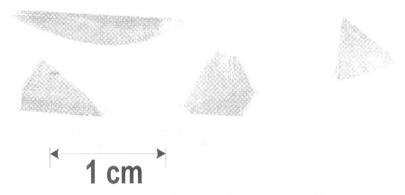

FIG. 23. Free-standing pieces of ~20-μm-thick GaN.

ration of the GaN from the sapphire substrate by using a chemically soluble, sputter-deposited ZnO buffer layer (Detchprohm et al., 1992, 1993). They reported that crystallites ~2 × 4 mm in size could be removed from the sapphire substrate by wet etching the ZnO in *aqua regia*. In order to generate larger areas of GaN, the thermal cracking, which limits the maximum crystallite size, needs to be suppressed. To do this, one needs either to develop techniques for separating or removing the substrates at the elevated growth temperatures prior to cooldown or to inhibit cracking so that post-growth dissolution is possible. As already mentioned, we have succeeded in growing films $>70\,\mu$m thick without thermal cracks. However, because the ZnO thermochemically desorbs early on in the film growth, separation of the GaN from the sapphire is not possible. It has been found, however, that by carefully controlling the initial stages of growth, a significant fraction of the ZnO layer can be buried by the GaN growth, resulting in a film that delaminates upon cooldown and yields free-standing GaN pieces, as shown in Fig. 23. Efforts are under way to develop thicker GaN films under these conditions, which will be less susceptible to fragmentation.

VI. Conclusions

The GaN films grown by HVPE can be uniformly deposited over large (2-in. diameter) areas. Film nucleation on the sapphire substrates can be enhanced by utilizing either a GaCl or ZnO pretreatment, thereby eliminating the need for a low-temperature nucleation step. While we believe that the ZnO is usually thermochemically desorbed early in the growth, this ZnO layer does drastically modify the sapphire surface properties and enhance nucleation, thus resulting in superior film morphology and electrical properties. The structural, optoelectronic, and electrical properties of these films compare with state-of-the-art OMVPE-grown films. Additionally, these films have been grown to thicknesses of 74 μm without optically visible thermally induced cracking. Structural characterization of these films shows that the film quality continues to improve with thickness, even for thicknesses greater than 15 μm. The high growth rate, large crack-free film thickness, and low-defect density demonstrate that HVPE is viable for the growth of low defect GaN thick films for use as substrates for device overgrowths.

Acknowledgments

I would like to acknowledge W. Götz from Hewlett-Packard Corporation and L. Romano from Xerox PARC for providing much of the characterization data presented, as well as the technical contributions of D. Bour from Xerox PARC and P. A. Maki, R. L. Aggarwal, Z. L. Liau, and I. Melngailis from MIT Lincoln Laboratory. I also would like to acknowledge the technical assistance of D. Hovey, J. Daneu, and B. S. Krusor. And finally I would like to thank I. Melngailis for reviewing this manuscript.

References

Aggarwal, R. L., Maki, P. A., Molnar, R. J., Liau, Z.-L., and Melngailis, I. (1996). Optically pumped GaN/Al$_{0.1}$Ga$_{0.9}$N double-heterostructure ultraviolet laser. *J. Appl. Phys.*, **79**, 2148–2150.

Akasaki, I., Naniwae, K., Itoh, K., Amano, H., and Hiramatsu, K. (1991). Growth and properties of single crystalline GaN films by hydride vapor phase epitaxy. *Cryst. Proper. Prep.*, **32–34**, 154–157.

Amano, H., Kito, M., Hiramatsu, K., and Akasaki, I. (1989). P-type conduction in Mg-doped GaN treated with low-energy electron beam irradiation (LEEBI). *Jpn. J. Appl. Phys.*, **28**, L2112–L2114.

Ban, V. S. (1972). Mass spectrometric studies of vapor-phase crystal growth: II. GaN. *J. Electrochem. Soc.*, **119**, 761.

Baranov, B., Däweritz, L., Gutan, V. B., Jungk, G., Neumann, H., and Raidt, H. (1978). Growth and properties of Al$_x$Ga$_{1-x}$N epitaxial layers. *Phys. Stat. Sol.*, **49**, 629–636.

Born, P. J. and Robertson, D. S. (1980). The chemical preparation of gallium nitride layers at low temperatures. *J. Mater. Sci.*, **15**, 3003–3009.

Crouch, R. K., Debnam, W. J., and Fripp, A. L. (1978). Properties of GaN grown on sapphire substrates. *J. Mater. Sci.*, **13**, 2358–2364.

Detchprohm, T., Hiramatsu, K., Amano, H., and Akasaki, I. (1992). Hydride vapor phase epitaxial growth of a high quality GaN film using a ZnO buffer layer. *Appl. Phys. Lett.*, **61**, 2688–2690.

Detchprohm, T., Amano, H., Hiramatsu, K., and Akasaki, I. (1993). The growth of thick GaN film on sapphire substrate by using ZnO buffer layer. *J. Cryst. Growth*, **128**, 384–390.

Detchprohm, T., Hiramatsu, K., Sawaki, N., and Akasaki, I. (1994). The homoepitaxy of GaN by metalorganic vapor phase epitaxy using GaN substrates. *J. Cryst. Growth*, **137**, 170–174.

Dismukes, J. P., Yim, W. M., Tietjen, X. X., and Novak, R. E. (1970). Vapor deposition of semiconducting mononitrides of scandium, yttrium, and the rare-earth elements. *RCA Rev.*, **31**, 680–691.

Dismukes, J. P., Yim, W. M., and Ban, V. S. (1972). Epitaxial growth and properties of semiconducting ScN. *J. Cryst. Growth*, **13/14**, 365–370.

Hagen, J., Metcalfe, R. D., Wickenden, D., and Clark, W. (1978). Growth and properties of $Ga_xAl_{1-x}N$ compounds *J. Phys. C: Solid State Phys.*, **11**, L143–L146.

Huang, K. H., Yu, J. G., Kuo, C. P., Fletcher, R. M., Osentowski, T. D., Stinson, L. J., and Craford, M. G. (1992). Twofold efficiency improvement in high performance AlGaInP light-emitting diodes in the 555–620 nm spectral region using a thick GaP window layer. *Appl. Phys. Lett.*, **61**, 1045–1047.

Ilegems, M. J. (1972). Vapor epitaxy of gallium nitride. *J. Cryst. Growth*, **13/14**, 360–364.

Jacob, G., Boulou, M., and Furtado, M. (1977a). Effect of growth parameters on the properties of GaN:Zn epilayers. *J. Cryst. Growth*, **42**, 136–143.

Jacob, G. and Bois, D. (1977b). Efficient injection mechanism for electroluminescence in GaN. *Appl. Phys. Lett.*, **30**, 412–414.

Jacob, G., Boulou, M., and Bois, D. (1978). GaN electroluminescent devices: preparation and studies. *J. Lumin.*, **17**, 263–282.

Karpinski, J., Porowski, S., and Miotkowska, S. (1982). High pressure vapor growth of GaN. *J. Cryst. Growth*, **56**, 77–82.

Khan, M. A., Shur, M. S., Kuznia, J. N., Chen, Q., Burm, J., and Schaff, W. (1995). Temperature activated conductance in GaN/AlGaN hetero-junction field effect transistors operating at temperatures up to 300 °C. *Appl. Phys. Lett.*, **66**, 1083–1085.

Lee, H., Yuri, M., Ueda, T., and Harris, J. S. (1996). Thermodynamic analysis and growth characterization of thick GaN films grown by chloride VPE using $GaCl_3/N_2$ and NH_3/N_2. *Mater. Res. Soc. Symp. Proc.*, **423**, 233–238.

Lester, S. D., Ponce, F. A., Craford, M. A., and Steigerald, D. A. (1995). High dislocation densities in high efficiency GaN-based light-emitting diodes. *Appl. Phys. Lett.*, **66**, 1249–1251.

Liau, Z. L., Aggarwal, R. L., Maki, P. A., Molnar, R. J., Walpole, J. N., Williamson, R. C., and Melngailis, I. (1996). Light scattering in high-dislocation density GaN. *Appl. Phys. Lett.*, **69**, 1665–1667.

Liu, S. S. and Stevenson, D. A. (1978). Growth kinetics and catalytic effects in the vapor phase epitaxy of gallium nitride. *J. Electrochem. Soc.*, **125**, 1161–1169.

Maruska, H. P. and Tietjen, J. J. (1969). The preparation and properties of vapor-deposited single-crystalline GaN. *Appl. Phys. Lett.*, **15**, 327–329.

Maruska, H. P., Rhines, W. C., and Stevenson, D. A. (1972). Preparation of Mg-doped GaN diodes exhibiting violet electroluminescence. *Mater. Res. Bull.* **7**, 777–782.

Maruska, H. P., Stevenson, D. A., and Pankove, J. I. (1973). Violet luminescence of Mg-doped GaN. *Appl. Phys. Lett.*, **22**, 303–305.

Melnik, Y. V., Nikitina, I. P., Zubrilov, A. S., Sitnikova, A. A., Musikhin, Y. G., and Dmitriev, V. A. (1996). High-quality GaN grown directly on SiC by halide vapour phase epitaxy.

Inst. Phys. Conf. Ser., **142**, 863–866.
Molnar, R. J., Lei, T., and Moustakas, T. D. (1993). Electron transport mechanism in gallium nitride. *Appl. Phys. Lett.*, **62**, 72–74.
Molnar, R. J., Nichols, K. B., Maki, P., Brown E. R., and Melngailis, I. (1995). The role of impurities in hydride vapor phase epitaxially grown gallium nitride. *Mater. Res. Soc. Symp. Proc.*, **378**, 479–484.
Molnar, R. J., Aggarwal, R., Liau, Z. L., Brown, E. R., Melngailis, I., Götz, W., Romano, L. T., and Johnson, N. M. (1996a). Optoelectronic and structural properties of high-quality GaN grown by hydride vapor phase epitaxy. *Mater. Res. Soc. Symp. Proc.*, **395**, 189–194.
Molnar, R. J., Maki, P., Aggarwal, R., Liau, Z. L., Brown, E. R., Melngailis, I., Götz, W., Romano, L. T., and Johnson, N. M. (1996b). Gallium nitride thick films grown by hydride vapor phase epitaxy. *Mater. Res. Soc. Symp. Proc.*, **423**, 221–226.
Molnar, R. J., Götz, W., Romano, L. T., and Johnson, N. M. (1997). Growth of gallium nitride by hydride vapor-phase epitaxy. *J. Cryst. Growth*, **178**, 147–156.
Morimoto, Y., Uchiho, K., and Ushio, S. (1973). Vapor phase epitaxial growth of GaN on GaAs, GaP, Si, and sapphire substrates from $GaBr_3$ and NH_3. *J. Electrochem. Soc.*, **120**, 1783–1785.
Nakamura, S. (1991). GaN growth using GaN buffer layer. *Jpn. J. Appl. Phys.*, **30**, L1705–L1707.
Nakamura, S., Senoh, M., Nagahama, S., Iwasa, N., Yamada, T., Matsushita, T., Kiyoku, H., Sugimoto, Y., Kozaki, T., Umemoto, H., Sano, M., and Chocho, K. (1998). InGaN/GaN/AlGaN-based laser diodes with modulation-doped strained-layer superlattices grown on an epitaxially laterally overgrown GaN substrate. *Appl. Phys. Lett.*, **72**, 211–213.
Naniwae, K., Itoh, S., Amano, H., Itoh, K., Hiramatsu, K., and Akasaki, I. (1990). Growth of single crystal GaN substrate using hydride vapor phase epitaxy. *J. Cryst. Growth*, **99**, 381–384.
Nickl, J. J., Just, W., and Bertinger, R. (1974). Preparation of epitaxial gallium nitride. *Mater. Res. Bull.*, **9**, 1413–1420.
Nikolaev, A., Melnik, Y., Kuznetsov, N., Strelchuk, A., Kovarsky, A., Vassilevski, K., and Dmitriev, V. (1998). GaN pn-structures grown by hydride vapor phase epitaxy. *Mater. Res. Soc. Symp. Proc.*, **482**, 251–256.
Ohki, Y., Toyoda, Y., Kobayasi, H., and Akasaki, I. (1982). Fabrication and properties of a practical blue-emitting GaN m-i-s diode. *Inst. Phys. Conf. Ser.*, **63**, 479–484.
Pankove, J. I., Miller, E. A., and Berkeyheiser, J. E. (1971). GaN electroluminescent diodes. *RCA Rev.*, **32**, 383–392.
Pankove, J. I. (1973). Blue-green numeric display using electroluminescent GaN. *RCA Rev.*, **34**, 336–343.
Pankove, J. I. and Hutchby, J. A. (1976). Photoluminescence of ion-implanted GaN. *J. Appl. Phys.*, **47**, 5387–5390.
Perkins, N. R., Horton, M. N., and Kuech, T. F. (1996). Halide vapor phase epitaxy of gallium nitride films on sapphire and silicon substrates. *Mater. Res. Soc. Symp. Proc.*, **395**, 243–248.
Reed, T. B. (1971). *Free Energy of Formation of Binary Compounds*. Cambridge, MA: The MIT Press.
Romano, L. T., Krusor, B. S., Anderson, G. A., Bour, D. P., Molnar, R. J., and Maki, P. (1996). Structural characterization of thick GaN films grown by hydride vapor phase epitaxy. *Mater. Res. Soc. Symp. Proc.*, **423**, 245–250.
Romano, L. T., Krusor, B. S., and Molnar, R. J. (1997). Structure of GaN films grown by hydride vapor phase epitaxy. *Appl. Phys. Lett.*, **71**, 2283–2285.
Safvi, S. A., Perkins, N. R., Horton, M. N., Thon, A., Zhi, D., and Kuech, T. F. (1996). Optimization of reactor geometry and growth conditions for GaN halide vapor phase

epitaxy. *Mater. Res. Soc. Symp. Proc.*, **423**, 227–232.
Sano, M. and Aoki, M. (1976). Epitaxial growth of undoped and Mg-doped GaN. *Jpn. J. Appl. Phys.*, **15**, 1943–1950.
Saparin, G. V., Obyden, S. K., Chukichev, M. V., and Popov, S. I. (1984). Cathodoluminescent contrast of direct writing patterns in the scanning electron microscope. *J. Lumin.*, **31 & 32**, 684–686.
Seifert, W., Fitzl, G., and Butter, E. (1981). Study on the growth rate in VPE of GaN. *J. Cryst. Growth*, **52**, 257–262.
Seifert, W., Franzheld, R., Butter, E., Sobotta, H., and Reide, V. (1983). On the origin of free-carriers in high-conducting n-GaN. *Cryst. Res. Technol.*, **18**, 383–390.
Shintani, A. and Minagawa, S. (1974). Kinetics of the epitaxial growth of GaN using Ga, HCl, and NH_3. *J. Cryst. Growth*, **22**, 1–5.
Tsuchiya, H., Okahisa, T., Hasegawa, F., Okumura, H., and Yoshida, S. (1994). Homoepitaxial growth of cubic GaN by hydride vapor phase epitaxy on cubic GaN/GaAs substrates prepared with gas source molecular beam epitaxy. *Jpn. J. Appl. Phys.*, **33**, 1747–1752.
Tsuchiya, H., Akamatsu, M., Ishida, M., and Hasegawa, F. (1996). Layer-by-layer growth of GaN on GaAs substrates by alternate supply of $GaCl_3$ and NH_3. *Jpn. J. Appl. Phys.*, **35**, L748–L750.
Tsuchiya, H., Sunaba, K., Yonemura, S., Suemasu, T., and Hasegawa, F. (1997). Cubic dominant GaN growth on (001) GaAs substrates by hydride vapor phase epitaxy. *Jpn. J. Appl. Phys.*, **36**, L1–L3.
Usui, A., Sunakawa, H., Sakai, A., and Yamaguchi, A. A. (1997). Thick GaN epitaxial growth with low dislocation density by hydride vapor phase epitaxy. *Jpn. J. Appl. Phys.*, **36**, L899–L902.
Wickenden, D. K., Faulkner, K. R., Brander, R. W., and Isherwood, B. J., (1971). Growth of epitaxial layers of gallium nitride on silicon carbide and corundum substrates. *J. Cryst. Growth*, **9**, 158–164.
Yamaguchi, A., Manak, T., Sakai, A., Sunakawa, H., Kimura, A., Nido, M., and Usui, A. (1996). Single domain hexagonal GaN films on GaAs (100) vicinal substrates grown by hydride vapor phase epitaxy. *Jpn. J. Appl. Phys.*, **35**, L873–L875.
Yim, W. M., Stofko, E. J., Zanzucchi, P. J., Pankove, J. I., Ettenberg, M., and Gilbert, S. L. (1973). Epitaxially grown AlN and its optical band gap. *J. Appl. Phys.*, **44**, 292–296.

CHAPTER 2

Growth of III-V Nitrides by Molecular Beam Epitaxy

T. D. Moustakas

PHOTONICS CENTER
DEPARTMENT OF ELECTRICAL AND COMPUTER ENGINEERING
BOSTON UNIVERSITY
BOSTON, MASSACHUSETTS

I. INTRODUCTION	33
II. BACKGROUND OF MOLECULAR BEAM EPITAXY TECHNIQUES	34
III. NITROGEN SOURCES USED FOR THE GROWTH OF III-V NITRIDES BY MOLECULAR BEAM EPITAXY	40
1. Ammonia as a Nitrogen Source	40
2. Nitrogen Plasma Sources	41
3. Nitrogen Ion Sources	50
IV. GaN FILMS	51
1. Film Growth	51
2. Film Structure and Microstructure	60
3. Doping Studies	70
4. Electronic Structure and Optoelectronic Properties	80
V. InGaAlN ALLOYS	89
1. Growth	89
2. Phase Separation and Long-range Atomic Ordering	89
VI. MULTIQUANTUM WELLS	101
1. $In_xGa_{1-x}N/Al_yGa_{1-y}N$ Multiquantum Wells	101
2. GaN/AlGaN Multiquantum Wells	105
VII. DEVICE APPLICATIONS	110
1. Device Processing	110
2. Devices	111
VIII. CONCLUSIONS	119
REFERENCES	121

I. Introduction

The full potential of III-V nitrides for various optoelectronic applications requires significant progress in the areas of heteroepitaxial growth, crystal structure and microstructure, impurity doping, formation of defects, alloying phenomena, and the formation of homojunction and heterojunction structures. The systematic study of these issues is important in order to develop predictive and reproducible methods of growing these materials with desired physical properties.

The various thin-film deposition methods employed in the growth of GaN and other III-V nitrides have a number of advantages for addressing the issues discussed previously. The chloride transport method, which was developed in the late 1960s (Maruska and Tietjen, 1969), is characterized by high growth rates, which leads to thick films with smaller concentrations of misfit-related defects close to the free surface and potentially can be useful as a substrate material (Molnar, 1998). The metallorganic chemical vapor deposition (MOCVD) method has been used by Nichia, Inc. for the fabrication of III-V nitride LEDs and lasers (Nakamura, 1998), and upon optimization has some attractive features for the manufacturing of these devices (Denbaars and Keller, 1998). The molecular beam epitaxy (MBE) method has a number of advantages for the study of new materials and in particular for the study of epitaxial phenomena, because it is equipped with a number of *in situ* probes that monitor the growth in real time. In addition, the MBE process has emerged as a practical growth method for the fabrication of GaAs microwave and optoelectronic devices. This method has been employed recently for the growth of the entire family of III-V nitrides and has led to materials with physical properties equivalent to those produced by the MOCVD method. In this chapter we present a detailed description of the MBE method as used for the growth of III-V nitrides.

Section II gives a general background of the MBE techniques as practiced for the growth of traditional III-V compounds and devices. Section III addresses issues related to nitrogen sources suitable for the growth of III-V nitrides. Section IV covers growth, structure, doping and properties of GaN films. Section V addresses the growth and structure of InGaAlN alloys with particular emphasis on phase separation and ordering phenomena. Section VI discusses InGaN/AlGaN and GaN/AlGaN multiquantum wells (MQWs). Finally, Section VII deals with device applications of III-V nitrides grown by MBE.

II. Background of Molecular Beam Epitaxy Techniques

Molecular beam epitaxy (MBE) is a thin-film deposition process in which thermal beams of atoms or molecules react on the clean surface of a single-crystalline substrate that is held at high temperatures under ultrahigh-vacuum conditions to form an epitaxial film. Thus, contrary to the chemical vapor deposition (CVD) processes where chemical reactions play an important role, the MBE process is a physical method of thin-film deposition. This method of thin-film deposition and its applications to the growth of III-V compounds has been reviewed by a number of authors (Ploog, 1980; Cho, 1983; Gossard, 1982; Chang and Ploog, 1985; Parker, 1986; Davies and Williams, 1986; Foxon and Harris, 1987; Foxon and Joyce, 1990).

The vacuum requirements for the MBE process are typically better than 10^{-10} torr. This makes it possible to grow epitaxial films with high purity and excellent crystal quality at relatively low substrate temperatures. Additionally, the ultrahigh-vacuum environment allows the study of surface, interface and bulk properties of the growing films in real time by employing a variety of structural and analytical probes. Although the MBE deposition process was first proposed by Gunther (1958), its implementation had to await the development of ultrahigh-vacuum technology, when it was successfully applied for the growth of epitaxial GaAs films (Davey and Pankey, 1968).

The development of the MBE process in its present state was primarily motivated by the desire to study new quantum phenomena in semiconducting superlattice structures (Esaki and Tsu, 1970). The demonstration of such phenomena required the growth of superlattice structures with atomically abrupt and perfect interfaces, and control of the layer thicknesses down to a single monolayer. The development of crystal-growth techniques, which led to such a degree of lattice and interface perfection, has been attained by the contributions of many disciplines of science and technology.

Modern MBE deposition systems are designed to produce high-quality materials and devices at high throughput. The requirement of maintaining an ultrahigh-vacuum environment while simultaneously improving the throughput was addressed through the design of MBE systems consisting of multiple chambers separated by gate valves. All commercially available equipment is constructed with at least three such chambers. The first chamber serves for sample introduction and is capable of medium-high vacuum (10^{-6}–10^{-8} torr). The second chamber is capable of ultrahigh vacuum and acts principally as a buffer between the introduction and the growth chambers. This chamber is also used for substrate preparations, such as outgassing or sputter etching, and for accommodation of surface analytical facilities such as auger electron spectroscopy (AES), secondary ion mass spectroscopy (SIMS), x-ray photoelectron spectroscopy (XPS), and ultraviolet photoelectron spectroscopy (UPS). The third chamber, the growth chamber, is capable of ultrahigh vacuum ($<10^{-10}$ torr) and its design criteria depend greatly on the nature of the materials being deposited.

The facilities in the third chamber are capable of forming and monitoring the ultrahigh-vacuum environment; heating and monitoring the temperature of the substrate fairly accurately; generating and determining the intensity of the molecular or atomic beams; controlling composition profiles through beam interruption; and studying surface and interface phenomena during film growth.

The primary pumping of the growth chamber is accomplished with a combination of either storage pumps (ion, titanium sublimation, and cryopumping facilities) or throughput pumps (diffusion or turbomolecular pumps) depending on the nature and vapor pressure of the evaporants. Additionally, the sources and the substrate, which employ a large heating load, are surrounded by a cryopanel, usually cooled by liquid nitrogen. This secondary pumping minimizes the unintentional incorporation of impurities into the growing film. Such a combination of pumping, after a typical system bakeout of approximately 24 h at 250 °C, results in base pressure below the x-ray limit of the ionization gauge ($<2 \times 10^{-11}$ torr). The principal impurities, as monitored by a residual gas analyzer, are H_2, H_2O, CO, and CO_2 at partial pressures of typically less than 5×10^{-13} torr.

The design of the substrate holder allows substrate rotation during film growth and additional motions to facilitate surface analysis and beam-flux monitoring. The substrate, which is usually held to a molybdenum block with indium solder, is heated radiatively and its temperature monitored with an optical pyrometer. All parts of the substrate holder are fabricated with ultrahigh purity and refractory materials. The uniform substrate heating in such designs results in thickness and doping uniformity between 0.5 to 1% over a 2-inch wafer.

The most common method of creating molecular beams for MBE growth is through the use of Knudsen effusion cells. In ideal Knudsen cells the orifice should be less than the mean free path of the vapor molecules within the cell, and the beam flux can be calculated from the equilibrium vapor pressure using kinetic theory. In practice, however, the molecular beam sources are not ideal Knudsen cells because they employ large apertures, which are necessary to achieve enhanced growth rates and better compositional uniformities. Thus, the beam fluxes are usually measured with a nude ionization gauge placed at the location of the substrate. The crucibles employed in Knudsen cells can be made from a variety of materials. Pyrolytic boron nitride (PBN) appears to be the preferred material for the growth of III-V compounds. The temperature of the crucible is controlled to within ± 1 °C.

There are a number of alternative sources for creating molecular beams. Most prominent among them are electron-beam evaporation and gas sources. Electron-beam evaporation is commonly used for low vapor-pressure materials such as silicon and refractory metals. Gas sources have been developed in several laboratories and implemented in commercial MBE systems. Such gas sources include, for example, AsH_3 and PH_3 to produce the group V elements (Panish, 1980; Calawa, 1981). These gases are thermally "cracked" to be dimers As_2 and P_2 before they reach the substrate. Thus, the process is not a chemical vapor deposition. The current tendency is to develop gas sources for the group III elements using metallorganic sources. The process of using such gas sources is called metallorganic molecular beam epitaxy (MOMBE) in analogy to MOCVD.

Control over the film composition and doping profile is attained by incorporating a mechanical shutter in front of each source. As the flow of molecules or atoms from the source to the substrate is in the molecular rather than the hydrodynamic flow regime, positioning a shutter in front of a source will effectively stop the beam from reaching the substrate. In fact it is because the flow regime is molecular that the process is called molecular beam epitaxy. Because the growth rate of the MBE process is inherently slow (~ 1 monolayer/s), shutter actuation times of a fraction of a second are required to produce compositionally modulated materials with interface smoothness of one atomic layer.

While the majority of surface analytical probes are accommodated in the preparation chamber to avoid possible contamination by the evaporants, reflection high energy electron diffraction (RHEED) is routinely used in the growth chamber to monitor and control the growth process. It is generally recognized that this method has played a major role in the development of the MBE process.

RHEED consists of a well-collimated monoenergetic electron beam, which is directed at a grazing angle of about 1° toward the substrate. The primary electron beam has an energy of between 10 to 20 keV, resulting in an energy component perpendicular to the substrate of about 100 eV. Thus, the penetration depth of the incident electron beam is approximately equal to a few atomic layers. As a result, a smooth crystal surface acts as a two-dimensional grating and diffracts the electrons. The diffraction pattern is formed on a fluorescent screen placed diametrically opposite to the electron gun. RHEED is routinely used to study thermal desorption of oxides prior to growth, to control the initial stages of epitaxial growth, and to study surface reconstruction as a function of growth parameters.

The removal of oxides from a substrate is the first step prior to epitaxial growth. In some cases this step takes place in the preparation chamber in order to avoid contaminating the growth chamber just before the initiation of the epitaxy. However, for III-V compounds, such as GaAs, oxide desorption takes place in the growth chamber — in the presence of arsenic overpressure — in order to prevent surface fractionation after the removal of the oxides. A clear diffraction RHEED pattern indicates that all oxides have been removed.

Another potential application of RHEED is the study of surface topography. The RHEED pattern from a smooth-single crystalline surface is expected to have the form of a series of streaks running perpendicular to the surface of the crystal, which is consistent with a two-dimensional diffraction. On the other hand, the RHEED pattern from a rough surface is expected to be spotty, because the penetration of electrons through surface asperities results in three-dimensional diffraction. RHEED also is used to study surface reconstruction. For example, the (100) GaAs surface used for MBE growth of GaAs reconstructs to different configurations in order to lower its free energy. As surface reconstruction leads to a lower symmetry than that of the bulk crystal, extra diffraction lines are expected in the RHEED pattern.

Besides these RHEED applications, it has been observed that when the growth is initiated, the intensity of the RHEED features shows an oscillatory behavior. Current thinking is that these intensity fluctuations are related to both crystal growth and electron diffraction phenomena. The influence of diffraction in the intensity enhancement is probably due to the multiple scattering originating from the beam penetration into the solid. Early interpretation of the RHEED intensity oscillations is that the thin-film growth proceeds in a layer-by-layer mode and thus the period of the oscillations corresponds to a monolayer growth (Foxon and Harris, 1987). These results can be used to calculate growth rates and the composition of ternary and quaternary III-V compounds. Figure 1 shows how the composition of AlGaAs can be calculated from the growth rates of their components of GaAs and AlAs (Moustakas, 1988).

From the previous discussion it is apparent that, compared to the other methods of thin-film epitaxial growth, molecular beam epitaxy has some unique advantages that can be summarized as follows: (a) The growth rate is generally low, approximately 1 monolayer/s. This allows compositional and doping profile changes to within atomic dimensions through the actuation of mechanical shutters; (b) the growth temperature is relatively low and thus interdiffusion between layers of different composition is negligible; (c) the MBE growth mechanism leads to atomically smooth surfaces; (d) the ability to study growth phenomena in real time provides opportunities for scientific innovation and quality control in the production environment; and (e) all steps of the MBE deposition process can be fully automated.

These favorable features, together with significant progress in the design of modern MBE systems, have led to the growth of films with excellent thickness and doping uniformities, and excellent crystal quality over large

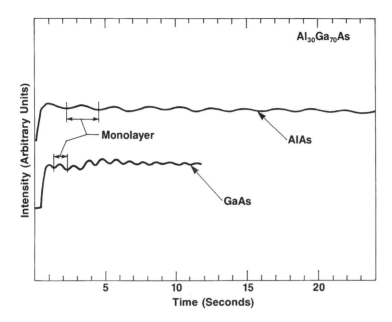

Fig. 1. Determination of the composition of AlGaAs from the growth rates of the GaAs and AlAs components, as determined from RHEED oscillations. (Reprinted from Moustakas, 1988.)

area substrates. Currently, the commercial MBE systems are designed for 3-in wafers.

The successful use of these features has led to the fabrication of GaAs microwave devices such as varactor diodes, IMPATT diodes, mixer diodes, and Schottky barrier field-effect transistors (FETs), as well as to the fabrication of optoelectronic devices such as optical waveguides, light-emitting diodes, and heterostructure-injection lasers.

Furthermore, progress has been made in the growth of epitaxial structures with lateral dimensional control using mechanical masks made of either refractory metals or silicon. Linewidths down to 1 μm have been reported by Tsang and Ilegems (1977). Additionally, by appropriate motion of the masks with respect to the substrate, three-dimensional patterns and tapered structures were fabricated (Tsang and Cho, 1978; Tsang and Ilegems, 1979). This type of lateral dimensional control is required in optoelectronic devices made of GaAs/AlGaAs multilayers.

The multichamber design of the MBE systems also facilitates the formation of metallic contacts. Ideal semiconductor-metal interfaces have been formed by epitaxial growth of single-crystalline Al onto (001) GaAs (Cho and Dernier, 1978). Progress has also been made in the fabrication of ohmic contacts.

The area of superlattice structures and device concepts based on such structures is one of the most active fields in scientific research today. The MBE process has contributed significantly to this new class of artificially modulated materials. Such structures have been fabricated from semiconductors, metals, insulators, and combinations of these materials (Chang and Giessen, 1985).

In conclusion, the MBE process has emerged as a practical growth method for a variety of materials and devices and as a unique scientific tool to study thin-film growth phenomena *in situ*. The application of this method to the growth of III-V nitrides is discussed in the following sections.

III. Nitrogen Sources Used for the Growth of III-V Nitrides by Molecular Beam Epitaxy

The primary precursor gases employed during the growth of GaN by the MOCVD method are $Ga(CH_3)_3$ or $Ga(CH_5)_3$ and NH_3 while $GaCl_3$ and NH_3 are used for the growth of GaN by the halide vapor-phase epitaxy (VPE) method. Typical epitaxial temperatures in these methods are 1000–1100 °C, which are sufficient for the thermal decomposition of ammonia and the required chemical reactions for the synthesis of III-V nitrides (Denbaars and Keller, 1998; Molnar, 1998).

Deposition of GaN and other III-V nitrides by MBE requires the development of appropriate nitrogen sources, as molecular nitrogen (N_2) does not chemisorb on Ga due to its large binding energy of 9.5 eV. Three approaches to this problem are currently being developed. The first utilizes ammonia as a nitrogen source, the second utilizes plasma-activated molecular nitrogen, and the third utilizes low-energy ionic nitrogen. In the following we describe these various approaches and point out their relative merits and problems.

1. AMMONIA AS A NITROGEN SOURCE

A number of groups have been developing the growth of GaN by MBE using ammonia as the source of nitrogen for the past fifteen years (Yoshida *et al.*, 1975; Yoshida *et al.*, 1979; Yoshida *et al.*, 1982; Yoshida *et al.*, 1983a, 1983b; Powell *et al.*, 1992; Yang et al., 1995; Moriyasu *et al.*, 1995; Kim *et al.*, 1996; Kamp *et al.*, 1996; Kamp *et al.*, 1997; Grandjean *et al.*, 1997a, 1997b; Grandjean *et al.*, 1998). In this method, ammonia is decomposed on the surface of the substrate by pyrolysis at 700–900 °C, a temperature that is much lower than that used for the growth of GaN by MOCVD. Detailed studies of the growth kinetics during growth of GaN by MBE with ammonia as the nitrogen source were recently presented by a number of researchers (Evans *et al.*, 1995; Kim *et al.*, 1996; Kamp *et al.*, 1997). These studies showed also a correlation between growth conditions and the structure and properties of the film. According to these studies, the growth mechanism of GaN by this method is believed to consist of the thermal activation of ammonia and surface reaction mitigated dissociation followed by reaction of N with Ga to form GaN.

The general trends in this method of GaN growth are that the growth rate increases with substrate temperature (700–800 °C), while simultaneously the structure and optoelectronic proaperties of the films improve. This implies that the growth rate is controlled not by the Ga flux but by the density of thermally activated reactive ammonia molecules on the surface. At 800 °C, the growth rate was found to be 1.2 μm/hr. Films grown at such temperatures were found to be semi-insulating with excellent optical properties (Kim *et al.*, 1996; Kamp *et al.*, 1997; Grandjean *et al.*, 1998).

Metallorganic MBE (MOMBE) that employs metallorganic precursors for group III sources is still in the early stages of development (Tsuchiya *et al.*, 1995; Abernathy *et al.*, 1997).

In conclusion, this deposition method leads to high-quality GaN films at a growth rate equivalent to that of the MOCVD method without the high consumption of ammonia, which is required in the latest method. Furthermore, this method demonstrated the fabrication of a high-transconductance modulation doped field effect transistor (MODFET) (Aktas *et al.*, 1995) and GaN/AlGaN separate confinement heterostructures (Schmidt *et al.*, 1996). One potential problem with this method is the reaction of ammonia with various filaments used to heat the cells and the substrate or employed in various diagnostic studies (mass spectroscopy, RHEED, ionization gauges).

2. NITROGEN PLASMA SOURCES

Two plasma methods have been developed over the past several years for the activation of molecular nitrogen. The one involves microwave plasmas (2.45 GHz) assisted by an electron cyclotron resonance (ECR) condition, while the second utilizes RF plasmas (13.5 MHz).

a. Microwave Plasma-assisted ECR Sources

Electron cyclotron resonance (ECR) plasma sources have been used by a number of groups to provide plasma-activated nitrogen for the growth of

III-V nitrides by molecular beam epitaxy (Paisley et al., 1989; Lei et al., 1991; Strite et al., 1991; Lei et al., 1992; Moustakas et al., 1992; Moustakas et al., 1993; Moustakas and Molnar, 1993; Eddy et al., 1993; Lin et al., 1993; Molnar and Moustakas, 1994; Beresford, 1994; Molnar et al., 1995a; Ohtani et al., 1995; Moustakas, 1996; Moustakas et al., 1996; Singh and Moustakas, 1996a, 1996b; Vaudo et al., 1996; Singh et al., 1996; Singh et al., 1997; Korakakis et al., 1997; Korakakis, 1998; Korakakis et al., 1998a; Moustakas et al., 1998). The design and operation of the compact ECR source manufactured by ASTeX Corporation was discussed in detail by Molnar et al. (1995a) and Ohtani et al. (1995). In the following, we summarize the design criteria and modifications that led to high-quality III-V nitride films, structures, and devices.

The ASTeX compact-ECR source uses an axial electromagnet to stimulate a resonant coupling of the microwave energy (2.45 GHz) with the

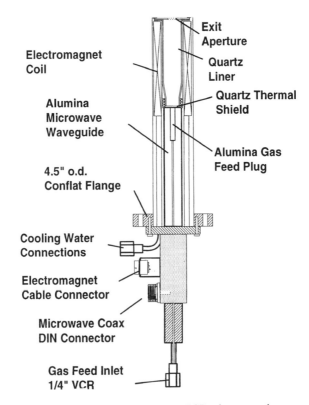

FIG. 2. Schematic drawing of a compact ECR microwave plasma source.

electrons in the discharge as shown in Fig. 2. The axial magnetic field in this system was calculated by the expression (Haus and Melcher, 1989)

$$B(z) = \mu_0 \frac{Ni}{2d} \cdot \left[\frac{\frac{d}{2a} - \frac{z}{a}}{\sqrt{1 + \left(\frac{d}{2a} - \frac{z}{a}\right)^2}} + \frac{\frac{d}{2a} + \frac{z}{a}}{\sqrt{1 + \left(\frac{d}{2a} + \frac{z}{a}\right)^2}} \right] \quad (1)$$

where Ni is the number of amp-turns, d is the length of the solenoid (6 in), and a is the radius of the solenoid (~ 1 in). Figure 3 indicates the strength of the magnetic field as a function of position along the axis of the ECR solenoid.

The cyclotron acceleration promotes electron/gas collisions and results in a high-density plasma, with gas ionization efficiencies as high as 10%. This efficient resonant coupling also allows the source to be operated stably with growth chamber pressures as low as 10^{-5} torr. The lower process pressures and high degree of ionization make these sources desirable candidates for the excitation of nonreactive gases, such as nitrogen, in plasma-assisted growth processes.

Due to the large mean free path of the gaseous species at these low background pressures (~ 1 m at 10^{-4} torr) the growth is carried out in the molecular flow regime, where the transport of atoms or molecules in both thermal beams from the effusion cells as well as the beam of activated

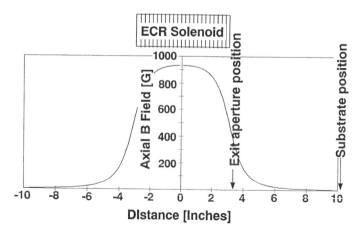

FIG. 3. Magnetic field strength along the axis of the ECR source solenoid. (Reprinted from Molnar et al., 1995a.)

nitrogen from the ECR source occurs in a collisionless manner. This low collision rate coupled with magnetic field effects on the charged species (electrons, ions), however, can have a profound effect on the operation of this source. In particular, in sources with axial magnetic fields, the acceleration of charged species down the divergence of the magnetic field can lead to plasma species with highly anisotropic energies (Chen, 1984). Generally, the electrons in the plasma are well confined to magnetic field lines by their Larmor gyration around these field lines and are guided down the divergence of the magnetic field by a relaxation of this gyration (Chen, 1984). The ions, on the other hand, are poorly confined due to their comparatively large mass. However, the diffusion of the forementioned electrons induces an electric field that accelerates the ionic species along the magnetic field lines by coulombic interaction, leading to the so-called ambipolar diffusion process (Matsuoka and Ono, 1988).

The gain in ion translational kinetic energy along the magnetic field lines corresponds to an acceleration toward the substrate and can, therefore, promote damage to the film through ion bombardment effects. This anisotropic kinetic energy component has been shown to increase with distance from the source (Köhler et al., 1993) and can be on the order of several tens of electronvolts at the lower pressure typically used in these sources (Haus and Melcher, 1989), above the ~ 24-eV damage threshold for GaN (Böer, 1990). Such ion energies together with the high-power densities employed in these compact ECR plasma sources can also result in significant sputtering of impurities from the source and chamber walls.

To moderate these phenomena, the original source was modified by the incorporation of exit apertures to control the gas pressure within the source (Molnar et al., 1995a). The aperture consists of an ~ 1-mm thick quartz disc with a 1-cm diam. hole in its center, which was installed at the lip of the source liner (see Fig. 2). Another exit aperture having a single 1/16-in diam. hole was also studied in a test chamber where optical emission studies were carried out to investigate trends in plasma chemistry with pressure inside the source.

As the diameter of the discharge zone in the compact ECR source is smaller than the diameter for stable mode propagation, microwave launching into this region occurs below cutoff, confining the ECR condition to a region close to the thermal shield of the source. In order to estimate the 1-cm aperture's effect on the pressure in the discharge region the flow conductances for both the exit aperture and the quartz liner were calculated. The quartz liner was approximated as a 2.54-cm diam. cylinder and 10 cm in length. The flow conductance of the quartz liner in the molecular flow regime can then be written as (Leybold-Heraues, 1993)

$$C = 12.1(d^3 l)[l/s] = 19.8[l/s] \qquad (2)$$

2 GROWTH OF III-V NITRIDES BY MOLECULAR BEAM EPITAXY

Similarly, the flow conductance for the aperture in the molecular flow regime is given by

$$C = 11.6 \, A[l/s] = 9.1[l/s] \tag{3}$$

The pressure in the ECR zone without the exit aperture (P_{NA}) can now be determined from the pressure in the growth chamber (P_{GC}) and for a nitrogen flow of 6 sccm

$$P_{NA} = P_{GC} + \frac{F_{N2}}{C_{liner}} = 1.2 \cdot 10^{-4}$$
$$+ \frac{6[\text{atm} \cdot cc/\text{min}]}{19.8[l/s]} \cdot \frac{1[\text{min}]}{60[s]} \cdot \frac{1[l]}{1000[cc]} \cdot \frac{760[T]}{1[\text{atm}]} = 4.0[\text{mT}] \tag{4}$$

The pressure in the ECR zone with the exit aperture can also be determined for a nitrogen flow of 6 sccm by first computing the effective conductance of both the liner and the aperture

$$\frac{1}{C_{eff}} = \frac{1}{C_{aper}} + \frac{1}{C_{liner}}, \quad C_{eff} = 6.2[l/s], \quad P_{aper} = P_{GC} + \frac{F_{N2}}{C_{eff}} = 12.4[\text{mT}] \tag{5}$$

Therefore, the introduction of the 1-cm exit aperture results in a pressure increase at the ECR discharge area by at least a factor of three. As discussed previously (Molnar et al., 1995a), the aperture also serves to suppress several pressure-lowering phenomena suggesting that this factor of three is very conservative.

The increase of the pressure inside the ECR source affects the mechanism of nitrogen activation. At low pressures, the electrons and molecules are not in thermal equilibrium. Due to the difference in mass between electrons and ions, the electrons accelerate rapidly, while the ions move slowly. Due to the large electron mean free path, there is no redistribution of energy between electrons and molecules and the gas temperature remains relatively cool. Thus, activated nitrogen is produced primarily by the high-energy electrons, whose concentration is relatively small. As the pressure in the discharge region increases, redistribution of energy between electrons and molecules is expected due to the smaller electron mean free path and this leads to the generation of a higher concentration of activated nitrogen. Furthermore, the increase in the pressure inside the source should lead to an improvement in the properties of films grown with such a source by reducing the kinetic energy of the charged species that collide with both the source and chamber walls. Therefore, it is expected that the exit aperture should also suppress the generation of impurities during the growth.

The ECR source was characterized by Langmuir probe and optical emission spectroscopy. The nude ionization gauge, which is used to measure the beam-equivalent pressure from the various Knudsen effusion cells, was also employed as an electrostatic probe to perform plasma diagnostic studies. This is accomplished by connecting a dc power supply in series with an ammeter to the collector of the gauge. Figure 4 shows a schematic of the measuring circuit and I-V characteristics obtained with the ECR source operating at 30-W microwave power, both with and without the exit aperture. The grid and filaments of the ionization gauge were left electrically floating during these measurements to minimize their perturbation of the plasma.

Relative ion densities were estimated by assuming that the plasma is collisionless, the electron energy distribution is Maxwellian, the plasma is quasi-neutral ($N_i = N_e$), the electron temperature is much greater than the ion temperature ($T_e \gg T_i$) and the ions are singly charged. Then I^2 and V are related by the expression (Heidenreich et al., 1987)

$$\frac{\partial I_i^2}{\partial V_P} = -\frac{3A^2 e^3 N_i^2}{4\pi m_i} \qquad (6)$$

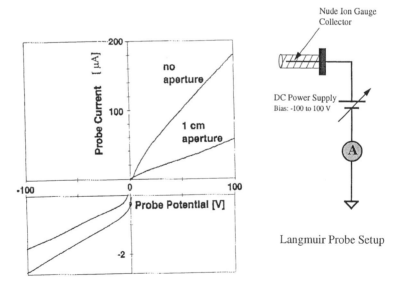

FIG. 4. Langmuir probe circuit with typical I-V characteristics. (Reprinted from Molnar et al., 1995a.)

where I_i is the probe current in the saturated ion density (highly negative) regime, V_p is the applied probe potential, A is the area of the probe, N_i is the ion density, and m_i is the ion mass. The square of the relative ion density, N_i^2 was determined from the slope of the line as shown in Fig. 5. The determination of the actual concentration of N_i requires the area of the probe to be known accurately and due to uncertainty of the effect of components surrounding the probe, this expression was only used to compute relative ion densities. By employing such an interpretation on the data in Fig. 4, we calculated that the ion density is reduced by $\sim 30\%$ by the introduction of the 1-cm exit aperture.

An optical emission spectra typical for a nitrogen plasma generated in the compact electron cyclotron resonance (CECR) source without an aperture is shown in Fig. 6. The measurement was performed using a UV fiber optic bundle to collect light from a viewport of the growth chamber in line-of-sight with the source. The light was dispersed through a 0.25-m monochrometer and imaged onto a diode detector array. A number of nitrogen-plasma emission peaks associated with neutral nitrogen molecules are shown in the spectrum, which involves the first-positive and second-positive series of molecular transitions (Pearse and Gayton, 1963; Wright and Winkler, 1968). Shown also are features related to the molecular N_2^+ ion. Vaudo et al. (1994) have demonstrated that when the source is operated at 250 W, the primary emission peaks are those due to atomic nitrogen occurring in the 600- to 900-nm spectral region.

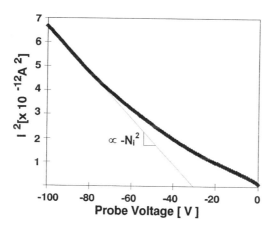

FIG. 5. I^2 vs V in ion-saturation (highly negative) region. (Reprinted from Molnar et al., 1995a.)

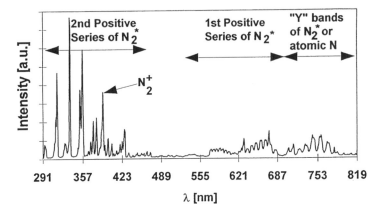

FIG. 6. Typical optical emission spectra obtained for ECR nitrogen plasma. (Reprinted from Molnar et al., 1995a.)

In order to study the effect of pressure on the plasma species generated, optical emission spectra were measured for various flow rates and exit apertures. The emission peak at 391.4 nm is associated with N_2^+ and the peak at 337.1 nm is associated with N_2^*. The ratio of these two peaks is used as a quantitative measure of the relative ion to excited neutral generation and is shown in Fig. 7 for a variety of nitrogen flows and exit apertures.

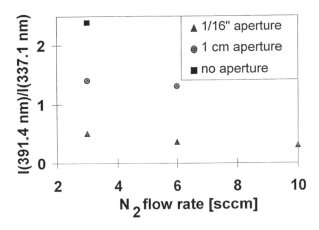

FIG. 7. Ratio of ionic (391.4 nm) to metastable neutral molecular (337.1 nm) nitrogen with various apertures/gas flow rates. (Reprinted from Molnar et al., 1995a.)

While the ratio depends somewhat on the flow, this dependence is weak compared to the effect of using an exit aperture. In fact, with the 1/16-in exit aperture, there were no ions or electrons detectable downstream with a Langmuir probe. This elimination may be at least partly due to Debye screening of the hole when its diameter is of the order of several Debye lengths (Chen, 1984).

GaN films grown using various exit apertures in the ECR source were found to have smooth surface morphologies, with excellent transport and photoluminescence properties (Moustakas and Molnar, 1993; Molnar and Moustakas, 1994; Molnar et al., 1995a; Ng et al., 1997; Ng et al., 1998a, 1998b; Iliopoulos et al., 1998a, 1998b). Furthermore, this method was successfully used for the fabrication of both optical and electronic devices as discussed in Section VII. The only problem with current design ECR sources is that the density of activated nitrogen is capable of GaN growth rates of only 200–300 nm/h. However, we have shown recently in our laboratory that by employing exit apertures with a large number of small holes growth rates of about 1 μm/h can be obtained.

b. Radio Frequency (RF) Plasma Sources

A schematic of the Oxford Applied Research RF plasma source is shown in Fig. 8. The discharge tube and the beam exit plate can be fabricated from

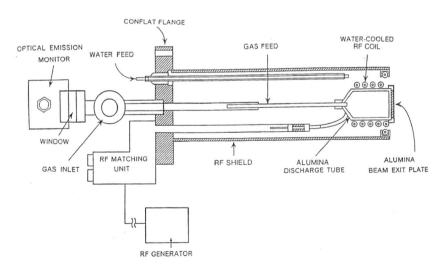

FIG. 8. Schematic drawing of an RF plasma source.

various materials such as quartz, boron nitride, or alumina. The RF coil around the discharge tube is either water- or liquid-nitrogen cooled. This source was investigated by a number of researchers (Park, 1992; Vaudo et al., 1993; Vaudo, 1994) and is currently employed by a large number of researchers studying the growth of III-V nitrides (Liu et al., 1993; Schetzina, 1995; Hooper et al., 1995; Brandt et al., 1995; Brandt et al., 1996; Guha et al., 1996; Hooper et al., 1997; Riechert et al., 1997).

The RF source is excited with inductively coupled 13.56-MHz power supplied by an RF generator capable of up to 600 W of RF power. The emission spectra from this source are dominated by atomic nitrogen lines occurring in the 700- to 900-nm spectral region (Vaudo, 1994). It has been demonstrated recently that those sources produce enough activated nitrogen to lead to GaN growth rates close to 1 μm/h (Riechert et al., 1997; Beresford et al., 1997). Such sources have become commercially available by a number of vendors and have demonstrated the production of high-quality GaN films.

3. NITROGEN ION SOURCES

The growth of GaN films by MBE using ionic N_2^+ species was also reported (Powell et al., 1992; Powell et al., 1993; Fu et al., 1995; Leung et al., 1997). In the report of Powell et al. (1993) the ionic N_2^+ flux is produced by a hot single tungsten grid capable of providing 3×10^{14} to 3×10^{15} ions/cm$^2 \cdot$s. The ion flux is controlled by varying the discharge current at a constant acceleration potential, typically 35 V. Energy and mass analysis of the emitted ion beam shows that it consists primarily of N_2^+ ($N_2^+/N^+ \cong 6$) with an energy spread of less than 5 eV; N_2^+ ions accelerated to energies greater than their molecular binding energy of 9.8 eV dissociate in the surface of the film. These authors reported that as the ion N_2^+ flux increases, the resistivity of the GaN films also increases. Films with resistivity up to 10^6 ohm·cm were produced by this method.

Fu et al. (1995) reported that the nitrogen ion beam was produced in a Commonwealth Scientific Kauffman-type ion source. They have demonstrated that GaN films grown by this method show strong band-edge luminescence when the kinetic energy of the ions is about 10 eV, while luminescence is not detectable when the kinetic energy of the ions exceeds 18 eV. They also found that the use of conductive SiC substrates leads to films with more uniform luminescence than the use of insulating sapphire substrates, due to surface charging effects.

Finally, Leung et al. (1997) produced ionic nitrogen using a hollow anode ion source. In this source, a dc voltage generates a glow discharge in a hollow anode ion source that is constricted to an area in the plasma chamber. It is the pressure difference between the plasma chamber and the MBE growth chamber that extracts the activated nitrogen with energies around 5 eV. The authors report that this method leads to films with good photoluminescence properties but relatively poor transport properties, which they attribute to the presence of second phase (cubic domains) and impurities.

IV. GaN Films

1. FILM GROWTH

Since III-V nitride substrates are unavailable, films must be grown heteroepitaxially on foreign substrates. The majority of GaN films reported have been grown on the basal plane of sapphire (Al_2O_3). Early work indicates that growth on such substrates tends to be three-dimensional. To alleviate this problem, Amano et al. (1986) and Akasaki et al. (1989) developed a low-temperature AlN-buffer followed by the growth of GaN at higher temperatures using the MOCVD method. This led to significant improvement in the films. The Boston University group (Menon, 1990; Lei et al., 1991; Lei et al., 1992; Moustakas et al., 1992; Moustakas et al., 1993; Moustakas and Molnar, 1993) and the Nichia group (Nakamura, 1991) reported independently the development of a GaN buffer layer instead of AlN using the MBE and MOCVD methods, respectively. Moustakas et al. (1993) have shown that this two-step growth process leads to lateral growth that is approximately 100 times higher than the vertical growth.

For the growth of GaN on sapphire substrates (0001, 11-20, 10-12), the Boston University group also reported the importance of surface nitridation (conversion of Al_2O_3 to AlN) using activated nitrogen from an electron-cyclotron resonance (ECR) source. The surface nitridation was confirmed with RHEED (Moustakas et al., 1992; Moustakas et al., 1993; Moustakas and Molnar, 1993). This process was also found to be important during growth of GaN by the MOCVD process (Keller et al., 1996). The nitridation process during growth by MBE with ammonia or during growth by the MOCVD method is done by exposing the surface of the substrate to ammonia at high temperatures. A recent study of such nitridation was reported by Grandjean et al. (1996). These authors also confirmed the formation of an AlN relaxed layer, and later (Grandjean et al., 1997a) showed that when the substrate is nitridated prior to the growth of GaN, the crystal is oriented with the *c*-axis perpendicular to that of sapphire and $[1\text{-}100]_{GaN}//[11\text{-}20]_{Al_2O_3}$. On the other hand, if the substrate is not nitridated the GaN film has a different orientation: $([11\text{-}20]_{GaN}//[1\text{-}100]_{Al_2O_3}$ and $[1\text{-}103]_{GaN}//[11\text{-}20]_{Al_2O_3})$. It has been shown recently (Korakakis, 1998; Ng et al., 1998c) that the AlN grown by the plasma nitridation process is under compressive stress with in-plane lattice constant of 3.01 Å instead of the relaxed value of 3.11 Å. Thus, the AlN tends to acquire the lattice constant of the smaller Al_2O_3 unit cell that is 2.75 Å (Lei et al., 1993). A number of authors (Uchida et al., 1996; Korakakis, 1998; Ng et al., 1997; Heinlein et al., 1997) reported that excessive nitridation leads to rough surfaces. The use of AlN or GaN buffers is also the subject of recent studies involving the growth of GaN by MBE with NH_3. A number of workers use AlN-buffers grown at 800 °C (Yang et al., 1995; Kim et al., 1996) while still others use a low-temperature GaN-buffer (Grandjean et al., 1997b).

The heteroepitaxial growth of these materials is being complicated further by the fact that the III-V nitrides can exist in various polymorphs. All three binaries (InN, GaN, AlN) were found to exist in the wurtzite and zincblende structures (Strite and Morkoc, 1992). It has been found that the wurtzite structures grow primarily on substrates with hexagonal symmetry while the zinc-blende structures grow on substrates with cubic symmetry. The two structures have either the hexagonal close-packed (HCP) stacking sequence (along the (0001) direction) or the face-centered cubic (FCC) stacking sequence (along the (111) direction) with approximately the same cohesive energies (Wright, 1997; Baudic et al., 1996). Therefore, the formation of stacking faults and the nucleation of second phases during the growth of either of the two phases is highly probable (Lei et al., 1992; Moustakas et al., 1993; Moustakas, 1996).

In this section we review the epitaxial growth of GaN in the wurtzite and zinc-blende structures and discuss evidence that certain experimental results may be related to the coexistence of both phases in the same film. Examples will be drawn from the author's own work.

GaN films were grown by the electron cyclotron resonance microwave-plasma assisted molecular beam epitaxy method (ECR-MBE). The deposition system is schematically illustrated in Fig. 9. It consists of a Varian Gen II MBE unit with an ASTeX compact ECR source mounted in one of the effusion cell ports. The base pressure of the overall system is 10^{-11} torr. A reflection high-energy electron diffraction (RHEED) setup is an integral part of the apparatus. Conventional Knudsen effusion cells are used for the evaporation of the group III elements as well as the dopants (Si and Mg). Active nitrogen is produced by passing molecular nitrogen through the ECR source at a flow rate, which produces a downstream pressure of 10^{-5} to 10^{-4} torr.

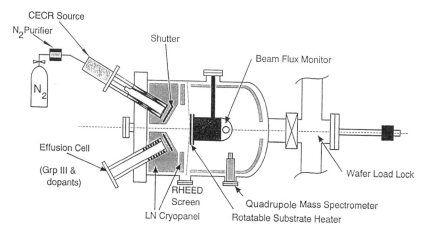

FIG. 9. Schematic of ECR-MBE system.

a. Growth of Wurtzite GaN Films

GaN films having the wurtzite structure (a-GaN) were grown on (0001) sapphire substrates. The substrates were degreased and etched in $H_3PO_4:H_2SO_4$ (1:3) for the removal of surface contaminants and mechanical damage due to polishing, and finally rinsed in deionized water. As discussed previously, the growth on sapphire involves three steps. The first step is the conversion of the surface of the Al_2O_3 substrate to AlN by exposing the substrate to an ECR nitrogen plasma at 800 °C for approximately 10 min. The exact mechanism of this nitridation is not well understood yet. However, we believe that this is related to ion-assisted removal of oxygen and replacement by nitrogen. The thin AlN film is a single crystal with atomically smooth surface morphology as indicated by the streakiness of the RHEED pattern in Fig. 10a. The broadness of the diffraction lines suggests that the AlN film is very thin and highly strained as discussed previously. It is also very likely that this film is not continuous, but that it consists of isolated AlN islands. We believe that the nitridated sapphire substrate is a better substrate for the growth of GaN because of the smaller lattice mismatch between AlN and GaN. It may also act as a diffusion barrier for oxygen and other impurities. The second step involves the deposition of a thin GaN-buffer (200–300 Å thick) at 550 °C. This thin GaN film is single crystal with atomically smooth surface morphology as indicated by the RHEED pattern of Fig. 10b. The diffraction lines of this film are sharper than those of the AlN film. The third step involves the

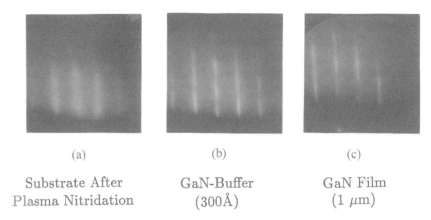

FIG. 10. RHEED data (a) after substrate nitridation; (b) after deposition of 300 Å GaN buffer; (c) after the growth of 1-μm film at 800 °C. (Reprinted from Moustakas et al., 1992.)

deposition of the rest of the GaN film at 700–800 °C. The growth during this step is close to homoepitaxy. The RHEED pattern upon the completion of approximately 1-μm thick film is shown in Fig. 10c. The streakiness and sharpness of the diffraction lines suggest that the film has atomically smooth surface morphology with excellent crystalline quality.

In our earlier work, the conversion of the surface of sapphire from Al_2O_3 to AlN was confirmed by RHEED studies conducted using a 10-kV electron gun (Moustakas et al., 1992). Recently, by using a 30-kV electron gun, we were able to observe both the AlN and the Al_2O_3 RHEED patterns simultaneously (Korakakis, 1998). From the analysis of these data, we concluded that the unit cell of the AlN layer formed on the surface of the sapphire by the nitridation process is rotated in the c-plane by 30° with respect to the unit cell of sapphire (Korakakis, 1998). In general, RHEED experiments are not accurate enough to be used in the determination of lattice parameters. However, in this case, the lattice parameter of the formed AlN was estimated by comparing it with the known lattice parameter of Al_2O_3. The AlN formed after 5 min of nitridation is found to have a lattice constant of 3.05 Å and that formed after 20 min of nitridation is 3.015 Å. Thus the AlN formed by this nitridation process is under extreme compressive stress. This is also consistent with results of deposited AlN layer on sapphire, which was reported to be 3.09 Å instead of 3.11 Å (Bedair, 1996).

The effect of nitridation on the surface morphology of the substrate was also investigated by atomic force microscopy (AFM) and the results are shown in Fig. 11 (Korakakis, 1998; Ng et al., 1997). According to these data,

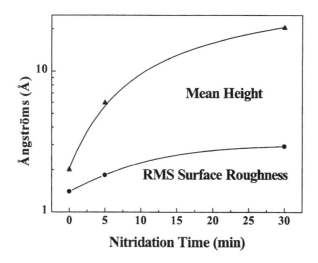

Fig. 11. Effect of nitridation on the roughness of the sapphire substrates. (Reprinted from Korakakis, 1998.)

the substrate becomes significantly rougher as a function of nitridation time. Our studies of the structure, electron mobility and photoluminescence of GaN films grown at various nitridation times show that optimum nitridation occurs at about 10 min.

The surface morphology of the GaN films was found to depend strongly on the microwave power in the ECR discharge (Moustakas and Molnar, 1993). More specifically, films grown at relatively low microwave power have atomically smooth surface morphology, while those grown at higher microwave powers have a rough surface morphology. However, if the Ga-flux increases in proportion to the power in the ECR discharge the surface morphology of the films remain smooth. Thus, the observed roughening is not directly related to the power in the ECR discharge, it is related to the ratio of group III to group V fluxes.

The observed surface roughening with the power in the discharge may result from two different processes. In one process the extra kinetic energy supplied by the energetic plasma increases the surface diffusion of the adatoms, which migrate and form clusters on the top of other clusters and this give rise to three-dimensional growth. The second process involves chemistry. Growth in nitrogen-rich environments increases the nucleation rate due to the strength of the Ga-N bond. In other words, the Ga atoms will be bonded to N atoms before they complete their migration to step edges. The fact that we do not observe surface roughening when growth proceeds under Ga-rich conditions supports the chemical interpretation. Thus, Ga-rich conditions appear to promote step-flow growth in which condensing material must reach the growth ledges by surface diffusion before new islands nucleate. GaN films grown under Ga-rich conditions on (0001) sapphire were found to have very high crystalline quality. A similar conclusion on the role of the ratio of group III to group V fluxes in the growth of GaN by MBE was also reached by other researchers (Yu *et al.*, 1996; Held *et al.*, 1997; Tarsa *et al.*, 1997). Their studies tend to suggest that the optimal film growth occurs close to the boundary between Ga- and N-stable growth regimes. However, some researchers report that N-stable conditions are preferable for growth (Morkoc *et al.*, 1995; Botchkarev *et al.*, 1995; Hughes *et al.*, 1995; Hacke *et al.*, 1996; and Feuillet *et al.*, 1996). Some of these workers also have argued that nitrogen-rich conditions lead to 2×2 surface reconstruction.

The origin of the observed layer-by-layer growth is not understood yet. According to Bauer (1958), if the deposited material has larger surface energy than the substrate, the growth mode will be three-dimensional. More recently, Bruinsma and Zangwill (1987) developed a phenomenological analysis based on continuous elasticity, which predict that in the presence of lattice misfit the layer-by-layer growth is never the equilibrium morphology but rather is metastable with respect to cluster formation on a thin wetting layer. Atomistic simulations by Grabow and Gilmer (1986) support the conclusion that layer-by-layer growth is the ground state only for zero misfit. The observation of a layer-by-layer growth of GaN on (0001) sapphire is probably related to the existence of the GaN buffer, which acts as the wetting layer.

Wurtzite GaN was also grown on other substrates, such as Si(111) (Lei *et al.*, 1991; Lei *et al.*, 1993; Basu *et al.*, 1994; Guha and Bjarczuk, 1998), and on MOCVD-grown GaN thick films (Johnson *et al.*, 1996; Tarsa *et al.*, 1997). Films grown on such substrates were found to have physical properties equivalent to those grown by the MOCVD method. In particular, the homoepitaxial growth leads to films that replicate the underlying GaN films although grown at a significantly lower temperature.

b. Growth of Zinc-blende GaN Films

Zinc-blende GaN (β-GaN) films were grown on a variety of cubic substrates. These include, for example, β-SiC (Paisley *et al.*, 1989; Liu *et al.*, 1993), GaAs (Kikuchi *et al.*, 1994; Kikuchi *et al.*, 1995; Yoshida *et al.*, 1997; Trampert *et al.*, 1997; Brandt *et al.*, 1997), MgO (Powell *et al.*, 1993) and

Si(100) (Lei *et al.*, 1991; Lei *et al.*, 1992; Lei *et al.*, 1993; Basu *et al.*, 1994). Next, we discuss the growth of zinc-blende GaN films on Si(100).

GaN films having the zinc-blende structure were grown on both *n*- and *p*-type (001) silicon substrates using the ECR-MBE method. The Si substrates were degreased and etched in buffered HF and outgased to 800 °C in the MBE preparation chamber. RHEED studies of such an Si wafer at 400 °C indicate that the Si (001) surface is unreconstructed (i.e., 1 × 1). We found that such an unreconstructed surface is required for the growth of cubic GaN. As discussed previously, such films were grown in two temperature steps (Lei *et al.*, 1991; Lei *et al.*, 1992). More specifically, a GaN buffer layer about 300–900 Å thick was deposited at a temperature of 400 °C; it was found by RHEED studies to have the zinc-blende structure with the [001] direction perpendicular to the substrate. The temperature of the substrate was then raised to 600 °C and the rest of the GaN film was grown. The zinc blende structure of these films was confirmed by RHEED and *ex situ* x-ray diffraction (XRD) and transmission electron microscopy (TEM) studies (Lei *et al.*, 1992; Lei *et al.*, 1993; Basu *et al.*, 1994).

The surface morphologies of the GaN films, grown under identical conditions, were found to be smoother on *n*-type rather than *p*-type Si-substrates. This may suggest that GaN wets to the *n*-type substrate better than to the *p*-type substrate (Lei *et al.*, 1992). This conclusion is also consistent with the work of Morimoto *et al.* (1973), who reported that in vapor phase growth, GaN adheres to *n*-type silicon but not to *p*-type silicon substrates. The origin of this phenomenon is not understood at this time. Figure 12 shows two characteristic surface morphologies observed on *p*-type

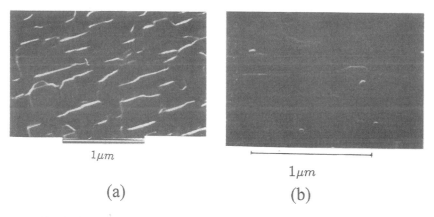

FIG. 12. Surface morphologies of GaN thin films on *p*-type Si substrates grown as described in the text. (Reprinted from Lei *et al.*, 1992.)

substrates. The sample with the smoother surface was grown at lower nitrogen pressure. The surface of the sample shown in Fig. 12a is roughened with many rectangular-shaped "tiles" oriented along the [110] and [1-10] directions. This presumably arises because the GaN surface corresponding to these directions is more closely packed than the (100) and (010) surfaces and therefore have lower surface energies. The size of these "tiles" reveals the size of the domains. The smoother surface morphology of the films grown at lower pressure is consistent with higher III-V ratio during the growth of this film.

From the data of Fig. 12b we can estimate the two-dimensional nucleation rate and the lateral growth rate (Moustakas et al., 1993). Let J be the nucleation rate, S the average area of the plateaus, h the height of the plateaus, and t the time for the plateaus to grow, which is limited by the nucleation rate. Then we have

$$\sqrt{S} = v_l t = \frac{v_l}{JS} \qquad (7)$$

$$h = v_n t = \frac{v_n}{JS} \qquad (8)$$

where v_l and v_n are the lateral and vertical growth rates, respectively. Using data from Fig. 12b ($\sqrt{S} \approx 1\ \mu m$, $h \approx 100\ \text{Å}$) and the known growth rate ($v_n = 2000\ \text{Å/h}$) we find from Eqs. (7) and (8) $v_l \approx 100\ v_n$ and $J = 20$ nuclei/($\mu m^2 \cdot h$).

These data clearly indicate that the two-step growth method leads to quasi-layer-by-layer growth with very small two-dimensional nucleation rate and high lateral growth rate. The buffer layer, which was employed both for the growth of GaN on sapphire and silicon, is much thicker than the critical thickness for strain layer epitaxy. Therefore, the buffer layer is expected to be relieved of the strain and thus be very defective due to the introduction of dislocations. Nevertheless, it is a GaN layer, which is found to be single crystalline in our case. As a result, the subsequent epitaxy of GaN at higher temperature is similar to homoepitaxy on a defective substrate. Therefore, with the two-step process, the growth is engineered into two stages controlled by a different mechanism.

In the first stage, the nucleation of GaN is heteroepitaxial, which involves the absorption of monolayers of GaN and the formation of critical nuclei. Growth in this stage at low temperatures leads to a continuous film since the sticking coefficient of the evaporant species is much higher and the nucleation process is much faster. On the other hand, at high temperatures, the adatoms have a very short lifetime before they reevaporate. This tends to lead to incomplete coverage of the surface with three-dimensional clusters. The subsequent growth of the films would then tend to be columnar. For example, Fig. 13 shows the surface morphology of a GaN film grown on Si(100) in one step at 600 °C. Contrary to the smoothness of the film grown with the two-step process, this film is rough with poorly connected small cubic crystals. It should be pointed out, however, that RHEED indicates that this film also has the zinc-blende structure in agreement with the observed surface morphology. By lowering the substrate temperature during the first stage, the adhesion of the GaN film is improved. However, if the temperature is too low, the condensation would be too fast with respect to the mobility of the adatoms on the substrates, hence the buffer layer would not be crystalline, which would make further epitaxy impossible. Therefore, we need a moderate temperature for the formation of the GaN buffer. Lei *et al.* (1992) reported that the X-ray rocking curve for a film grown in one step at 400 °C on Si(100) has a full width at half maximum (FWHM) as wide as 4 degrees, suggesting very poor crystalline quality. The growth during the second stage is homoepitaxial, which has no energy barrier. Therefore, it can and should be carried out at higher temperature, so that the adatoms would have enough mobility to reach their equilibrium configuration and thus improve the crystallinity of the film.

Conventionally, epitaxial thin-film growth is classified into three categories: (i) layer-by-layer growth mode or Frank-Van der Merwe mode; (ii) island growth mode or Volmer-Weber mode; and (iii) layer-by-layer followed by island growth mode or Stranski-Krastanov mode. It is understood that the mode of nucleation and growth of epitaxial films is strongly governed by the interfacial free energies between the deposit and substrate

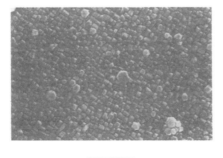

$1 \mu m$

FIG. 13. SEM surface morphology of a film grown in one step at 600 °C. (Reprinted from Lei *et al.*, 1992.)

σ_{sd}, the deposit and the vapor σ_{dv}, the vapor and the substrate σ_{sv}. It has been shown (Landau and Lifshitz, 1980) that at equilibrium, the deposit forms a hemispherical cap on top of the substrate and the contact angle θ satisfies

$$\cos \theta = (\sigma_{sv} - \sigma_{sd})/\sigma_{dv} \tag{9}$$

If $\sigma_{sv} - \sigma_{sd} = \sigma_{dv}$, then $\theta = 0$, the deposit is said to wet the substrate, that is, the growth is two-dimensional. In this case, it is expected that growth will follow the layer-by-layer mode.

In the case of homoepitaxy, the deposit has exactly the same structure as the substrate; then by definition, the interfacial free energy between the deposit and the substrate $\sigma_{sd} = 0$. Thus, the contact angle is zero since $\sigma_{sv} = \sigma_{dv}$ and this under equilibrium conditions leads to a layer-by-layer growth mode. Therefore, in the second stage of the growth, if the surface mobility of the deposited GaN is large enough, the growth of GaN would be at a quasi-equilibrium, in which the GaN would be able to achieve its equilibrium configuration. Therefore, if the buffer layer is relatively smooth, we would expect the subsequent GaN growth to follow the layer-by-layer mode, which would result in a smooth surface morphology.

Growth of the GaN during the second stage at elevated temperatures is desirable to promote high surface mobility of the adsorbed adatoms. However, too high temperatures may not be desirable for epitaxy due to the roughening transition (Weeks, 1980). Lei et al. (1992) provided evidence that the surface of β-GaN on Si(100) is completely roughened by thermal fluctuations at about 800 °C. X-ray diffraction in such films also detected a large portion of (0002)- or (111)-oriented domain.

2. Film Structure and Microstructure

In this section, we discuss the structure and microstructure of GaN films grown by MBE either on (0001) sapphire (wurtzite) or on (100) Si (zinc blende). A discussion also is presented on methods developed to detect the coexistence of the two phases on the same film.

a. Structure of GaN Films on (0001) Sapphire

The structure and microstructure of GaN films grown on (0001) sapphire substrates by MBE was found to be affected very significantly by the nucleation steps (substrate nitridation and low-temperature buffer) as well as by the ratio of group III to group V fluxes. If the surface of the substrate is not converted into AlN, as discussed in the previous section, the grown GaN films tend to be polycrystalline with a high concentration of misoriented domains or second phases (Doppalapudi et al., 1998). Excessive nitridation appears to also be undesirable since it leads to rough initial surfaces (Korakakis, 1998; Ng et al., 1998c). Similarly, the presence of a low-temperature buffer was found to affect the film microstructure (Lei et al., 1991; Lei et al., 1992; Moustakas et al., 1992; Moustakas et al., 1993; Lei et al., 1993; Basu et al., 1994; Fuke et al., 1998; Romano et al., 1997; Kim et al., 1997; Sugiura et al., 1997).

The effect of the ratio of III/V fluxes on the structure and microstructure of GaN films grown by MBE was studied by Romano et al. (1997). These authors investigated two types of films (A and B) grown under identical conditions except that the ratio of fluxes (III/V) for films A were 30% higher than for films B during the growth of the buffer layer and of the high temperature layer. This was achieved by changing the Ga beam flux while maintaining the nitrogen flux constant for both films. In the following we discuss the results of Romano et al. (1997).

The structural properties of these films were characterized by high-resolution X-ray diffraction (XRD) and transmission electron microscopy (TEM). The XRD was performed with a Philips four-circle diffractometer using a Cu K_α wavelength source. Reciprocal space maps were taken in triple axis (TA) mode near the (0002) reflection and rocking curve measurements were made of both the (0002) and the (1-102) reflections. In TA mode, the X-ray detector limit is 12 arc s, resulting in a sampling volume that is small compared to the (0002) reciprocal lattice point. Therefore, broadening of the XRD peak due to variations in the lattice parameter can be distinguished from broadening due to mosaic. The experimental setup for the TA mode requires an analyzer crystal in front of the detector. The distribution of tilts in the crystal is measured by scanning the sample with the detector at a fixed position. In this configuration, only regions of the crystal with the same Bragg plane spacing will be analyzed. The mosaic spread of the crystal can be measured by scanning the sample at twice the speed of the detector. In a standard rocking curve (SRC) the integrated intensity of the Bragg reflection is over a relatively large area of reciprocal space and therefore variations of lattice parameter and crystal mosaic both contribute to the broadening of the peak.

High-resolution (HREM) and conventional and multiple dark-field imaging were carried out in a JEOL 3010 at 300 kV. HREM image simulations were performed at different values of defocus and tilts of the crystal near the $\langle 11\text{-}20 \rangle$ orientation using the MaCTempas multislice program (Kilaas, 1987). Samples for cross-sectional TEM (XTEM) analysis were prepared by cutting bars perpendicular to the ⟨11-20⟩ direction, followed by flat polishing to <5 μm and then ion milling to achieve electron transparency. Plan-view TEM (PTEM) samples were prepared by polishing the substrate side of ⟨0001⟩ oriented specimens and ion milling.

The XRD spectrum from a TA measurement taken along the (0002) direction with the detector at a fixed position is shown in Fig. 14 for films A and B. The FWHM of the peaks are 54 arc s for film A and 41 arc s for film B indicating small variations in interplanar spacing along the c-axis. Scherrer's formula can be used to calculate a coherence length from these measurements. It was found to be 336 nm for film A and 442 nm for film B. These lengths correspond to the film thickness determined by TEM measurements for films A and B, respectively. The presence of Pendellosung fringes in this type of measurement indicates that the GaN/sapphire interface must be sharp and of good crystalline quality along the c-axis. No Pendellosung fringes were observed in the SRC measurements. The FWHM of the (0002) rocking curve measurements was 69 arc s for film A and 174 arc s for film B. The difference in the width of the (0002) peaks between

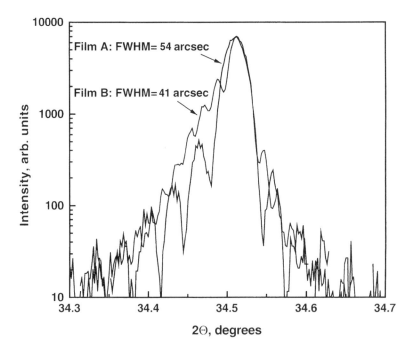

FIG. 14. XRD spectra of a cut through reciprocal space in a direction perpendicular to the c-axis for films A and B. (Reprinted from Romano et al., 1997.)

the SRC measurements and the TA measurements is greater for film B than for film A. It was found from the TEM measurements that this was due to a greater density of mixed and screw dislocations, as discussed by Heying *et al.* (1996) as well as crystallites with local misorientations of the *c*-axis in film B compared to film A. The FWHM of rocking curves for the asymmetric (1-102) reflection was 52 arc min for film A and 30 arc min for film B. This extensive broadening of the asymmetric scans suggests a high density of edge threading dislocations. From PTEM the majority of dislocations in both of these films were found to be edge dislocations with a density of 10^{11} cm^{-2} for film A and 10^{10} cm^{-2} for film B.

Conventional XTEM micrographs of films A and B are shown in Figs. 15 and 16, respectively. The images are taken near the (11-20) zone axis of the GaN with $g = 0002$ (Fig. 15a and Fig. 16a) or $g = 10\text{-}10$ (Fig. 15b and Fig. 16b). For $g = 0002$, all edge dislocations will be invisible, whereas screw and mixed screw/edge dislocations would be invisible only if their Burger's

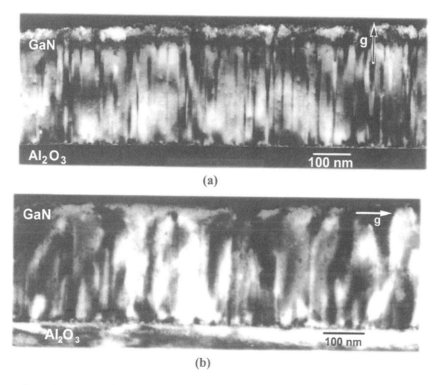

FIG. 15. XTEM micrographs taken near the [11-20] zone axis of the GaN for film A with (a) $g = (0002)$ and (b) $g = (10\text{-}10)$. (Reprinted from Romano *et al.*, 1997.)

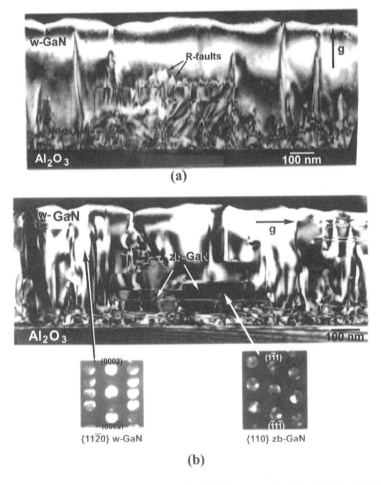

FIG. 16. XTEM micrographs taken near the [11-20] zone axis of the GaN for film B with (a) $g = (0002)$ and (b) $g = (10$-$10)$. R-plane facets labeled in (a) are parallel to the (1-101) planes. Convergent beam diffraction patterns from regions in the GaN film containing the zinc-blende (zb) and wurtzite (w) phases are shown in (b). (Reprinted from Romano et al., 1997.)

vector $\boldsymbol{b}$ is such that $\boldsymbol{b} \cdot g = 0$. With $g = 10$-10, screw dislocations would be invisible, whereas edge and mixed dislocations would be invisible only if $\boldsymbol{b} \cdot g = 0$. In Fig. 15a, a sharp black/white columnar-like contrast is observed that is not present in Fig. 15b or Fig. 16. The high density of edge dislocations makes it difficult to resolve individual dislocations in Fig. 15b.

However, the contrast in Fig. 15a cannot be due to screw or mixed screw/edge dislocations because they were not observed in PTEM. Instead, the contrast is due to fringes from overlapping inversion domains (IDs) (Romano et al., 1996). The IDs were also imaged in plan-view samples by using a nonzero reflection along the inversion axis. Fringe contrast occurs at the inversion domain boundaries (IDBs) under these diffracting conditions since there is a phase difference in the structure factor between the inverted region of the crystal and the host matrix as discussed by Serneels et al. (1973). The fringes are not visible when imaging with (hkil) reflections with $1 = 0$.

Film B was found to have a very different microstructure than film A. Figure 16 shows that film B contains crew and mixed line dislocations and facets on the (1-101) planes (R-plane). Edge dislocations and slabs of second-phase material parallel to the interface are visible in Fig. 16. It was found by microdiffraction and HREM that the second phase consisted of zinc-blende (zb) GaN with the (1-11) axis parallel to the (0001) axis of wurtzite GaN. Diffraction patterns from each of the phases are also shown in Fig. 16. The (110) cubic diffraction pattern is taken from the zb regions and the (11-20) diffraction pattern is taken from the surrounding wurtzite matrix.

Figure 17 is an HREM image near the interface of sample B showing the zb- and w-GaN phases. The image is taken along the (11-20) zone axis of the GaN. The (111) planes of zb-GaN are parallel to the (0001) planes of the w-GaN phase. The zb phase was found near the film/substrate interface and appeared as slabs about 10–200 nm thick and about 100–400 nm in length along the interface. The total amount of zb-GaN was $<2\%$ of the film area and therefore unable to be detected by XRD. It is likely that the growth of zb-GaN was nucleated during the growth of the buffer layer (at 550 °C) because it was found near the film/substrate interface. However, when the substrate temperature was raised to 800 °C, it became favorable to form the more energetically stable w-GaN phase. Both R-plane facets and c-plane defects can be observed in the HREM image in film B (Fig. 17) that were not found in A films. The R-facets separate regions of w-GaN with small misorientations. These misorientations would contribute to the broader width in the (0002) X-ray rocking curve for film B compared to A.

In summary, the microstructure of GaN films grown by ECR-MBE was found to depend on the ratio of the Ga and N flux. Inversion domains of wurtzite GaN were found for A films that were grown with a 30% higher Ga to N ratio compared to B films. In B films, both the zinc-blende and wurtzite GaN phase were present. This indicates that the ratio of the III/V species can influence the type of microstructure that is formed and that the growth of zb-GaN may be favorable by controlling the amount of Ga at the surface.

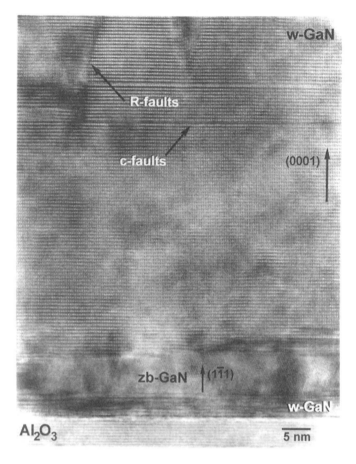

Fig. 17. HREM image taken along the [11-20] zone axis of the GaN near the interface of sample B showing the zinc-blende (zb) and wurtzite (w) phases. (Reprinted from Romano et al., 1997.)

b. Structure of GaN Films on Silicon Substrates

The structure and microstructure of the zinc-blende GaN films grown on (001) Si substrates were investigated by Lei et al. (1991, 1992, 1993) and Basu et al. (1994).

Plan-view specimens suitable for TEM examination at 120 kV were prepared by removing the Si substrate chemically and then thinning regions approximately 5×5 μm^2 in the resulting GaN film using a Gatan precision

ion milling system (PIMS) employing Ar$^+$ ions. When the film is milled from the face containing the buffer layer, a rectangular tile-like morphology is again clearly evident and is found to be orientated along the (110) and (1-10) directions in agreement with the data in Fig. 12a (Lei et al., 1992).

The cubicity of the GaN films may be conveniently verified by determining the crystal symmetry using convergent beam electron diffraction (CBED); CBED patterns taken with a small probe from a relatively unbuckled region of the film are shown in Fig. 18. Both the zero-layer detail and higher-order Laue zone (HOLZ) ring show both whole pattern and bright-field 4 mm symmetry as expected from a crystal with m3m point group down (001) direction (Lei et al., 1992). Technically, however, these symmetries would also be seen from the tetragonal point group 4/mmm down [001] direction. Measurements of the HOLZ ring diameter indicate that, within experimental error, there is no tetragonal distortion and that the material has the zinc-blende structure with a measured lattice parameter of 4.52 ± 0.05 Å.

The buffer layer was also imaged by removing material from the growth face with the PIMS. The layer appears to be highly faulted with some very weak texturing visible, which is more apparent in the selective area diffraction (SAD) patterns as shown in Fig. 19. Figure 19a shows single-crystal reflections expected from the (001) zone axis and additionally arcs or incomplete rings. These features can all be indexed in terms of β-GaN and show that the buffer layer consists of highly defective single crystal, and to a lesser extent, highly textured and possibly twinned polycrystalline material. On moving into a thicker region of the specimen containing both buffer and growth layers, the single-crystal nature of the layer becomes more emphasized (Fig. 19b) with numerous zero-layer reflections and streaking of the spots along the (110) directions evidently due to specimen buckling and faulting on the (110) planes. Figure 19c shows the SAD pattern taken from a thick region of the film that had been milled from the buffer layer face, that is, buffer layer and adjacent growth layer were removed. Although many zero-layer reflections are still present, the reduction in streaking implies a lower level of faulting.

X-ray diffraction shows a strong peak at $2\theta = 40.1°$, whose d spacing is 2.25 Å, which is due to (002) reflection from β-GaN. Hence the lattice constant for β-GaN was found to be 4.50 Å in good agreement with the electron-diffraction data. A small peak was also observed at $2\theta = 34.6°$, which is due to the d spacing of (111) β-GaN or (0002) α-GaN. This suggests that the GaN film has some misoriented domains. These domains may have developed in the early stage of the buffer-layer deposition as revealed by the TEM imaging. Similar structural faults have also been observed in the interface of GaN epitaxy on GaAs (001) (Strite et al., 1991). We have

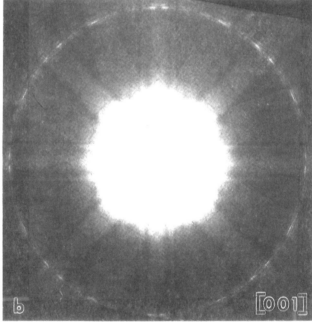

FIG. 18. CBED patterns from [001] β-GaN using 200-Å probe: (a) at high camera length showing zero-layer detail (bright field and adjacent reflections); (b) at low camera length showing first HOLZ ring. (Reprinted from Lei *et al.*, 1992.)

2 GROWTH OF III-V NITRIDES BY MOLECULAR BEAM EPITAXY

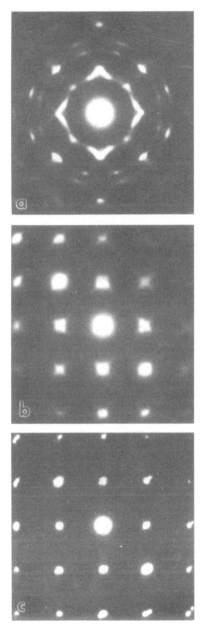

FIG. 19. SAD patterns down [001] from various layers within the β-GaN film: (a) buffer layer; (b) buffer layer plus adjacent growth layer; (c) layer near growth surface. (Reprinted from Lei *et al.*, 1992.)

observed these misoriented domains in almost all of our samples. The consistent appearance of the misoriented domains in GaN-Si and GaN-GaAs heteroepitaxy is very likely to be related to the large lattice mismatch between the GaN and the substrate. Because of the large lattice mismatch, the interface of GaN and the substrate is under significant strain and, therefore, it might be favorable for the system to introduce dislocations or misorientations of GaN to reduce the interfacial energy. In zinc-blende or wurtzite structures, the (111) or (0002) planes are the most closely packed, and hence have the lowest surface energies. Therefore, the introduction of (111)- or (0002)-oriented GaN would lower the surface energy to the GaN vapor-interface, and would not necessarily increase the energy for the GaN substrate interface, as it was strained significantly. However, the (111)- and (0002)-oriented GaN domains would grow slower than the (001)-oriented GaN grains stabilized by introduction of dislocations. As a result, misoriented domains would be buried at the interface region as the film grows thicker.

Thus, the XRD and TEM studies on these films indicate that the films have the zinc-blende structure with approximately 1% wurtzite phase occurring primarily close to the Si-GaN interface. This is to be contrasted with wurtzite GaN films grown on Si(111), which were found to have an approximately 25% zinc-blende component (Lei *et al.*, 1993; Basu *et al.*, 1994). This clearly suggests that stacking faults may be introduced much more easily when growth proceeds with the basal plane parallel to the substrate. We believe that this high concentration of stacking faults is instigated by the high interfacial strain between Si and GaN.

In Katsikini *et al.* (1996, 1997a, 1997b, 1998) methods were developed to determine the percentage of the cubic and hexagonal phases in GaN using the angular dependence of the near-edge X-ray absorption fine structure (NEXAFS) spectra recorded at the N-K edge. Thus, these spectra probe the partial density of unfilled states with p component, which is characteristic of the symmetry of the material. This method is capable of providing a quantitative estimate of the percentage of the wurtzite and cubic phases present in a mixed crystal.

3. DOPING STUDIES

a. n-Type Doping

A number of groups have doped GaN *n*-type by MBE using Si as a dopant (Doverspike and Pankove, 1998). In the following, we describe the *n*-type doping studies from the author's laboratory. A large number of GaN films have been doped n-type with Si using the ECR-MBE method. The carrier concentration in these films was varied from 10^{15} to above 10^{20} cm^{-3} (Ng et al., 1997, 1998a, 1998b, 1998c). These films were grown using various nitridation and buffer conditions, which, as discussed previously, affect the epitaxial growth and the incorporation of defects in the films. To confirm that the doping is related to silicon rather than nitrogen vacancies or accidental impurities, we plot in Fig. 20 the net carrier concentration vs the reciprocal temperature of the silicon cell. The observed activation energy of 4.5 eV is in good agreement with the activation energy of the silicon vapor pressure vs the silicon reciprocal temperature. Samples with a carrier concentration less than 10^{17} cm^{-3} fall below the straight line, suggesting that in these samples there is significant compensation.

The electron mobility vs the net carrier concentration in these films, as determined by Hall effect measurements, is shown in Fig. 21. Samples with carrier concentration 3×10^{17} cm^{-3} were found to have the highest electron mobilities (320 cm^2/V·s). Such mobility values are identical to those re-

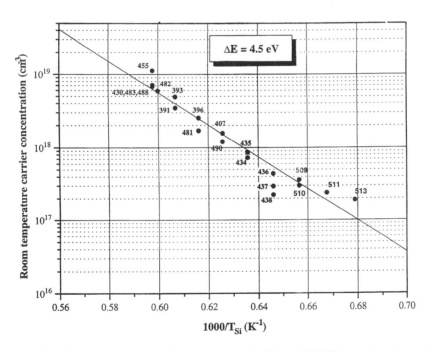

FIG. 20. Room-temperature carrier concentration for Si-doped GaN films vs the reciprocal of the temperature of the silicon cell.

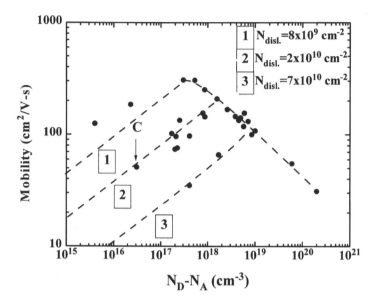

FIG. 21. Electron mobility vs carrier concentration in n-GaN films. The curves in the low-carrier concentration regions are theoretical curves fitted to Eqs. (10) and (11) with the indicated dislocation densities. (Reprinted from Ng et al., 1997.)

ported for MOCVD-grown GaN with the same carrier concentration (Doverspike and Pankove, 1998). These data do not follow the traditional behavior characteristic of ionized impurity scattering. The bell-shaped curves of the data can be accounted for by considering the effect of dislocations. It is well known that dislocations with an edge component introduce acceptor-like centers along the dislocation line (dangling bonds) which capture electrons from the conduction band in an *n*-type semiconductor (Read, 1954a; Pödör, 1966). Thus, the dislocation lines become negatively charged and a space charge is formed around them that scatters electrons crossing the dislocations and, thus, reduces their mobility. This phenomenon has been studied both experimentally and theoretically in *n*-type Ge (Read, 1954b; Pearson et al., 1954) and the electron mobility due to such scattering by dislocations is given by the expression (Pödör, 1966)

$$\mu_{disl} = \frac{30\sqrt{2\pi}\,\varepsilon^2 a^2 (kT)^{3/2}}{N_{disl} e^3 f^2 \lambda_d \sqrt{m}} \quad (10)$$

where *d* is the distance between acceptor centers (dangling bonds) along the dislocation line, f is the occupation rate of the acceptor centers, N_{disl} is the density of dislocations and λ_d is the Debye screening length

$$\lambda_d = \left(\frac{\varepsilon kT}{e^2 n}\right)^{1/2} \qquad (11)$$

where n is the net carrier concentration. Thus, the electron mobility due to scattering by dislocations should increase monotonically with net carrier concentration. As a result, the combined effect of ionized impurity and dislocation scattering would lead qualitatively to the observed experimental results in Fig. 21.

The results of Fig. 21 have recently been accounted for quantitatively by Weimann et al. (1998). These authors argued that a highly dislocated wurtzite crystal can be pictured as consisting of hexagonal columns rotated relatively to each other by a small angle, with inserted atomic planes to fill the space between two columns (Ponce, 1997). Structural studies in GaN indicate single dislocation lines between two prisms with approximately one threading dislocation per hexagonal prism. These edge dislocations introduce dangling bond-type states whose vertical spacing is given by the c-lattice constant of hexagonal GaN (5.18 Å). Thus, for a GaN film with 10^{10} cm^{-2} dislocation density the volume dangling bond density of states is expected to be 2×10^{17} cm^{-3}. In n-GaN films, the dislocations become negatively charged and act as Coulomb scattering centers in lateral transport. At high carrier concentrations the charged dislocations are screened out by the excess free carriers and thus the dominant scattering mechanism arises from ionized impurities. On the other hand, at low carrier concentrations scattering at charged dislocations becomes the dominant mechanism.

It should be emphasized that the previous model predicts that dislocations affect only the lateral transport. Vertical devices are unaffected by the scattering of electrons at threading dislocation lines due to the repulsive band bending around dislocations. This may explain the observed performance of LEDs and lasers in the presence of high dislocation densities.

b. *p-Type Doping*

p-Type doping studies of GaN by the various deposition methods were discussed by Doverspike and Pankove (1998). Here we report on a systematic study of p-type doping of GaN by MBE (Moustakas and Molnar, 1993; Brandt et al., 1994a; Ng et al., 1997). A large number of GaN films have been doped p-type with Mg. As reported in these references, the samples show p-type conductivity without any post-growth annealing. Figure 22 shows the resistivity for two series of films grown at 750 and 700 °C,

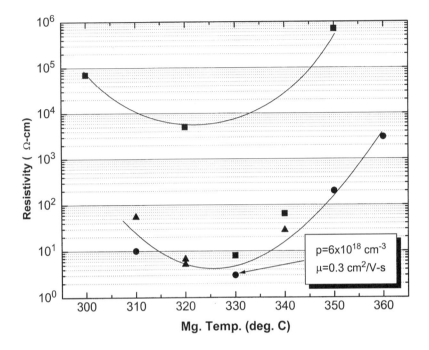

Fig. 22. Resistivity of Mg-doped p-GaN films as a function of the Mg cell temperature. (Reprinted from Ng et al., 1997.)

respectively, as a function of the temperature of the Mg cell. Thus, both the substrate temperature and the temperature of the Mg cell influence the incorporation of Mg. At lower temperatures of the Mg cell, the incorporated Mg is not sufficient to overcompensate the native defects or impurities, which tend to dope the films n-type. At higher Mg-cell temperatures, we believe that the incorporation of a high concentration of Mg causes clustering of Mg atoms or incorporation in interstitial or nitrogen sites instead of substitutional sites. The effect of substrate temperature can be accounted for by the reevaporation of Mg before it incorporates into the lattice. A similar conclusion has also been reached by Kamp et al. (1997) who reported that Mg incorporation in GaN films decreases exponentially with temperature with an activation energy of 2.56 eV. The sample with the lowest resistivity was found by Hall effect measurements to have a carrier concentration of 6×10^{18} cm^{-3} and electron mobility of 0.3 cm^2/V-s (Moustakas and Molnar, 1993). This value of the Hall mobility is rather low suggesting that the holes move by a hopping mechanism. We propose that

Mg incorporation in our method of growth induces a certain degree of disorder in the material, which introduces band-tails in the various energy bands. In general, disorder can be partially removed by annealing to high temperatures. We conducted such experiments for one of our films and the results are shown in Fig. 23 (Ng et al., 1997).

These data indicate that upon annealing up to 850 °C in a nitrogen ambient, the carrier concentration of the film was reduced by roughly a factor of two while the hole mobility was increased by more than an order of magnitude. The origin of this annealing effect was investigated by studying the temperature dependence of the resistivity before and after annealing to 850 °C. Before annealing, the acceptor activation energy was 140 meV while after annealing the activation energy became 176 meV. This change in activation energy accounts qualitatively for the reduction in hole concentration This result together with the improvement of the mobility upon annealing can be accounted for in a model, which assumes that in the as-grown films, Mg incorporation induces a significant degree of static disorder. In such a model the density of states for the as-grown and annealed films is shown in Fig. 24.

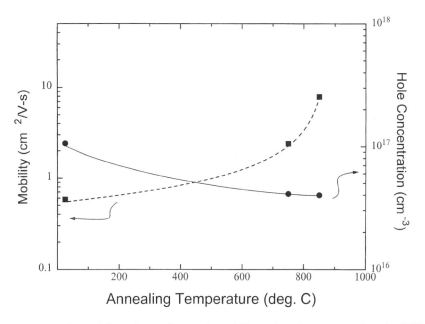

FIG. 23. Effects of thermal annealing on the mobility and carrier concentration of p-GaN films. (Reprinted from Ng et al., 1997.)

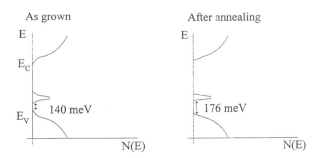

FIG. 24. Model for the density of states of Mg-doped GaN films before and after annealing. (Reprinted from Ng et al., 1997.)

It is understood that the incorporation of Mg in MOCVD growth of GaN is facilitated by the formation of Mg-H complexes. It was found that low-energy-electron irradiation (Amano et al., 1989) or thermal annealing (Nakamura et al., 1991) are required to activate the p-type doping, a result which was attributed to hydrogen evolution (Van Vechten et al., 1992). In the case of GaN growth by the ECR-MBE method, we believe that the Mg incorporation is facilitated by the abundance of electrons arriving at the surface of the film from the ECR source. Under such conditions of growth the Mg dopants enter the lattice in their charged state, which requires much less energy than the formation of a neutral acceptor (Ng et al., 1997).

To investigate the role of hydrogen in MBE-grown films we conducted a study of post-growth hydrogenation of both n-type (Si-doped or auto-doped) and p-type (Mg-doped) GaN films (Brandt et al., 1994a; Brandt et al., 1994b). Hydrogenation was performed with a remote microwave plasma operating at 2 torr with the sample held in a holder whose temperature was varied up to 675 °C. At the end of each annealing experiment the sample was cooled to room temperature and its carrier concentration was determined by Hall effect measurements. The results of these hydrogenation experiments are shown in Fig. 25. These data indicate that while the n-type samples are unaffected by the hydrogenation up to hydrogenation temperatures of 600 °C, the p-type samples show a reduction in carrier concentration by more than an order of the magnitude by hydrogenation at $T > 500\,°C$. This reduction in hole concentration was attributed to the formation of Mg-H complexes (Brandt et al., 1994b).

The introduction of hydrogen (deuterium) into the material was verified by secondary ion mass spectroscopy (SIMS) measurements. Figure 26 shows the depth profiles for both Mg and deuterium. The Mg concentration together with the hole concentration shown in Fig. 25 permits an estimation

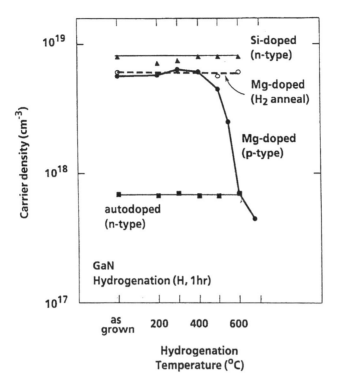

FIG. 25. Dependence of carrier density on hydrogenation temperature in MBE-grown GaN films. (Reprinted from Brandt et al., 1994a.)

of the doping efficiency that was achieved with ECR-assisted MBE. This efficiency is about 6% at room temperature. However, if one takes into account that the acceptor ionization energy for this sample was estimated to be 150 MeV (Moustakas and Molnar, 1993), only 1% of the acceptors should be ionized at 300 K. The considerably higher apparent ionization of about 6% suggests that the dominant transport process is impurity-band conductivity.

The luminescence spectra of Mg-doped GaN films grown by MBE were found to differ significantly from those of films grown by the MOCVD method (Nakamura et al., 1991). The MOCVD films show a broad peak at 2.75 eV. The luminescence spectra of MBE films at 77 K are shown in Fig. 27. Such spectra have been reported by others as well (Strite and Morkoç, 1992). These spectra show a small peak at 3.46 eV associated with neutral donor-bound excitons and a strong broad peak at 3.26 eV. At the

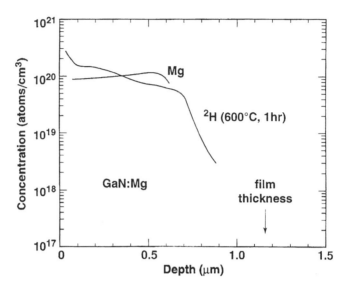

FIG. 26. Depth profile of deuterium and magnesium in p-type GaN determined by SIMS. (Reprinted from Brandt et al., 1994a).

low-energy side of this peak one can distinguish two other peaks at 3.17 eV and 3.08 eV, which can be accounted for as one optical and two optical phonon replicas of the 3.26-eV transition. The transition at 3.26 eV was also observed in undoped GaN films and was attributed to donor-acceptor (DA) recombinations, and the lower-energy peaks were attributed to phonon replicas (Lagerstedt and Monemar, 1979; Ilegems and Dingle, 1973).

Luminescence studies of undoped GaN film grown on β-SiC substrates are shown in Fig. 28. Structural studies of this film indicate that the film has primarily the zinc-blende structure; however, a small diffraction peak associated with the wurtzite structure is also present. The data of Fig. 28 show the transitions at 3.45 eV, which is associated with the small concentration of the wurtzite structure and a broad feature with three peaks. These three peaks are identical with those of Fig. 27. (The relative strength of these peaks was slightly modified due to the use of a glass filter during the collection of these data.) This raises the question of whether the peak at 3.26 eV, which was associated with DA recombination, is due instead to a transition across the gap of the zinc-blende structure, whose gap at 77 K is about 3.26 eV (Moustakas, 1996).

To account for the similarity of the data of Figs. 27 and 28, it was proposed (Moustakas, 1996) that Mg doping of wurtzite-GaN by MBE instigates the formation of stacking faults, which are nucleation sites for zinc-

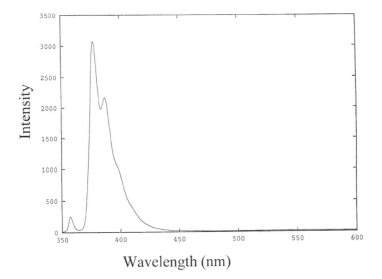

FIG. 27. Photoluminescence spectra at 77 K of an Mg-doped wurtzite-GaN film. (Reprinted from Moustakas, 1996.)

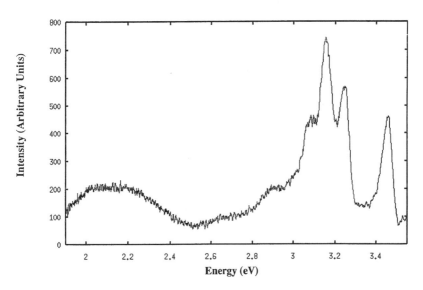

FIG. 28. Photoluminescence spectra at 77 K of an undoped zinc-blende GaN film. (Reprinted from Moustakas, 1996.)

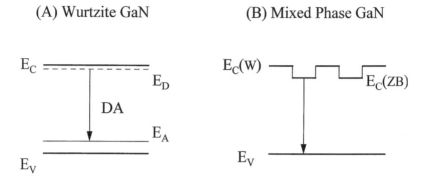

FIG. 29. Luminescence transitions in Mg-doped GaN in (A) wurtzite structure; (B) mixed phase structure. (Reprinted from Moustakas, 1996.)

blende GaN domains. This is probably related to the incorporation of a group II element in a wurtzite structure. The composite material of wurtzite GaN with randomly distributed microscopic domains of zinc-blende GaN or isolated stacking faults has a band diagram as illustrated in Fig. 29. The quantum wells in Fig. 29b represent domains of the zinc-blende structure. Transitions in these quantum wells give rise to the peak at 3.26 eV of the Mg-doped films. Doping of wurtzite GaN with high concentrations of Mg will introduce a high concentration of cubic domains and the Mg doping of these domains will give rise to transitions in the blue part of the spectrum. This may account for the observation that GaN p-n junction LEDs emit in the blue rather than the UV part of the spectrum. This model needs to be verified by conducting structural studies on both lightly and heavily Mg-doped GaN samples.

4. ELECTRONIC STRUCTURE AND OPTOELECTRONIC PROPERTIES

The films discussed in the previous sections were characterized by studying their bulk and surface electronic structure using angle-resolved photoemission, soft X-ray emission and soft X-ray absorption, their transport properties by Hall effect measurements, and their optical properties by absorption, reflectance and photoluminescence measurements.

a. Electronic Structure

The bulk and surface electronic structure of GaN films produced by plasma-assisted MBE was investigated over the past few years (Stagarescu et al., 1996; Smith et al., 1997; Dhesi et al., 1997). The occupied partial density and the unoccupied partial density of states of GaN were determined by a combination of soft X-ray absorption and emissions spectroscopies (Stagarescu et al., 1996). The elementally and orbitally resolved GaN valence and conduction bands were measured by recording the GaL and NK spectra. The experimental spectra were found to be in good agreement with the partial density of states from *ab initio* calculations (Hu and Ching, 1993). The X-ray spectra show that the bottom of the valence band is primarily Ga derived, the top of the valence band is primarily of N_{2P} character, and the bottom of the conduction band is a mixture of Ga_{4s} and N_{2P} states.

The previous spectroscopies do not directly provide information on the dispersion of the bands and they are not surface-sensitive. To obtain a direct measurement of the band dispersion of surface and bulk states, high resolution, angle-resolved photoemission spectroscopy was employed (Smith et al., 1997; Dhesi et al., 1997). The bulk band dispersion along the $\Gamma\Delta A$, $\Gamma\Sigma M$, and $\Gamma T K$ directions of the bulk Brillouin zone was measured. The experimental results are in good agreement with local-density-approximation (LDA) band-structure calculations over a wide region of the Brillouin zone (Rubio et al., 1993). Furthermore, a nondispersive feature was identified near the valence-band maximum in a region of k-space devoid of bulk states. This feature is identified as emission from a surface state on GaN (0001)- (1×1). The symmetry of this surface state is even with respect to the mirror planes of the surface and polarization measurements indicate that it is of SP_z character consistent with the dangling-bond state. The angle-resolved photoemission spectra indicate that this state is destroyed by adsorption of O_2 or activated H_2 or by ion bombardment of the surface.

The identification that the surface states in wurtzite GaN are resonant with the valence band is in agreement with the lack of Fermi level pinning (Foresi and Moustakas, 1993) and the strong evidence that dislocations are not recombination centers in GaN.

In zinc-blende GaN, the conduction electron spin resonance was measured and the observed resonance has an isotropic g value of 1.9533 (Fanciulli et al., 1993). This value was compared to theoretical predictions based on the five-band model $k \cdot p$ calculations (Chadi et al., 1976). Using the same parameters the five-band model predicts an electron effective mass for the zinc-blende structure of $m^* = 0.15\,m_0$. This value is close to that measured experimentally in wurtzite GaN ($m^* = 0.20\,m_0$) (Barker and Ilegems, 1993).

b. *Transport Properties*

GaN films produced heteroepitaxially on various substrates were generally found to be heavily doped *n*-type, a result that was originally attributed to nitrogen vacancies. This conclusion was supported by empirical calculation (Jenkins and Dow, 1989), which show that nitrogen vacancies contribute donor-like states close to the conduction band edge. Recently, *ab initio* calculations (Neugebauer and Van de Walle, 1994; Boguslawski *et al.*, 1995) show that nitrogen vacancies introduce a resonant state inside the conduction band. Free carrier absorption measurements under high pressure tend to support this hypothesis (Perlin *et al.*, 1995). More recently, *ab initio* calculations (Van de Walle and Neugebauer, 1997) suggest that because of energetics, nitrogen vacancies may not be the predominant donors but instead accidental impurities such as oxygen and silicon may be predominant. However, the kinetically controlled concentration of vacancies could be made to be the dominant factor. Recently, GaN films with background carrier concentration of the order of 10^{17} cm^{-3} have been produced in a number of laboratories, and such films have been intentionally doped *n*- or *p*-type with silicon or magnesium, respectively (Doverspike and Pankove, 1998). Previously, we discussed the importance of dislocations in determining the lateral transport in GaN (Ng *et al.*, 1997; Weiman *et al.*, 1998; Ng *et al.*, 1998) and it was argued that dislocations with an edge component introduce acceptor centers (dangling bonds) along the dislocation line. For a GaN film with 10^{10} cm^{-2} dislocation density the concentration of dangling bonds at dislocations are 2×10^{17} cm^{-3}. These acceptor-like centers set a limit on our ability to dope the material intentionally *n*-type to values below 10^{17} cm^{-3}.

The mechanism of transport in *n*-type GaN films was investigated by studying the temperature dependence of the transport coefficients. With reference to the data of Fig. 21, we studied the temperature dependence of the carrier concentration and electron mobility for samples belonging clearly in the impurity- or dislocation-scattering regimes (Ng *et al.*, 1998a, 1998b). The samples falling in the region where ionized impurity scattering is the dominant factor in determining their mobility show the typical temperature dependence (Fig. 30). In the high-temperature range, the mobility approaches a $T^{-3/2}$ dependence due to phonon scattering. In the low-temperature range, the mobility varies as $T^{3/2}$, which is characteristic of ionized impurity scattering. The peak mobility for sample #1 is 425 cm^2/V·s. The net carrier concentration as a function of temperature is shown in Fig. 31. The activation energy of the Si donors, extracted from the slope of the carrier concentration vs $1/T$ is 13–15 meV, in agreement with published values for Si-doped GaN films grown by the MOCVD method (Götz *et al.*, 1996).

Samples that clearly belong to the dislocation-scattering regime were found to have thermally activated carrier concentration and electron mobility (Ng *et al.*, 1998a, 1998b). Figure 32 shows the results for sample C of

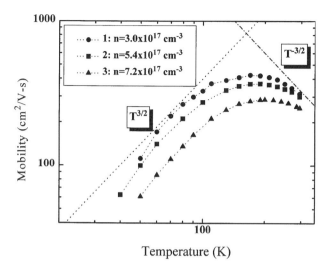

FIG. 30. Temperature dependence of electron mobility of a number of Si-doped GaN films. (Reprinted from Ng et al., 1998a.)

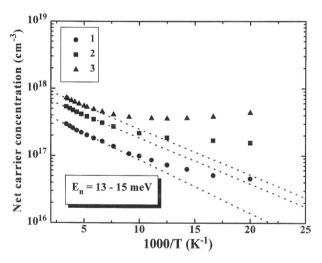

FIG. 31. The net carrier concentration vs $1/T$ for the GaN samples discussed in Fig. 30. (Reprinted from Ng et al., 1998a.)

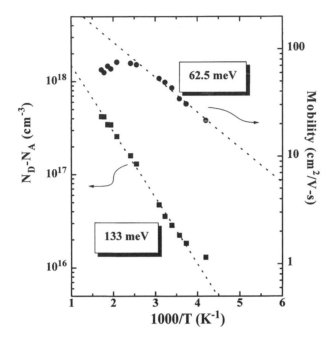

FIG. 32. Temperature dependence of electron mobility and carrier concentration for sample C of Fig. 21. (Reprinted from Ng et al., 1998a.)

Fig. 21. It is interesting to note that the mobility is thermally activated with an activation energy about half that of the carrier concentration. This result is in agreement with the dislocation-scattering model, which according to Eqs. (10) and (11) predicts that $\mu_{disl} \sim n^{1/2}$.

c. *Optical Properties*

The optical properties of cubic and hexagonal GaN thin films, grown by the ECR-MBE method on Si(100) and (0001) sapphire substrates, respectively, have been studied at photon energies up to 25 eV with conventional and synchrotron-radiation spectroscopic elipsometry (Logothetidis et al., 1994; Logothetidis et al., 1995). The fundamental gaps of the two polytypes were found to be at 3.25 and 3.43 eV for cubic and wurtzite GaN. The dielectric function of the two phases in the spectral region 3–10 eV was determined and the results were compared with band-structure calculations. The same authors also reported the reflectivity of both GaN phases and compared their results to early studies by Bloom et al. (1974). They observed that the spectra in this early reference bear resemblance not only with the hexagonal polytype (main peak at 7.0 eV) but also with the cubic polytype. One possible explanation is that early methods of film growth led to films with a significant cubic component in the material.

The photoluminescence (PL) spectra at low temperatures for n-type GaN films generally exhibit a strong and sharp near bandgap (NBG) luminescence at about 3.47–3.48 eV, a broad luminescence peak at about 2.2 eV, and sometimes luminescence peaks in the range of 3.0–3.3 eV (Monemar, 1998). In this section, we discuss the luminescence at 77 K for a large number of GaN films, produced by ECR-MBE with variable carrier concentration $(N_D - N_A)$ from 10^{15} to 10^{20} cm^{-3} (Iliopoulos et al., 1998a, 1998b). The transport coefficients of these films were discussed in Fig. 21. The photoluminescence of the investigated GaN films was excited with a 10-mW He-Cd laser ($\lambda = 325$ nm) and the spectra were collected with a 0.5-m spectrophotometer equipped with a holographic grating blazed at 250 nm and detected with a photomultiplier tube.

Typical photoluminescence spectra (measured at 77 K) for three samples whose carrier concentrations are $1.0 \cdot 10^{20}$, $1.5 \cdot 10^{19}$, and $5.3 \cdot 10^{16}$ cm^{-3} are shown in Fig. 33. As seen from these spectra, the main recombination

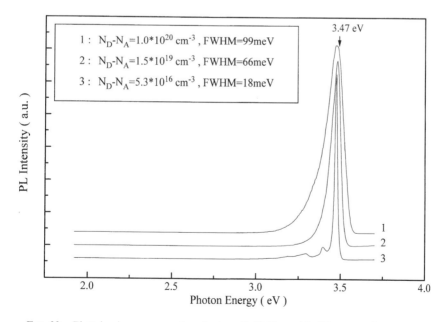

FIG. 33. Photoluminescence spectra of n-type GaN films with different carrier concentration. (Reprinted from Iliopoulos et al., 1998a.)

transition occurs at 3.47 eV with no detectable transitions in the yellow part of the spectrum. For the heavily doped samples the near bandgap peak is broadened toward the Stokes part of the line, characteristic of bound electron states. The main difference between these samples is the width of the photoluminescence line, which in this case varies from 18 meV for the lightly doped sample to 99 meV for the heaviest doped sample.

Figure 34 shows the FWHM of the 3.47-eV photoluminescence peak vs the net carrier concentration ($n = N_D$-N_A) in a log-log scale. It is apparent from these data that the FWHM for samples with $n < 10^{17}$ cm^{-3} is approximately constant (~ 20 meV) and the FWHM for heavily doped samples ($n > 10^{18}$ cm^{-3}) increases as a function of doping approximately as a power of net carrier concentration. The values of dopant concentration in semiconductors above which the impurity electronic states, introduced in the bandgap, evolve into impurity bands can be approximated by the values of free carrier concentration for which the Debye-Hückel screening length (which is appropriate for a nondegenerate case) with a Yukawa-type potential of an impurity equals the first excitonic Bohr radius (Bonch-Bruyevich, 1966). For GaN, with electron effective mass $m_n^* = 0.2 \cdot m_0$ and static dielectric constant $\varepsilon_0 = 10$, the density is calculated to be of the order 10^{17} cm^{-3}. This value is in good agreement with the net carrier concentra-

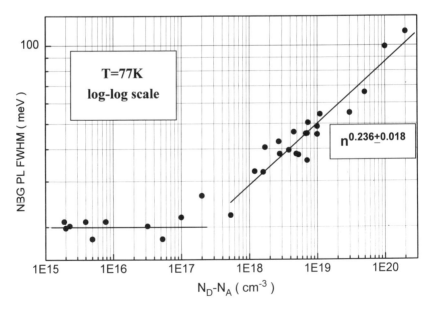

FIG. 34. FWHM of near-bandgap photoluminescence peak (at 77 K) vs net carrier concentration in log-log scale. (Reprinted from Iliopoulos et al., 1998a.)

tion above which our data show a strong dependence of the photoluminescence FWHM on N_D-N_A, indicating that the broadening is due to the width of the donor band. The width of the PL line at carrier concentrations less than 10^{17} cm^{-3} was found to be about 20 meV. If the width of the line was due solely to thermal broadening, it would be expected to be about $3K_BT/2 \sim 10$ meV. We attribute the additional broadening to static disorder in these films, arising from the fact that they were grown at a temperature significantly lower than half the melting point of GaN. An additional cause of that broadening could be homogeneous and inhomogeneous strain, which arise from stoichiometric fluctuations and thermal expansion differences between substrate and film.

The observed functional dependence of the FWHM at high carrier concentrations has been accounted for by using a semiclassical model regarding the broadening of impurity bands in heavily doped semiconductors (Morgan, 1965). In this model ionic impurities are distributed randomly throughout the crystal, with average density N and probability $p = N/N_s$, where N_s is the density of available ionic sites in the crystal. The electrostatic potential at a given point differs from that of a corresponding point in a pure crystal by the total potential produced at the point by all ions in the crystal. The ionic potentials of impurities are screened (with screening length λ) Coulomb ones, and the randomness of ionic positions results in potential fluctuations. The moments of the resulting electronic states probability density function have been calculated in the case of several species of ions being present, differing in charge Z and average density N_Z. The resulting probability distribution is approximated by a Gaussian function with second moment σ given by (Morgan, 1965)

$$\sigma^2 = \langle (E - \bar{E})^2 \rangle_{av} = 1.07 \cdot \frac{2\pi e^4}{(4\pi\varepsilon\varepsilon_0)^2} \cdot \lambda \cdot \sum_Z N_Z \cdot Z^2 \cdot \exp\left(-\frac{2R_Z}{\lambda}\right) \quad (12)$$

where the parameters R_Z are model parameters, used to exclude from the calculation lattice sites, which, if occupied by an ion would contribute an energy of more than 2.3σ, because the probability of finding two ions in this range of distance is generally very small.

By including the screening length and using Thomas-Fermi approximation and the compensation ratio in Eq. (12) we find that the Gaussian-shaped impurity band's second moment, as a function of net carrier concentration and compensation ratio, is given by (Iliopoulos et al., 1998a, 1998b)

$$\sigma_i^2 = 1.138 \cdot 10^{-13} \text{ meV}^2 \cdot \text{cm}^{5/2} \cdot c \cdot n^{5/6} \cdot \exp\left(-8.14 \cdot 10^4 \cdot \sqrt{\frac{2}{c+1}} \cdot n^{-1/4}\right)$$

(13)

We believe that the observed transition at 3.47 eV is due to neutral donor-bound exciton recombination. The broadening of this transition can be accounted for by the broadening of the valence band edge due to potential fluctuations introduced by the impurities. More specifically, the valence band edge potential fluctuations dominate the recombination due to density-of-states effects. The width of this potential fluctuation distribution is the same as that calculated by Eq. (13). Including the broadening σ_τ due to thermal and static disorder, equal to 20 meV as explained before, we expect the FWHM of the PL lines to be

$$\text{FWHM} = \left((20\,\text{meV})^2 + (2^{3/2} \ln 2 \cdot \sigma_i)^2\right)^{1/2} \qquad (14)$$

where $\sigma_i(n, c)$ is given by Eq. (13) and the factor $2^{3/2} \cdot \ln 2$ comes from the relation of the FWHM of a Gaussian distribution to its second moment.

Figure 35 presents the measured FWHM of the near bandgap (NBG) PL peaks of our samples as well as the parametric curves, described by Eq. (14), for compensation ratio values 1 to 5. Thus, according to these data, the observed scattering in similarly doped samples is due to compensation arising from acceptor impurities or dislocations (Weiman et al., 1998).

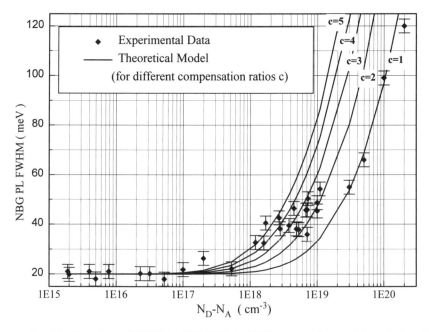

FIG. 35. Experimental FWHM of NBG PL (at 77 K) compared to model predictions for different compensation ratios. (Reprinted from Iliopoulos et al., 1998a.)

In conclusion, the PL spectra of GaN films grown by plasma-assisted MBE and doped n-type with Si to net carrier concentrations 10^{15} to 10^{20} cm^{-3} were investigated at 77 K. In all cases we observed a near bandgap luminescence at 3.47–3.48 eV which we attributed to donor-bound excitonic recombination. The FWHM for samples with carrier concentration less than 10^{17} cm^{-3} was found to be constant of ~ 20 meV. The width of this transition was attributed to thermal, static and strain-induced broadening. The FWHM of the NBG luminescence for samples with carrier concentration higher than 10^{18} cm^{-3} was found to increase monotonically with carrier concentration consistent with the impurity band-broadening model (Morgan et al., 1965). An expression was derived to take into account the effect of compensation on the width of this transition.

V. InGaAlN Alloys

1. Growth

Both the InGaN and AlGaN alloys were grown by plasma-assisted molecular beam epitaxy (MBE) using the deposition system described in Fig. 9. The films were grown on (0001) sapphire substrates, which were first subjected to plasma nitridation (conversion of the surface of Al_2O_3 to AlN) and then coated with 30-nm low-temperature GaN buffer (Moustakas et al., 1992). The InGaN films were grown at 650 to 700 °C at growth rates between 10–30 Å/min, which is of the same order as those of InGaN films grown by the MOCVD method (Nakamura et al., 1994). The AlGaN films were grown at 750 °C at growth rates of 25–35 Å/min. During the growth of these films, a number of deposition parameters, such as the ratio of the III/V fluxes and the degree of n-type doping, were varied.

The structure of the films was determined by X-ray diffraction (XRD) using Cu Kα_1 in a four-circle diffractometer with a Ge(111) single-crystal monochromator that forbids X-ray reflection of the $\lambda/2$ harmonic. The composition of the films was determined from the XRD peaks assuming that Vegard's law is applicable in these ternary alloy systems. The InGaN films were also characterized by optical absorption and photoluminescence studies as reported earlier (Singh and Moustakas, 1996a, 1996b; Singh et al., 1997).

2. Phase Separation and Long-range Atomic Ordering

The solid solutions of III-V compounds were originally considered to be made of a random substitution of the mixed atoms. However, it was found over the past several years that there are deviations from perfect randomness due to either phase separation or atomic ordering during the growth of nominally homogeneous epitaxial layers (Zunger and Mahajan, 1994). In fact, the majority of III-V ternary and quaternary alloys are predicted to be thermodynamically unstable at the low-growth temperature and show a tendency toward clustering and phase separation (Zunger and Mahajan, 1994; Norman et al., 1993). Thus, atomic ordering was not expected to occur during the epitaxial growth of these materials. However, atomic long-range ordering has been observed in a number of epitaxial III-V alloy semiconductors. Such include, for example, InGaAs grown on InP (Nakayama et al., 1986), AlGaAs grown on GaAs (Kuan et al., 1985), and InGaP grown on GaAs (Gomgo et al., 1986) and many other III-V semiconductor alloys. Such phenomena were primarily observed in epitaxial films and the degree of ordering was found to depend strongly on growth conditions, such as growth method, substrate nature and orientation, surface step structure of the starting substrate plane, substrate temperature, III/V flux ratio, and so on. The fact that the degree of ordering and the structure of the ordered alloys depend on the growth conditions strongly suggests that growth kinetics are the dominant factors, which lead to the long-range ordering rather than intrinsic thermodynamic stability of the long-range ordered phases (Nakayama et al., 1994). Evidence has recently been presented that these phenomena have a significant effect on the optical and electronic properties of the alloys (Zunger and Mahajan, 1994).

Similarly, the solid solutions of III-V nitrides were considered until recently to be homogeneous random alloys. However, there is ample recent evidence of phase separation and atomic long-range ordering in InGaN and AlGaN alloys (Singh and Moustakas, 1996a; Singh and Moustakas, 1996b; Singh et al., 1997; Korakakis et al., 1997; Moustakas et al., 1998; Doppalapudi et al., 1998). The study of such phenomena in these alloys is important since these materials are currently employed in various optical and electronic devices.

a. Phase Separation and Ordering in InGaN Alloys

Phase separation in InGaN alloys was studied in thick (300–400 nm) InGaN films or thinner (<50 nm) InGaN films in the form of InGaN/GaN double heterostructures or InGaN/AlGaN multiquantum wells (MWQs). The phenomenon was studied directly using XRD or indirectly by determining the optical properties of the films through transmission and photoluminescence measurements. These InGaN films or structures were grown on top of thick GaN films on (0001) sapphire substrates.

All InGaN films were examined by scanning electron microscopy (SEM), found to have mirror-like surfaces, and show no evidence of indium droplets. X-ray diffraction of the thick $In_xGa_{1-x}N$ films with InN mole fraction greater than 30% show an additional peak that corresponds to In concentration very close to that of pure InN (Singh and Moustakas, 1996a; Singh and Moustakas, 1996b; Singh et al., 1997). Figure 36 shows the θ-2θ XRD pattern for an InGaN film with 37% InN mole fraction. This result represents the first observed evidence of phase separation in InGaN films. Recently, a similar result was reported for films grown by the MOCVD method (Wakahara et al., 1997).

X-ray diffraction studies of InGaN/GaN double heterostructures with the thickness of the InGaN less than 50 nm show no evidence of phase separation as described previously. Figure 37 shows the XRD patterns for two such double heterostructures with InN mole fraction 53 and 81%, respectively. This result can be accounted for by taking into account that phase separation in any alloy requires long-range diffusion and thus a correlation should exist between phase separation and length of time required for the growth of the film. We believe that this is one of the reasons for the nonobservable phase separation in GaN/InGaN/GaN double heterostructures with thin InGaN layers. Strain associated with thin InGaN quantum wells could also stabilize the alloy against phase separation.

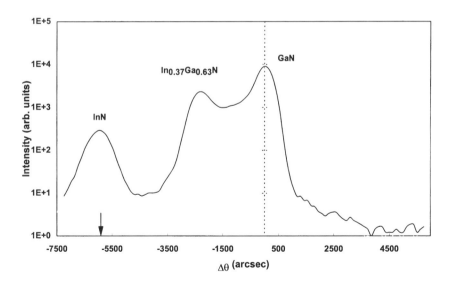

FIG. 36. XRD data for an $In_{0.37}Ga_{0.63}N$ sample. (Reprinted from Singh et al., 1997.)

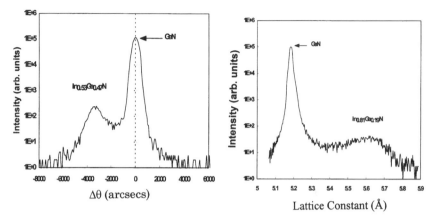

FIG. 37. XRD data for two InGaN/GaN double heterostructures. (Reprinted from Singh et al., 1997.)

In addition to direct observation of phase separation by XRD, we also studied these phenomena indirectly by examining the effect of phase separation on the optical properties of the films. Figure 38 shows the optical absorption constant vs wavelength for three films with InN mole fraction of 0, 18, and 37%, respectively. It is obvious from these data that while the low In concentration films have a single absorption edge, the film with 37% In has two distinct absorption edges, characteristic of a composite structure. However, from the data of Fig. 38, it is difficult to determine the optical gap because the optical absorption constant of any direct-gap semiconductor consists of the region in which the energy of the photon is greater than the bandgap (Kane region) and that in which the energy of the photon is less than the bandgap (Urbach edge). The optical absorption constant in the two regions is given by the following expressions, respectively (Ariel et al., 1995)

$$\alpha \propto \sqrt{E - E_g} \qquad (15)$$

$$\alpha = \alpha_0 \exp\left(\frac{E - E_1}{E_0}\right) \qquad (16)$$

where α_0 and E_1 are independent of either thermal or structural disorder.

An accurate determination of the optical bandgap can be accomplished by using the derivative of the optical absorption constant (Ariel et al., 1995). The derivatives of Eqs. (15) and (16) are

$$\frac{d\alpha}{dE} \propto \frac{1}{\sqrt{E - E_g}} \qquad (17)$$

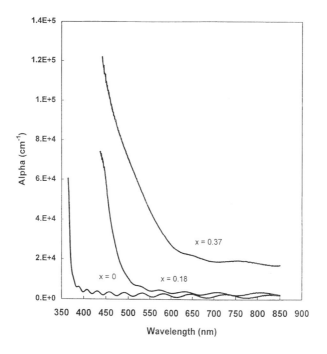

FIG. 38. Optical absorption vs wavelength for (a) GaN; (b) $In_{0.18}Ga_{0.82}N$; and (c) $In_{0.37}Ga_{0.63}N$ film. (Reprinted from Singh et al., 1997.)

$$\frac{d\alpha}{dE} \propto \exp\left(\frac{E - E_1}{E_0}\right) \qquad (18)$$

From Eqs. (17) and (18), the derivative of the absorption constant at the Kane regime depends on $1/\sqrt{E - E_g}$, while the derivative of the absorption constant at the Urbach edge has the same exponential dependence on E. A plot of the derivative of the absorption constant vs the energy, for the samples described in Fig. 38, is shown in Fig. 39. According to these data the bandgap of pure GaN is 3.37 eV. The bandgap of the InGaN film with 18% indium is 2.74 eV with a broad Urbach edge, and the film with 37% indium appears to have multiple bandgaps, suggesting phase separation in such films.

Evidence of phase separation in InGaN films, based on optical absorption data has also been reported in films grown by the MOCVD method (Matsuoka et al., 1990). In Fig. 40, we reproduced the optical bandgap data from this reference. The solid line represents the relationship between optical

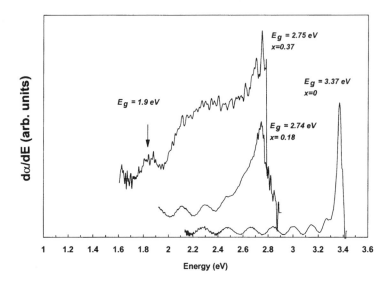

FIG. 39. The $d\alpha/dE$ vs photon energy for the samples discussed in Fig. 38. (Reprinted from Moustakas et al., 1998.)

bandgap vs InN mole fraction, assuming the validity of Vegard's law. It is important to note that the data points for the bulk films fall significantly below the straight line, a result which can be accounted for by phase separation in such films. For example, the film with 42% indium has approximately the same optical gap as that of pure InN, which is consistent with the data of Fig. 36. Plotted in Fig. 40 are also the optical gaps, determined from the peaks of the photoluminescence, for two MQWs of InGaN/AlGaN in which the InGaN wells are about 100 Å thick (Singh et al., 1996). The fact that these points fall closer to the straight line supports our earlier conclusion that phase separation increases with the film thickness or duration of film growth.

Phase separation in the InN-GaN pseudo-binary system is thermodynamically expected because of the large difference in lattice parameter between GaN and InN (11%) (Singh and Moustakas, 1996a, 1996b; Singh et al., 1997). The maximum value of the critical temperature for miscibility in this system was estimated using Stringfellow's delta lattice parameter model (DLP) (Stringfellow, 1982)

$$T_C = \frac{8.75K}{4R} \frac{(\Delta a)^2}{(\bar{a})^{4.5}} \tag{19}$$

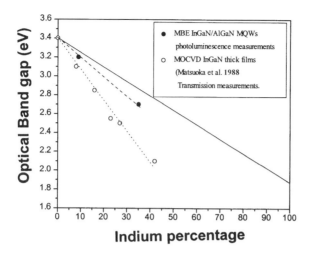

FIG. 40. Optical bandgap vs InN mole fraction for the InGaN films and InGaN/AlGaN MQWs described in the inset of the figure. (Reprinted from Moustakas et al., 1998.)

where Δa is the difference in the lattice constants of GaN and InN, $\bar{a}$ is the average lattice of GaN and InN, and R is the gas constant; K is the proportionality constant between atomization enthalpy (bonding energy) and lattice constant in the Philips and Van Vechten formula (Philips and Van Vechten, 1970)

$$\Delta H^{at} = K a^{-2.5} \tag{20}$$

The constant K was found by Stringfellow by fitting Eq. (20) to experimental data for III-V compounds to have a value of 1.17×10^7 cal/mol $\text{Å}^{2.5}$.

Equation (19) was derived for zinc-blende structure, while GaN and InN, in their equilibrium phases, have the wurtzite structure. However, the cohesive energies of these two allotropic forms must differ only slightly. Using the lattice constants for zinc-blende GaN and InN, we find from Eq. (19) that $T_C = 2457$ K. Based on this estimate, the growth of InGaN alloys at about 1000 K occurs within the miscibility gap of this system. More accurate calculations (Ho and Stringfellow, 1996) based on the valence-force-field (VFF) model with a number of nearest neighbors allowed to relax, have arrived at much lower temperature ($T_C = 1500$ K). However, even such a value is much higher than the typical growth temperature of InGaN alloys. Thus, thermodynamically phase separation is expected in InGaN alloys, although it is very likely that growth kinetics also play a very important role in these phase-separation phenomena.

Recently, Doppalapudi et al. (1998) reported additional observations of phase separation and long-range atomic ordering in InGaN alloys. Specifically, films grown by MBE at substrate temperatures of 700–750 °C with indium concentration higher than 35% show phase separation in good agreement with thermodynamic predictions (Ho and Stringfellow, 1996). Films grown at lower substrate temperatures (650–675 °C) revealed composition inhomogeneities when the In content is larger than 25%. Upon annealing of such films to 725 °C, the material undergoes similar phase separation as films grown at the same temperatures. The InGaN films were also found to show long-range atomic ordering of the Cu-Pt type. The ordering parameter was found to increase monotonically with growth rates of the films, a result which is consistent with the notion that ordering is induced at the surface of the growing films where it is thermodynamically stable and it is then subsequently "frozen in" during further growth. The ordered phase was found to be stable to annealing up to 725 °C.

In conclusion, experimental results on phase separation in InGaN alloys produced by MBE were presented. Phase separation was directly observed by XRD in thick films with InN mole fraction greater than 30%. However, such phenomenon was not observed in InGaN/GaN double heterostructures or InGaN/AlGaN MWQs (InGaN thickness < 50 nm) with InN mole fraction up to 80%. Phase separation was found to affect the optical properties of the films. We propose that the optical gaps in these composite films should be determined by plotting the derivative of the absorption constant vs energy. Films grown at lower temperatures were found to undergo the same type of phase separation upon annealing as those grown at higher temperatures. These findings are consistent with thermodynamic predictions. The same alloys were found to show ordering of the Cu-Pt type.

b. Long-range Order in AlGaN Alloys

Because AlGaN alloys crystallize in the wurtzite structure, the geometrical structure factor for the Bragg reflection (hkl) is given by the expression

$$F_{hkl} = f_A + f_B \exp\left\{2\pi i\left(\frac{h + 2k}{3} + \frac{l}{2}\right)\right\} \tag{21}$$

where f_A and f_B are the average atomic scattering factors of the (Al, Ga) atoms occupying the (000) and (1/3, 2/3, 1/2) sublattice sites of the hexagonal cell, respectively. When the two sites are occupied by the same atomic species or by random mixtures of the two species, Bragg reflections with $l = $ odd and $(h + 2k) = 3n$ are forbidden according to Eq. (21). Thus, in pure GaN, pure AlN and random $Al_xGa_{1-x}N$ alloys, diffraction peaks such as (0001), (0003), (0005), and so on, do not occur. However, if one of the sublattice sites is preferentially occupied by Al or Ga in an $Al_xGa_{1-x}N$ alloy, the two terms no longer cancel and superlattice peaks result.

Korakakis et al. (1997) have recently reported that in $Al_xGa_{1-x}N$ alloys, grown by MBE, the superlattice peaks (0001), (0003), and (0005) do exist. Figure 41 shows the θ-2θ XRD pattern for an $Al_{0.5}Ga_{0.5}N$ alloy, in which the existence of superlattice peaks, in addition to the allowed diffractions, provide clear evidence of ordering in this sample.

In general, the superlattice structure factor of an ordered alloy can be written in terms of the long-range order parameter $S = p_{Al}^1 - p_{Al}^2$, where p_{Al}^1 and p_{Al}^2 are the probabilities of finding aluminum on sublattice (000) or sublattice (1/3, 2/3, 1/2) sites, respectively:

$$F_{hkl} = S(f_{Al} - f_{Ga}) \tag{22}$$

The long-range order parameter must vanish at the compositional endpoints of pure GaN and pure AlN; thus the superlattice peak scattering is expected to be largest in alloys with compositions near 50%. Indeed, if the ordering was perfect, S would be proportional to x, where x is the composition factor of the minority component, and therefore in the

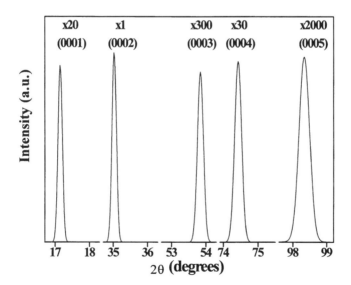

FIG. 41. θ-2θ XRD scan from an $Al_{0.5}Ga_{0.5}N$ sample with particularly strong ordering grown on c-plane sapphire. (Reprinted from Korakakis et al., 1997.)

kinematical scattering limit (Guinier, 1963) the superlattice peak intensity would increase with composition as x^2. Figure 42 shows that the relative superlattice peak intensity is largest for films with relatively high concentrations of each component. Here the superlattice (0001) intensity has been normalized with respect to the (0004) fundamental intensity to account for differences in film thickness and orientational quality. The solid line in this figure was calculated from the expression

$$\frac{I_{(0001)}}{I_{(0004)}} = \left[\frac{\frac{1 + \cos^2 2\theta_1 \cos^2 2\alpha}{\sin 2\theta_1}}{\frac{1 + \cos^2 2\theta_4 \cos^2 2\alpha}{\sin 2\theta_4}}\right] \cdot \frac{(2x(f_{GaN}^{(1)} - f_{AlN}^{(1)}))^2}{(2(1-x)f_{GaN}^{(4)} + 2xf_{AlN}^{(4)})^2} \cdot \left\{\frac{\sin \theta_4}{\sin \theta_1}\right\} \quad (23)$$

which represents the kinematical limit for perfectly ordered alloys, taking into account the Lorentz-polarization factor (in square brackets) and differences in film volume arising from illuminating the sample at different angles (in curly brackets) but neglecting the effect of absorption and the Debye-Waller temperature factor. In this expression, θ_k is the Bragg angle at the $(000k)$ peak, $f_A^{(k)}$ is the scattering factor of molecule A at the $(000k)$

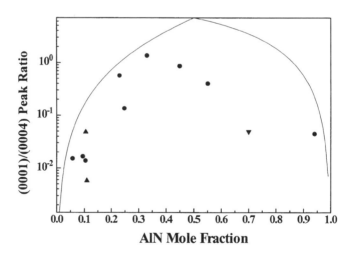

FIG. 42. Relative superlattice peak-integrated intensity for a number of samples grown as a function of AlN mole fraction. Samples grown on sapphire are shown as circles and those grown on 6H-SiC are shown as triangles. The solid line is the theoretical prediction in the kinematic limit. (Reprinted from Korakakis et al., 1997.)

peak and α is the Bragg angle of the monochromator. The experimental data are in qualitative agreement with the theoretical prediction. However, the determination of the long-range order parameter S from the experimental data of Fig. 42 is not accurate because of dynamical scattering effects, surface roughness, and so on.

In addition, the domain size D and the inhomogeneous strain ε_{in} of the grown films was investigated by examining the FWHM wavenumber δk of the fundamental and superlattice XRD peaks. The relationship between these quantities is given in reciprocal space by (Lei et al., 1993)

$$\delta k = \frac{2\pi}{D} + \varepsilon_{in} k \qquad (24)$$

Figure 43 shows a plot of δk vs k for an $Al_{0.55}Ga_{0.45}N$ film. To within the accuracy of our measurements the fundamental peaks indicate that the grain size of the alloy is 6400 Å, which is comparable to the thickness of the film. From the same plot we observe that the superlattice peaks fall on a different line. This broadening of the superlattice peaks, compared to the fundamental peaks, can be attributed to either size-effect broadening or formation of antiphase domains. In either case, we deduce that the average size of the ordered domains is somewhat less than the film thickness — approximately 1700 Å. As we do not have independent determinations of

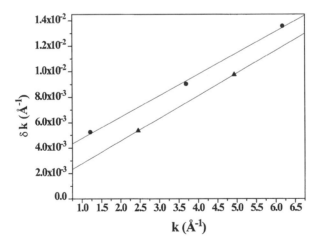

FIG. 43. Plot of δk vs k for an $Al_{0.55}Ga_{0.45}N$ film grown on c-plane sapphire. The triangles indicate the fundamental peaks and the circles indicate the superlattice peaks. (Reprinted from Korakakis et al., 1997.)

the Al composition we cannot determine homogeneous strain in those films utilizing the shift of the peaks. Therefore we cannot explore any dependence of lattice constant on the degree of atomic order.

The long-range ordering is expected to depend on a number of thermodynamic, growth and kinetic factors. In agreement with the findings reported in traditional III-V compounds, we found that the degree of ordering is affected by the ratio of III/V fluxes (Fig. 44) (Korakakis, 1998; Moustakas et al., 1998). More specifically, films grown at lower III/V flux ratios were found to have higher long-range order. Furthermore, we observed that Si doping has a small effect in increasing the degree of ordering. Another parameter that most likely affects the long-range ordering is the step structure of the substrate. However, we did not investigate this parameter systematically (Korakakis, 1998).

In conclusion, long-range atomic ordering in AlGaN alloys produced by MBE has been observed by studying the (0001), (0003), and (0005) superlattice peaks by XRD. We found that the ordering is largest for AlN mole fraction close to 50% in qualitative agreement with the kinematical scattering theory. The ordered domains in these films were found to be approximately one-third of the thickness of the films. From the various kinetic factors, we found that the III/V ratio has a significant effect on the degree of ordering in these films.

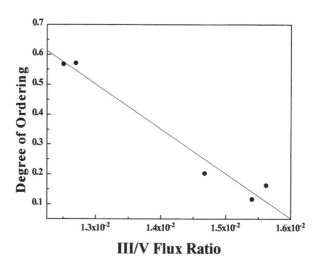

FIG. 44. Dependence of the degree of ordering on the III/V flux ratio for $Al_xGa_{1-x}N$ films grown under otherwise identical conditions. (Reprinted from Moustakas et al., 1998.)

VI. Multiquantum Wells

1. $In_xGa_{1-x}N/Al_yGa_{1-y}N$ Multiquantum Wells

The development of InGaN quantum wells (QWs) is important for the fabrication of LEDs, diode laser structures and detectors operating in the visible and near-UV part of the electromagnetic spectrum. Single-quantum wells (SQWs) and multiquantum wells (MQWs) of InGaN have already been used as the active layer for both blue-green LEDs and blue laser diodes (Nakamura et al., 1996a, 1996b). In spite of the rapid progress by the Nichia group in the development of such devices, the growth and properties of the InGaN alloys are poorly investigated. We discussed in the previous section that InGaN alloys undergo phase separation, a result accounted for by both thermodynamic and kinetic factors. These studies indicate that while phase separation occurs readily in bulk (thick) InGaN films, it is generally not observable in thin InGaN films incorporated in quantum well structures. This suggests that the efficiency, monochromaticity and stability of devices based on InGaN will require the fabrication of the active region in the form of quantum well structures.

There are only limited reports on the growth of InGaN SQW and MQW structures. Nakamura et al. (1993) reported quantum effects in $In_xGa_{1-x}N/In_yGa_{1-y}N$ superlattices grown by the MOCVD method. Koike et al. (1996) reported on $In_{0.08}Ga_{0.92}N/GaN$ MQWs grown also by the MOCVD method. Finally, Keller et al. (1996) reported the growth and properties of $In_{0.16}Ga_{0.84}N/In_xGa_{1-x}N$ (graded) SQW also by the MOCVD method. In this section we discuss the growth by MBE and characterization of InGaN/GaN and InGaN/AlGaN MQWs where both the wells and the barriers were grown at the same temperature.

Figure 45 shows a schematic of the two types of MQW structures investigated. The first structure (Fig. 45a) consists of a 2-μm GaN film and five periods of $In_{0.09}Ga_{0.91}N/GaN$. The thicknesses of the quantum wells and barriers were 80 and 90 Å, respectively. Finally, a 200-Å cap layer was deposited. The whole structure is doped with silicon. The second structure (Fig. 45b) consists of a 400-nm GaN film followed by 50 nm of AlGaN and seven periods of $In_{0.35}Ga_{0.65}N/Al_{0.1}Ga_{0.9}N$. The thicknesses of the wells and barriers were both estimated to be 120 Å.

Epitaxial growth of the various MQW structures was carried out on C-plane sapphire substrates using the deposition system described in Fig. 9 and the procedures outlined earlier. The substrates were first exposed to a nitrogen plasma for surface nitridation at a temperature of about 800 °C and then cooled to 550 °C for the deposition of a 300-Å thick GaN buffer. Finally, the temperature was raised to 750 °C for the growth of a GaN film

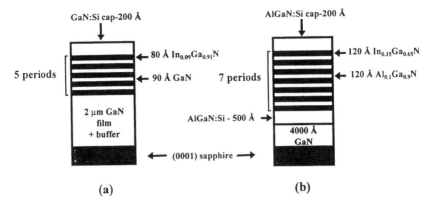

FIG. 45. Schematic of MQW structures (a) $In_{0.09}Ga_{0.91}N/GaN$ having five periods; and (b) $In_{0.35}Ga_{0.65}N/Al_{0.1}Ga_{0.9}N$ having seven periods. (Reprinted from Singh et al., 1996.)

(approximately 0.5 μm or thicker). In the case of the $In_{0.35}Ga_{0.65}N/Al_{0.2}Ga_{0.8}N$ multiquantum well structure, the substrate temperature was maintained at 660 °C during the growth of the InGaN wells as well as the AlGaN barrier layers. The gallium beam equivalent pressure (BEP) was kept constant, while the aluminum and indium cell shutters were opened alternately for the growth of the 7-period structure. No changes were made in the nitrogen plasma power, which was kept constant at 100 W with a nitrogen flow rate of 10 sccm. The $In_{0.09}Ga_{0.81}N/GaN$ structure was grown at a substrate temperature of 670 °C. The gallium BEP was kept constant while the nitrogen plasma power was kept at 100 W during the InGaN layer growth and was reduced to 80 W for the GaN barrier layers.

The samples were characterized by room-temperature photoluminescence which was excited by a 10-mW He-Cd laser with a wavelength of 325 nm. The luminescence spectra were dispersed through a 0.5-m Acton Research spectrometer and measured with a photomultiplier tube (Hamamatsu R928). The indium and aluminum mole fractions were determined from bulk InGaN and AlGaN samples grown under identical beam fluxes of indium, gallium, and aluminum. Cross-sectional TEM studies were conducted on the InGaN/AlGaN MQW structure, using a JEOL 200CX transmission electron microscope operated at 200 kV.

Figure 46 is a TEM bright field image of a cross section of the structure discussed in Fig. 45b, taken near the $GaN[1100]/Al_2O_3[11-20]$ zone axis with a two-beam condition ($g = [0002]$) to obtain the best contrast between the layers. The bending of the layers seen in the micrograph is due to ion

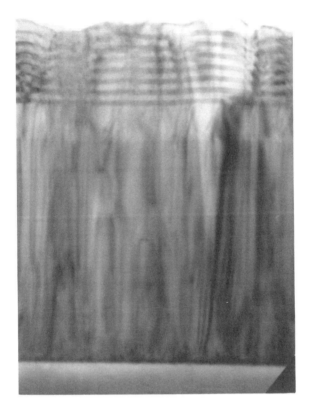

FIG. 46. Cross-sectional TEM micrograph of the InGaN/AlGaN MQW described in Fig. 45b. (Reprinted from Singh *et al.*, 1996.)

damage and uneven thinning resulting from the TEM sample preparation procedure. The micrograph shows sharp interfaces between adjacent layers indicating abrupt wells and barriers. The thicknesses of the wells and the barriers calculated from this micrograph are 130 Å each, in close agreement with those calculated from beam-flux measurements.

Figure 47 shows the room-temperature photoluminescence from the structure discussed in Fig. 45a. The spectra show two peaks: a small peak at 365 nm, which is luminescence from the GaN cap, GaN barriers and the GaN film underneath the MQW; and a second peak, centered at 387 nm, which is luminescence emitted by the $In_{0.09}Ga_{0.91}N/GaN$ quantum wells. The FWHM of this peak is 16 nm and the PL intensity is significantly higher and narrower than that of bulk InGaN films having the same In concentration (Singh and Moustakas, 1996b). These data resemble the work

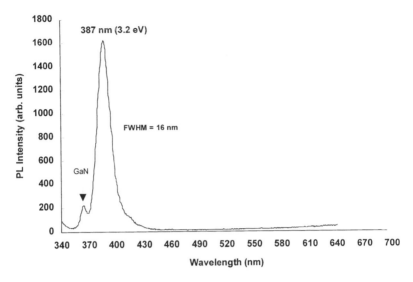

FIG. 47. Room-temperature photoluminescence from a five-period ($In_{0.09}Ga_{0.91}N$/GaN) MQW structure the schematic of which is shown in Fig. 45. (Reprinted from Singh et al., 1996.)

of Koike et al. (1996) who studied $In_{0.08}Ga_{0.92}N$(7 nm)/GaN(9 nm) MQWs with six periods. In this work the properties of the MQWs were evaluated by cathodoluminescence (CL) measurements. The CL emission peak occurs at 386 nm and its FWHM is 11 nm. High-resolution submicron microscopy was used to assess the radiative efficiency and spatial uniformity of these MQWs by illuminating and collecting the luminescence from a facet perpendicular to the layers (Herzog et al., 1997). These studies reveal that the radiative recombination for the InGaN quantum wells is 50–60 times more efficient than that of the underlying GaN films.

Figure 48 shows the photoluminescence from the structure discussed in Fig. 45b. The spectra here show two peaks: a small peak at 345 nm, which is luminescence from the AlGaN barriers and the AlGaN film underneath the MQW; and a second peak, centered at 463 nm, which is luminescence emitted by the $In_{0.35}Ga_{0.65}N$/$Al_{0.1}Ga_{0.9}N$ quantum wells. The FWHM of this peak is 28 nm and the PL intensity is significantly higher than that of bulk InGaN films having the same In concentration (Singh and Moustakas, 1996b). The spectrum of this MQW is broader than the one described in Fig. 47, which is consistent with a wider well (120 Å) employed in the present structure.

In conclusion, multiquantum wells of InGaN/GaN and InGaN/AlGaN were fabricated by the MBE method, using the same substrate temperature

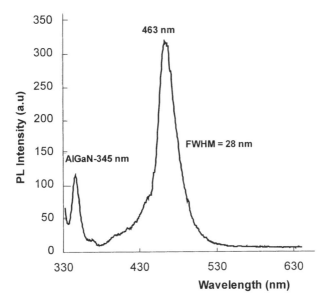

Fig. 48. Room-temperature photoluminescence from a seven-period ($In_{0.35}Ga_{0.65}N$/ $Al_{0.1}Ga_{0.9}N$) MQW structure the schematic of which is shown in Fig. 45. (Reprinted from Singh et al., 1996.)

(660–670 °C) for both wells and barriers. The interfaces between wells and barriers were found to be abrupt by TEM studies and the luminescence was found to be significantly narrower and more intense than that emitted from bulk InGaN films. These studies clearly indicate that $In_xGa_{1-x}N$/ $Al_yGa_{1-y}N$ MQWs have superior photoluminescence properties compared to bulk InGaN films and for this reason are likely to be used in the active region of emitters operating in the visible part of the electromagnetic spectrum.

2. GaN/AlGaN Multiquantum Wells

Laser diodes, having InGaN MQWs as an active region, have recently been demonstrated to emit in the violet-blue part of the electromagnetic spectrum (Nakamura et al., 1996). There is significant technological incentive to develop UV laser diodes from this class of materials. The fabrication of UV laser diodes using III-V nitrides requires that the active region be made of GaN/AlGaN or $Al_xGa_{1-x}N/Al_yGa_{1-y}N$ MQWs. In this section we discuss epitaxial growth by MBE and crystal structure of GaN/Al$_{0.2}$Ga$_{0.8}$N MQWs grown onto an Al$_{0.25}$Ga$_{0.75}$N cladding layer to simulate a laser-diode structure (Korakakis et al., 1998a).

The investigated laser-diode structure was grown by plasma-assisted MBE, using the deposition system described in Fig. 9. The (0001) sapphire substrates were subjected first to plasma nitridation at 750 °C. Following this step, a low-temperature (550 °C) GaN buffer of approximately 200 Å was grown. To simulate a laser-diode structure, we grew next 4500 Å of n-GaN as the bottom contact layer and 4000 Å of n-Al$_{0.25}$Ga$_{0.75}$N as the corresponding cladding layer. These films were doped n-type with silicon to levels of 1×10^{19} cm^{-3} and 5×10^{18} cm^{-3}, respectively. The undoped GaN/Al$_{0.2}$Ga$_{0.8}$N MQW consisting of 15 pairs of 50 Å GaN and 50 Å Al$_{0.2}$Ga$_{0.8}$N (periodicity $\Lambda = 100$ Å) was grown next. The contact, cladding and MQW layers were grown at 750 °C. A schematic of the investigated structure is shown in Fig. 49.

The structure described in Fig. 49 was characterized by XRD studies, using a four-circle diffractometer with a Ge(111) crystal as a monochromator to isolate the Cu-Kα_1 radiation. Specifically, the reciprocal lattice map of off-axis diffraction peaks, as well as the θ-2θ pattern around the (0002) reflection were examined. The chemical composition of the films was determined from the measured c-lattice parameter by XRD and assuming the validity of Vegard's law.

Figure 50 shows the intensity contour plot around the ($\bar{2}$024) reciprocal lattice point. The diffraction peak, which originates from the GaN film below the MQW (labeled A) corresponds to lattice parameters of bulk GaN ($a = 3.189$ Å and $c = 5.185$ Å). The diffraction pattern around points labeled

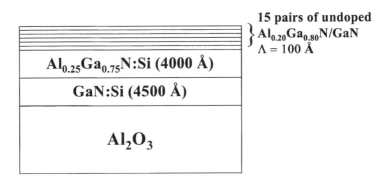

FIG. 49. A schematic of the investigated GaN/AlGaN MQW structure. (Reprinted from Korakakis et al., 1998.)

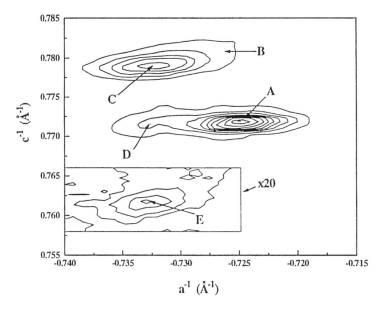

FIG. 50. Intensity contour plot of the (20-24) reciprocal lattice area of the structure described in Fig. 49. (Reprinted from Korakakis et al., 1998.)

B and C is due to the cladding $Al_{0.25}Ga_{0.75}N$ film below the MQW. This peak is skewed toward larger a- and smaller c-lattice constants (point B). We attribute this skewness to the fact that the first few monolayers of the $Al_{0.25}Ga_{0.75}N$ are strained and assume the in-plane lattice parameter of GaN underneath. However, as the $Al_{0.25}Ga_{0.75}N$ film grows thicker it relaxes to a larger c-lattice parameter ($c = 5.16$ Å) and to a smaller a-lattice parameter ($a = 3.154$ Å). The c/a ratio for this film is the ideal for a hexagonal closed-packed structure (HCP) $c/a = 1.63$. The observed smaller peak (labeled D) with almost the same c-lattice parameter ($c = 5.182$ Å) as GaN and smaller a-lattice parameter is the primary diffraction peak of the superlattice structure. This identification of the peak D is based on the observation of a satellite peak (labeled E) at a distance $1/\Lambda$ ($1/100$ Å^{-1}) parallel to the c^{-1} axis and with the same a-lattice parameter. The fact that we observe only one first-order satellite peak is evidence that the superlattice structure is coherent; otherwise a satellite peak below the primary GaN peak (point A) would have been observed. The symmetric first-order satellite peak in this ($\bar{2}024$) mapping is expected to be at a distance $1/\Lambda$ from point D parallel to the c^{-1} axis toward smaller c. However, this brings the peak underneath the $Al_{0.25}Ga_{0.75}N$ peak (point C), and thus cannot be independently resolved. It is important to note that peak C, which corresponds to the underlying $Al_{0.25}Ga_{0.75}N$ thick film, and peaks D and E, which correspond to the superlattice structure, all have the same in-plane lattice parameter.

Thus, the superlattice is strained and has assumed the a-lattice parameter of the layer on which it was grown. The contraction of the a-lattice constants of the layers forming the superlattice causes an expansion of the c-lattice constant predicted by the Poisson ratio (Arfken, 1970). This phenomenon is well investigated in metallic structures and in other alloy semiconductor superlattices (McWhan, 1985). It has not been reported in the wurtzite III-nitrides where the lattice mismatch among the binary alloys of the family is larger than that of the arsenides. However, this result is expected because for the difference in the a-lattice constants of GaN and $Al_{0.2}Ga_{0.8}N$ the Matthews critical thickness is expected to be 120 Å (Amano et al., 1997).

Figure 51 shows the θ-2θ experimental and theoretical diffraction patterns around the (0002) reflection of the structure described in Fig. 49. The mathematical expression for the diffraction pattern of superlattices with

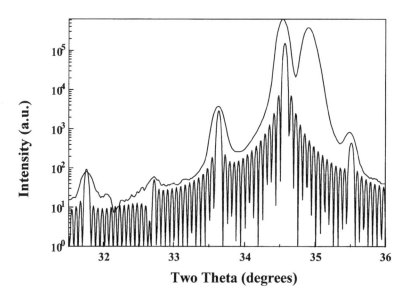

FIG. 51. θ-2θ XRD scan around the (0002) reflection of the structure described in Fig. 49. (Reprinted from Korakakis et al., 1998.)

2 GROWTH OF III-V NITRIDES BY MOLECULAR BEAM EPITAXY

ideally abrupt interfaces is given by the expression (McWhan, 1985)

$$I(s) = \left(\frac{\sin(\pi s k \Lambda)}{\sin(\pi s \Lambda)}\right)^2 \left[C_A^2 \left(\frac{\sin(\pi s n_A a)}{\sin(\pi s a)}\right)^2 + C_B^2 \left(\frac{\sin(\pi s n_B (a - \varepsilon))}{\sin(\pi s(a - \varepsilon))}\right)^2 \right.$$
$$\left. + 2 C_A C_B \cos(\pi s \Lambda) \frac{\sin(\pi s n_A a)}{\sin(\pi s a)} \frac{\sin(\pi s n_B (a - \varepsilon))}{\sin(\pi s(a - \varepsilon))} \right] \qquad (25)$$

where k is the number of pairs of layers in the superlattice, n_A, a, n_B, a-ε are the number of planes and plane spacing of material A and material B in the respective layers. Λ is the superlattice periodicity, and s is the position in reciprocal space. The coefficients C_A, C_B include the atomic, or molecular, properties of the constituent materials, atomic scattering factor, and Debye temperature factor. The Lorentz-polarization factor can be neglected over a small angular range and therefore is not included in Eq. (25). From the discussion on the data presented in Fig. 50 we have established that the superlattice is coherent with $a = 3.154$ Å and $c = 5.185$ Å. Using this value of a we can determine the coefficients C_A and C_B in Eq. (25). Using the value of c and assuming that the superlattice layers have an equal number of planes we can derive $n_A = n_B = 9.5$. In the hexagonal close-packed (HCP) unit cell there are two monolayers and thus n_A and n_B are allowed to have half-integer values. Substituting these values into Eq. (25), the theoretical curve shown in Fig. 51 is constructed.

The experimental diffraction pattern in Fig. 51, in addition to the superlattice features, shows also two peaks due to GaN and $Al_{0.20}Ga_{0.80}N$ layers. The GaN coincides with the primary superlattice peak while the $Al_{0.25}Ga_{0.75}N$ peak occurs at slightly higher 2θ angle. The satellite peaks of the superlattice are symmetric at $\pm 1/\Lambda$ around the primary peak. In order to evaluate the quality of the interfaces we compared the experimentally derived integrated intensity peak ratios to the theoretically expected peak ratios. As the FWHM of the satellite peaks are almost equal, the peak height can be used for the comparison. These data show that there is an excellent agreement between experimental and theoretical data in the relative intensities of up to the third order of satellite peaks. This is evidence that the superlattice layers are well formed with smooth and abrupt interfaces (compared to 50 Å). Apparently there was little interdiffusion of Ga and Al atoms between the layers. This is a surprising result considering that Ga and Al are miscible at the growth temperature employed, that is, 750 °C. Bearing in mind, however, that $Al_xGa_{1-x}N$ alloys are strongly bonded, the activation energy for the diffusion of atoms in their lattice is expected to be higher than the energy provided by the substrate temperature.

In conclusion, GaN/Al$_{0.2}$Ga$_{0.8}$N MQWs were grown by plasma-assisted MBE on an Al$_{0.25}$Ga$_{0.75}$N cladding layer simulating a laser-diode structure. The MQW was found to be coherent with the a-lattice parameter of the underlying Al$_x$Ga$_{1-x}$N layer. Up to the third-order satellite peaks were observed with relative intensities in very good agreement with theoretical predictions for a superlattice with abrupt interfaces. This strong evidence of smooth and abrupt interfaces indicates negligible interdiffusion of Ga and Al atoms at the growth temperature of 750 °C, a result that is consistent with the strength of the Al-N and Ga-N bonds.

VII. Device Applications

In this section, we present a brief review of progress reported in the fabrication of GaN devices by MBE. An excellent review of nitride-based devices was recently presented by Mohammad *et al.* (1995).

1. Device Processing

A number of devices were fabricated on GaN and AlGaN films grown by MBE. Standard photolithography, and lift-off techniques were used to pattern metal contacts on the devices (Foresi and Moustakas, 1993; Misra *et al.*, 1995; Vaudo *et al.*, 1996; Sampath *et al.*, 1998). Reactive ion etching (RIE) with SiCl$_4$ gas was employed for the definition of the mesas (Manfra *et al.*, 1994; Vaudo *et al.*, 1996).

Various contacts to n-GaN (Foresi and Moustakas, 1993; Sampath *et al.*, 1998) were investigated. These investigations provided the first evidence that surface states in GaN do not pin the Fermi level of metals and thus the ideal Schottky limit is applicable to GaN. Based on this, it was proposed (Foresi and Moustakas, 1993) that metals with work function close to that of the electron affinity of GaN (4.1 eV) (Pankove and Schade, 1974) should form good ohmic contacts to n-GaN films. The first metal to be investigated was Al, which has a reported work function of 4.08 eV (Boer, 1990). The specific contact resistivity for this metal was determined by transfer length measurements (TLM) to be 10^{-3}–10^{-4} Ω cm^2 (Foresi and Moustakas, 1993). Upon annealing to 575 °C for 10 min, the contacts became slightly nonohmic, a result which was attributed to the formation of an interfacial AlN layer during the annealing process. The authors proposed that such a phenomenon can be avoided by the deposition first of a thin layer of Ti or Cr to act as a diffusion barrier. Indeed, Lin *et al.* (1994) employed Ti/Al bilayers on n-GaN, which upon rapid thermal annealing to 900 °C for 30 s yielded contacts with contact resistivity $8 \times 10^{-6} \Omega \text{cm}^2$. Such bilayers are currently employed routinely in commercial LEDs and experimental laser and transistor structures (Nakamura, 1998; Pankove et al., 1994).

Subsequent to the original proposal by Foresi and Moustakas (1993), a number of workers investigated the dependence between work function and barrier height of various metals to n-GaN (Hacke et al., 1993; Binari et al., 1994; Schmitz et al., 1996; Kalinina et al., 1997). These studies indicate a monotonic increase of the barrier height ϕ_B with the work function of the metal. However, the increase does not scale with the work function as expected from the relationship $q\phi_B = q\phi_M - qX$, where ϕ_m and X are the metal work function and the semiconductor electron affinity, respectively. Whether this is due to surface states or process-induced defects requires further investigation. Recently, Smith et al. (1997) and Dhesi et al. (1997) produced evidence from photoemission studies that the surface states are close to the top of the valence band, and thus they should not contribute to the pinning of the Fermi level. Also Kalinina et al. (1997) produced evidence that the chemical reactivity of the metal affects the characteristics of the Schottky barriers.

Similar conclusions regarding a direct relation between barrier height and work function of the metal has recently been presented for p-GaN films (Ishikawa et al., 1997) and n-$Al_xGa_{1-x}N$ films (Sampath et al., 1998). More specifically, it was found that the resistance at the p-GaN/metal interface decreased exponentially with increasing the metal work function. The lowest contact resistance was obtained with nickel and platinum metals. Similarly, it was observed that Ti/Al bi-metals, which form excellent ohmic contacts to n-GaN, shows rectifying behavior in contact with $Al_xGa_{1-x}N$ (for $x > 10\%$). Thus, high work-function metals are appropriate to form low-resistivity contacts to p-GaN and Ni/Au bilayers are routinely used in commercial LED structures (Nakamura et al., 1996). Also, low work-function metals should be used to contact $Al_xGa_{1-x}N$ alloys with $x > 10\%$.

2. DEVICES

a. Emitters

Molecular-beam-epitaxy grown *p-n* junction LEDs were first reported by Molnar et al. (1995b). Such structures were fabricated into mesa-etched LEDs (Vaudo et al., 1996) as shown schematically in Fig. 52. Circular mesas, 330 µm in diameter, were formed by reactive ion etching through the *p*-type GaN:Mg layer using $SiCl_4$ as an etching gas. At a system pressure of

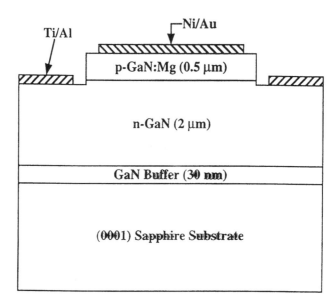

FIG. 52. Schematic representation of the GaN p-n junction LED structure. (Reprinted from Vaudo et al., 1996.)

30-m torr and substrate bias of 500 V etch rates of 330 Å/min were reproducibly achieved. Such etching rates are similar to those reported for MOCVD-grown n-GaN films (Pearton and Shul, 1998).

Figure 53 shows the I-V characteristics of one of such devices, with a 330-μm diameter p-metal contact. The device exhibits good rectifying behavior with on-series resistance of 14 Ω, which is comparable to that reported for commercial III-V nitride LEDs (Nakamura et al., 1994).

The electroluminescence spectra of the same device at two dc drive currents is shown in Fig. 54. The emission peak at 396 nm (violet) is the narrowest (FWHM = 30 nm) reported in the literature for any GaN p-n junction LED. We attribute the dominant emission peak to donor-acceptor pair recombination between shallow donors and Mg-acceptors on the p-side of the junction. This electroluminescence peak corresponds exactly to the photoluminescence peak measured from bulk Mg-doped films grown by MBE. The location of this peak is at a notably shorter wavelength than the 430-nm emission peak, which is commonly observed from MOCVD-grown p-n junction LEDs (Akasaki et al., 1991; Nakamura et al., 1991; Goldenberg et al., 1993).

Similar results were reported with an RF plasma source for activating the nitrogen during the MBE process by Riechert et al. (1997), who investigated

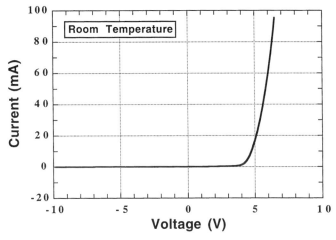

FIG. 53. Current-voltage characteristics for a GaN p-n junction LED with a 300-μm diam. p contact. This device was rapid thermally annealed at 700 °C for 20 s before device processing. (Reprinted from Vaudo et al., 1996.)

both GaN homojunctions as well as InGaN/GaN heterojunctions. Using similar sources, Guha et al. (1998) have demonstrated the fabrication of AlGaN/GaN double heterostructure LEDs on conducting Si(111) substrates. LED structures using ammonia as a nitrogen source were reported by Kamp et al. (1997).

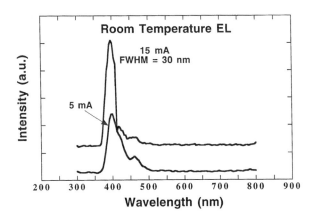

FIG. 54. Electroluminescence spectra for a GaN p-n junction LED operated at dc currents of 5 and 15 mA. (Reprinted from Vaudo et al., 1996.)

Electrically pumped nitride laser structures have not yet been fabricated by the MBE method. Schmidt et al. (1996) have fabricated by MBE GaN/AlGaN separate confinement heterostructures (SCH), which exhibited stimulated emission threshold pumping powers as low as 90 kW/cm^2 at room temperature. Similar results were recently obtained in our laboratory by optically pumping the GaN/AlGaN MQW structure shown in Fig. 49 (Korakakis et al., 1998b). These authors report that at low excitation power the spectra consist of a relatively broad emission with the peak position at 366 nm. As the excitation power density increases a sharp, narrow emission feature (FWHM = 1.5 nm) appears at 363 nm.

b. *Detectors*

Ultraviolet photoconducting detectors fabricated on films grown by MBE were first reported by Misra et al. (1995). The metal contacts consist of interdigitated electrodes as shown schematically in Fig. 55. These devices were characterized by measuring the gain-quantum efficiency product, spectral response and response time.

The spectral response of the GaN detectors was measured and the results are shown in Fig. 56. These results are similar to those reported on GaN grown by MOCVD (Khan et al., 1992). It is important to note that the

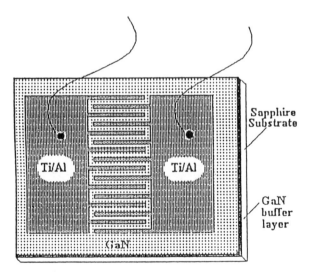

FIG. 55. Schematic illustration of the GaN photoconductive UV detector. (Reprinted from Misra et al., 1995.)

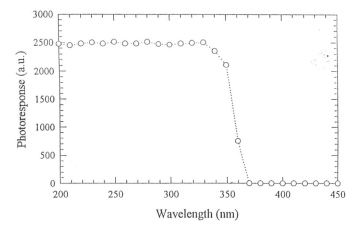

FIG. 56. Spectral response of the photoconducting UV detector made from GaN film grown by ECR-MBE. (Reprinted from Misra et al., 1995.)

photoresponse remains constant at wavelengths shorter than the gap of the semiconductor. This is to be contrasted with the photoconducting spectral response of other semiconductors such as GaAs, where the response falls abruptly at shorter wavelengths due to surface recombination. This is further evidence that the surface states in GaN do not occur at the middle of the gap (Foresi and Moustakas, 1993).

A large number of GaN photoconducting detectors were fabricated on MBE films on which resistivity was varied from 10^1 to 10^9 (ohm · cm) (Misra et al., 1997; Misra et al., 1998). The resistivity in these films was varied by changing the ratio of group-III to group-V fluxes. Figure 57 shows the $(\mu\tau)$ product, calculated from the gain measurements vs the resistivity in the films. Thus, the internal photoconductive gain varies by almost five orders of magnitude with the resistivity of the films. Our evidence suggests that the variation of the $(\mu\tau)$ product with the resistivity is due mainly to the variation of the response time (τ). As a result, one can fabricate out of GaN both high sensitivity but slow photodetectors or lower sensitivity but faster photodetectors (Misra et al., 1997; Misra et al., 1998).

Similar photoconducting detectors were also fabricated from AlGaN alloys (Misra et al., 1998). Again, the resistivity of the films was varied from 10^1 to 10^7 (ohm · cm) by varying the Al content in the films. The $(\mu\tau)$ product for these detectors was found to be higher than those of pure GaN. Figure 58 shows the $(\mu\tau)$ product vs resistivity for a number of AlGaN photoconducting detectors. We attribute the improvement

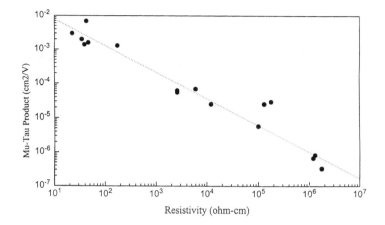

FIG. 57. Variation of the mobility-lifetime product with GaN film resistivity in GaN photoconductive detectors, as determined from measurements of photoconductive gain. (Reprinted from Misra et al., 1998.)

in ($\mu\tau$) product in AlGaN detectors to atomic long-range ordering in these materials (Korakakis et al., 1997; Moustakas et al., 1998; Misra et al., 1998).

PIN photodiodes grown by MBE were also reported by a number of researchers (Xu et al., 1997; Van Hove et al., 1997).

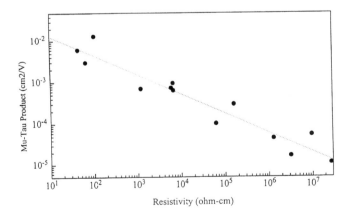

FIG. 58. Variation of the mobility-lifetime product with AlGaN film resistivity in AlGaN photoconductive detectors, as determined from measurements of photoconductive gain. (Reprinted from Misra et al., 1998.)

c. Transistors

Heterojunction bipolar transistors (HBTs) based entirely on III-V nitrides have not been reported in the literature. Pankove *et al.* (1994) have reported the fabrication of the first heterojunction bipolar transistor by combining 6H-SiC and GaN as indicated schematically in Fig. 59. In this report, the n-GaN emitter was deposited by the ECR-MBE method. The base and the collector of the transistor doped at the indicated levels were made of 6H-SiC. The base was formed by liquid-phase epitaxy. The two materials have ideal properties for the fabrication of such transistors. The energy bandgaps of GaN and 6H-SiC are 3.4 and 2.9 eV, respectively. Both materials have high thermal conductivities (for GaN 1.3 W/cm · °C and for SiC 5.0 W/cm · °C). In addition, the lattice constants of the two materials are closely matched. The high doping concentration in the base as well as in the collector led to negligible early voltage and very small breakdown voltage. The I-V characteristics of this device at 300 K in the common-base mode are shown in Fig. 60. It is interesting to note that in the saturation regime, the collector current is almost identical to the emitter current, which means that the base current is negligibly small. Soft breakdown of the base-to-collector junction was observed around 10 V. The large area (0.25 cm^2) of the base-to-collector junction also contributed to the leakage current at

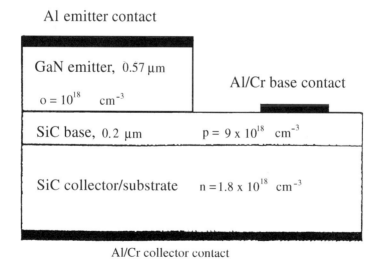

FIG. 59. Cross section of the GaN/SiC transistor. (Reprinted from Pankove *et al.*, 1994).

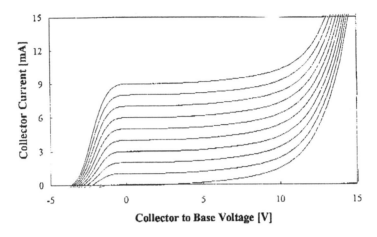

FIG. 60. Common-base I-V characteristics at 300 K of HBT discussed in Fig. 59. The curves correspond to an emitter current of 0 mA for the bottom curve, 1, 2, 3, 4, 5, 6, 7, 8, and 9 mA for the top curve. The emitter area is 75 μm by 75 μm. (Reprinted from Pankove et al., 1994.)

higher voltages. To eliminate the effect of the collector-base leakage current, the common emitter-current gain was extracted from the common base mode. The current gain was found to vary with the emitter current (I_E) and the collector-base voltage (V_{CB}) with a peak value of 108,701 at $I_E = 100$ mA and $V_{CB} = 2$ V.

The emitter efficiency of this device was calculated from the equation

$$\gamma = \frac{I_{nE}}{I_{nE} + I_{PE}} \qquad (26)$$

and found to be 0.999999. The base transport factor was calculated from the equation

$$a_T = 1 - 0.5 \left[\frac{X_B}{L_n} \right]^2 \qquad (27)$$

where X_B is the width of the base and L_n is the diffusion length of electrons. Using a mobility of 110 cm^2/V·s and a lifetime of 5 μs, the diffusion length was found to be $L_n = 37.7$ μm and the base transport factor 0.999987. The current gain was computed by the expression

$$\beta = \frac{a_T \gamma}{1 - a_T \gamma} \qquad (28)$$

and found to be 80,409, in reasonable agreement with the experiment.

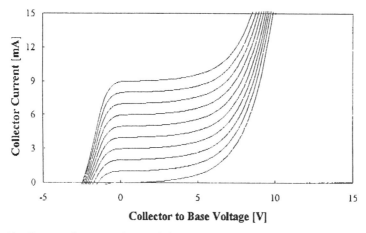

FIG. 61. Common-base I-V characteristics at 260 °C of HBT discussed in Fig. 59. The curves correspond to an emitter current of 0 mA at the bottom increasing in 1-mA increments to 9 mA at the top. (Reprinted from Pankove et al., 1994.)

The device was evaluated up to 260 °C. The I-V characteristics at this temperature are shown in Fig. 61. The curves in this figure are essentially the same as those in Fig. 60, except that the soft breakdown of the base-collector junction occurs at much lower voltage ($V_{CB} = 5$ V). Thus, the main problem with this HBT is the leakage through the SiC base-collector junction. Because of this high leakage, the HBT could not operate in the common emitter mode. In order for this device to operate in the common emitter mode, the leakage current must be reduced by either reducing the concentration of doping in the collector or by improving the base-collector junction.

Molecular beam epitaxy has also been used for the fabrication of a high-performance n-channel normally off GaN MODFET by Aktas et al. (1995). This device was fabricated on sapphire substrates and was found to have a room-temperature sheet carrier concentration of 1.2×10^{13} cm^{-2}, which is about one order of magnitude higher than that obtained from GaAs/AlGaAs MODFET structures. This MODFET structure was reported to have a room-temperature transconductance of $g_m = 120$ mS/mm.

VIII. Conclusions

The MBE process was proven to be a practical growth method for the deposition of III-V compounds and devices as well as a unique scientific tool for the study of thin-film growth phenomena *in situ*. The application of this method to the growth of GaN and its alloys with InN and AlN required the development of appropriate nitrogen sources. Such sources, based on nitrogen plasmas or NH_3 decomposition, were developed over the past few years and led to the deposition of GaN films of the highest quality in terms of their electronic and optical properties. Furthermore, the growth rate was increased to above 1 μm/h without the high consumption of ammonia required in the MOCVD process.

The heteroepitaxial growth of GaN on sapphire substrates follows the conversion of the surface of the substrate from Al_2O_3 to AlN and the deposition of either a GaN or an AlN buffer. The optimization of these steps was found to be important for the control of the structure, microstructure and propagation of defects in the films. The method has also been successfully applied for the growth of zinc-blende GaN on a number of substrates with cubic symmetry such as GaAs, (0001) Si, and β-SiC.

Controlled n-type doping of GaN with carrier concentration from 10^{15} to above 10^{20} cm^{-3} was accomplished by the incorporation of Si. The electron mobility, determined from lateral transport measurements, was found to be limited by scattering from charged dislocations rather than background impurities. The problem is more pronounced at lower carrier concentration and less pronounced at higher carrier concentrations due to screening effects.

Controlled p-type doping with carrier concentration above 10^{18} cm^{-3} has been accomplished by the incorporation of Mg. The samples are p-type as grown without requiring post-growth annealing as films grown by the MOCVD method. Annealing studies of p-type films grown by the MBE method have shown the opposite effect than MOCVD films. Specifically, the carrier concentration decreases upon annealing by about a factor of three while correspondingly, the hole mobility increases by more than a factor of ten. These novel annealing phenomena were attributed to the removal of static disorder in the films.

InGaN and AlGaN alloys were developed by MBE and phenomena such as phase separation and long-range atomic ordering were investigated. Furthermore, MQW structures of the InGaN/AlGaN and GaN/AlGaN configuration were developed. These structures were found to have excellent luminescence properties and show lasing upon optical pumping.

Finally, the MBE method has been successfully used to fabricate devices such as LEDs, optically pumped lasers, photodetectors and transistors. The results on optically pumped lasers are the best reported in the literature. Similarly, the application of MBE III-V nitrides to electronic devices has also led to some of the best results in the literature. In the area of photoconductive detectors, the MBE method proved to be versatile in the fabrication of detectors capable of both high sensitivity and high speed.

In view of the recent developments in the fabrication of efficient nitrogen sources and the successful application of MBE in the fabrication of unique and efficient electronic and optoelectronic devices we believe that this method of nitride growth is going to play a major role in the development of the nitride technology.

ACKNOWLEDGMENTS

The author would like to thank former and current students as well as post-doctoral fellows whose work is featured in this chapter. This includes: Dr. M. Fanciulli, Dr. J. Foresi, Dr. S. Jin, Dr. D. Korakakis, Dr. T. Lei, Mr. Menon, Dr. R. Molnar, Dr. R. Singh, Dr. R. Vaudo, Ms. G. Bunea, Mr. D. Doppalapudi, Mr. E. Iliopoulos, Ms. M. Misra, Mr. K. Nam, Mr. H. Ng, Mr. A. Sampath, and Mr. R. Singh. The author would also like to thank his fellow faculty members at Boston University for their support. These include Prof. S. N. Basu, Prof. C. Eddy, Prof. B. Goldberg, Prof. K. F. Ludwig, Prof. W. Skocpol, Prof. K. Smith, and Prof. S. Unlu. Many thanks are also due to his collaborators at Xerox Laboratories: Dr. N. Johnson, Dr. L. Romano, Dr. M. Brandt, and Dr. D. Hofstetter. At the University of Thessaloniki he would like to thank Prof. Stoimenos, Prof. E. Paloura, Prof. S. Logethetidis, Dr. Petalas and Ms. Katsikini. From the University of Colorado he would like to thank Prof. J. Pankove, Prof. Van Zeghbroeck, and Dr. J. Torvik. From Hewlett-Packard he would like to thank Dr. G. Crawford and Dr. R. Flesher for their help. Thanks are due to Dr. R. J. Graham at Arizona State University and to Dr. P. Perlin, Dr. S. Porowski, and Dr. T. Suzski at the Polish Academy of Sciences. Thanks are also due to many colleagues for discussions and reprints of their work. The author would also like to thank Dr. A. Hussein of DARPA and Dr. C. Wood of ONR, who monitored the bulk of the work reported in this chapter for their respective funding agencies.

REFERENCES

Abernathy, C. R., Mackenzie, J. D., and Donovan, S. M. (1997). *J. Crys. Growth*, **178**, 74.
Akasaki, I., Amano, H., Kiode, Y., Hiramatsu, H., and Sawaki, N. (1989) *J. Crys. Growth*, **89**, 209.
Akasaki. I., Amano, H., Kito, M., and Hiramatsu, K. (1991). *J. Lumin.*, **48 & 49**, 666.
Aktas, O., Kim, W., Fan, Z., Botchkorev, A., Salvador, A., Mohammad, S. N., Sverdlov, B., and Morkoc, H. (1995). *Electron. Lett.*, **31**, 1389.
Amano, H., Sawaki, N., Akasaki, I., and Toyoda, Y. (1986). *Appl. Phys. Lett.*, **48**, 353.
Amano, H., Kito, M., Hiramatsu, K., and Akasaki, I. (1989). *Jpn. J. Appl. Phys.*, **28**, L2112.

Amano, H., Takeuchi, T., Sota, S., Sakai, H., and Akasaki, I. (1997). *Mat. Res. Soc. Proc.*, **449**, 1143.
Arfken, G. (1970). *Mathematical Methods for Physicists*, 2nd edition, New York: Academic Press.
Ariel, V., Garber, V., Rosenfeld, D., and Bahir, G. (1995). *Appl. Phys. Lett.*, **66**, 2101.
Baudic, Z. Z., McGill, T. C., and Ikonic, Z. (1996). Preprint.
Barker, A. S. and Ilegems, M. (1973). *Phys. Rev. B*, **7**, 743.
Basu, S. N., Lei, T., and Moustakas, T. D. (1994). *J. Mater. Res.*, **9**, 2370.
Bauer, E. (1958). *Krist Z.*, **100**, p. 372.
Bedair, S. M. (1996). The GaN Workshop, St. Louis, MO.
Beresford, R. (1994). *JOM*, **46**, 54.
Beresford, R., Stevens, K. S., Cui, Q., Schwartzman, A., and Cheng, H. (1997). In *III-V Nitrides*. F. A. Ponce, T. D. Moustakas, I. Akasaki, and B. A. Monemar, eds., Pittsburgh, PA: *Mater. Res. Soc. Proc.*, **449**, p. 361.
Binari, S. C., Dietrich, H. B., Kelner, G., Rowland, L. B., Doverspike, K., and Gaskill, D. K. (1994). *Electron Lett.*, **30**, 909.
Bloom, S., Harbeke, G., Meier, E., and Ortenburger, L. B. (1974). *Phy. Stat. Solid. B*, **66**, 161.
Boer, K. W. (1990). *Survey of Semiconductor Physics*, Vol. I, New York: Van Nostrand Reinhold.
Boguslawski, P., Briggs, E., and Berholc, J. (1995). *Phys. Rev.*, **B51**, 17255.
Bonch-Bruyevitch, V. L. (1966). *The Electronic Theory of Heavily Doped Semiconductors*, New York: Elsevier, pp. 16–17.
Botchkarev, A., Salvador, A., Sverdlov, B., Myoung, J., and Morkoc, H. (1995). *Appl. Phys. Lett.*, **77**, 4455.
Brandt, M. S., Johnson, N. M., Molnar, R. J., Singh, R., and Moustakas, T. D. (1994a). *Appl. Phys. Lett.*, **64**, 2264.
Brandt, M. S., Ager, III, J. W., Gotz, W., Johnson, N. M., Harris, Jr., J. S., Molnar, R. J., and Moustakas, T. D. (1994b). *Phys. Rev. B*, **49**, p. 14758.
Brandt, O., Yang, H., Jenichen, B., Suzuki, Y., Daweritz, L., and Ploog, K. H. (1995). *Phys. Rev.*, **B52**, R2253.
Brandt, O., Yang, H., Trampert, A., Wassermeier, M., and Ploog, K. H. (1997). *Appl. Phys. Lett.*, **71**, 473.
Bruinsman, R. and Zangwill, A. (1987). *Europhys. Lett.*, **4**, 729.
Calawa, A. R. (1981). *Appl. Phys. Lett.*, **38**, 701.
Chadi, D. J., Clark, A. H., and Burnham, R. (1976). *Phys. Rev.*, **B13**, 4466.
Chang, L. L. and Giessen, B. C. (eds.) (1985). *Synthetic Modulated Structures*, New York: Academic Press.
Chang, L. L. and Ploog, K. (eds.) (1985). *Molecular Beam Epitaxy and Heterostructures*, Boston: Martinus Nijhoff Publishers.
Chen, F. F. (1984). *Introduction to Plasma Physics and Controlled Fusion*. **1**, 2nd edition, Boston: Plenum Publishing.
Cho, A. Y. and Dernier, P. D. (1978). *J. Appl. Phys.*, **49**, 3328.
Cho, A. Y. (1983). *Thin Solid Films*, **10**, 291.
Davey, J. E. and Pankey, T. (1968). *J. Appl. Phys.*, **39**, 1941.
Davies, G. J. and Williams, D. (1986). In *The Technology and Physics of Molecular Beam Epitaxy*, E. H. C. Parker, ed., Boston: Plenum Publishing Corp., **2**, 15.
Denbaars, S. P. and Keller, S. (1998). In *Gallium Nitride (GaN) I*, J. I. Pankove and T. D. Moustakas, eds., New York: Academic Press, Chapter 2.
Dhesi, S., Stagarescu, B., Smith, K., Doppalupudi, D., Singh, R., and Moustakas, T. D. (1997). *Phys. Rev.*, **B56**, 10271.

Dingle, R., Sell, D. D., Stokowski, S. E., and Illegems, M. (1971). *Phys. Rev. B*, **4**, 1211.
Doppalapudi, D., Iliopoulos, E., Basu, S. N., and Moustakas, T. D. (1998). *MRS Symp. Proc.*, **482**, 51.
Doppalapudi, D., Basu, S. N., Ludwig, K., and Moustakas, T. D. (1998). *J. Appl. Phys.*, **84**, 1384.
Doverspike, K. and Pankove, J. I. (1998). In *Gallium Nitride (GaN) I*, J. I. Pankove and T. D. Moustakas, eds., New York: Academic Press, pp. 259–269.
Eddy, C. R., Moustakas, T. D., and Scanlon, J. (1993). *J. Appl. Phys.*, **73**, 448.
Esaki, L. and Tsu, R. (1970). IBM Research Development, **14**, 61.
Evans, K. R., Lei, T., Kaspi, R., and Jones, C. R. (1995). *Proc. of the Topical Workshop on III-V Nitrides*. I. Akasaki and K. Onabe, eds. (Nagoya, Japan), 363.
Fanciulli, M., Lei, T., and Moustakas, T. D. (1993). *Phys. Rev. B*, **48**, 15144.
Feuillet, G., Hacke, P., Okumura, H., and Yoshida, S. (1996). *Proc. of the International Symp. on Blue Laser and Light-Emitting Diodes*, paper LN-4.
Foresi, J. S. and Moustakas, T. D. (1993). *Appl. Phys. Lett.*, **62**, 2859.
Foxon, C. T. and Harris, J. J., eds. (1987). *Molecular Beam Epitaxy*. Amsterdam: North Holland.
Foxon, C. T. and Joyce, B. A. (1990). *Growth and Characterization of Semiconductors*. R. A. Stradling and P. C. Klipstein, eds., New York: Adam Hilger, p. 35.
Fu, T. C., Newman, N., Jones, E., Chan, J. C., Liu, X., Rubin, M. D., Cheung, N. W., and Weber, E. R. (1995). *J. Electron. Mat.* **24**, 249.
Fuke, S., Teshigawara, H., Kuwahara, K., Takano, Y. (1998). *J. Appl. Phys.*, **83**, 764.
Goldenberg, B., Zook, J. D., and Ulmer, R. J. (1993). *Appl. Phys. Lett.*, **62**, 381.
Gomgo, A., Kobayashi, K., Kawata, S., Hino, I., Suski, T., and Yudasa, Y. (1986). *J. Crys. Growth*, **77**, 367.
Gossard, A. C. (1982). In *Treatise on Material Science and Technology, Preparation and Properties of Thin Films*. K. N. Tu and R. Rosenberg, eds., New York: Academic Press, **24**, p. 13.
Götz, W., Johnson, N. M., Chen, C., Liu, H., Kuo, C., and Inler, N. (1996). *Appl. Phys. Lett.*, **68**, 22.
Grabow, M. H. and Gilmer, G. H. (1986). *Layered Structures and Epitaxy*. J. M. Gibson, G. C. Osbourn, and R. M. Thompson, eds., Pittsburgh, PA: *Mater. Res. Soc.*, p. 13.
Grandjean, N., Massies, J., and Leroux, M. (1996). *Appl. Phys. Lett.*, **69**, 2071.
Grandjean, N., Massies, J., Vennegues, P., Laügt, M., and Leroux, M. (1997a). *Appl. Phys. Lett.*, **70**, 643.
Grandjean, N., Leroux, M., Laugt, M., and Massies, J. (1997b). *Appl. Phys. Lett.*, **71**, 240.
Grandjean, N., Massies, J., Vennegues, P., and Leroux, M. (1998). *J. Appl. Phys.* **83**, 1379.
Guha, S., Bojarczuk, N. A., and Kisker, D. W. (1996). *Appl. Phys. Lett.*, **69**, 2879.
Guha, S. and Bojarczuk, N. A. (1998). *Appl. Phys. Lett.*, **72**, 415.
Guinier, A. (1963). *X-Ray Diffraction*. San Francisco: Freeman.
Gunther, K. Z. (1958). *Z. Naturforsch*, **13a**, 1081.
Hacke, P., Detchprohm, T., Hiramatsu, K., and Sawaki, N. (1993). *Appl. Phys. Lett.*, **63**, 2676.
Hacke, P., Feuillet, G., Okumura, H., and Yoshida, S. (1996). *Appl. Phys. Lett.*, **69**, 2507.
Haus, H. A., and Melcher, J. R. (1989). *Electromagnetic Fields and Energy*. Englewood Cliffs, N.J.: Prentice-Hall.
Heidenreich, J. E., Paraszczak, J. R., Moisan, M., and Sauve, G. (1987). *J. Vac. Sci. Tech.*, **B5**, 347.
Heinlein, C., Grepstad, J., Berge, T., and Riechert, H. (1997). *Appl. Phys. Lett.*, **71**, 341.
Held, R., Crawford, D. E., Joshnston, A. M., Dabiran, A. M., and Cohen, P. I. (1997). *J. Electron Mat.*, **26**, 272.

Herzog, W. D., Singh, R., Moustakas, T. D., Goldberg, B. B., and Unlu, M. S. (1997). *Appl. Phys. Lett.,* **70**, 1333.
Heying, B., Wu, X. H., Keller, S., Li, Y., Kapolnek, D., Keller, B. P., DenBaars, S. P., and Speck, J. P. (1996). *Appl. Phys. Lett.,* **68**, 643.
Ho, I. and Stringfellow, G. B. (1996). *Appl. Phys. Lett.,* **69**, 2701.
Hooper, S. E., Foxon, C. T., Cheng, T. S., Jenkins, L. C., Lacklison, D. E., Orton, J. W., Bestwick, T., Kean, A., Dawson, M., and Duggan G. (1995). *J. Cryst. Growth,* **155**, 157.
Hooper, S. E., Foxon, C. T., Cheng, T. S., Jeffs, N. L. C., Lacklison, D. E., Orton, J. W., Dawson, M., and Duggan G. (1997). In *III-V Nitrides.* F. A. Ponce, T. D. Moustakas, I. Akasaki, B. A. Monemar, eds. Pittsburgh, PA: *Mater. Res. Soc. Proc.* 449, p. 325.
Hu, Y. N. and Ching, W. Y. (1993). *Phys. Rev. B,* **48**, 4335.
Hughes, W. C., Rowland, Jr., W. H., Johnson, M. H. L., Fujita, S., Cook, J. M., Schetzina, J. F., Ren, J., Edmond, J. A. (1995). *J. Vac. Sci. Tech.,* **B13**, 1571.
Ilegems, M. and Dingle, R. (1973). *J. Appl. Phys.,* **39**, p. 4234.
Iliopoulos, E., Doppalapudi, D., Ng, H. M., and Moustakas, T. D. (1998a). *MRS Symp. Proc.,* **482**, 655.
Iliopoulos, E., Doppalapudi, D., Ng, H. M., and Moustakas, T. D. (1998b). *Appl. Phys. Lett.,* **73**, 315.
Ishikawa, H., Kobayashi, S., Koide, Y., Yamasaki, S., Nagai, S., Umezaki, J., Koike, M., and Murakami, M. (1997). *J. Appl. Phys.* **81**, 1315.
Jenkins, D. W. and Dow, J. D. (1989). *Phys. Rev. B,* **39**, 3317.
Johnson, M. A. L., Fujita, S., Rowland, W. H., Hughes, W. C., He, Y. W., El-Masry, N. A., Cook, J. W., and Schetzina, J. F. (1996). *J. Electronic Mat.,* **25**, 793.
Kalinina, E. V., Kuznetsov, N. I., Babanin, A. I., Dmitriev, V. A., Shchukarev, A. V. (1997). *Diamond and Related Materials,* **6**, 1528.
Kamp, M., Mayer, M., Pelzmann, A., Thies, A., Chung, H. Y., Sternschulte, H., Marti, O., and Ebeling, K. J. (1996). *Mat. Res. Soc. Symp. Proc.,* **395**, 135.
Kamp, M., Mayer, M., Pelzmann, A., and Ebeling, K. J. (1997). *Mat. Res. Soc. Symp. Proc.,* **449**, 161.
Katsikini, M., Paloura, E. C., and Moustakas, T. D. (1996). *Appl. Phys. Lett.,* **69**, 4206.
Katsikini, M., Paloura, E. C., Fieber-Erdmann, M., Kalomiros, J., and Moustakas, T. D. (1997a). *Phys. Rev. B,* **56**, 13380.
Katsikini, M., Paloura, E. C., Moustakas, T. D., Holub-Krappe, E., and Antonopoulos, J. (1997b). In *III-V Nitrides.* F. A. Ponce, T. D. Moustakas, I. Akasaki, and B. A. Monemar, eds., *Mat. Res. Soc. Proc.,* **449**, 411.
Katisikini, M., Paloura, E. C., and Moustakas, T. D. (1998). *J. Appl. Phys.,* **83**, 1437.
Keller, S., Keller, B. P., Wu, Y. F., Heying, B., Kapolnek, D., Speck, J. S., Mishra, U. K., and DenBaars, S. P. (1996). *Appl. Phys. Lett.,* **68**, 1525.
Khan, M. A., Kuznia, J. N., Olson, D. T., Van Hove, J. M., Blasingame, M., and Reitz, L. F. (1992). *Appl. Phys. Lett.,* **60**, 2917.
Kikuchi, A., Hoshi, H., and Kishino, K. (1994). *Jap. J. Appl. Phys.,* **33**, 688.
Kikuchi, A., Hoshi, H., and Kishino, K. (1995). *J. Crys. Growth,* **150**, 897.
Kilaas, R. (1987). *Proc 45th EMSA,* Baltimore, MD. G. W. Bailey, ed., San Francisco, CA: San Francisco Press, 66.
Kim, W., Aktas, O., Botchkarev, A. E., Salvador, A., Mohammad, S. N., and Morkoc, H. (1996). *J. Appl. Phys.,* **79**, 1.
Kim, M. H., Sone, C., Yi, J. H., and Yoon, E. (1997). *Appl. Phys. Lett.,* **71**, 1228.
Kohler, W. E., Romheld, M., Seebock, R. J., and Scaberna, S. (1993). *Appl. Phys. Lett.,* **63**, 2890.
Koike, M., Yamasaki, S., Nagai, S., Koide, N., Asami, S., Amano, H., and Akasaki, I. (1996). *Appl. Phys. Lett.,* **68**, 1403.

Korakakis, D., Ludwig, K. F., and Moustakas, T. D. (1997). *Appl. Phys. Lett.*, **71**, 72.
Korakakis, D. (1998). Ph.D. Dissertation, Boston University.
Korakakis, D., Ludwig, K. F., and Moustakas, T. D. (1998a). *Appl. Phys. Lett.*, **72**, 1004.
Korakakis, D., Hofstetter, D., Iliopoulos, E., and Moustakas, T. D. (1998b). (Unpublished).
Kuan, T. S., Kuech, T. F., Wang, W. I., and Wilkie, E. L. (1985). *Phys. Rev. Lett.*, **54**, 201.
Lagerstedt, O. and Monemar, B. (1979). *J. Appl. Phys.*, **45**, 2266.
Landau, L. D. and Lifshitz, I. M. (1980). *Statistical Physics*. 3rd ed. Oxford: Pergamon.
Lei, T., Fanciulli, M., Molnar, R. J., Moustakas, T. D., Graham, R. J., and Scanlon, J. (1991). *Appl. Phys. Lett.*, **58**, 944.
Lei, T., Moustakas, T. D., Graham, R. J., He, Y., and Berkowitz, S. J. (1992). *J. Appl. Phys.*, **71**, 4933.
Lei, T., Ludwig, K. F., and Moustakas, T. D. (1993). *J. Appl. Phys.*, **74**, 4430.
Leung, M. S. H., Klockenbrink, R., Kisielowski, C., Fujii, H., Kruger, J., Sudhir, G. S., Anders, A., Liliental-Weber, Z., Rubin, M., and Weber, E. R. (1997). In *III-V Nitrides*. F. A. Ponce, T. D. Moustakas, I. Akasaki, and B. A. Monemar, eds., Pittsburgh, PA: *Mater. Res. Soc. Proc.*, **449**, p. 221.
Leybold-Heraues Vacuum Products (1993). *Vacuum Technology, Its Foundations, Formulae and Tables* (Catalog).
Lin, M. E., Strite, S., Agarwal, A., Salvador, A., Zhou, G. L., Teraguchi, N., Rockett, A., and Morkoc, H. (1993). *Appl. Phys. Lett.*, **62**, 702.
Lin, M. E., Sverdlov, B., Zhou, G. L., and Morkoc, H. (1993). *Appl. Phys. Lett.*, **62**, 3479.
Lin, M. E., Ma, Z., Huang, F. Y., Fan, Z., Allen, L. H., and Morkoc, H. (1994). *Appl. Phys. Lett.*, **64**, 1003.
Liu, H., Frenkel, A. C., Kim, J. G., and Park, R. M. (1993). *J. Appl. Phys.*, **74**, 6124.
Logothetidis, S., Petalas, J., Cardona, M., and Moustakas, T. D. (1994). *Phys. Rev. B*, **50**, 18017.
Logothetidis, S., Petalas, J., Cardona, M., and Moustakas, T. D. (1995). *Mat. Sci. Eng. B*, **29**, 65.
McWhan, D. B. (1985). In *Synthetic Modulated Structures*. L. L. Chang and B. C. Giessen, eds., New York: Academic Press, Chapter 2.
Manfra, M., Berkowvitz, S., Molnar, R., Clark, A., Moustakas, T. D., and Skocpol, W. J. (1994). *Mat. Res. Soc. Proc.*, **324**, 477.
Maruska, H. P. and Tietjen, J. J. (1969). *Appl. Phys. Lett.*, **15**, 367.
Matsuoka, T. and Ono, K. (1988). *J. Vac. Sci. Tech.*, **A6**, 25.
Matsuoka, T., Sasaki, T., and Katsui, A. (1990). *Optoelectron. Dev. and Tech.*, **5**, 53.
Menon, G. (1990). M.S. Thesis, Boston University, Boston, MA.
Misra, M., Moustakas, T. D., Vaudo, R. P., Singh, R., and Shah, K. S. (1995). *Proc. SPIE Conference on X-ray and UV Sensors and Applications*, **2519**, 78.
Misra, M., Korakakis, D., Singh, R., Sampath, A., and Moustakas, T. D. (1997). *Mater. Res. Soc. Proc.*, **449**, 597.
Misra, M., Korakakis, D., and Moustakas, T. D. (1998). *Appl. Phys. Lett.*
Mohammad, S. N., Salvador, A. A., and Morkoc, H. (1995). *Proc. IEEE*, **83**, 1306.
Molnar, R. J. and Moustakas, T. D. (1994). *J. Appl. Phys.*, **76**, 4587.
Molnar, R. J., Singh, R., and Moustakas, T. D. (1995a). *J. Electron Mat.*, **24**, 275.
Molnar, R. J., Singh, R., and Moustakas, T. D. (1995b). *Appl. Phys. Lett.*, **66**, 268.
Molnar, R. J. (1998). In *Semiconductors and Semimetals*, Vol. 57. J. Pankove and T. Moustakas, eds., New York: Academic Press.
Monemar, B. A. (1998). In *Gallium Nitride (GaN) I*. J. I. Pankove and T. D. Moustakas, eds., New York: Academic Press, pp. 311–334.
Monemar, B. A., Bergman, J. P., Buyanova, I. A., Li, W., Amano, H., and Akasaki, I. (1996). *Mater. Res. Soc. Internet J. Nitride Semicond. Res.*, **1**, 2.
Morgan, T. N. (1965). *Phys. Rev.*, **139**, A343.

Morimoto, Y., Uchiho, K., and Ushio, S. (1973). *J. Electrochem. Soc. Solid State and Technol.*, **120**, 1783.
Moriyasu, Y., Goto, H., Kuze, N., and Matsui, M. (1995). *J. Cryst. Growth*, **150**, 916.
Morkoc, H., Botchkarev, A. E., Salvador, A., and Sverdlov, B. (1995). *J. Crys. Growth*, **150**, 887.
Moustakas, T. D. (Nov. 1988). *MRS Bulletin XIII*, p. 29.
Moustakas, T. D., Molnar, R., Lei, T., Menon, G., and Eddy, C. R. (1992). In *Wide Band Gap Semiconductors*. T. D. Moustakas, J. I. Pankove, and Y. Hamakawa, eds., Pittsburgh, PA: *Mater. Res. Soc. Proc.*, **242**, pp. 427–432.
Moustakas, T. D., Lei, T., and Molnar, R. J. (1993). *Physica B*, **185**, p. 36.
Moustakas, T. D. and Molnar, R. J. (1993). *Mat. Res. Soc. Symp. Proc.*, **281**, 753.
Moustakas, T. D. (1996). In *Gallium Nitride and Related Materials*. F. A. Ponce, R. D. Dupuis, S. Nakamura, and J. A. Edmond, eds., Pittsburgh, PA: *Mater. Res. Soc. Proc.*, **395**, pp. 111–122.
Moustakas, T. D., Vaudo, R. P., Singh, R., Korakakis, D., Misra, M., Sampath, A., and Goepfert, I. D. (1996). *Inst. Phys. Conf. Ser.* No. 42, Chapter 5, p. 833. Paper presented at Silicon Carbide and Related Materials 1995 Conference (Kyoto, Japan).
Moustakas, T. D., Singh, R., Korakakis, D., Doppalapudi, D., Ng, H. M., Sampath, A., Iliopoulos, E., and Misra, M. (1998). *MRS Symp. Proc.*, **482**, 193.
Nakamura, S. (1991). *Jpn. J. Appl. Phys.*, **30**, L1705.
Nakamura, S., Senoh, M., and Mukai, T. (1991). *Jpn. J. Appl. Phys.*, **30**, L1708.
Nakamura, S., Mukai, T., Senoh, M., Nagahama, S., and Iwasa, N. (1993). *J. Appl. Phys.*, **74**, 3911.
Nakamura, S. (1994). *Microelectron. J.*, **25**, 651.
Nakamura, S., Senoh, M., Nagahama, S., Iwasa, N., Yamada, T., Matsushita, T., Kiyoku, H., and Sugimoto, Y. (1996a). *Appl. Phys. Lett.*, **68**, 3269.
Nakamura, S., Senoh, M., Nakahama, S., Iwasa, N., Yamada, T., Matsushita, T., Kiyoku, H., and Sugimoto, Y. (1996b). *Appl. Phys. Lett.*, **68**, 2105.
Nakamura, S. (1998). In *Gallium Nitride (GaN) I*. J. I. Pankove and T. D. Moustakas, eds., New York: Academic Press, Chapter 14.
Nakayarua, H. and Fujitsu, H. (1986). *Inst. Phys. Conf. Ser.*, **79**, 289.
Nakayama, H., Tochigi, M., Maeda, H., and Mishino, T. (1994). *Appl. Surf. Science*, **82/83**, 214.
Neugebauer, J. and Van de Walle, C. G. (1994). *Phys. Rev. B*, **50**, 8067.
Ng, H. M., Doppalapudi, D., Korakakis, D., Singh, R., and Moustakas, T. D. (1997). *Proc. Second International Conference on Nitride Semiconductors* (Tokushima, Japan) p. 10.
Ng, H. M., Doppalapudi, D., Singh, R., and Moustakas, T. D. (1998a). *MRS Symp. Proc.*, Vol. 482, 507.
Ng, H. M., Doppalapudi, D., Moustakas, T. D., Weimann, N. G., and Eastman, L. F. (1998b). *Appl. Phys. Lett.*, **73**, 000.
Ng, H. M., Doppalapudi, D., Korakakis, D., Singh, R., and Moustakas, T. D. (1998c). *J. Cryst. Growth*, **189–190**, 349.
Norman, A. G., Seong, T. Y., Ferguson, L. T., Booker, G. R., and Joyce, B. A. (1993). *Semicond. Sci. Tech.*, **8**, 9.
Ohtani, A., Stevens, K. S., Kinniburg, M., and Beresford, R. (1995). *J. Cryst. Growth*, **150**, 902.
Paisley, M., Sitar, Z., Posthil, J. B., and Davis, R. F. (1989). *J. Vac. Sci. Technol.*, **7**, 701.
Panish, M. B. (1980). *J. Electrochem. Soc.*, **127**, 2729.
Pankove, J. I. and Schade, H. E. P. (1974). *Appl. Phys. Lett.*, **25**, 53.
Pankove, J. I., Chang, S. S., Lee, H. C., Molnar, R. J., Moustakas, T. D., and Van Zeghbroeck, B. (1994). *IDEM-94*, **389**.
Park, R. M. (1992). *J. Vac. Sci. Tech.*, **A10**, 701.
Parker, E. H. C., ed. (1986). *The Technology and Physics of Molecular Beam Epitaxy*. New

York: Plenum Publishing Corp.
Pearse, R. W. B. and Gayton, A. G. (1963). *The Identification of Molecular Spectra.* New York: Wiley, pp. 209–220.
Pearson, G. L., Read, W. T., and Morin, F. J. (1954). *Phys. Rev.*, **93**, 666.
Pearton, S. J., and Shul, R. J. (1998). In *Gallium Nitride (GaN) I.* J. I. Pankove and T. D. Moustakas, eds., New York: Academic Press, Chapter 5.
Perlin, P., Suzki, T., Teisseyre, H., Lesczynski, M., Grzegory, I., Jun, J., Porowski, S., Boguslawski, P., Berholc, J., Chervin, J. C., Polian, A., and Moustakas, T. D. (1995). *Phys. Rev. Lett.,* **75**, 296.
Philips, J. C. and Van Vechten, J. A. (1970). *Phys. Rev. B*, **2**, 2147.
Ploog, K. (1980). In *Crystal Growth, Properties and Applications.* H. C. Freyhard, ed., Berlin: Springer-Verlag, Vol. 3, p. 73.
Pödör, B. (1966). *Phys. Stat. Solidi,* **16**, K167.
Ponce, F. A. (1997). *MRS Bulletin*, **22**, 51.
Powell, R. C., Lee, N. E., and Greene, J. E. (1992). *Appl. Phys. Lett.,* **60**, 2505.
Powell, R. C., Lee, N. E., Kim, Y. W., and Greene, J. E. (1993). *J. Appl. Phys.*, **73**, 189.
Read, W. T. (1954a). *Phil. Mag.,* **45**, 775.
Read, W. T. (1954b). *Phil. Mag.,* **46**, 111.
Riechert, H., Averbeck, R., Graber, A., Schienle, M., Straub, V., and Tews, H. (1997). In *III-V Nitrides.* F. A. Ponce, T. D. Moustakas, I. Akasaki, and B. A. Monemar, eds., Pittsburgh, PA: *Mat. Res. Soc. Proc.*, **449**, 149.
Romano, L. T., Northrup, J. E., and O'Keefe, M. A. (1996). *Appl. Phys. Lett.,* **69**, 2394.
Romano, L. T., Krusor, B. S., Singh, R., and Moustakas, T. D. (1997). *J. Electron. Mater.*, **26**, 285.
Rubio, A., Corkhill, J. L., Cohen, M. L., Shirley, E. L., and Louie, S. G. (1993). *Phys. Rev. B.*, **48**, 11810.
Sampath, A., Ng, H. M., Korakakis, D., and Moustakas, T. D. (1998). *MRS Symp. Proc.*, **482**, 1095.
Schetzina, J. F. (1995). *Mat. Res. Soc. Proc.*, **395**, 123.
Schmidt, J. J., Shan, W., Song, J. J., Salvador, A. A., Kim, W., Atakas, O., Botchkarev, A., and Morkoc, H. (1996). *Appl. Phys. Lett.,* **68**, 1820.
Schmitz, A. C., Ping, A. T., Kahn, M. A., Chen, Q., Yang, J. W., and Adesida, I. (1996). *Semicond. Sci. Technol.*, **11**, 1464.
Serneels, R., Synkers, M., Delavignette, P., Gevers, R., and Amelinckx, S. (1973). *Phys. Stat. Sol.* (*b*), **58**, 277.
Singh, R. and Moustakas, T. D. (1996a). *Mat. Res. Soc. Symp. Proc.*, **395**, 163.
Singh, R. and Moustakas, T. D. (1996b). *ECS Proc.*, **96-11**, 186.
Singh, R., Doppalapudi, D., and Moustakas, T. D. (1996). *Appl. Phys. Lett.,* **69**, 2388.
Singh, R., Doppalapudi, D., Moustakas, T. D., and Romano, L. T. (1997). *Appl. Phys. Lett.,* **70**, 1089.
Smith, K. E., Dhesi, S. S., Duda, L. C., Stagerescu, C. B., Guo, J. H., Nordgren, J., Singh, R., and Moustakas, T. D. (1997). *MRS Proc.*, **449**, 787.
Stagrescu, C. B., Duda, L. C., Smith, K. E., Guo, J. H., Nordgren, J., Singh, R., and Moustakas, T. D. (1996). *Phys. Rev. B*, **54**, 17,335.
Stringfellow, G. B. (1982). *J. Cryst. Growth,* **58**, 194.
Strite, S., Ruan, J., Li, Z., Manning, N., Salvador, A., Chen, H., Smith, D. J., Choyke, W. J., and Morkoc, H. (1991). *J. Vac. Sci. Technol.*, **B9**, 1924.
Strite, S. and Morkoc, H. (1992). *J. Vac. Sci. Tech.*, **B10**, 1237.
Sugiura, L., Itaya, K., Nishio, J., Fujimoto, H., and Kokubun, Y. (1997). *J. Appl. Phys.*, **82**, 4877.

Tarsa, E. J., Heying, B., Wu, X. H., Fini, P., DenBaars, S. P., and Speck, J. S. (1997). *J. Appl. Phys.*, **82**, 5472.
Trampert, A., Brandt, O., Yang, H., and Ploog, K. H. (1997). *Appl. Phys. Lett.*, **70**, 683.
Tsang, T. S. and Ilegems, M. (1977). *Appl. Phys. Lett.*, **31**, 301.
Tsang, T. S. and Cho, A. Y. (1978). *J. Appl. Phys.*, **32**, 491.
Tsang, T. S. and Ilegems, M. (1979). *Appl. Phys. Lett.*, **35**, 792.
Tsuchiya, H., Takeuchi, A., and Kurihara, M. (1995). *J. Cryst. Growth*, **152**, 21.
Uchida, K., Watanabe, A., Yano, F., Kouguchi, M., Tanaka, T., and Misagawa, S. (1996). *J. Appl. Phys.*, **79**, 3487.
Van de Walle, C. G. and Neugebauer, J. (1997). In *III-V Nitrides*. F. A. Ponce, T. D. Moustakas, I. Akasaki, and B. A. Monemar, eds., Pittsburgh, PA: *Mater. Res. Soc. Proc.*, **147**, pp. 861–870.
Van Hove, J. M., Chow, P. P., Hickman, R., Wowchak, A. M., Klaassen, J. J., and Polley, C. J. (1997). *Mat. Res. Soc. Symp. Proc.*, **449**, 1227.
Van Vechten, J. A., Zook, J. D., Horning, R. D., and Goldenberg, B. (1992). *Jpn. J. Appl. Phys.*, **31**, 3662.
Vaudo, R. P., Yu, Z., Cook, J. W., and Schetzina, J. F. (1993). *Opt. Lett.*, **18**, 1843.
Vaudo, R. P., Cook, J. W., and Schetzina, J. F. (1994). *J. Vac. Sci. Technol.*, **B12**, 1232.
Vaudo, R. P. (1994). Ph.D. Thesis, North Carolina State University.
Vaudo, R. P., Goepfert, I. D., Moustakas, T. D., Beyea, D. M., Frey, T. J., and Meehan, K. (1996). *J. Appl. Phys.*, **79**, 2779.
Wakahara, A., Tokuda, T., Dang, Z., Noda, V., and Sasaki, A. (1997). *Appl. Phys. Lett.*, **71**, 906.
Weimann, N. G., Eastman, L. F., Doppalapudi, D., Ng, H. M., and Moustakas, T. D. (1998). *J. Appl. Phys.*, **83**, 3656.
Weeks, J. D. (1980). *Ordering in Strongly Fluctuating Condensed Matter Systems.*
Wright, A. N. and Winkler, A. (1968). *Active Nitrogen*. New York: Academic Press.
Wright, A. F. (1997). *J. Appl. Phys.*, **82**, 5269.
Xu, G. Y., Salvador, A., Kim, W., Fan, Z., Lu, C., Tang, H., Morkoc, H., Smith, G., Estes, M., Goldenberg, B., Yang, W., and Krishnankutty, S. (1997). *Appl. Phys. Lett.*, **71**, 2154.
Yang, Z., Li, L. K., and Wang, W. I. (1995). *Appl. Phys. Lett.*, **67**, 1686.
Yoshida, S., Misawa, S., and Itoh, A. (1975). *Appl. Phys. Lett.*, **26**, 461.
Yoshida, S., Misawa, S., Fuji, Y., Tanaka, S., Hayakawa, H., Gonda, S., and Itoh, A. (1979). *J. Vac. Sci. Tech.*, **16**, 990.
Yoshida, S., Misawa, S., and Gonda, S. (1982). *J. Appl. Phys.*, **53**, 6844.
Yoshida, S., Misawa, S., and Gonda, S. (1983a). *Appl. Phys. Lett.*, **42**, 427.
Yoshida, S., Misawa, S., and Gonda, S. (1983b). *J. Vac. Sci. Tech.*, **B1**, 250.
Yoshida, S., Okumura, H., Feuillet, G., Hacke, P., Balakrishnan, K. (1997). In *III-V Nitrides*. F. A. Ponce, T. D. Moustakas, I. Akasaki, and B. A. Monemar, eds., Pittsburgh, PA: *Mater. Res. Soc. Proc.*, **449**, p. 173.
Yu, Z., Buczkowski, S. L., Gilis, N. C., Meyers, T. H., and Richards-Babb, M. R. (1996). *Appl. Phys. Lett.*, **69**, 2731.
Zunger, A. and Mahajan, S. (1994). *Handbook of Semiconductors*, **3**, Second edition, T. S. Moss, ed., Amsterdam: Elsevier.

CHAPTER 3

Defects in Bulk GaN and Homoepitaxial Layers

Zuzanna Liliental-Weber

CENTER FOR ADVANCED MATERIALS
MATERIALS SCIENCE DIVISION
LAWRENCE BERKELEY NATIONAL LABORATORY
BERKELEY, CALIFORNIA

I. INTRODUCTION		129
II. POLARITY OF THE CRYSTALS		130
III. DEFECT DISTRIBUTION		135
IV. NANOTUBES		140
V. PL AND POINT DEFECTS		142
VI. INFLUENCE OF ANNEALING		145
VII. LARGER–DIMENSION BULK GaN CRYSTALS		146
VIII. HOMOEPITAXIAL LAYERS		147
	1. Influence of Polarity	147
	2. Pinholes and Growth Rate	152
IX. SUMMARY		153
	REFERENCES	154

I. Introduction

In recent years wide-bandgap semiconductors have attracted much attention because of the potential to produce high-temperature power devices (Kahn et al., 1993), bright blue LEDs (Nakamura et al., 1994; Nakamura et al., 1995), and pulse and continuous wave (cw) lasers (Nakamura et al., 1996). Improvements in crystal-growth techniques triggered this dynamic development (Moustakas and Molnar, 1993; Lin et al., 1993; Wang and Davis, 1993; Newman et al., 1993). Large high-quality crystals of III-V nitrides have not yet been obtained. This is because the thermodynamic properties of these nitrides eliminate standard growth methods like Czochralski or Bridgman growth from the melt. For epitaxial growth, both the lack of lattice-matched substrates and the difference in thermal expansion coefficient hamper good structural quality of nitride films. For example, GaN films are commonly grown on 14% lattice-mismatched sapphire, which leads to large strain during growth, as well as during the cooling-down process, and therefore leads to high defect density (Lester et al., 1995; Liliental-Weber et al., 1995; Liliental-Weber et al., 1996b; Qian et al., 1996; Ruvimov et al., 1996; Ponce et al., 1996). Currently available bulk GaN is grown Ga-rich (from a Ga melt) under a nitrogen hydrostatic pressure of about 15 kbars at temperatures ranging from 1500–1800 K (Porowski et al., 1989; Porowski and Grzegory, 1994). The high stability of the nitrides results in very high melting temperatures, close to 3000 K. In addition, GaN has a very high decomposition pressure, making growth of bulk GaN difficult. Bulk GaN grown by this technique crystallizes in the form of platelets or rod-like crystals. X-ray studies using (004) Cu K_α reflection of the plate-like crystals show that quality is related to the size of the crystal. The experimental x-ray rocking curve full-width at half maximum (FWHM) is about 20–30 arc s for crystals not larger then 1 mm. For larger crystals (1–3 mm), the FWHM of the rocking curve is broader, and for crystals from 3–10 mm the rocking curve splits (Leszczyñski et al., 1995). For epitaxial GaN grown on SiC, the x-ray rocking curves have a FWHM of 2.5–4 arc min and for GaN grown on sapphire the rocking curves have a FWHM of 6–15 arc min (Leszczyñski et al., 1995).

In this chapter plate-like bulk crystals will be characterized using transmission electron microscopy (TEM) and x-ray diffraction. The results of cathodoluminescence (CL) and TEM studies will be compared using the same samples. Two types of crystals will be described: those grown in 8-mm diameter and 16-mm diameter crucibles. As another chapter of this book deals with the subject of crystal growth, this issue will be discussed only marginally in this chapter. Particular types of defects present in these plate-like crystals will be characterized and the relevance of these defects as relates to using these crystals as substrates for homoepitaxial growth will be discussed. Defects in homoepitaxial layers and growth-rate hierarchy measured from bulk and homoepitaxial films will be characterized toward the end of this chapter.

II. Polarity of the Crystals

For this investigation, bulk GaN crystals were grown from a Ga melt under nitrogen hydrostatic pressure (Porowski et al., 1989; Porowski and Gizegory, 1994). Detailed studies using transmission electron microscopy

(TEM) in plan-view [0001] direction and cross section along [11$\bar{2}$0] and [1$\bar{1}$00] of small plates of about 1 mm in length obtained in an 8-mm crucible will be described. Much larger crystals were obtained in this crucible, but only the small crystals were chosen for TEM characterization to learn about crystal quality for the smallest FWHM of the x-ray rocking curves. The highest perfection was expected for these crystals. The larger bulk crystals with broader FWHM will be described in a later part of the chapter.

The GaN crystal plates had the shape of elongated hexagons (Fig. 1) and crystallized with the wurtzite structure. The longest axis was frequently along [11$\bar{2}$0]. In some cases the samples were less elongated and the dimensions along [11$\bar{2}$0] and [1$\bar{1}$00] were similar, but in all cases the smallest dimension was in the c-axis direction. The ratio of plate length to thickness along the c axis was as large as 100. This shows that growth along the c axis is very slow. Study of cross-section samples showed that one side of the plate is almost atomically flat (Fig. 2a, b) (with fluctuations not larger than 10–15 Å) while the opposite side is rough. In an extreme case a high density of pyramids, with heights of about 100 nm was present on the rough side of the plate. Our earlier studies showed that crystal polarity plays an extremely important role in the growth. The first report on this issue was an abstract to the MSA meeting (Liliental-Weber and Kisielowski, 1995),

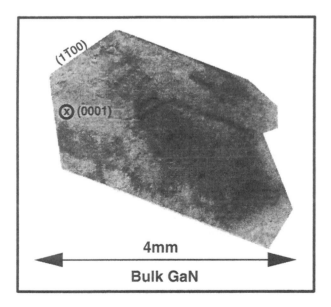

FIG. 1. Optical micrograph of a typical GaN platelet crystal obtained in a small cavity with indicated (1$\bar{1}$00) cleavage planes.

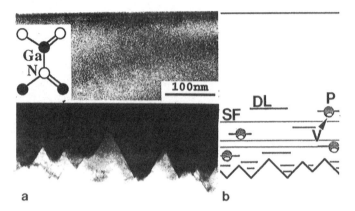

FIG. 2. (a) Cross section through the GaN plate with indicated polarity (only areas close to the plate surfaces normal to the **c** axis are shown). Note different surface roughnesses on the two opposite polar surfaces. (b) Schematic drawing of the same sample showing presence of stacking faults (SF) and dislocation loops (DL) decorated by Ga precipitates (P). The voids (V) located toward the rough surface attached to these precipitates are indicated.

but a full report on the importance of polarity on growth came in an invited talk at the International School of Semiconductor Physics (Liliental-Weber, 1995).

Crystal polarity was determined using convergent beam electron diffraction (CBED), where instead of a parallel electron beam a convergent beam was focused on the sample with a beam spot size of 1–2 nm. As a result each diffracted spot is formed as a disc. Each diffracted disc carries information that can be used to determine point group of the crystal, lattice parameter, and other crystal properties (Buxton *et al.*, 1976; Steeds and Morniroli, 1992; Tafto and Spence, 1982). This method is known to be especially useful for polarity determination in noncentrosymmetric crystals (Spence and Zuo, 1992; Liliental-Weber and Parachenian-Allen, 1986). The dynamic approach, taking the full CBED pattern of the GaN cross-section samples of the ($1\bar{1}00$) pole, was used in the 002B Topcon microscope. As a reference, computer-simulated CBED patterns were calculated along the [$1\bar{1}00$] zone axis for the acceleration voltage (200 kV) used in the experiment. Because the intensity pattern within the (0002) and (000$\bar{2}$) discs is different, this information can be used to determine unambiguously the direction of the N to Ga **c**-axis bond in the crystal. It was determined that the long bonds along the **c**-axis in the direction from N to Ga atoms point toward the smooth surface. This direction with N at the origin and Ga in +**c** direction will be defined in this chapter as **B** polarity. This information is very important, because there is a clear dependence of crystal quality and growth rate on polarity. The results of this study were reported at conferences and in journal publications (Liliental-Weber *et al.*, 1996b; Liliental-Weber, 1995; Liliental-Weber *et al.*, 1997b; Liliental-Weber *et al.*, 1996a). Polarity of the mechanically polished bulk GaN obtained from the same crystal grower was also reported by another group, which appeared to show that the results were 180° rotated compared to our reported results (Ponce *et al.*, 1996). This result was reported in spite of the fact that the interpretation of CBED discs in both cases was identical (Liliental-Weber *et al.*, 1996a). Because 180° rotations can be a common mistake in microscopy (so-called "180° ambiguity") an effort was undertaken to eliminate this possibility for our reported results. Two independent calibrations using GaAs wafers grown in $\langle 111 \rangle$ direction along with etching studies revealing either Ga or As surfaces as well as Z—contrast studies of our standards confirmed that our Topcon microscope manual was correct and our original assignment of the GaN polarity was correct. This original polarity assignment applied as well to homoepitaxial layers, which were grown on the original bulk GaN crystals without additional polishing, where the "smooth" and "rough" surfaces can be easily recognized. The polarity issue in homoepitaxial films will be discussed in the last part of this chapter.

The polarity appeared to be extremely important for GaN crystal growth, for example, the bond direction between N and Ga influences growth rate and crystal quality near the two opposite surfaces. One possible explanation for different surface morphology is to consider bonding between Ga and N at the surface. Because growth of bulk GaN platelets is governed by arrival of nitrogen atoms at the growing surface, one can consider that a newly arrived nitrogen atom is likely to be attached weakly to Ga on the N side by a single bond (Fig. 2). However, a nitrogen atom is attached strongly to Ga on the Ga side (Fig. 2) by strong triple bonding. This side of the crystal might be expected to grow more smoothly. In addition, one needs to consider that liquid Ga surrounds the crystal. It is likely that clusters of Ga atoms (as exist in liquid Ga) may attach simultaneously to the N side, making grown-in stacking faults more probable.

For some larger crystals with broader FWHM, it was observed that two or three near parallel plates were connected with each other, as shown in Fig. 3. It is not easy to recognize several almost parallel plates in plan-view, but they can be seen when cross-sectional samples are prepared. As each individual plate originally had both a smooth and a rough surface, the parallel plates would first grow together at the most extended parts of the rough surface (Fig. 3). Each individual plate in these double- or multilayer crystals had the typical structure with respect to the distribution of extended defects as was observed in single plates, but the individual plates might be

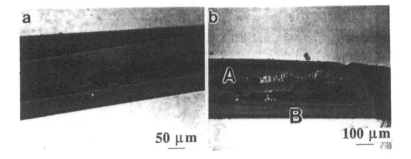

Fig. 3. Low-magnification images of two cross-sectional GaN bulk samples: (a) double-plate crystal; and (b) multiplate crystal.

slightly tilted with respect to each other. Each individual plate in this structure is a single monocrystalline "domain" with a particular polarity, but the arrangement of these "domains" and their polarity in a multiplate sample can be arbitrary. The sample shown in Fig. 3a had a reverse polarity for the two plates, leaving both the remaining outside surfaces chemically identical. The upper and lower crystal with the outside smooth surfaces had the long bonds along the c-axis in the direction from N to Ga atom pointed toward the smooth surfaces. The interface area between these connected plates can have empty spaces, or have some imbedded Ga. Discovery of the existence of these multilayer plate crystals might explain the broadening of x-ray rocking curves (Leszczyñski *et al.*, 1995), since small angle boundaries were never found by TEM in the individual crystals. The arrangement of two plates with reverse polarity also explains the recent results of x-ray studies (Leszczyñski *et al.*, 1996), which show that the lattice parameter on the two sides of the sample can occasionally be identical, because chemically identical surfaces could be present on both sides of such a plate. Each individual plate is monocrystalline, without any small-angle boundaries, as was suggested by x-ray studies (Leszczyñski *et al.*, 1996). Such boundaries are possible only where two plates are connected to each other. Such multiple crystals are expected to broaden FWHM of the x-ray rocking curve if the thickness of one subplate is smaller than the x-ray penetration depth, but such multiple crystals should not create a problem for homoepitaxial growth as long as one plate is not interrupted. One can assume that growth on either of the two surfaces of a double crystal as shown in Fig. 3a should not create a problem, but for the crystal shown in Fig. 3b growth on the surface of a crystal marked by "B" would be expected to be good in spite of the fact that the crystal contains four plates, but growth on the opposite side "A" would not bring good results.

III. Defect Distribution

Small individual bulk plates observed in thin cross-sectional samples under the optical microscope appeared to consist of two or sometimes more "sublayers" of different colors: "transparent" (Fig. 2a–upper part) and "yellow" (Fig. 2a–rough lower part). The "yellow" areas contained a high density of planar defects. These planar defects were always observed close to the rough surface of the crystal as shown schematically in Fig. 2b and in detail in Fig. 4a. The planar defects did not extend more than one-tenth to one-fourth of the plate thickness away from the rough side. The rougher the surface of the sample, the more such defects were observed. The opposite side of the plate, which appeared to be transparent, was structurally perfect, with no extended defects present (see Fig. 2a–upper part, and Fig. 4b) and had an almost atomically flat surface. It is important to notice that bulk plate GaN crystals do not have threading dislocations along the c axis that would end at the (0001) surfaces. This is very important with respect to application of these plates as substrates for homoepitaxial growth, because threading dislocations propagate into the epitaxial layers.

Four types of extended defects were observed near the rough side of the crystal: stacking faults (SF); large dislocation loops attached to stacking faults (DL—with a diameter in the range of 100–1000 Å); small dislocation loops with a diameter about 30 nm; and precipitates (P), which were found

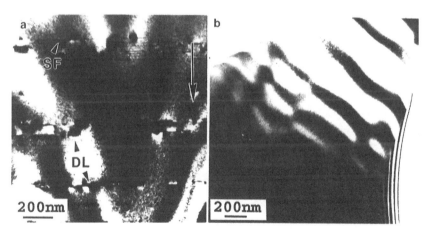

FIG. 4. (a) TEM micrograph showing in higher magnification defects present close to the rough "yellow" side of the crystal (a portion from the lower surface shown on Fig. 2a); (b) lack of extended defects in the area close to the smooth surface "transparent side" (a portion from the upper surface shown on Fig. 2a). The arrow indicates crystal polarity in the direction Ga to N along c-axis (A polarity).

to be associated with both large and small dislocation loops. Studies of these precipitates were very difficult and required special precautions because melting of the precipitates under the electron beam can easily occur (as one can see in Fig. 5a). Careful studies under the least possible exposure to electrons showed that these precipitates are pure Ga and very often are associated with small voids (Fig. 5b). It was interesting to observe that the void associated with each precipitate was always located closer to the rough side of the crystal. The polarity of the crystal also appears to be responsible for this preferential location. The strain field at the surrounding dislocation core helps to nucleate excess Ga precipitation. (All bulk crystals were grown in an atmosphere of excess Ga.) For crystals with a smaller density of dislocation loops small microhomogeneities (as shown in Fig. 5c) were often observed.

The most common defects in these crystals were long extended stacking faults, which are probably nucleated at an early stage of crystal growth by growth mistakes. These stacking faults often extend through the entire plate from edge to edge. In the cross-sectional micrographs observed in both the [11$\bar{2}$0] and [1$\bar{1}$00] directions, they are visible as long straight lines (Fig. 4a) located preferentially near the rough side of the plates. Therefore, it is assumed that they were nucleated during the early stages of growth and propagated during rapid growth in the nonpolar directions.

There are three possible types of stacking faults for the wurtzite structure (Liliental-Weber *et al.*, 1996a): (I) — one can be formed by removal of a basal layer followed by a shear of 1/3 $\langle 10\bar{1}0 \rangle$ of the crystal above the fault to reduce the energy; (II) — one which results from a shear of 1/3 $\langle 10\bar{1}0 \rangle$ in an originally perfect crystal; (III) — a fault of the highest energy, which is formed by inserting an extra plane. These three types of stacking faults were found in the bulk GaN crystals (Fig. 6a–c). These stacking faults in the wurtzite structure can be described as the existence of layers having sphalerite unit cells, which requires rotation of the bonds; for example, the

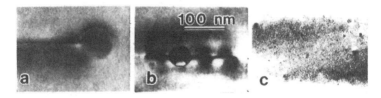

FIG. 5. Ga precipitates found close to the rough side of the crystal: (a) a precipitate melted under electron-beam illumination; (b) a precipitate on a dislocation loop with a void located toward the rough side (as schematically shown in Fig. 2b); and (c) microinhomogeneities in the "yellow" part of the crystal (rough side).

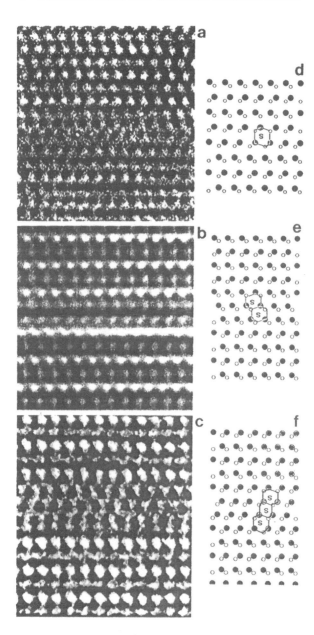

FIG. 6. Different types of stacking faults determined in the bulk GaN crystals: (a–c) indicate faults with increasingly higher energies; (d–f) indicate hard ball models of the atom arrangement for the faults shown on the left-hand side. Note that the number of sphalerite units inserted into the wurtzite structure increases with increase of fault energy.

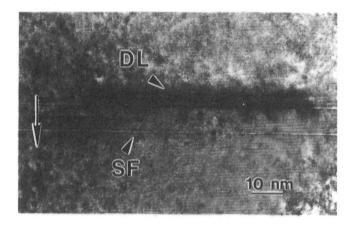

FIG. 7. High-resolution image of stacking faults (SF) and a dislocation loop (DL-darker contrast) attached to the fault. Note lattice plane bending around the dislocation loop. The arrow indicates crystal polarity in the direction Ga to N along c-axis (A polarity).

mirror symmetry for bond arrangement typical of the wurtzite structure is lost and it is locally converted to the zinc-blende bond arrangement where bonds are rotated 60° with respect to the nearest neighbor bonds. Analysis of the electron micrographs showed that the number of zinc-blende "unit cell layers," shown schematically in (Fig. 6d–f), is related to the fault energy. The faults with the highest energy were found very often in the bulk GaN crystals. Therefore, formation of an extrinsic dislocation loop located on the right atomic layer within such a high-energy stacking fault is energetically favorable, because insertion of the additional layer can convert locally the highest-energy fault configuration into the low-energy type of fault.

Therefore, associated with the extended stacking faults, large interstitial-type dislocation loops with a diameter of several hundred angstroms are often present (Fig. 7). Following an individual stacking fault, several such dislocation loops, separated laterally from each other, can usually be found (Fig. 4a). These dislocation loops, decorated by precipitates, can be seen easily in a plan-view micrograph (Fig. 8). Formation of such a loop results in a different strain within and outside the loop because of the surrounding dislocation line. At the edges of these loops, the dislocation core appears to act as the nucleation site for the observed precipitation of excess Ga.

Smaller dislocation loops, which are not connected with stacking faults, are very often decorated by Ga precipitates. They are also interstitial types with a Burger's vector $\langle \bar{2}203 \rangle$ (Fig. 9).

3 DEFECTS IN BULK GaN AND HOMOEPITAXIAL LAYERS 139

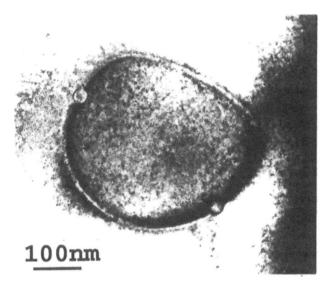

FIG. 8. Plan-view image showing a dislocation loop decorated by Ga precipitates. Density of these loops changes in different areas of the crystal.

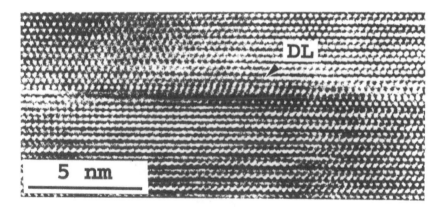

FIG. 9. TEM micrograph showing a cross section through an interstitial-type dislocation loop (DL) with a Burger's vector of $\langle \bar{2}203 \rangle$. Note an inserted extra plane.

IV. Nanotubes

As in SiC and Al_2O_3 the most detrimental defects that have precluded application of these materials are nanotubes. The nanotubes in SiC have received great attention because they are known to be the defects that limit breakdown voltage of high-power devices (Powell *et al.*, 1994). The original explanation for the formation of nanotubes was given by Frank (1951), who suggested hollow-core screw dislocation. The explanation was given that a state of local equilibrium exists in which a dislocation of a large Burger's vector should have less energy when its core is hollow compared to a core filled with the highly strained lattice. The open-core size, therefore, should be an equilibrium between the added surface free energy and the decreased lattice strain energy. According to the Frank theory, the radius of the hole is proportional to the square of its Burger's vector. The density of the nanotubes was estimated to be in the range of 10^5–10^7 cm^{-2}. The radii of the pipes are in the range of 3–500 nm and they appear to propagate along the c-axis of the film. These defects are usually revealed using atomic force microscopy (AFM) or plan-view TEM (Fig. 10). Based on these methods it is not easy to determine the length and location of these defects within the sample. Only in some circumstances can proper image defocus in the microscope give some information about defect depth. Our studies of plan-view bulk GaN samples showed a distribution of dark round spots on the sample (Fig. 11a), which were different from the images of the pinholes or "nanotubes" observed in heterolayers of GaN. These dark spots were determined to be Ga precipitates. When the high-resolution TEM image is

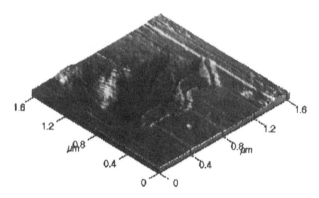

FIG. 10. An example of a pinhole formed in metallorganic chemical vapor deposition (MOCVD)-grown GaN layer shown by AFM.

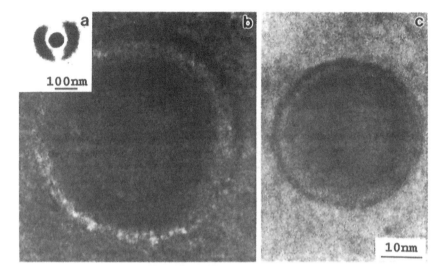

FIG. 11. Hollow defects filled by Ga: (a) a Ga precipitate on the sample surface (dark disc); (b) a high-resolution image of a dark disc. Note amorphous material (probably melted Ga); (c) the same area as above taken under different defocus value showing crystalline GaN at the bottom of the hollow.

focused on the sample surface these dark areas appeared to be amorphous (Fig. 11b), confirming our earlier observation that Ga precipitates melt under the electron beam. However, by defocusing of these images one can obtain the GaN lattice image (Fig. 11c), suggesting that a nanotube of some length along the electron beam axis can be located below the Ga precipitate. From the defocus value it can be estimated that the depth of a particular tube was about 200–300 nm. It appears that these nanotubes differ from the ones observed in SiC (Qian et al., 1996; Qian et al., 1995), or heteroepitaxial GaN (Hobgood et al., 1994; Takasu and Shimanuki, 1974; Heindl and Strunk 1996), because they are not hollow, but instead are filled by Ga.

Another type of nanotube was observed in annealed bulk GaN crystals for which optical microscopy shows hexagonal islands similar to those observed in heteroepitaxial GaN layers (Liliental-Weber et al., 1997a). The TEM studies on cross-sectional samples from such areas show a defect with V-shape in the center of the hexagonal island (shown by the arrow in Fig. 12). Starting from the V-shape defect a nanotube (tubular and hollow) runs through the entire crystal from the center of the hexagonal island (Fig. 12). This is similar to the observation of empty hollow tubes in the center of hexagonal islands in heteroepitaxial GaN layers.

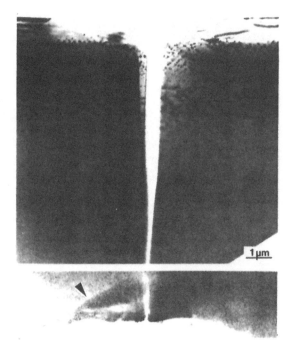

FIG. 12. A tubular defect formed in an annealed bulk GaN crystal. Note a V-shaped defect (marked by arrow) at a starting point of the nanotube defect (shown in the center of the micrograph).

V. PL and Point Defects

As already mentioned, the bulk GaN crystals consist of sublayers, one of which is defect-free (Figs. 2a and 4b), while the other has a high density of stacking faults, dislocation loops, and Ga precipitates (Figs. 2b and 4a). These sublayers appear as transparent or yellow, respectively, under the optical microscope. The CBED studies for the same sample thickness taken in the transparent and yellow part of the layer show different intensity distribution for particular discs in $[1\bar{1}00]$ projection. The difference can be observed in the central discs. Comparing the [0002] disc taken in the yellow area with the $[000\bar{2}]$ disc of the transparent one, the patterns almost seemed to be interchanged, although they are not identical (Fig. 13). It is believed that this change is introduced by the presence of different point defects (or their different concentration) in the transparent and yellow parts of the crystal (Kisielowski *et al.*, 1996). However, at this point it is not easy to

Fig. 13. CBED patterns obtained for the same sample thickness (130 nm) in the transparent (a) and yellow (b) parts of the plate crystal.

calculate the concentration of these defects based on CBED simulation, because many necessary parameters are not accurately known for GaN. In addition, very recent studies have found a high concentration of oxygen in these crystals (Perlin et al., 1996) and it is not clear at this point how to introduce oxygen presence for CBED pattern simulation. One conclusion from this study is clear: there must have been a substantial supersaturation of intrinsic point defects in the crystal that caused Ga precipitation and the formation of voids in the defective (yellow) area. In the defect-free (transparent) area, much higher point-defect concentration must remain in interstitial or substitutional positions responsible for the difference in the CBED pattern in this area of the crystal. This difference in CBED pattern and presumably different point defects (or their concentration) would explain the different lattice parameter on both sides of the platelet crystals (Leszczyñski et al., 1996).

The CL spectra were obtained in both the transparent and the yellow parts of the samples (Fig. 14). The CL was mapped at two different wavelengths. One was chosen to be close to the band-to-band transition at

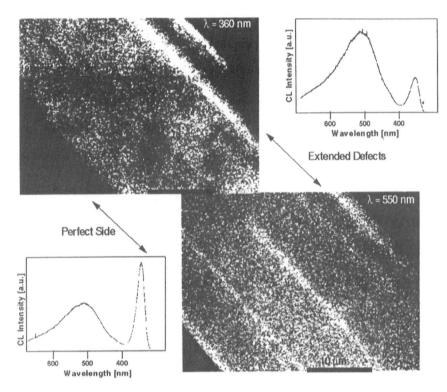

FIG. 14. Cathodoluminescence spectra and images taken at the wavelength of 360 nm (band-to-band luminescence) and 550 nm (yellow luminescence). Note interchangeable contrast for these two wavelengths. The areas of the crystals from which spectra were taken are indicated.

360 nm (eV), and the other one was centered on the yellow luminescence band at 550 nm (eV). Complementary images were obtained: the areas that appeared dark at one wavelength were bright if mapped with the other wavelength (Fig. 14). A comparison of CL and TEM images of the two crystals showed a consistent trend: in transparent areas the bandedge CL dominates over the yellow CL, while in the yellow areas this trend is inverted; for example, the yellow CL dominates over the bandedge CL. Especially in the sample, which had a high density of stacking faults and dislocation loops decorated by Ga precipitates, the yellow luminescence was large. The exact ratio of the yellow luminescence to the bandedge luminescence depended strongly on the sample position.

VI. Influence of Annealing

Annealing of bulk GaN crystals under hydrostatic pressure of N (GPa) at a temperature of 1500 K for half an hour leads to a decrease of the lattice parameter (Leszczyński, 1997). As was mentioned in the previous section, unannealed crystals show a difference in the lattice parameter on the two sides of the plates (Leszczyński *et al.*, 1996). This difference disappeared after annealing and the lattice parameter of annealed crystals is in the same range as was measured for the rough side of unannealed crystals that show high concentration of dislocation loops decorated by Ga precipitates (Leszczyński, 1997). First annealing experiments under high nitrogen pressure revealed additional dislocation loops decorated by Ga precipitates (Fig. 15). Two sizes of dislocation loops were observed in annealed crystals: one with a diameter of 20–50 nm and the second with a diameter of 200–400 nm. In the unannealed samples only dislocation loops with smaller diameter were

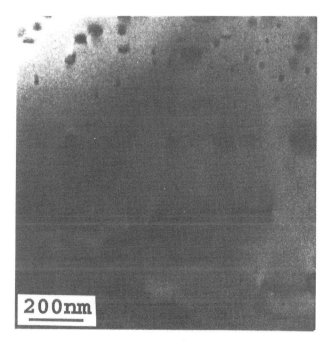

FIG. 15. Distribution of dislocation loops in annealed GaN crystals. Note two different sizes of loops. Smaller loops were most probably formed after annealing of the transparent part of the crystal free of extended defects before annealing.

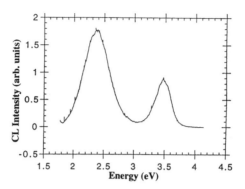

FIG. 16. A cathodoluminescence spectra of annealed GaN crystals. Note high intensity of yellow luminescence, similar to that observed in defected areas of unannealed crystals.

observed. This is consistent with an assumption of high concentrations of point defects present in these crystals as suggested by the CBED studies. Annealing of bulk crystals leads to the formation of additional dislocation loops in crystal areas, which were free of these defects before annealing, and increase of the diameter of dislocation loops in the areas where these loops were present before annealing. Formation of extrinsic dislocation loops and precipitation of the extra Ga atoms, which cause expansion of the lattice parameter in unannealed crystals, would explain the decrease of the lattice parameter on annealing, which was observed by x-ray studies.

Cathodoluminescence studies of these annealed crystals show consistently much higher intensity of yellow luminescence compared to unannealed crystals free of extended defects (Fig. 16), suggesting that higher density of dislocation loops decorated by Ga precipitates correlates with this luminescence. This would support earlier studies showing enhanced yellow luminescence near dislocations (Ponce *et al.*, 1996).

VII. Larger-Dimension Bulk GaN Crystals

Sizes of platelets grown in the 16-mm diameter crucible were much larger (Porowski *et al.*, 1997) compared to those obtained from the smaller 8-mm diameter crucible. First of all, these crystals are transparent. They appear to have less excess Ga. TEM cross-sectional studies of these crystals show much higher perfection compared to the smaller crystals. On one side the subsurface area contains dislocation loops (Fig. 17b). Their diameter is in the range of 50–200 nm. These loops are only within the distance of 200–300 nm from the surface. Additional ion milling, which removes this

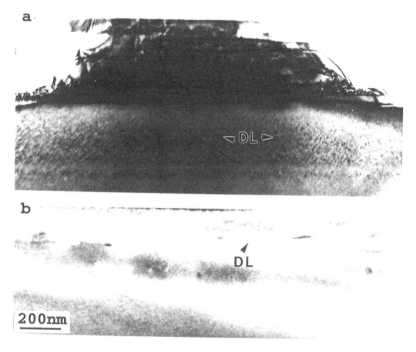

FIG. 17. (a) A crystallite formed on the GaN surface. Note wide band of darker contrast in the subsurface area indicating high density of small dislocation loops (DL) formed on this side of the crystal; (b) high density of small dislocation loops (DL) present only in the subsurface area of the crystals grown in the larger cavity.

surface area, results in a platelet of high perfection essentially without defects. The opposite surface has high density of small dislocation loops and some crystallites with a diameter of 1–1.5 μm on the surface (Fig. 17a). The presence of such crystallites on the surface would lead to defect formation in a homoepitaxial layer; these crystallites should be removed prior to homoepitaxial growth.

VIII. Homoepitaxial Layers

1. INFLUENCE OF POLARITY

Homoepitaxial layers were grown on bulk GaN either on the smooth or on the rough surface, thereby in two opposite polar directions. Based on CBED studies of bulk GaN plates it was previously concluded (Liliental-

Weber et al., 1996b; Liliental-Weber, 1995; Liliental-Weber et al., 1997b; Liliental-Weber et al., 1996a) that the long bonds parallel to the c-axis, in the direction N to Ga point toward the smooth surface in the bulk GaN platelets (defined as B polarity). Studies on homoepitaxial layers showed that the homoepitaxial layers grow with the same polarity as the substrate. Therefore, by growth on the two opposite surfaces of bulk crystals homoepitaxial layers with opposite polarity of the growth direction could be studied (Baranowski et al., 1996).

TEM studies showed that the layer quality and type of defects are different depending on the substrate polarity. Figure 18a shows the layer grown on the smooth surface of the substrate and Fig. 18b shows the substrate with the rough side opposite to the layer. In the epilayer grown on the smooth surface threading dislocations and inversion domains were found (Fig. 19a). The inversion boundaries originate from dislocation loops formed at the interface and their density was not larger than 5×10^5 cm^{-2}. A dark contrast can be observed on the TEM micrographs along the

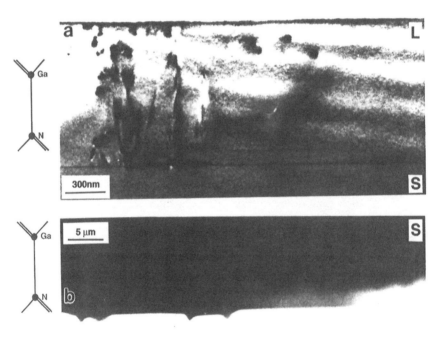

FIG. 18. (a) TEM cross-sectional micrograph showing a homoepitaxial layer (L) grown on the smooth side of the substrate (S) (B polarity indicated); (b) the substrate with the rough side opposite to the layer.

3 DEFECTS IN BULK GaN AND HOMOEPITAXIAL LAYERS 149

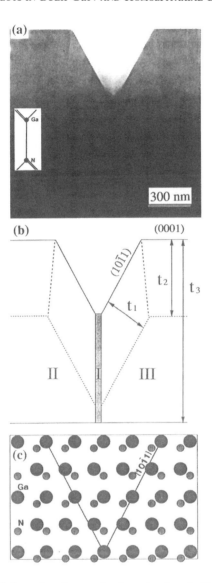

FIG. 19. (a) A pinhole formed on the top of an inversion domain (I). Note two dislocations originating from the interface bending toward the inversion domain; (b) the above micrograph (a) is shown schematically. Note that the height of the inversion domain (I) is about half the surrounding layer thickness t_3, showing different growth rates for the two polar directions. Two subgrains (marked in b as II and III) are growing on the two sides of the inversion domain. Growth proceeds on top of these grains with thickness t_2 along the **c**-axis and t_1 perpendicular to the (10$\bar{1}$1) planes. (c) Ball model of atom arrangement on the (10$\bar{1}$1) planes. Note the angle between these two planes fits to the angle between V arms of a pinhole.

interface for this sample observed in two perpendicular projections [11$\bar{2}$0] and [1$\bar{1}$00]. Detailed studies showed that this contrast was not related to any structural defect. Bright- and dark-field images taken with different diffraction conditions showed that some inhomogeneity, segregation, takes place at this interface. All threading defects started from dislocation loops formed at the interface. In most cases threading dislocations were grouped into clusters within the islands and the number of threading dislocations was related to the island size. Considering one dislocation per island their number would be in the range of $5 \times 10^6 \, \text{cm}^{-2}$, but the density of dislocations within one island can vary from 1 to 20. Dislocation density can drastically increase if the substrate surface is not cleaned properly. Any impurities and inhomogeneities at the interface can be a source of a threading dislocation, as is shown in Fig. 20.

In addition to these defects, interstitial-type dislocation loops and pinholes were occasionally found in this sample. The pinholes were associated with inversion boundaries, which were associated with dislocations. Figure 19 shows a pinhole observed in a homoepitaxial layer. It is clear from this micrograph that the pinhole started on the top of an inversion domain (confirmed by CBED). Two dislocations were also attracted to this inversion domain. This inversion domain and the dislocations were independently originated at the interface but the dislocations terminated on the facet of the pinhole. The dislocations were bent toward the hole during layer

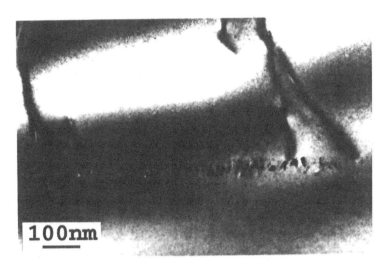

FIG. 20. Threading dislocations initiated from inhomogeneities (due to imperfect surface preparation) formed at the smooth surface of GaN.

growth to minimize their length and, therefore, reduce the total free energy of the system. Diffraction contrast used for the characterization of these dislocations showed that they were of a different Burger's vector: The lower dislocation closer to the inversion domain had a near screw character, while the upper one was lying at a larger angle to its Burger's vector.

The difference between the homoepitaxial layers grown with different polarity developed during crystal growth (Baranowski and Porowski, 1996). For homoepitaxy on the "smooth" surface the initiation of the growth had to be done at lower temperature than in the case of growth on the "rough" surface. In the GaN sample grown on the rough surface (A polarity) no threading defects were observed (Fig. 21). Only interstitial-type dislocation loops mostly formed on c-planes with a density in the range of $1 \times 10^7 \text{cm}^{-2}$ were found. No dark contrast was present at the interface between the bulk GaN substrate and the epilayer. The initial growth on this surface was surprisingly good, making it nearly impossible to determine conclusively the exact position of the substrate-epilayer interface. However, a high density of pinholes (in the range of 10^6-10^7cm^{-2}) was seen in these samples. These

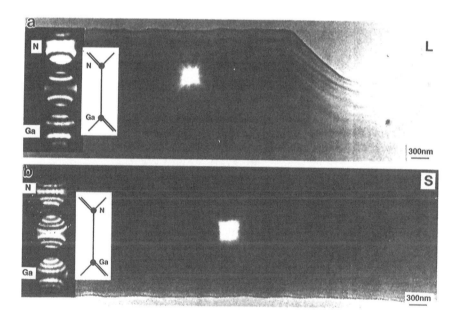

FIG. 21. (a) Homoepitaxial layer (L) grown on the rough side of a GaN bulk crystal (S). CBED patterns indicate polarity of the substrate and the layer. Note formation of a pinhole in the layer. (b) The substrate (S) on the opposite side from the layer. CBED patterns were taken in the positions indicated by the electron-beam white marks.

pinholes were not attached to inversion domains as observed in the layers with B polarity. Some of them originated from the interface; and others appeared to nucleate during layer growth. Point-defect clustering and substrate-growth surface roughness could be the cause of pinhole initiation.

2. PINHOLES AND GROWTH RATE

From careful study of the contrast around the inversion domain shown in Fig. 19a, one can distinguish three subgrains of GaN that were formed at the interface (with the inversion domain in the center). The growth of the inversion domain was much slower than in the two surrounding subgrains of opposite polarity. A schematic of the growth sequence is shown in Fig. 19b. Measurements of the layer thickness and the height of the inversion domain, which has the opposite polarity compared to the surrounding layer, show that the growth rates for the two opposite polarities have a ratio of about 2:1. These relative growth rates are in agreement with measurements of layer thicknesses of homoepitaxial layers grown on surfaces of opposite polarity. It is assumed that the "deepest" pinholes start at the substrate interface. A layer thickness of $0.7\,\mu m$ for growth on the rough surface compared to $1.2\,\mu m$ for growth on the smooth surface can be estimated. Studying the contrast on the two sides of the inversion domain observed in the sample grown on the smooth side of GaN sideways growth of the two faceted grains can also be followed. These facets were formed on $[10\bar{1}1]$ polar planes (Fig. 19c), and the growth rate on these planes was 2 to 2.5 times slower in comparison to the growth on the c-planes in the direction N to Ga (B polarity). The details of this study are described elsewhere (Liliental-Weber *et al.*, 1997c).

From the measured growth rates for these homoepitaxial films as well as from the growth of the bulk GaN platelike crystals (Liliental-Weber *et al.*, 1996a; Liliental-Weber *et al.*, 1996b) it is clear that the highest growth rates for these crystals are along nonpolar directions (about 100 times higher), the growth rate along the [0001] polar direction is approximately 60% higher in one polarity (on the smooth surface of the bulk GaN) compared to the opposite $[000\bar{1}]$ direction and the growth on $[10\bar{1}1]$ planes is the slowest, and not equal for A and B polarity. This suggests that the slow growth rate on $[10\bar{1}1]$ polar facets can be the origin of "pinhole" formation. This would also explain the observation that pinholes are formed with higher density for growth in one polarity compared to the other. The rough side of the bulk plate is initially faceted on $[10\bar{1}1]$ planes (Liliental-Weber *et al.*, 1996a) and further growth on this faceted surface does not planarize these facets.

IX. Summary

This chapter decribes the crystal perfection of both bulk GaN crystals grown in the form of plates from liquid Ga under high hydrostatic pressure of N and homoepitaxial GaN layers grown on top of these bulk crystals. It was shown that the bulk crystal quality is exceptionally good, and the crystal quality of the platelets grown in the larger 16-mm crucible is superior to those grown in the small 8-mm crucible. The smaller crystals were not stoichiometric; excess Ga is present in the form of Ga inclusions or point defects. The platelet crystals show different surface morphology on the opposite [0001] surfaces, which is related to crystal polarity. It was shown that the side of the plate perpendicular to the **c** axis in the direction N to Ga has an almost atomically flat surface and the opposite surface is rough. This rough side is also associated with a high density of stacking faults of the high-energy type formed during the crystal growth. Extrinsic dislocation loops are often formed on the adjacent atomic layer to these stacking faults, locally converting the fault to the lower-energy configuration. The dislocation line surrounding these areas then provides nucleation sites for precipitation of excess Ga. The Ga precipitates were always associated with a smaller void, suggesting that this precipitation is also associated with a net contraction in volume. As the stacking faults and dislocation loops found in these crystals are parallel to the basal plane and are only near the rough side, these defects would not be expected to have any effect on quality of an epilayer of GaN grown on the smooth side of the plates, assuming that the substrate surface is properly prepared for the growth.

The crystals grown in the larger crucible have better quality than those grown in the smaller one. These crystals are transparent. However, some small crystallites are attached to the platelets. Therefore, polishing of these crystals will be necessary for subsequent homoepitaxial growth.

Homoepitaxial layers grown on bulk platelets repeat the polarity of the substrate. Different types of defects are formed in the layers grown with different polarity. For the layers grown in the N to Ga direction (smooth side of the bulk platelet-B polarity) growth is difficult and much lower growth temperature must be used (Baranowski and Porowski, 1996). Inhomogeneities were present in the interfacial area. The characteristic defects in these layers are threading dislocations and a small density (10^5 cm^{-2}) of inversion domains. All these defects originate at the interface with the substrate.

Growth in the Ga to N direction (A polarity) is so good that it is almost impossible to determine the position of the substrate/epilayer interface. Typical defects in these layers are extrinsic dislocation loops along **c**-planes and pinholes. The density of these pinholes is almost two orders of magnitude higher than that observed for growth in the opposite polar direction. No structural defects were found to be associated with these pinholes and they were most probably a result of the growth on the faceted substrate.

Growth rates of GaN in different crystallographic direction were estimated; growth in the nonpolar direction [11$\bar{2}$0] is the fastest (Liliental-Weber et al., 1996a). Growth of epilayers in the N to Ga direction (B polarity) along [0001] direction is about 50–100 times slower compared to the nonpolar directions and growth is even slower in the [000$\bar{1}$] direction. Growth on [10$\bar{1}$1] planes is 2 to 2.5 times slower than on [0001]. This extremely slow growth rate on [10$\bar{1}$1] was suggested to be responsible for the formation of pinholes in GaN (Liliental-Weber et al., 1997c).

Based on TEM studies of the defects in bulk GaN it appears that this material is suitable as a substrate for epitaxial growth. Optimization of gas flow, gas purity, and also cleaning of the bulk surface are necessary to obtain high crystal quality of the homoepitaxial layers.

ACKNOWLEDGMENT

This work was supported by the Director, Office of Basic Science, Materials Science Division, United States Department of Energy, under the Contract No. DE-AC03-76SF00098. The use of the facility at the National Center for Electron Microscopy at Lawrence Berkeley National Laboratory is greatly appreciated. The author wishes to thank W. Swider for excellent sample preparation and Dr. I. Grzegory and Dr. S. Porowski for bulk GaN crystals, Dr. J. Baranowski and Dr. K. Pakuła for the homoepitaxial layers grown on the bulk GaN, Dr. K. Kisielowski for the calculation of CBED diffraction patterns, L. Schloss for CL mesurements, Dr. J. June and Dr. T. Suski for the annealing of bulk GaN crystals, Dr. S. Ruvimov for the cooperation in analyses of stacking faults in bulk GaN crystals and Dr. J. Washburn and Dr. E. R. Weber for the discussion of point and extended defects in bulk GaN.

REFERENCES

Baranowski, J. M., Liliental-Weber, Z., Korona, K., Pakuła, K., Stepniewski, R., Wysmolek, A., Grzegory, I., Novak, G., Porowski, S., Monemar, B., and Bergman, P. (1997). *Math. Ass. Soc. Symp. Proc.* (1997). **449**, 393.

Baranowski, J. and Porowski, S. (1996). *Proc. 23rd Int. Conf. on the Physics of Semiconductors*, Berlin, 1996, M. Scheffler, R. Zimmerman, eds., Singapore: World Scientific, p. 497.

Buxton, B. F., Eades, J. A., Steeds, J. W., and Rackham, G. M. (1976). *Phil. Trans. R. Soc. London*, **281**, 171.
Frank, F. C. (1951). *Acta Cryst.*, **4**, 497.
Heindl, J. and Strunk, H. P. (1996). *Phys. Stat. Sol. (B)*, **193**, K1.
Hobgood, H. M., Barret, D. L., McHugh, J. P., Clarke, R. C., Sriram, S., Burk, A. A., Greggi, J., Brandt, C. D., Hopkins, R. H., and Choyke, W. J. (1994). *J. Cryst. Growth*, **137**, 181.
Kahn, M. A., Bhattarai, A., Kuznia, J. N., and Olsen, D. T. (1993). *Appl. Phys. Lett.*, **63**, 1214.
Kisielowski, C., Liliental-Weber, Z., and Weber, E. R. (1996). *Brazilian Journal of Physics*, **26**, 83.
Lester, S. D., Ponce, F. A., Craford, M. G., and Steigerwald, D. A. (1995). *Appl. Phys. Lett.*, **66**, 1249.
Leszczyński, M., Suski, T., Perlin, P., Grzegory, I., Bočkowski, M., Jun, J., Porowski, S., and Major, J. (1995). *J. Phys. D. Appl. Phys.*, **28**, A 149.
Leszczyński, M., Teisseyre, H., Suski, T., Grzegory, I., Bočkowski, M., Jun, J., Porowski, S., Pakuła, K., Baranowski, J. M., Foxon, C. T., and Cheng, T. S. (1996). *Appl. Phys. Lett.*, **69**, 73.
Leszczyński, M. (1997). Private communication.
Liliental-Weber, Z., Sohn, H., Newman, N., and Washburn, J. (1995). *J. Vac. Sci. Technol.*, **B13**, 1578.
Liliental-Weber, Z., Ruvimov, S., Suski, T., Ager, III, J. W., Swider, W., Washburn, J., Amano, H., Akasaki, I., Imler, W. (1996b). *Mat. Res. Soc. Symp. Proc.* **423**, 487.
Liliental-Weber, Z. and Kisielowski, C. (1995). *Proc. Microscopy and Microanalysis*. Bailey, G. W., Ellisman, M. H., Henningar, R. A., and Zaluzec, N. J., eds., New York: Jones and Begell Publ., p. 148.
Liliental-Weber, Z. (1995). 24th International School of Physics of Semiconducting Compounds, Jaszowiec, Poland, 27 May–2 June, 1995. Invited talk.
Liliental-Weber, Z. and Parachenian-Allen, L. (1986). *Appl. Phys. Lett.*, **49**, 1190.
Liliental-Weber, Z., Kisielowski, C., Liu, X., Schloss, L., Washburn, J., Weber, E. R., Grzegory, I., Bockowski, M., Jun, J., Suski, T., and Porowski, S. (1997b). *Proc. of Topical Workshop on III-V Nitrides (TWN95)*. I. Akasaki and K. Onabe, eds. Amsterdam: Elsevier Science, p. 167.
Liliental-Weber, Z., Kisielowski, C., Ruvimov, S., Chen, Y., Washburn, J., Grzegory, I., Bockowski, M., Jun, J., and Porowski, S. (1996a). *J. Electr. Mat.* **25**, 9, 1545–1550.
Liliental-Weber, Z., Chen, Y., Ruvimov, S., and Washburn, J. (1997a). *Proc. Mat. Res. Soc. Boston*, **449**, 417.
Liliental-Weber, Z., Chen, Y., Ruvimov, S., and Washburn, J. (1997c). *Phys. Rev. Lett.* **79**, 2835.
Lin, M. E., Xue, G., Zhou, G. L., Green, J. E., and Morkoc, H. (1993). *Appl. Phys. Lett.*, **63**, 932.
Moustakas, T. D. and Molnar, R. J. (1993). *Mat. Res. Soc. Conf. Proc.* **281**, 753.
Nakamura, S., Mukai, T., and Senoh, M. (1994). *J. Appl. Phys.*, **76**, 8189.
Nakamura, S., Senoh, M., Iwasa, N., and Nagahama, S. (1995). *Jpn. J. Appl. Phys.*, **4**, L797.
Nakamura, S., Senoh, M., Nagahama, S., Iwasa, N., Yamada, T., Matsushita, T., Kiyoku, H., and Sugimoto, Y. (1996). *Jpn. J. Appl. Phys.*, **35**, L217.
Newman, N., Ross, J., and Rubin, M. (1993). *Appl. Phys. Lett.*, **62**, 1242.
Perlin, P., Polian, A., Chervin, J. E., Knapp, W., Suski, T., Grzegory, I., Porowski, S., and Erickson, J. W. (1997). *Proc. Mat. Res. Soc. Boston*, **449**, 519.
Ponce, F. A., Cherns, D., Yong, W. T., and Steeds, J. W. (1996). *Appl. Phys. Lett.*, **69**, 770.
Ponce, F. A., Bour, D. P., Young, W. T., Saunders, M., and Steeds, J. W. (1996). *Appl. Phys. Lett.*, **69**, 337.
Ponce, F. A., Bour, D. P., Götz, W., and Wright, P. J. (1996). *Appl. Phys. Lett.*, **68**, 57.
Porowski, S., Grzegory, I., and Jun, J. (1989). In *High Pressure Chemical Synthesis*. J. Jurczak and B. Baranowski, eds., Amsterdam: Elsevier Science Publishers, B.V. p. 21.

Porowski, S. and Grzegory, I. (1994). In *Properties of Group III Nitrides*. James H. Edgar, ed., EMIS Data-reviews Series No. 11, p. 76.

Porowski, S., Bockowski, M., Lucznik, B., Wroblewski, M., Kurkowski, S., Grzegory, I., Leszczyński, M., Novak, G., Pakula, K., and Baranowski, J. (1997). *Mat. Res. Soc. Symp. Proc.*, **449**, 35.

Powell, J. A., Neudeck, P. G., Larkin, D. J., Yang, J. W., and Pirouz, P. (1994). *Inst. Phys. Conf. Ser.* **137**, 161.

Qian, W., Skowronski, M., and Rohrer, G. S. (1996). *Mat. Res. Soc. Symp. Proc.*, **423**, 475.

Qian, W., Rohrer, G. S., Skowronski, M., Doverspike, K., Rowland, L. B., and Gaskill, D. K. (1995). *Appl. Phys. Lett.*, **67**, 2284.

Ruvimov, S., Liliental-Weber, Z., Suski, T., Ager, J. W., Washburn, J., Krueger, J., Kisielowski, C., Weber, E. R., Amano, H., and Akasaki, I. (1996). *Appl. Phys. Lett.*, **69**, 1454.

Spence, J. C. H. and Zuo, J. M. (1992). In *Electron Microdiffraction*, New York and London: Plenum Press, p. 131.

Steeds, J. W. and Morniroli, J. P. (1992). In *Minerals and Reactions at the Atomic Scale: Transmission Electron Microscopy*, Vol. 27, P. R. Buseck, ed., Chelsea, MI: Book Crafters, pp. 37–83.

Tafto, J. and Spence, J. C. H. (1982). *J. Appl. Crystallogr.*, **15**, 60.

Takasu, S. and Shimanuki, S. (1974). *J. Cryst. Growth*, **24/25**, 641.

Wang, C. and Davis, R. F., (1993). *Appl. Phys. Lett.*, **63**, 990.

CHAPTER 4

Hydrogen in III-V Nitrides

Chris G. Van de Walle and Noble M. Johnson

XEROX PALO ALTO RESEARCH CENTER
3333 COYOTE HILL ROAD
PALO ALTO, CALIFORNIA

I.	INTRODUCTION	157
	1. *General Features of Hydrogen in Semiconductors*	158
	2. *Presence of Hydrogen During Nitride Growth and Processing*	160
II.	THEORETICAL FRAMEWORK	161
	1. *Theoretical Approaches Used for Hydrogen in Nitrides*	161
	2. *Isolated Interstitial Hydrogen in GaN*	164
	3. *Hydrogen Molecules*	168
	4. *Interaction of Hydrogen with Shallow Impurities*	168
III.	EXPERIMENTAL OBSERVATIONS	175
	1. *Intentional Hydrogenation and Diffusion in GaN*	175
	2. *Effects of Hydrogen During Growth/Hydrogen on GaN Surfaces*	176
	3. *Passivation and Activation of Mg Acceptors*	177
	4. *Passivation of Other Acceptor Dopants*	179
	5. *Local Vibrational Modes of the Mg—H Complex in GaN*	180
IV.	CONCLUSIONS AND OUTLOOK	181
	REFERENCES	183

I. Introduction

While the technological potential of nitride semiconductors has been appreciated for several decades, current rapid progress in the field resulted from breakthroughs that occurred at the end of the 1980s. One such breakthrough consisted of the achievement of *p*-type doping in GaN. It was found that mere incorporation of acceptor impurities during metal organic chemical vapor deposition (MOCVD) growth was not sufficient; a post-growth processing step, either involving low-energy electron beam irradiation (LEEBI) (Amano et al., 1989) or thermal annealing (Nakamura et al., 1992) was required. It is now understood that hydrogen plays an important role in this process, providing strong motivation for understanding the behavior of hydrogen in the group III-V nitrides. Based on current knowledge about the behavior of hydrogen in other semiconductors, we expect hydrogen to be involved in other important processes as well.

The rest of this Introduction will be devoted to a brief review of the highlights of this knowledge; references to the literature can be found in review volumes by Pankove and Johnson (1991) and by Pearton *et al.* (1992), and in a review paper by Estreicher (1995).

In Section II we will describe what is known from theory and computations about hydrogen in GaN. While many features are similar to the behavior of hydrogen in other semiconductors, a number of important differences occur; the physical reasons for these differences will be discussed. At this stage, experimental knowledge about hydrogen in the nitrides lags the theoretical understanding: the number of physical parameters that have been determined is much lower than for Si or GaAs. A summary of the experimental results obtained to date is given in Section III. Section IV concludes the chapter, with an emphasis on future directions.

1. GENERAL FEATURES OF HYDROGEN IN SEMICONDUCTORS

a. Charge States of Isolated Interstitial Hydrogen

Isolated interstitial hydrogen can assume different charge states in the semiconductor. The impurity introduces a level in the bandgap, and the charge state depends on the occupation of that level. The charge state determines the most favorable location of hydrogen in the semiconductor lattice.

In the positive charge state (H^+, essentially a proton) hydrogen seeks out regions of high electronic charge density. In most covalent semiconductors, the maximum charge density is found at the bond-center (BC) site, and in Si and GaAs H^+ is indeed known to be located at this site. Hydrogen in GaN will turn out to behave differently. This charge state is most favorable in *p*-type material.

In the negative charge state (H^-) hydrogen prefers regions of low electronic charge density. In semiconductors with zinc-blende structure, H^- can be found at the tetrahedral interstitial site; H^- is the preferred charge state in *n*-type material.

The lattice location of neutral hydrogen (H^0), finally, is quite sensitive to the details of the charge density distribution. In all cases that have been theoretically investigated, however, the energy of H^0 is always higher than that of H^+ or H^-, that is, the neutral charge state is never the most stable state, under equilibrium conditions. This is the defining characteristic of a "negative-U" center. We will see that H in GaN is a negative-U center with a very large magnitude of U.

b. *Diffusion*

Hydrogen, being a small and light impurity, is expected to move quite easily through the perfect lattice. This is indeed true for the positive and neutral charge states. The migration barriers for the negative charge state are usually somewhat higher.

c. *Compensation and Passivation of Shallow Impurities*

Based upon the characteristics of isolated interstitial hydrogen, one can immediately predict that hydrogen will interact with shallow dopants. Hydrogen always counteracts the electrical activity of the dopants. The presence of hydrogen in *p*-type material leads to formation of H^+, a donor, which compensates acceptors. When H^+ forms a *complex* with the acceptor impurity (to which it is coulombically attracted), we say that the acceptor is *passivated* (*or neutralized*). *Passivation* is characterized by the presence of neutral complexes (which cause little scattering of carriers), whereas *compensation* involves donors and acceptors that are spatially separated (and hence contribute separately to ionized impurity scattering); the difference can be observed in carrier mobilities.

The *binding energy* of hydrogen to the shallow-level is defined as the energy difference between the initial state (a complex) and the final state (hydrogen and the shallow dopant far apart). This binding energy is usually of order 1 eV. The *dissociation energy* is the activation energy required to break up the complex; it is somewhat higher than the binding energy, due to the presence of a barrier in the potential energy surface. Dissociation of the complex usually requires annealing at a temperature of a few hundred degrees Centigrade. Studies of the dissociation kinetics are most reliably performed with the complex situated in an electric field, such as that present in the space charge region of a reverse biased diode; the electric field removes the dissociated hydrogen from the vicinity of the dopant and prevents retrapping. In the absence of an electric field, higher annealing temperatures are required to activate the dopants due to retrapping. It has also been found that minority carriers can enhance the dissociation process.

Just as H^+ passivates acceptors, we expect H^- to passivate donors. The passivation efficiency and activation kinetics depend on the relative stability of the various hydrogen charge states.

d. Interactions with Deep Levels

Hydrogen can of course also interact with impurities or defects that form deep levels in the bandgap. A prime example is the interaction of hydrogen with dangling bonds in amorphous silicon, a process that significantly improves the electronic quality of the material. Hydrogen interacts with native defects, the passivation of vacancies being most extensively studied. Interactions with impurities that introduce deep levels have been observed but have not been studied in great detail. Finally, one expects hydrogen to interact with extended defects as well.

e. Hydrogen Molecules and Molecular Complexes

Theory predicts that H_2 molecules readily form in many semiconductors; the binding energy is somewhat smaller than its value in free space, but still large enough to make interstitial H_2 one of the more favorable configurations hydrogen can assume in the lattice. Experimental observation of these molecules has proven challenging, due mainly to the lack of a dipole moment in the molecule, rendering it difficult to observe with IR absorption spectroscopy. The H_2 molecules seem to form often after H has been released from shallow dopants. We will see in Section III that H_2 molecules are not nearly as favorable in GaN as they are in the other semiconductors.

Another configuration involving two hydrogen atoms is the so-called H_2^* complex, in which a host-atom bond is broken and one hydrogen atom is inserted between the two atoms, in a BC position, while the other hydrogen occupies an antibonding (AB) site. This configuration is somewhat higher in energy than the H_2 molecule, at least in diamond, silicon, and GaAs, but it may play an important role in diffusion as well as in nucleation of larger hydrogen-induced defects.

2. PRESENCE OF HYDROGEN DURING NITRIDE GROWTH AND PROCESSING

Hydrogen can be incorporated in the semiconductor either intentionally or unintentionally. Techniques for intentional hydrogenation are reviewed by Pankove and Johnson (1991). The preferred technique of introducing hydrogen in a controlled fashion is hydrogenation in a remote plasma. Shielding the sample from the plasma prevents introduction of plasma-induced defects. Atomic hydrogen, generated in the plasma, can be introduced in the semiconductor at moderate temperatures, unlike molecular hydrogen, which penetrates only at high temperatures. Isotope studies can be performed by using deuterium (^2H); deuterium also lends itself to detection by secondary ion mass spectroscopy (SIMS), where the presence of hydrogen in the background limits the sensitivity of SIMS to ^{1}H.

Hydrogen can also be introduced using proton implantation. Implantation always carries the risk of creating damage in the solid; such damage is very difficult to anneal out in a hard material such as GaN. The effects of hydrogen introduced by implantation are therefore often hard to separate from those caused by the implantation damage. Sometimes the implantation damage itself is actually the sought-after effect of proton implantation, for instance for device isolation (Binari *et al.*, 1995).

Unintentional incorporation of hydrogen can occur during a host of processes. First and foremost, MOCVD growth is bound to introduce large concentrations of hydrogen, both from the source gases and from the use of H_2 as a carrier gas. This also applies to hydride vapor phase epitaxy (HVPE). Growth by molecular beam epitaxy (MBE) is often considered to be free of hydrogen; in reality, however, measurable concentrations of hydrogen have been found in MBE-grown samples. Hydrogen is usually the prevailing contaminant in MBE chambers; it can be introduced through water vapor, from sources, and so on. Recently, Yu *et al.* (1996) have intentionally introduced hydrogen during MBE growth of GaN, finding that hydrogen favorably affects growth under Ga-rich conditions. These findings will be discussed in Section III. Finally, post-growth processing of samples can also introduce hydrogen; among the examples are annealing in forming gas and chemomechanical polishing.

II. Theoretical Framework

1. THEORETICAL APPROACHES USED FOR HYDROGEN IN NITRIDES

Most of the theoretical studies carried out for hydrogen in GaN to date are based on density-functional theory in the local-density approximation, using pseudopotentials, a plane-wave basis set, and a supercell geometry (Neugebauer and Van de Walle, 1995a; Bosin *et al.*, 1996; Okamota *et al.*, 1996). Only one study has been based on a different approach, namely, Hartree-Fock based cluster calculations (Estreicher and Maric, 1996). A thorough discussion of these methods has been given by Estreicher (1994, 1995). Here we provide only a brief overview of the techniques.

a. Pseudopotential-Density-Functional Calculations

This computational approach is now regarded as a standard for performing first-principles studies of defects in semiconductors. Density functional theory (DFT) in the local density approximation (LDA) (Hohenberg and

Kohn, 1964; Kohn and Sham, 1965) allows a description of the ground state of the many-body system in terms of a one-electron equation with an effective potential. The total potential consists of an ionic potential due to the atomic cores, a Hartree potential, and a so-called exchange and correlation potential that describes the many-body aspects. This approach has proven very successful for a wide variety of solid-state problems. One shortcoming of the technique is its failure to produce reliable excited-state properties, widely referred to as the "bandgap problem." Many useful results of the calculations depend on ground-state properties and are thus not affected by this shortcoming. In cases where the bandgap enters the calculations either directly or indirectly, judicious inspection of the results still allows extraction of reliable information. Situations where the bandgap error could potentially affect the results will be discussed where appropriate (see Neugebauer and Van de Walle, 1995b).

Most properties of molecules and solids are determined by the valence electrons; the core electrons can usually be removed from the problem by representing the ionic core (i.e., nucleus plus inner shells of electrons) with a pseudopotential. State-of-the-art calculations employ nonlocal norm-conserving pseudopotentials (Hamann et al., 1979) that are generated solely based on atomic calculations and do not include any fitting to experiment.

The case of the nitride semiconductors poses two challenges for the use of pseudopotentials. First, nitrogen is a first-row element characterized by a fairly deep potential, which requires a rather large cutoff in the plane-wave basis set that is commonly used to expand wave functions and potentials in a variational approach. Second, the $3d$ states of Ga (and $4d$ states of In) cannot strictly be considered as core states. Indeed, the energetic position of these states is fairly close to the N $2s$ states, and the interaction between these levels may not adequately be described if the d states are treated as part of the core. Including the d electrons as valence states significantly increases the computational burden, because their wave functions are much more localized than those of the s and p states. Sometimes explicit inclusion of the d states as valence states is the only way to obtain accurate results (Neugebauer and Van de Walle, 1994a). In many cases, however, the effect of the d states can be adequately included by a correction to the exchange and correlation potential, commonly referred to as the "nonlinear core correction" (Louie et al., 1982). Specifically, for the case of the hydrogen impurity in GaN, tests have shown that use of the nonlinear core correction provides a reliable description of the system (Neugebauer and Van de Walle, 1995b).

The last ingredient commonly used in pseudopotential-density-functional calculations for defects or impurities is the supercell geometry. Ideally, one would like to describe a single isolated impurity in an infinite crystal. In the supercell approach, the impurity is surrounded by a finite number of semiconductor atoms, and this structure is periodically repeated. Maintaining periodicity allows continued use of algorithms such as fast Fourier transforms (FFT). One can also be assured that the band structure of the host crystal is well described (which may not be the case in a cluster approach). For sufficiently large supercells, the properties of a single, isolated impurity can be derived. Convergence tests have indicated that the energetics of impurities and defects are usually well described by using 32-atom supercells. Specific cases, such as description of highly charged states or detailed investigations of wave functions of shallow impurities, may require larger supercells, but at present that is still computationally challenging.

Error bars on the values derived from first-principles calculations depend on the situation. When comparing energies of an impurity in different positions in the lattice, the accuracy is quite high and the error bar is smaller than 0.1 eV. When looking at formation energies, or comparing energies of different charge states, the limitations of density-functional theory may play a role, and the error bar could increase to a few tenths of an electron volt. These error bars are still small enough not to affect any of the qualitative conclusions discussed here.

b. *Hartree-Fock Based Cluster Calculations*

Hartree-Fock based models are usually based on quantum-chemistry approaches that have been successfully applied to atoms and molecules. The main problem with the technique is the computational demands: *ab initio* Hartree-Fock methods can only be applied to systems with small numbers of atoms, because they require evaluation of a large number of multicenter integrals. Various semiempirical approaches have been developed that either neglect or approximate some of these integrals. The accuracy and reliability of these methods are hard to assess.

In the cluster approach, the crystalline environment is simulated by a cluster of host atoms, typically terminated by hydrogen atoms. The impurity is then embedded in such a cluster. Size–convergence tests should ensure that the cluster is large enough to suppress any spurious interactions with the surface of the cluster, and to provide a reasonable description of the host band structure. The latter is usually challenging, because for cluster sizes that are computationally tractable the confinement of the electronic states will lead to significant modifications of the bulk band structure.

2. Isolated Interstitial Hydrogen in GaN

a. Configurations in the Lattice

Comprehensive calculations of the energetics, atomic geometry, and electronic structure for isolated hydrogen in a variety of different interstitial positions in GaN were carried out by Neugebauer and Van de Walle (1995a, 1995b) and by Bosin et al. (1996). Both groups used the pseudopotential-density functional approach, and their conclusions are very similar. Some quantitative differences will be discussed where appropriate. Estreicher and Maric (1996) reported preliminary Hartree-Fock based calculations on clusters of zinc-blende–GaN containing 44 host atoms. Their results will also be included in what follows.

Neugebauer and Van de Walle actually mapped out complete total-energy surfaces by placing the hydrogen atom in a variety of positions in the crystal and allowing the host atoms to relax. The resulting energies as a function of hydrogen position form an adiabatic total energy surface, and include information abut diffusion barriers. The calculations of Neugebauer and Van de Walle (1995a, 1995b) were carried out for GaN in the zinc-blende structure; those of Bosin et al. (1996) for the wurtzite structure, but only exploring sites with C_{3v} symmetry. As the zinc-blende and wurtzite structures only differ beyond third-nearest neighbors, the differences with respect to atomic and electronic structure of the hydrogen impurity are expected to be small.

Hydrogen in the positive charge state: H^+. In the positive charge state, hydrogen as usual seeks out regions of high electronic charge density; in the GaN crystal, this means positions close to the nitrogen atom. Indeed, the first-principles calculations show that essentially all positions in which the H–N bond length is about 1.02 Å are approximately equal in energy, this bond length being representative of H–N bonds in molecules. This situation is different from more covalent semiconductors such as Si and GaAs; there, the covalent nature of the bonding leads to a significant buildup of charge in the bonds between the atoms. In a more ionic crystal such as GaN this concentration of charge in the bonds is less pronounced, and the charge density is distributed more spherically around the anion. Still, the charge density peaks in the bonds between the atoms, and H^+ would probably still prefer this site, were it not for another effect that counterbalances the energy gained by immersing the proton in a region of high electronic density. This effect is the lattice relaxation that is required to accommodate the H^+ at the BC site. Indeed, the Ga–N bond length of 1.95 Å is too short to allow for insertion of a H atom; the Ga–N bond has to stretch by more than 40% of the bond length. This relaxation obviously costs energy, and it is particularly costly in a hard material such as GaN. In spite of this high cost, the BC site is still only a few 0.1 eV higher in energy than the AB_N site (antibonding site behind nitrogen atom), where virtually no relaxation is necessary, and which provides the most stable site for H^+. Estreicher and Maric (1996), in their Hartree-Fock based calculations, actually found H^+ to reside at the BC site.

The total energy surface for H^+ (Neugebauer and Van de Walle, 1995a) shows that H^+ can move between equivalent sites in the lattice with a barrier of only 0.7 eV. This rather low value is indicative of a high diffusivity. Indeed, experiments on hydrogen diffusion (to be discussed in Section III) have confirmed this prediction of high mobility for H^+.

Hydrogen in the neutral charge state: H^0. The energy differences between different sites are much smaller for H^0 than they are for H^+, that is, the total energy surface is rather flat (except for regions within 1 Å of the nuclei, where the energy always increases sharply). This behavior is similar to the trend in Si (Van de Walle *et al.*, 1989). Neugebauer and Van de Walle (1995a) find the most stable site to be the antibonding site behind a Ga atom (AB_{Ga}). As the Ga–H bond length is only about 0.1 Å smaller than the Ga–N bond length, the AB_{Ga} site nearly coincides with the position of the tetrahedral interstitial site (T_d^{Ga}) in the zinc-blende structure. Bosin *et al.* find the BC site to have the lowest energy; a possible explanation for the difference will be given in what follows. Estreicher and Maric (1996) reported that for H^0 the BC site is nearly degenerate with the T_d^{Ga} site.

Hydrogen in the negative charge state: H^-. All calculations find that the most stable site for H^- is at the AB_{Ga} site (very close to T_d^{Ga}, as already discussed). The charge density of the GaN crystal has a global minimum at this site, rendering it favorable for incorporation of a negatively charged impurity. Other sites in the crystal are much higher in energy. Consequently, the energy barrier for H^- to move between equivalent sites is quite high; the total energy surface (Neugebauer and Van de Walle, 1995a) yields a migration barrier of 3.4 eV, a value high enough to render H^- essentially immobile.

b. *Relative Stability of Different Charge States and "Negative-U" Character*

The first-principles calculations also allow a comparison of the energy of hydrogen in different charge states. These energy differences necessarily depend on the position of the Fermi level; for instance, in order to convert

H^0 into H^+, an electron has to be removed from H^0, and the energy of this electron has to be taken into account in the comparison of energies. In a thermodynamic context the electrons are placed in the reservoir of electrons, with an energy corresponding to the Fermi energy (E_F). If the Fermi level is high in the gap, it will cost more energy to place an electron there, and consequently the energy difference between H^+ and H^0 will increase. The energy of the neutral species is, of course, independent of the Fermi energy. The *formation energy* of interstitial H^0 is defined as the energy difference between a free H atom (in vacuum) and H^0 at its stable site inside GaN. In thermodynamic equilibrium, the formation energy determines the concentration c of an impurity at temperature T through the following expression:

$$c = N_{\text{sites}} \exp[-E^f/k_B T] \quad (1)$$

where N_{sites} is the number of sites on which the defect can be incorporated, k_B the Boltzmann constant, and E^f the formation energy. The formation energies of interstitial hydrogen in various charge states in GaN are depicted in Fig. 1. Bosin et al. (1996) produced a similar figure, but with some

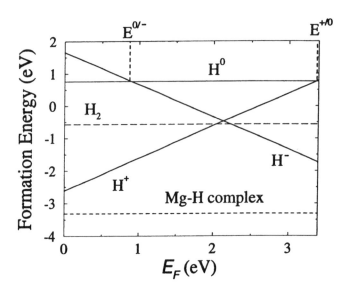

FIG. 1. Formation energies as a function of Fermi level for H^+, H^0, and H^- (solid lines), for a H_2 molecule (dashed line), and for a Mg-H complex (short-dashed line) in GaN. $E_F = 0$ corresponds to the top of the valence band. The formation energy is referenced to the energy of a free H atom. (Reprinted with permission from Neugebauer and Van de Walle, 1995a.)

quantitative differences. Most of these differences can be traced back to the difference in bandgaps that occur in the computational approaches. The small value of the gap in the calculations by Bosin et al. probably causes them to occupy states (for the case of H^0 and H^-) that are not H-induced impurity levels, but conduction-band states. This incorrect level occupation then affects the formation energies, and may also be responsible for the different conclusions regarding the stable position of the H^0 species. We now discuss several conclusions that can be drawn from Fig. 1.

Incorporation in the lattice and solubility. Figure 1 shows that the formation energy of H^+ can be significantly lower than that of H^-. According to Eq. (1), lower formation energies translate into higher concentrations that can be incorporated in the solid. We therefore see that conditions favoring formation of H^+ will lead to higher levels of hydrogen incorporation. Figure 1 shows that H^+ has a lower formation energy in *p*-type material (low values of E_F), while H^- is lower in energy in *n*-type GaN (high values of E_F). We therefore conclude that hydrogen is more likely to incorporate in *p*-type GaN than in *n*-type material. This prediction has been confirmed in a number of experiments, to be discussed in Section III. Values of E_F below ~ 2.1 eV favor incorporation of H^+; for E_F higher in the gap, H^- is more stable. Note that H^0 is not the lowest-energy species for *any* position of the Fermi level; this is characteristic of a "negative-U" center, to be discussed in the next section.

Transition levels and negative-U. Our discussion so far indicates that hydrogen is an amphoteric impurity, that is, it behaves as both a donor and an acceptor. The position of E_F for which the formation energies of H^+ and H^0 are equal is, by definition, the donor level of an impurity. Similarly, the E_F value for which the formation energies of H^0 and H^- coincide determines the acceptor level. These acceptor and donor levels can be determined from Fig. 1, the most obvious conclusion being that the donor level is located *above* the acceptor level, by a value of 2.4 eV. This is a rather unusual situation for impurities in semiconductors; more commonly, Coulomb repulsion (expressed by the parameter U) makes it more difficult to add additional electrons to an impurity and, therefore, the acceptor level (the $0/-$ transition) is expected to lie above the donor level ($+/0$) ("positive-U"). The energy difference between the acceptor and the donor level is negative here, characteristic of a "negative-U" center. This behavior can occur because hydrogen assumes different configurations in the lattice as the charge state changes.

The value of U predicted here, -2.4 eV, is larger in magnitude than any value predicted or measured for any impurity in any semiconductor! While smaller in magnitude, negative-U behavior is also observed for hydrogen in other semiconductors: $U = -0.4$ eV has been calculated (Van de Walle *et al.*, 1989) and measured (Johnson *et al.*, 1994) for Si, and in GaAs, a value of -0.7 eV has been calculated (Pavesi and Giannozzi, 1992). The tendency of hydrogen to form negative-U centers, and the large magnitude of U in GaN can be understood in the context of a simple model for the interaction between H and the semiconductor (Neugebauer and Van de Walle, 1995a).

3. Hydrogen Molecules

Neugebauer and Van de Walle (1995a) have studied the incorporation of H_2 molecules in the GaN lattice. Various interstitial positions were investigated, all yielding about the same formation energy, which is shown in Fig. 1. It is clear that H_2 is unstable with respect to dissociation into monatomic hydrogen. This is a distinct property of GaN, and very different from the case of Si and GaAs. The low stability of H_2 can be explained by the small lattice constant of GaN, which leaves little room for incorporation of a molecule in interstitial positions. Neugebauer and Van de Walle (unpublished) also studied H_2^* complexes (see Section I.1.e), and found them to be higher in energy than H_2 molecules.

4. Interaction of Hydrogen with Shallow Impurities

Knowledge of the behavior of isolated interstitial hydrogen already provides major clues to the interaction of hydrogen with shallow acceptors. Incorporation of acceptor impurities renders the material *p*-type, and we have seen that hydrogen prefers the positive charge state (H^+) in *p*-type GaN. Hydrogen therefore acts as a donor, counteracting the intended doping of the material. When hydrogen is spatially separated from the acceptor, *compensation* occurs: an electron is transferred from the hydrogen to the acceptor, eliminating a hole that could otherwise have been contributed by the acceptor. Because H^+ is coulombically attracted to negatively charged acceptors, complex formation is possible; this situation is referred to as *passivation*. The difference with compensation can be observed, for example, in carrier mobilities: passivation leads to neutral complexes that cause less scattering than isolated ionized donor and acceptor impurities. Another signature of complex formation is the occurrence of vibrational modes characteristic of the complex. Whether or not the hydrogen forms a complex with the acceptor depends on the binding energy of the complex, and on the temperature.

a. The Mg-H Complex

Magnesium is the most commonly used acceptor in GaN; investigations of hydrogen interactions with acceptors have therefore focused on Mg. Bosin et al. (1996) also studied the interaction of hydrogen with other acceptor impurities, in particular, Be_{Ga}, C_N, Ca_{Ga}, and Zn_{Ga}. However, the authors pointed out that their results were preliminary due to the limited size of the supercell (16 atoms) and an incomplete investigation of all possible sites.

Several groups (Neugebauer and Van de Walle, 1995a; Bosin et al., 1996; Okamoto et al., 1996) have performed detailed calculations of the interaction between hydrogen and a Mg acceptor. As in the case of isolated H^+, the sites around the N atom are energetically preferred, with the AB_N site apparently winning out over the BC site. Okamoto et al. find the BC site to be slightly more favorable than the AB site, but only by 0.1 eV. This small quantitative difference may be due to their use of a smaller wurtzite supercell (16 atoms, as opposed to 32 atoms used by the other groups). The binding energy of the Mg-H complex is 0.7 eV (see also Fig. 1) (Neugebauer and Van de Walle 1995a, 1996a). The binding in the Mg-H complex is clearly dominated by hydrogen interacting with a nitrogen atom, and this is reflected in the vibrational frequency of the stretch mode, which is representative of N—H bonds (such as in NH_3). A comparison of results from the various calculations is presented in Table I. While the qualitative

TABLE I

RESULTS OF COMPUTATIONAL STUDIES OF Mg-H COMPLEXES IN GaN BY VARIOUS GROUPS

Reference	H Site	Structure	Symmetry	ΔE (eV)	Frequency (cm^{-1})
Neugebauer and	AB	zb	C_{3v}	0.0	—
Van de Walle (1995a)	BC	zb	C_{3v}	0.3	3360
Bosin et al. (1996)	AB	w	C_s	0.00	2939
	AB	w	C_{3v}	1.13	3069
	BC	w	C_s	0.62	3917
	BC	w	C_{3v}	0.32	3611
Okamoto et al. (1996)	BC	w	C_s	0.0	3450
	BC	w	C_{3v}	0.0	3490
	AB	w	?	0.1	—

The H site is either AB, the antibonding site next to a N neighbor of the Mg atom, or BC, the bond center site; "w" stands for wurtzite, "zb" for zinc blende; for the wurtzite structure the complex can be either oriented along the c-axis (C_{3v} symmetry), or in the other bonds (C_s symmetry); ΔE is the energy difference with the lowest-energy configuration within each calculation; the frequency of the vibrational stretch mode is given.

conclusions are very similar, some quantitative differences occur. The value calculated by Neugebauer and Van de Walle (1995a) is 3360 cm^{-1}, which is somewhat larger than the experimental result (3125 cm^{-1}, see Section III.5). It should be kept in mind that the calculated value does not include anharmonic effects, which are sizable in the case of N—H vibrations; in NH$_3$ the anharmonicity lowers the frequency by 170 cm^{-1} (Johnson et al., 1993). Experimental results for vibrations of the Mg-H complex will be presented in Section III.

The interaction of hydrogen with Mg acceptors nicely explains the necessity of an activation procedure when Mg-doped GaN is grown by MOCVD. The MOCVD process allows ready incorporation of hydrogen in the p-type layer. During cooldown this hydrogen forms complexes with the Mg acceptors, eliminating the electrical activity. In order to activate the Mg, the complexes have to be dissociated and hydrogen neutralized or removed from the p-type layer.

Neugebauer and Van de Walle (1995a) estimated this dissociation barrier of the Mg—H complex to be 1.5 eV or slightly higher. This value should allow dissociation of the complex at temperatures of a few hundred degrees Centigrade. Experimentally, however, temperatures exceeding 600 °C have been found necessary to activate Mg-doped MOCVD-growth GaN. This indicates that the activation process does not merely consist of dissociating Mg-H complexes. When hydrogen leaves the Mg acceptor, it still behaves as a donor, and therefore can still compensate the acceptor as long as it remains in the p-type layer. The hydrogen therefore has to be removed from the p-type layer (e.g., into the substrate or through the surface); alternatively, the hydrogen can be neutralized, for example, by binding to extended defects. The process is schematically illustrated in Fig. 2. Formation of H$_2$ molecules, which is common after acceptor passivation in other semiconductors, is not possible in GaN because of the high formation energy of H$_2$ (see Section II.3).

b. *Role of Hydrogen in Doping of GaN*

n-type GaN. As discussed in Section II.2.b, the solubility of hydrogen in n-type GaN is quite low, and hence hydrogen will have little or no effect on n-type doping. *If* hydrogen were present in n-type GaN, it could bind to Si donors with essentially the same binding energy (0.7 eV) as calculated for Mg-H complexes (Neugebauer and Van de Walle, 1996a); however, due to the low solubility of H in n-type GaN the hydrogen concentration will be very small. This is fortunate, because the high diffusion barrier of H$^-$ in n-type GaN would prevent hydrogen from being removed from the material. This high diffusion barrier is also responsible for the fact that hydrogenation

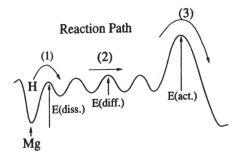

FIG. 2. Schematic illustration of the reaction path and energy barriers for the activation of Mg acceptors in GaN. At low temperatures hydrogen forms a neutral complex with the Mg acceptor. With increasing temperature the Mg-H complex dissociates (1), and the positively charged H can then easily migrate through the crystal (2). Further increase in the temperature allows the hydrogen to overcome an additional activation barrier (3), corresponding to removal from the *p*-type layer or neutralization at extended defects, clusters, and so on. (Reprinted with permission from Neugebauer and Van de Walle, 1996b.)

(exposure to atomic hydrogen) has no discernible effect on *n*-type GaN (Götz et al., 1995).

p-type GaN. In order to appreciate how hydrogen affects the concentrations of acceptor impurities and native defects in the system, we need to go back to the concept of formation energy that was introduced in Section II.2.b, and define formation energies for other species. For all of these species, Eq. (1) will apply, providing the link between formation energy and concentration. The formation energy of a Mg acceptor in GaN depends on various parameters. In addition to the dependence on Fermi level that occurs for a charged species (as discussed in Section II.2.b), the formation energy also depends on the relative abundances of Mg, Ga, and N. In thermodynamic equilibrium these abundances can be described by the chemical potentials μ_{Mg}, μ_{Ga}, and μ_N. A first-principles calculation can provide a value E_{tot} (GaN:Mg$_{Ga}^-$) for the energy of a GaN supercell in which one Ga atom has been replaced by a Mg atom. The expression for the formation energy then takes into account that a Ga atom had to be removed (and placed in the appropriate reservoir, with energy μ_{Ga}), and that a Mg atom has been brought in from a reservoir with energy μ_{Mg}

$$E^f(\text{GaN:Mg}_{Ga}^-) = E_{tot}(\text{GaN:Mg}_{Ga}^-) - E_{tot}(\text{GaN-bulk}) - \mu_{Mg} + \mu_{Ga} - E_F \quad (2)$$

Similar expressions apply to the hydrogen impurity, and to any defects that may be present in the system. Extensive investigations of native defects in

GaN (Neugebauer and Van de Walle, 1994b; Boguslawski et al., 1995) have indicated that self-interstitials and antisites are high-energy defects that are unlikely to occur in GaN; we can therefore focus on vacancies.

The chemical potentials are variables, reflecting the variety of conditions that can occur during growth. To facilitate the discussion, however, we focus on a specific set of conditions, namely, Ga-rich growth ($\mu_{Ga} = \mu_{Ga\text{-bulk}}$), and Mg incorporation limited by the formation of Mg_3N_2, which determines the solubility limit of Mg. The formation energy then becomes solely a function of the Fermi energy, as shown in Fig. 3. Results for other growth conditions can easily be derived by changing the chemical potential values.

Figure 3 summarizes the results of Neugebauer and Van de Walle (1996b) for magnesium and native defects in *p*-type GaN, in the absence of hydrogen. The nitrogen vacancy V_N is the dominant defect under *p*-type conditions [note that V_N has a *high* formation energy in *n*-type GaN, and hence is not the cause of *n*-type conductivity in as-grown GaN (Neugebauer and Van de Walle, 1994b; Van de Walle and Neugebauer, 1997)]. The transition between Mg^0 and Mg^- indicates the position of the Mg acceptor level, close to the experimental value of 0.16 eV (Akasaki et al., 1991). These formation energies can be used to calculate equilibrium concentrations, using Eq. (1) and taking charge neutrality into account (for details of this procedure, see Laks et al., 1991 and Van de Walle et al., 1993).

When no hydrogen is present in the system (see Fig. 3) the nitrogen vacancy is the dominant compensating defect, and due to charge neutrality the Fermi level will be located near the point where the formation energies

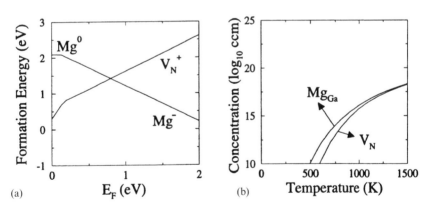

FIG. 3. (a) Formation energies as a function of Fermi level for the Mg acceptor (Mg_{Ga}) and the nitrogen vacancy (V_N). (b) Equilibrium concentrations of these defects and impurities as a function of temperature, in the absence of hydrogen. (Reprinted with permission from Neugebauer and Van de Walle, 1996b.)

of V_N^+ and Mg_{Ga}^- are equal. Significant compensation by nitrogen vacancies would therefore occur; this is a concern, for example, for MBE growth, where little or no hydrogen is present. However, MBE growth is performed at much lower temperatures than MOCVD, and therefore the assumption of thermodynamic equilibrium, which is implicit in Fig. 3(b), is likely not fulfilled. Formation of nitrogen vacancies may therefore be suppressed in MBE.

Hydrogen is highly abundant in many of the high-temperature growth techniques such as MOCVD or HVPE. Therefore, we also consider growth under H-rich conditions (determined by the hydrogen chemical potential being set equal to the energy of free H_2). Figure 4(a) shows that the formation energy of hydrogen is lower than that of the nitrogen vacancy; as a consequence, the formation of nitrogen vacancies is suppressed, and hydrogen becomes the dominant compensating center, as can be seen in Fig. 4(b). We also note that the Mg concentration is increased with respect to the hydrogen-free case. This can be understood by inspection of the formation energies: as the formation energy of hydrogen is lower than that of V_N, the Fermi level equilibration point is moved higher in the gap, leading to a lower formation energy and hence higher concentration of Mg. Incorporation of hydrogen is therefore beneficial in two respects: suppression of native defects and enhancement of the acceptor concentration.

Incorporation of hydrogen of course has the downside that complete compensation of the acceptors occurs. Note that hydrogen is found to be present in the acceptor-doped material *at the growth temperature*; that is,

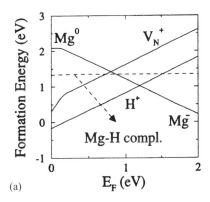
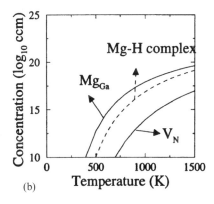

FIG. 4. (a) Formation energies as a function of Fermi level for the Mg acceptor (Mg_{Ga}), the nitrogen vacancy (V_N), and interstitial hydrogen in *p*-type GaN. The formation energy of the Mg-H complex is also shown. (b) Equilibrium concentrations of these defects and impurities as a function of temperature, for the H-rich limit. (Reprinted with permission from Neugebauer and Van de Walle, 1996b.)

hydrogen incorporation occurs during growth, and not merely from the NH_3 ambient during cooldown after growth. This incorporation process may differ from that in some other III-V compounds, where incorporation of hydrogen results mainly from indiffusion during the cooldown process. Hydrogen is thus already present in p-type GaN during growth, although not in the form of Mg—H complexes: at the high temperatures used during growth, these complexes are essentially dissociated. After growth, during the cooldown process, the highly mobile H^+ is coulombically attracted to Mg^- acceptors, and eventually forms Mg—H complexes. These complexes can be dissociated, and hydrogen removed from the p-type layer, during a post-growth activation process as already described.

General criteria for hydrogen to enhance doping. Van Vechten et al. (1992) pointed out that incorporation of hydrogen enables p-type doping by suppressing compensation by native defects. They went on to propose the incorporation (and subsequent removal) of hydrogen as a general method for improving p-type as well as n-type doping of wide-bandgap semiconductors. This model highlights the important role of hydrogen, but leaves various issues unexplained, such as the lack of hydrogen incorporation in n-type GaN, and the success of p-type doping (without post-growth treatments) in MBE. These issues have now been addressed, as has been described herein. More importantly, the incorporation of a compensating donor should not be considered as a universal solution to doping problems in semiconductors. Indeed, the investigation of H in p-type GaN allows us to derive the following general conditions that must be fulfilled for hydrogen (or any other compensating impurity) to improve doping:

(i) Hydrogen must be the dominant compensating defect (i.e., its formation energy must be lower than that of all native defects, and comparable to that of the dopant impurity).
(ii) The activation barriers to dissociate the H-impurity complex and to remove or neutralize H must be lower than the activation energies for native-defect formation, or the diffusion barrier of the impurity.
(iii) The dissociated hydrogen atom must be highly diffusive.

Whether or not these conditions are realized depends on the specific situation (semiconductor, dopant impurity, growth conditions). All of the conditions are met in p-type GaN, judging from the success of post-growth annealing. Failure to meet condition (ii) is the likely cause of the difficulties in obtaining p-type activity in MOCVD-grown nitrogen-doped ZnSe. Indeed, the N—H bond is very strong and would require high temperatures for dissociation; these temperatures exceed those at which the structural quality of the ZnSe crystal can be maintained.

III. Experimental Observations

1. INTENTIONAL HYDROGENATION AND DIFFUSION IN GaN

There is now a general consensus among experimentalists that at least in MOCVD-grown material the ability to diffuse hydrogen into GaN at moderate temperatures depends strongly on the electrical conductivity type. This is illustrated in Fig. 5 with results for specimens of n-type vs p-type GaN that were exposed to monatomic deuterium at specified temperatures for one hour in a remote microwave plasma (Götz et al., 1995). The p-type material was Mg doped with a hole concentration of 8×10^{17} cm^{-3} at room temperature, and the n-type material was Si doped with an electron

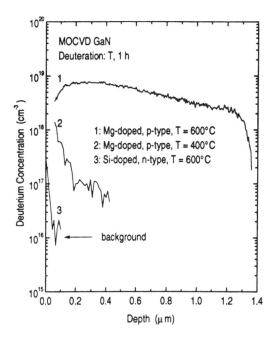

FIG. 5. Deuterium concentration vs depth below the sample surface measured with SIMS. Depth profiles were recorded for p-type, Mg-doped GaN after remote plasma deuteration at 600 °C (Curve 1), after deuteration at 400 °C (Curve 2) and for n-type, Si-doped GaN after deuteration at 600 °C (Curve 3). The deuterium background level of the SIMS measurement is $\sim 10^{16}$ cm^{-3}, as indicated by the arrow. (Reprinted with permission from Götz et al., 1995.)

concentration of 2×10^{17} cm^{-3} at room temperature. Both specimens were ~1-μm thick, and the Mg-doped specimen was thermally activated prior to hydrogenation. The figure shows depth profiles of deuterium that were obtained with SIMS. Exposure to monatomic deuterium at 600 °C leads to a substantial incorporation of deuterium in the *p*-type material and a negligible amount, beyond the immediate surface, in *n*-type GaN. Even a hydrogenation at 400 °C introduces more deuterium (e.g., volume concentration at any given depth or total depth integrated deuterium) into *p*-type material than is introduced by the 600 °C hydrogenation in the *n*-type material. The experimental results in Fig. 5 indicate that hydrogen readily diffuses and incorporates in *p*-type GaN at temperatures ≥ 600 °C but not in *n*-type GaN. These results are fully consistent with the total energy calculations of the properties of monatomic hydrogen in GaN, as summarized in Section II.2.

Hydrogen diffusion in GaN has been the focus of several studies in recent years. Pearton *et al.* (1996c, 1996d) noted that hydrogen is easily incorporated in GaN during many different process steps, including boiling in water, wet chemical etching, dielectric deposition using SiH_4, or dry etching. These authors reported an effective diffusivity of $> 10^{-11}$ cm^2 s^{-1} at 170 °C. Wilson *et al.* (1995) used SIMS to observe outdiffusion of deuterium from plasma-treated or implanted GaN, AlN, and InN. And hydrogen redistribution during anneal was reported by Zavada *et al.* (1994). A few early studies report penetration of hydrogen into *n*-type GaN [e.g., Pearton *et al.* (1996c, 1996d)]. One possible explanation may be that although hydrogen does not incorporate in bulk *n*-type GaN, it can penetrate heavily defected regions by diffusion through grain boundaries or along extended defects. Therefore, *n*-type samples with high densities of extended defects may exhibit effects after hydrogenation.

2. Effects of Hydrogen During Growth/Hydrogen on GaN Surfaces

Sung *et al.* (1996) reported surface studies of *n*-type GaN {000-1}-(1 × 1) surfaces grown by MOCVD, using time-of-flight scattering and recoiling spectrometry, low-energy electron diffraction, and thermal decomposition mass spectrometry. Elastic recoil detection was used to determine the bulk hydrogen concentration. Apart from various conclusions about the surface structure, they found that hydrogen atoms were bound to the outermost layer of N atoms, protruding outward from the surface with a coverage of three-fourths monolayer. Sung *et al.* proposed an explanation for this coverage in terms of autocompensation of the (1 × 1) structure. They also suggested that the presence of hydrogen during GaN growth is an important factor in the growth of material with high crystalline quality: H atoms can maintain sp^3 hybridization of the evolving surface, resulting in less ionic character. A high concentration of hydrogen was also found below the surface, within a depth of 5–10 monolayers from the surface.

Yu et al. (1996) investigated the effect of intentional introduction of atomic hydrogen during MBE growth of GaN. An RF plasma source was used for nitrogen, and atomic hydrogen was produced with a thermal cracker. The authors found that in the absence of hydrogen high-quality growth was obtained at 730 °C under Ga-rich conditions. It was found that the presence of atomic hydrogen increased the growth rate by as much as a factor of two. It was suggested that hydrogen increases the effective surface concentration of nitrogen, but the issue requires further investigation.

Chiang et al. (1995) studied H_2 desorption and NH_3 adsorption on polycrystalline GaN surfaces with time-of-flight detection of recoiled H^+ and D^+ ions. They found that hydrogen is released from Ga sites at temperatures between 250 and 450 °C. Some N—H species decompose at 500 °C, and hydrogen eventually all desorbs at 600 °C. The authors therefore concluded that during CVD growth of GaN at temperatures above 800 °C, the liberation of surface hydrides can be ruled out as the rate-limiting step. They also showed that hydrogen is mobile on the surface at roughly 250–500 °C, and desorbs in a recombinative fashion. Chiang et al. (1995) also observed that H/D exchange can occur rapidly during NH_3 exposure, even at room temperature.

3. PASSIVATION AND ACTIVATION OF Mg ACCEPTORS

The crucial role played by hydrogen in p-type GaN was established by Nakamura et al. (1992). They showed that MOCVD-grown Mg-doped films, which are highly resistive after growth, become p-type conductive after thermal annealing in an N_2 ambient. Earlier, Amano et al. (1989) had demonstrated that the activation can also be achieved by low-energy electron-beam irradiation (LEEBI). Nakamura et al. then annealed their samples in an NH_3 ambient at 600 °C, and found that the resistivity increased significantly. They concluded that atomic hydrogen produced by NH_3 dissociation was responsible for the passivation of acceptors. The correlation between hydrogen and magnesium concentrations was further investigated by Ohba and Hatano (1994).

The precise mechanism by which LEEBI treatment (Amano et al., 1989) activates acceptors is still controversial. The irradiation process can generate electron-hole pairs. These pairs could either directly provide the energy for releasing hydrogen from the acceptor; or the presence of minority carriers could lower the dissociation barrier. Alternatively, the irradiation process may cause local heating that effectively leads to thermal dissociation. Experiments by Li and Coleman (1996) monitored the LEEBI process with cathodoluminescence spectroscopy. The fact that the LEEBI process remained effective down to liquid helium temperature was interpreted as evidence for an athermal mechanism.

The influence of minority-carrier injection on the activation process was investigated by Pearton et al. (1996a). They found that acceptor activation takes place at a temperature as low as 175 °C under minority-carrier injection (accomplished by forward bias of a p-n junction). Conventional annealing under zero-bias conditions does not produce Mg-H dissociation until temperatures are above 450 °C. On the basis of these results the authors suggested that minority-carrier-enhanced dissociation of Mg—H complexes is the mechanism for activation during LEEBI treatment.

While Nakamura et al. (1992) introduced hydrogen through dissociation of NH_3, Götz et al. (1995) used remote plasma hydrogenation, as described in Section III.1. Götz et al. (1996a, 1997) went on to investigate the activation kinetics of Mg acceptors. They annealed Mg-doped samples grown by MOCVD incrementally for 5 min at temperatures ranging from 500 to 850 °C. Figure 6(a) demonstrates the resistivity measured as a function of temperature in the as-grown sample and after each annealing step. The resistivity decreases significantly after the 600 °C annealing step and reaches a value of about 3 Ω-cm at 300 K after annealing at 850 °C. In samples annealed above 700 °C the saturation of the resistivity for measurement temperatures below 200 K is due to hole conduction in an acceptor impurity band. Reliable Hall measurements were possible in samples annealed at temperatures of 600 °C and higher; the resulting hole concentrations are depicted in Fig. 6(b). The data for the anneal at 600 °C reveal acceptor activation over the entire measurement range, whereas the data for the anneals above 700 °C reveal ionization of a shallow acceptor only for temperatures above 200 K, with impurity-band conduction at lower temperatures. Fitting of the Hall-effect data indicates the presence of donor compensation, at a level of about 3×10^{18} cm^{-3}.

These results are consistent with the picture of Mg—H complexes being present in as-grown, Mg-doped GaN. Hydrogen is incorporated during growth, and Mg—H complexes are formed during cooldown. After growth only a fraction of Mg atoms act as acceptors and the material is semi-insulating. The absence of impurity-band conduction after the 600 °C anneal and its appearance after the 700 °C anneal clearly indicate that the activation of Mg acceptors is due to the generation of Mg-related acceptor states in the bandgap of GaN, and, therefore, strongly supports the existence of Mg—H complexes in the as-grown material.

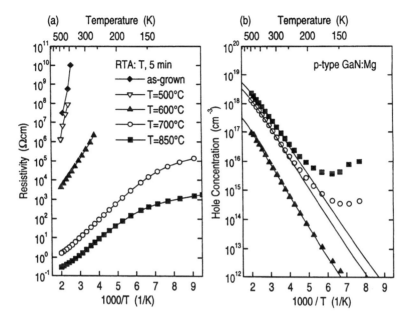

FIG. 6. (a) Resistivity and (b) hole concentration of a Mg-doped GaN sample as a function of reciprocal temperature. The resistivities (a) are shown for as-grown material and after annealing for 5 min at incremental temperatures T. The hole concentrations (b) are shown after acceptor activation at incremental temperatures T. The solid lines (b) represent calculated hole concentrations. (Reprinted with permission from Götz and Johnson, 1997.)

We note that Pearton et al. (1996d) reported an activation energy for dopant activation of order 2.5 eV in bulk samples, where hydrogen retrapping is expected to be significant. Pearton et al. (1996b) also reported that the apparent thermal stability of hydrogen-passivated Mg acceptors depends on the ambient employed during annealing: an H_2 ambient leads to a reactivation temperature approximately 150 °C higher than N_2. The authors attributed this to indiffusion of hydrogen from the H_2 ambients, causing a competition between passivation and reactivation. They also concluded that the hydrogen does not leave the sample after activation of Mg acceptors, but accumulates in regions of high defect density (in their case, the InGaN layer in a GaN/InGaN double heterostructure).

4. Passivation of Other Acceptor Dopants

Hydrogen passivation of Ca acceptors in GaN was reported by Lee et al. (1996). The Ca acceptors were introduced by implantation of Ca^+ or Ca^+ with coimplantation of P^+. Exposure to a hydrogen plasma at 250 °C led to a reduction in sheet carrier density of approximately an order of magnitude. The passivation could be reversed by annealing at 400 to 500 °C under N_2 ambient. The authors concluded that hydrogen passivation of acceptor dopants is a ubiquitous phenomenon, as it is in other semiconductors. Passivation of carbon acceptors was also reported by Pearton et al. (1996d).

5. LOCAL VIBRATIONAL MODES OF Mg—H COMPLEX IN GaN

Determination of the local vibrational modes (LVM) of the Mg—H complex in GaN provides satisfying confirmation of the significance of hydrogen in GaN and useful information, when compared with theory, on the structure of the complex. Currently, the stretch frequency is in fact the only reliably established physical parameter available from experiment for the Mg—H complex in GaN.

To provide convincing evidence in support of their spectroscopic identification Götz et al. (1996b) performed Fourier-transform infrared absorption spectroscopy on three differently treated specimens of Mg-doped GaN grown by MOCVD: the first specimen was as grown and electrically semi-insulating; the second specimen received a thermal anneal and displayed p-type conductivity; the third sample was exposed to monatomic deuterium at 600 °C for 2 hr, which increased the resistivity of the material. The infrared absorption spectra are shown in Fig. 7. The as-grown sample displays an LVM at $3125\,\text{cm}^{-1}$. After thermal activation of the Mg, the intensity of this absorption line is reduced. After deuteration (3), a new absorption line appears at $2321\,\text{cm}^{-1}$, which disappears after a thermal activation treatment (not shown in the figure). The isotopic shift clearly establishes the presence of hydrogen in the complex.

The spectroscopic evidence simultaneously establishes that hydrogen forms a complex with a shallow acceptor and that the acceptor activation process is due to the dissociation of Mg—H complexes. The LVM at $3125\,\text{cm}^{-1}$ appears only in Mg-doped GaN, and the isotopic shift is in excellent agreement with that observed for H-related LVMs in other semiconductors (Stavola and Pearton, 1991; Chevallier et al., 1991). Further, the decrease in the intensity of the LVM upon post-growth activation of the p-type conductivity is consistent with thermal dissociation of Mg—H complexes. Finally, the assignment of the LVM to the stretch mode of the Mg—H complex is in good agreement with the calculated H-stretch frequency (see Section II.4.a), which establishes that the Mg—H complex contains a strong N—H bond.

In light of the conclusive assignment produced by Götz et al. (1996b), we note that the local vibrational modes around $2200\,\text{cm}^{-1}$ observed by Brandt

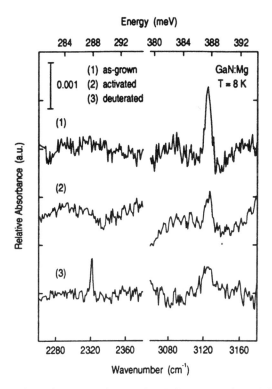

FIG. 7. Infrared absorption spectra for Mg-doped GaN grown by MOCVD. Spectra are shown for as-grown material (1), after RTA activation of the *p*-type conductivity (2), and after deuteration (3). The vertical bar indicates the magnitude of the absorbance scale. (Reprinted with permission from Götz et al., 1996b.)

et al. (1994) in Mg-doped material cannot be associated with Mg—H complexes. As already pointed out by Brandt *et al.*, modes in this frequency range could be due to Ga—H bonds, for instance, at or near extended defects.

IV. Conclusions and Outlook

Most of our current knowledge about hydrogen interactions with nitrides comes from computational studies. Experimental work has established the main qualitative features, but quantitative assessments are lacking. As pointed out in Section III.5, the vibrational frequency of the Mg—H complex is currently the only reliably established physical parameter available from experiment. A major problem on the experimental side, of course, has been the lack of samples with sufficiently good crystal quality to perform reliable impurity studies. This problem will hopefully be addressed when bulk crystals, as well as material with lower defect densities, become available. The computational work has definitely pointed out some extremely interesting features that should be explored experimentally: for instance, the huge negative-U value for interstitial hydrogen in GaN.

The aspect of hydrogen that is best understood so far is its tendency to passivate acceptors, and the need to perform a post-growth treatment to activate the acceptors. Nonetheless, various aspects of the activation process need to be clarified. For instance, no conclusions have been reached regarding the mechanisms operative during the LEEBI treatment. Also, thermal annealing gives rise to interesting changes in the photoluminescence spectrum (see Götz et al., 1996a) for which no explanation is available yet.

Only a few studies have been performed of the behavior of hydrogen on GaN surfaces, and its consequences for growth. Additional fundamental studies, both experimental and theoretical, would be very fruitful in this area and could potentially point the way toward improvements in the crystalline quality, suppression of defects, or incorporation of dopants.

The interaction between hydrogen and defects is just starting to be addressed. Regarding the native point defects, binding between hydrogen and vacancies is expected to occur and to play a role in the passivation (and potential activation upon annealing) of these defects. Computational work along these lines has recently been completed (Van de Walle, 1997), and experimental work on proton-implanted GaN has also been reported (Weinstein et al., 1998). Of course, hydrogen may also interact with extended defects, such as dislocations.

Finally, we point out that almost all the work so far has focused on GaN, and few studies have been performed on hydrogen interactions with other nitrides (InN and AlN) or nitride alloys. Although qualitatively similar behavior is expected, the quantitative details could turn out to be very important. For instance, the role played by hydrogen in p-type doping of GaN (as discussed in Section III.4.b) could undergo some changes in AlGaN alloys; and, of course, doping of such alloy layers is essential for device structures. Studies of hydrogen in InGaN and AlGaN alloys are therefore necessary.

Acknowledgment

Thanks are due to W. Götz and J. Neugebauer for very productive collaborations.

References

Akasaki, I., Amano, H., Kito, M., and Hiramatsu, K. (1991). *J. Lumin.*, **48&49**, 666.
Amano, H., Kito, M., Hiramatsu, K., and Akasaki, I. (1989). *Jpn. J. Appl. Phys.*, **28**, L2112.
Binari, S. C., Dietrich, H. B., Kelner, G., Rowland, L. B., Doverspike, K., and Wickenden, D. K. (1995). *J. Appl. Phys.*, **78**, 3008.
Boguslawski, P., Briggs, E. L., and Bernholc, J. (1995). *Phys. Rev. B*, **51**, 17 255.
Bosin, A., Fiorentini, V., and Vanderbilt, D. (1996). *Mat. Res. Soc. Symp. Proc.*, **395**, Pittsburgh: Materials Research Society, p. 503.
Brandt, M. S., Ager, J. W. III, Götz, W., Johnson, N. M., Harris, J. S., Molnar, R. J., and Moustakas, T. D. (1994). *Phys. Rev. B*, **48**, 14 758.
Chevallier, J., Clerjaud, B., and Pajot, B. (1991). In *Hydrogen in Semiconductors. Semiconductors and Semimetals*, **34**, J. I. Pankove and N. M. Johnson, eds. Boston: Academic Press, Chapter 13.
Chiang, C.-M., Gates, S. M., Bensaoula, A., and Schultz, J. A. (1995). *Chem. Phys. Lett.* **246**, 275.
Estreicher, S. K. (1994). In *Hydrogen in Compound Semiconductors*. S. J. Pearton, ed., Mat. Sci. Forum, **148–149**, p. 349. Aedermannsdorf: Trans Tech.
Estreicher, S. K. (1995). *Mat. Sci. Engr. Reports*, **14**, 319.
Estreicher, S. K. and Maric, D. M. (1996). *Mat. Res. Soc. Symp. Proc.* **423**, Pittsburgh: Materials Research Society, p. 613.
Götz, W., Johnson, N. M., Walker, J., Bour, D. P., Amano, H., and Akasaki, I. (1995). *Appl. Phys. Lett.*, **67**, 2666.
Götz, W., Johnson, N. M., Walker, J., Bour, D. P., and Street, R. A. (1996a). *Appl. Phys. Lett.*, **68**, 667.
Götz, W., Johnson, N. M., Bour, D. P., McCluskey, M. D., and Haller, E. E. (1996b). *Appl. Phys. Lett.*, **69**, 3725.
Götz, W. and Johnson, N. M. (1997). Unpublished.
Hamann, D. R., Schlüter, M., and Chiang, C. (1979). *Phys. Rev. Lett.*, **43**, 1494.
Hohenberg, P. and Kohn, W. (1964). *Phys. Rev.*, **136**, B864.
Johnson, B. G., Gill, P. M. W., and Pople, J. A. (1993). *J. Chem. Phys.*, **98**, 5612.
Johnson, N. M., Herring, C., and Van de Walle, C. G. (1994). *Phys. Rev. Lett.*, **73**, 130.
Kohn, W. and Sham, L. J. (1965). *Phys. Rev.*, **140**, A1133.
Laks, D. B., Van de Walle, C. G., Neumark, G. F., and Pantelides, S. T. (1991). *Phys. Rev. Lett.*, **66**, 648.
Lee, J. W., Pearton, S. J., Zolper, J. C., and Stall, R. A. (1996). *Appl. Phys. Lett.*, **68**, 2102.
Li, X. and Coleman, J. J. (1996). *Appl. Phys. Lett.*, **69**, 1605.
Louie, S. G., Froyen, S., and Cohen, M. L. (1982). *Phys. Rev. B*, **26**, 1739.
Nakamura, S., Iwasa, N., Senoh, M., and Mukai, T. (1992). *Jpn. J. Appl. Phys.*, **31**, 1258.
Neugebauer, J. and Van de Walle, C. G. (1994a). *Mat. Res. Soc. Symp. Proc.*, **339**, Pittsburgh: Materials Research Society, p. 687.
Neugebauer, J. and Van de Walle, C. G. (1994b). *Phys. Rev. B*, **50**, 8067.
Neugebauer, J. and Van de Walle, C. G. (1995a). *Phys. Rev. Lett.*, **75**, 4452.
Neugebauer, J. and Van de Walle, C. G. (1995b). *Mat. Res. Soc. Symp. Proc.*, **378**, Pittsburgh: Materials Research Society, p. 503.
Neugebauer, J. and Van de Walle, C. G. (1996a). *Appl. Phys. Lett.*, **68**, 1829.
Neugebauer, J. and Van de Walle, C. G. (1996b). *Mat. Res. Soc. Symp. Proc.*, **423**, Pittsburgh: Materials Research Society, p. 619.
Ohba, Y. and Hatano, A. (1994). *Jpn. J. Appl. Phys.*, **33**, L1367.
Okamoto, Y., Saito, M., and Oshiyama, A. (1996). *Jpn. J. Appl. Phys.*, **35**, L807.

Pankove, J. I. and Johnson, N. M., eds. (1991). *Hydrogen in Semiconductors. Semiconductors and Semimetals*, **34**, R. K. Willardson and A. C. Beer, Treatise eds. Boston: Academic Press.
Pavesi, L. and Giannozzi, P. (1992). *Phys. Rev. B,* **46**, 4621.
Pearton, S. J., Corbett, J. W., and Stavola, M. (1992). *Hydrogen in Crystalline Semiconductors.* Berlin: Springer-Verlag.
Pearton, S. J., Shul, R. J., Wilson, R. G., Ren, F., Zavada, J. M., Abernathy, C. R., Vartuli, C. B., Lee, J. W., Mileham, J. R., and Mackenzie, J. D. (1996a). *J. Electron. Mat.*, **25**, 845.
Pearton, S. J., Abernathy, C. R., Vartuli, C. B., Lee, J. W., Mackenzie, J. D., Wilson, R. G., Shul, R. J., Ren, F., and Zavada, J. M. (1996b). *J. Vac. Sci. Technol. A*, **14**, 831.
Pearton, S. J., Lee, J. W., and Yuan, C. (1996c). *Appl. Phys. Lett.*, **68**, 2690.
Pearton, S. J., Bendi, S., Jones, K. S., Krishnamoorthy, V., Wilson, R. G., Ren, F., Karlicek, R. F., Jr., and Stall, R. A. (1996d). *Appl. Phys. Lett.*, **69**, 1879.
Stavola, M. and Pearton, S. J. (1991). In *Hydrogen in Semiconductors, Semiconductors and Semimetals*, **34**, J. I. Pankove and N. M. Johnson, eds. Boston: Academic Press, Chapter 8.
Sung, M. M., Ahn, J., Bykov, V., Rabalais, J. W., Koleske, D. D., and Wickenden, A. E. (1996). *Phys. Rev. B,* **54**, 14 652.
Van de Walle, C. G., Denteneer, P. J. H., Bar-Yam, Y., and Pantelides, S. T. (1989). *Phys. Rev. B,* **39**, 10791.
Van de Walle, C. G., Laks, D. B., Neumark, G. F., and Pantelides, S. T. (1993). *Phys. Rev. B,* **47**, 9425.
Van de Walle, C. G. and Neugebauer, J. (1997). *Mat. Res. Soc. Symp. Proc.*, **449**, Pittsburgh: Materials Research Society, p. 861.
Van de Walle, C. G. (1997). *Phys. Rev. B,* **56**, R10 020.
Van Vechten, J. A., Zook, J. D., Horning, R. D., and Goldenberg, B. (1992). *Jpn. J. Appl. Phys.*, **31**, 3662.
Weinstein, M. G., Song, C. Y., Stavola, M., Pearton, S. J., Wilson, R. G., Shul, R. J., Killeen, K. P., and Ludowise, M. J. (1998). *Appl. Phys. Lett.*, **72**, 1703.
Wilson, R. G., Pearton, S. J., Abernathy, C. R., and Zavada, J. M. (1995). *J. Vac. Sci. Technol. A*, **13**, 719.
Yu, Z., Buczkowski, S. L., Giles, N. C., Myers, T. H., and Richards-Babb, M. R. (1996). *Appl. Phys. Lett.*, **69**, 2731.
Zavada, J. M., Wilson, R. G., Abernathy, C. R., and Pearton, S. J. (1994). *Appl. Phys. Lett.*, **64**, 2724.

CHAPTER 5

Characterization of Dopants and Deep Level Defects in Gallium Nitride

W. Götz

HEWLETT-PACKARD COMPANY
OPTOELECTRONICS DIVISION
SAN JOSE, CALIFORNIA

N. M. Johnson

XEROX PALO ALTO RESEARCH CENTER
PALO ALTO, CALIFORNIA

I. INTRODUCTION	185
II. MATERIALS PREPARATION	187
III. SHALLOW DOPANTS	187
1. *Hall-Effect Measurement*	187
2. *n-type, Si-doped GaN*	188
3. *p-type, Mg-doped GaN*	190
4. *Thickness Dependence of Electronic Properties for "Thick" GaN Films* . . .	193
IV. DEEP LEVEL DEFECTS	195
1. *Rectifying Devices and Capacitance Transient Methods*	195
2. *Conventional (Thermal) Deep Level Transient Spectroscopy*	197
3. *Optical Deep Level Transient Spectroscopy*	200
V. CONCLUSIONS	203
REFERENCES	205

I. Introduction

Impurities and native defects may affect the electrical conductivity of semiconductors. Shallow dopant impurities are intentionally introduced to control the type (i.e., *n*- vs *p*-type) and magnitude of the electrical conductivity for electronic or optoelectronic device applications. The species of dopant is chosen to best achieve the desired device performance, and controlled introduction of the dopant is required for reproducible characteristics. Consequently, the formative stage of any semiconductor device technology requires the selection and characterization of dopant impurities. The development of the III-V nitrides is currently at this stage.

Impurities and native defects that introduce deep levels are also important in semiconductors. Such defects are commonly classified as electron traps, hole traps, or recombination centers, dependent on a defect's predominant form of interaction with free carriers. The controlled introduction of deep level defects can be used to deliberately adjust the electronic or photoelectric properties of a semiconductor. For example, zinc, a deep acceptor in GaN, was utilized in the active InGaN region of the first commercially available III-V nitride double heterostructure (DH) LEDs to shift the emission wavelength into the visible range of the spectrum, because the growth of InGaN with a high InN composition was not possible at that time (Nakamura et al., 1994). On the other hand, the incorporation of deep-level defects can be detrimental to device performance. For example, deep level defects present in the active region of LEDs can act as nonradiative recombination centers and considerably reduce the quantum efficiency. For electronic devices, deep levels situated in the junction region of the device can act as carrier–generation centers and thus increase leakage currents.

The conventional Hall-effect technique is a well-established, powerful tool to investigate the electronic properties of shallow dopants in semiconductors, when performed over a range of temperatures. The analysis of the temperature dependence of the carrier concentration enables determination of important dopant parameters, such as the concentration and thermal activation energy for dopant ionization. For example, with variable-temperature Hall-effect measurements it was demonstrated that Mg-doping introduces a relatively deep acceptor level into the bandgap of GaN (Tanaka et al., 1994). For highly conductive unintentionally doped GaN, it was shown that the electron transport is due to hopping conduction (Ilegems and Montgomery, 1973; Molnar et al., 1993). In this chapter, the application of variable-temperature Hall-effect measurements is demonstrated for undoped GaN, Si-doped GaN, and Mg-doped GaN.

Capacitance transient spectroscopy encompasses a powerful set of techniques to detect and characterize deep levels in semiconductors. In particular, in III-V nitrides deep level defects have been investigated with deep level transient spectroscopy (DLTS) (Götz et al., 1994; Hacke et al., 1994, Haase et al., 1996), double correlation DLTS (DDLTS) (Götz et al., 1996c), and photoemission capacitance transient spectroscopy (ODLTS) (Götz et al., 1995, Yi and Wessels, 1996). These studies are reviewed in this chapter. Deep level defects for GaN were also characterized by isothermal capacitance transient spectroscopy (ICTS) and optical-ICTS (Hacke et al., 1997).

II. Materials Preparation

For optoelectronic applications, III-V nitride films are predominantly grown by organometallic chemical vapor deposition (OMVPE) (Akasaki and Amano, 1996). The OMVPE of GaN is usually conducted at sufficiently high temperatures (1000 to 1100 °C) to enable growth of films with good structural and electronic properties. On the other hand, the growth conditions are flexible enough to allow growth of the ternary alloys InGaN and AlGaN over a wide range of alloy compositions.

Most of the films to be discussed in this review were grown by OMVPE at ~ 1050 °C and doped with either Si or Mg during growth. Film nucleation on the sapphire substrates was initiated with a low-temperature GaN buffer layer. The films were typically between 2 and 3 μm thick. We also present electrical data for GaN films grown by hydride vapor phase epitaxy (HVPE). These films were grown to a thickness of 15 μm without a low-temperature buffer layer directly on the sapphire substrates (Molnar et al., 1996).

III. Shallow Dopants

1. HALL-EFFECT MEASUREMENT

Hall-effect measurements were performed with a computer-controlled system, which included a constant current source, an electrometer, and a high-impedance voltage meter. A temperature stage equipped with a Joule-Thompson high-pressure N_2 refrigerator was employed to vary the sample temperature in the range between 80–500 K. The magnetic field was 17.4 kG. The samples were configured in a square Van der Pauw geometry (Van der Pauw, 1958) and typically 5×5 mm^2. Small-area metal contacts were vacuum evaporated into the four corners of the samples to obtain ohmic contacts on n- and p-type GaN for the voltage ranges of interest. Contacts with good ohmic behavior are preferred for Hall-effect measurements in order to avoid depletion-layer effects, which would necessitate film thickness corrections.

The Hall-effect measurement yields the sheet resistance (r_{sq}) and the Hall coefficient ($R_H(n, p)$) of a GaN film. Under the assumptions of both uniformity of film thickness (d) and the transport properties within the films, the resistivity (ρ), the carrier concentration (n, p), and the carrier Hall mobility (μ_n, μ_p) can be determined from the measured quantities as follows (Schubert, 1993):

$$\rho = r_{sq} \cdot d \qquad (1)$$

$$n, p = \frac{r_H}{eR_H(n, p)} \tag{2}$$

and

$$\mu_n, \mu_p = \frac{R_H(n, p)}{\rho} \tag{3}$$

In Eq. (2) r_H is the Hall scattering factor, which is unknown for GaN. For the results presented in this chapter, r_H was assumed to be isotropic and of unity value. In case of inhomogeneities, Eqs. (1)–(3) are inapplicable, as will be discussed later in this section.

We used the temperature dependence of the electron and hole concentrations of the GaN films to derive parameters for shallow donors and acceptors, respectively. For n-type semiconductors, in the case where the temperature dependence of the charge transport is dominated by ionization of shallow donors, the electron concentration is given by (charge–neutrality condition):

$$n(T) = \sum_{i=1}^{m} \frac{N_i}{1 + \frac{n(T)g_i}{N_{C,\text{eff}}(T)} \exp\left(\frac{\Delta E_i}{kT}\right)} - N_{\text{comp}} \tag{4}$$

In Eq. (4), m is summed over the number of distinct donor species; N_i and ΔE_i are the concentrations and the activation (thermal ionization) energies of the donors, respectively; N_{comp} is the concentration of compensating acceptors, $N_{C,\text{eff}}(T)$ is the effective density of states for the conduction band, g_i is the degeneracy of the donor states, and k is the Boltzmann constant. A similar expression applies for acceptors in p-type semiconductors. A least-squares fit of Eq. (4) to the experimental $n(T)$ or $p(T)$ data yielded concentrations N_i and activation energies ΔE_i for donors or acceptors, respectively. For practical reasons the number of donors or acceptors was not allowed to exceed two ($m = 2$). The degeneracy factor for both donors and acceptors was taken as two. The effective density of states for the conduction band and the valence band were computed with an effective mass for electrons of $0.2 m_0$ and for holes of m_0 (m_0 = free electron mass).

2. n-TYPE, Si-DOPED GaN

Silicon is the donor species of choice for doping III-V nitrides n-type (Nakamura et al., 1992). It has been demonstrated that Si incorporates effectively on substitutional Ga sites in GaN. The Si doping is commonly achieved by flowing SiH$_4$ during OMVPE growth. It has also been shown that Si doping induces cracks into GaN films grown on sapphire substrates for Si-doping levels exceeding $\sim 1 \times 10^{19}$ cm^{-3} for a ~ 1 μm thick film. The cracking problem is most likely a consequence of the small ionic radius of Si$^+$ (0.41 Å) as compared to Ga$^+$ (0.62 Å). For growth on c-plane sapphire Si-doping introduces compression in the c-axis direction. As a consequence the basal plane of GaN is put in tension (Ager et al., 1997). If—for a given film thickness—the critical Si-doping concentration is exceeded, strain is relieved by film cracking. In addition, Si doping causes the GaN to embrittle, which further enhances the tendency of GaN to crack.

Results from variable-temperature Hall-effect measurements for n-type GaN are shown in Fig. 1. The experimental results are represented by unique symbols. Sample No. 1 was unintentionally doped and sample Nos. 2 to 4 were Si-doped. The electron concentrations are shown as functions of the reciprocal temperature in Fig. 1a, and the electron mobilities are shown as functions of temperature in Fig. 1b. The unintentionally doped GaN film (No. 1) exhibits the lowest electron concentration ($\sim 5 \times 10^{16}$ cm^{-3}, 300 K)

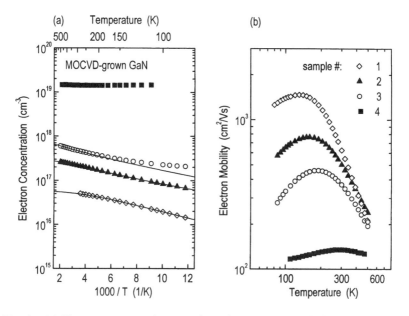

FIG. 1. (a) Electron concentration vs reciprocal temperature and (b) Hall mobility vs temperature for undoped (sample No. 1) and Si-doped (sample Nos. 2–4), n-type GaN films. The symbols refer to experimental data. The solid lines in Fig. 1a result from least-squares fits to the experimental data. The fits yield parameters for shallow donors that are given in the text.

and the highest electron mobility ($\sim 650\,\text{cm}^2(\text{V}\cdot\text{s})^{-1}$, 300 K). The Si-doped samples show higher electron concentrations and lower electron mobilities. The SiH_4 flux was gradually increased from sample Nos. 2 to 4. Consequently, the electron concentrations increase accordingly as determined by the Hall-effect measurements. Sample No. 4 shows an electron concentration of $\sim 2 \times 10^{19}\,\text{cm}^{-3}$, which is temperature independent. Such behavior is typically observed for degenerately doped semiconductors. The peak mobilities for the n-type GaN samples shown in Fig. 1 range from 135 (sample No. 4, 300 K) to 1450 cm^2(V·s)$^{-1}$ (sample No. 1, 130 K) and decrease with increasing electron concentration (Fig. 1b). The solid lines in Fig. 1a result from least-squares fits of Eq. (4) to the experimental $n(T)$ data of sample Nos. 1 to 3. The fits yield parameters for two donor levels that are dominant in the GaN films. The modeling yields activation energies for the shallower level of 16 meV (sample No. 1), 14 meV (No. 2), and 12 meV (No. 3). From electrical measurements, activation energies in this range have been reported for Si donors in GaN (Götz et al., 1996d). Activation energies for donor ionization as determined by variable temperature Hall-effect measurements depend on the donor concentration and are also influenced by the presence of ionized acceptors. Typically, the activation energies are smaller than the energy difference between the donor level in the bandgap and the conduction band minimum and further decrease with increasing donor concentration. The concentration of this donor is $6 \times 10^{16}\,\text{cm}^{-3}$ in sample No. 1, $2.3 \times 10^{17}\,\text{cm}^{-3}$ in No. 2, and $7.4 \times 10^{17}\,\text{cm}^{-3}$ in No. 3. The presence of Si in sample Nos. 2 to 4 was verified with secondary ion mass spectrometry (SIMS). The elemental Si concentration was found to agree within experimental uncertainties with the concentration of the shallowest donor level for sample Nos. 2, 3; for sample No. 4 the Si concentration was found to be $\sim 2 \times 10^{19}\,\text{cm}^{-3}$. For sample No. 1 Si could not be detected with SIMS above the background level of the SIMS apparatus ($\sim 1 \times 10^{17}\,\text{cm}^{-3}$).

The presence of a second donor level had to be assumed to model the temperature dependence of the electron concentrations of sample Nos. 1 to 3. The activation energy for this donor level ranges from 32–37 eV and its concentration from 3 to $6 \times 10^{16}\,\text{cm}^{-3}$, independent of the Si doping. It is possible that this second donor level is due to oxygen donors (Chung and Gershenzon, 1992). Oxygen is a potential impurity in nitride films grown with NH_3 as N precursor because water vapor is present in NH_3.

3. p-TYPE, Mg-DOPED GaN

The achievement of p-type doping was a critical step in the development of III-V nitrides for the fabrication of visible and ultraviolet LEDs and LDs.

Amano et al. (1989) and later Nakamura et al. (1992) demonstrated that post-growth treatments such as low–energy electron–beam irradiation (LEEBI) and thermal treatment in N_2 atmosphere, respectively, produce p-type conductivity in Mg-doped GaN. It has been demonstrated that H donors form electrically inactive complexes with uncompensated Mg acceptors during cooling after growth (Götz et al., 1996e). The presence of compensated donors and acceptors and the formation of neutral Mg-H complexes are responsible for the semi-insulating nature of as-grown Mg-doped GaN (Van Vechten et al., 1992; Neugebauer and Van de Walle, 1995). At temperatures below the growth temperature and in hydrogen-free ambients, the Mg-H complexes can be dissociated by either minority carrier injection (LEEBI treatment) or thermal annealing. The Mg acceptors substitute for Ga in the GaN lattice (Neugebauer and Van de Walle, 1995) and can be incorporated in concentrations as high as $\sim 10^{20}$ cm^{-3} (Götz et al., 1996b). However, the achievable hole concentrations at room temperature are usually at least two orders of magnitude below the acceptor concentrations because for Mg-doped GaN the thermal activation energy for acceptor ionization, as determined by variable temperature Hall-effect measurements, ranges up to ~ 180 meV (Götz et al., 1996b).

Results from variable-temperature Hall-effect measurements for three Mg-doped GaN samples (Nos. 5–7) are shown in Fig. 2. The p-type conductivity for all three samples was activated by thermal annealing. Figure 2a shows the hole concentration as a function of reciprocal temperature, and in Fig. 2b the hole mobility is shown as a function of temperature. Sample No. 5 exhibits the highest hole concentration and the lowest hole mobility. At 300 K, the hole concentration is $\sim 8 \times 10^{17}$ cm^{-3} and the mobility is only ~ 3 cm^2(V · s)$^{-1}$. The analysis of the temperature dependence of the hole concentration at temperatures above 280 K with the p-type version of Eq. (4) is consistent with the presence of a single acceptor with an activation energy of ~ 160 meV and a concentration of $\sim 8 \times 10^{19}$ cm^{-3} (solid line). The elemental Mg concentration as determined by SIMS was found to be $\sim 1 \times 10^{20}$ cm^{-3}. For Mg-doped GaN we typically find good agreement between the concentration of Mg and the concentration of the dominant acceptor level, the latter determined by analysis of the Hall-effect data. This indicates that most of the Mg atoms are incorporated on Ga lattice sites and act as acceptors. For sample No. 5 at temperatures below ~ 280 K the temperature dependence of the hole concentration is not accurately described by the ionization of an acceptor (Fig. 2a). In this temperature range the experimental hole concentration increases with decreasing temperature. This behavior, which is accompanied by a strong decrease of the hole mobility (Fig. 2b), is indicative of hopping conduction via closely spaced acceptor states. A similar behavior is observed for sample

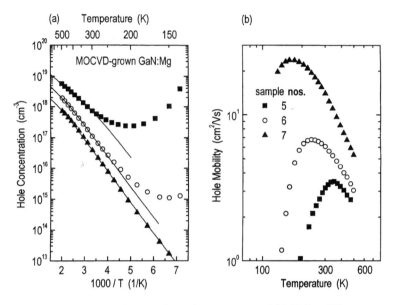

FIG. 2. (a) Hole concentration vs reciprocal temperature and (b) Hall mobility vs temperature for Mg-doped, p-type GaN films. The symbols refer to experimental data. The solid lines in Fig. 2a result from least-squares fits to the experimental data. The fits yield parameters for shallow acceptors that are given in the text.

No. 6. The temperature dependence of the hole concentration can be modeled with Eq. (4) only for temperatures above ~ 250 K. For sample No. 7, however, the temperature dependence of the hole concentration can be described over the entire temperature range of the Hall-effect data (Fig. 2a). The analysis reveals that the acceptor concentrations are lower than in sample No. 5. For sample Nos. 6 and 7 the acceptor concentration is $\sim 4 \times 10^{19}$ cm^{-3} and $\sim 2 \times 10^{19}$ cm^{-3}, respectively. The Mg concentration as measured by SIMS was found to be $\sim 5 \times 10^{19}$ cm^{-3} and $\sim 2 \times 10^{19}$ cm^{-3}, respectively. The acceptor activation energies for acceptors in sample Nos. 6 and 7 were found to be ~ 170 and 175 meV, respectively. Slightly higher activation energies such as these are usually obtained from samples with lower dopant concentrations (Schubert, 1993). Also, the observation of hopping conduction in sample Nos. 5 and 6 occurring at lower temperatures is consistent with the presence of higher acceptor concentrations in these samples. Hopping conduction is also an indication of the presence of compensation. The analysis of the Hall-effect data for the p-type samples with Eq. (4) allows the determination of the concentration of compensating donors. For sample Nos. 5, 6, and 7 the concentration of donors was determined to be $\sim 5 \times 10^{18}\,\text{cm}^{-3}$, $\sim 2 \times 10^{18}\,\text{cm}^{-3}$, and $\sim 2 \times 10^{8}\,\text{cm}^{-3}$, respectively.

As these concentrations are an order of magnitude higher than the concentration of residual donors in our unintentionally doped, n-type material, the formation energy of impurity or native donors must be considerably lower under p-type as compared to n-type growth conditions (Neugebauer and Van de Walle, 1996). Possible candidates for compensating donors are the nitrogen vacancy (V_N), isolated hydrogen, and substitutional oxygen (O_N), the latter exhibiting a strong affinity to Mg. However, the hole concentration is only marginally affected by the presence of compensation because the acceptor concentration in Mg-doped GaN is typically more than one order of magnitude higher than that of compensating donors.

The hole mobility is adversely affected by both the presence of compensating donors and high concentrations of acceptors. The 300 K hole mobility for sample Nos. 6 and 7 is $\sim 6\,\text{cm}^2(\text{V}\cdot\text{s})^{-1}$ and $15\,\text{cm}^2(\text{V}\cdot\text{s})^{-1}$, respectively. The 300 K resistivity for all three Mg-doped GaN films depicted in Fig. 2 is $\sim 3\,\Omega\text{-cm}$. Such a high resistivity constituted one of the obstacles in achieving low forward voltages in GaN light emitters. The device problem is magnified by the difficulty in forming low-resistance ohmic contacts to Mg-doped, p-type GaN.

A slightly higher activation energy was determined for Mg-doped AlGaN, which contained 8% AlN by Tanaka *et al.* (1994). For an acceptor concentration of $\sim 1.5 \times 10^{19}\,\text{cm}^{-3}$ an activation energy of 192 meV was derived from variable temperature Hall-effect measurements.

The ionization energy for Mg acceptors for samples with very low Mg concentrations are expected to be higher than the activation energies determined for sample Nos. 5 to 7 which contain Mg concentration in excess of $2 \times 10^{19}\,\text{cm}^{-3}$. Due to the high concentration the Coulomb potentials of ionized acceptors are likely to overlap, which leads to a lowering of the activation energy for ionization. In fact, from optical measurements the position of the Mg acceptor level was estimated to be ~ 208 meV above the valence band maximum (Götz *et al.*, 1996b).

4. THICKNESS DEPENDENCE OF ELECTRONIC PROPERTIES FOR "THICK" GaN FILMS

It has been shown for a 13-μm thick, n-type GaN film grown by HVPE on GaCl-pretreated sapphire that the concentrations of shallow and deep donors determined by Hall-effect measurements are inconsistent with results from both capacitance-voltage (C-V) measurements and deep level transient spectroscopy (DLTS) (Götz et al., 1996f; Götz et al., 1997). Both the background, shallow n-type doping concentration and the deep–level concentration were found to be constant as the film was thinned by mechanical polishing from the original film thickness to a thickness of $\sim 1.2\,\mu$m. However, Hall-effect data when analyzed under the assumption of thickness uniformity show an increase of the electron concentration from $\sim 2 \times 10^{17}\,\text{cm}^{-3}$ (300 K) at the original film thickness of $\sim 13\,\mu$m to $\sim 10^{20}\,\text{cm}^{-3}$ (300 K) at a film thickness of $\sim 1.2\,\mu$m. The increase of the electron concentration is accompanied by a decrease of the electron mobility from $452\,\text{cm}^2(\text{V}\cdot\text{s})^{-1}$ (300 K, 13 μm) to $2.1\,\text{cm}^2(\text{V}\cdot\text{s})^{-1}$ (300 K, 1.2 μm). The Hall-effect data suggest inhomogeneous transport properties for the 13 μm thick HVPE-grown GaN film. However, the C-V and DLTS data show that both the dopant and deep-level concentrations are uniform throughout most of the film thickness. At a film thickness of 1.2 μm the depletion region for the capacitance measurements reached $\sim 0.5\,\mu$m below the surface. Thus, the depth inhomogeneity of the transport properties must be limited to the first $\sim 0.7\,\mu$m of film growth.

For depth inhomogeneities in the free electron concentration and electron mobility, the Hall-effect measurement yields an effective areal density of free electrons $n_{S,\text{eff}}$ and an effective Hall mobility μ_{eff} (Herring, 1960). For simplicity, we consider here only the electron density-mobility product $n_{S,\text{eff}}\,\mu_{\text{eff}}$, which is related to the sample resistivity as follows ($r_H = 1$):

$$n_{S,\text{eff}}\,\mu_{\text{eff}} = \int_0^d dx/e \cdot \rho(x) \qquad (5)$$

The dependence of $n_{S,\text{eff}}\,\mu_{\text{eff}}$ on film thickness is displayed in Fig. 3. The experimental data are represented by symbols. For uniform film properties, a linear dependence of $n_{S,\text{eff}}\,\mu_{\text{eff}}$ on d is expected with $n_{S,\text{eff}}\,\mu_{\text{eff}}\,(d=0) = 0$. The data presented in Fig. 3 can be fitted by a straight line (solid line) given the experimental uncertainty of the data. However, the straight line intersects the $n_{S,\text{eff}}\,\mu_{\text{eff}}$ axis at $n_{S,\text{eff}}\,\mu_{\text{eff}}(0) \sim 2.6 \times 10^{16}\,(\text{cm V s})^{-1}$. Such nonzero intercept is indicative of the presence of two distinct conducting layers where the electron concentrations and mobilities are different but constant within each layer. For two layers with thicknesses d_1 and d ($d_1 \ll d$) and resistivities ρ_1 and ρ, Eq. (5) becomes

$$n_{S,\text{eff}}\,\mu_{\text{eff}}(d) = \frac{d_1}{e\rho_1} + \frac{d}{e\rho} \qquad (6)$$

The slope of the straight line yields the resistivity of the dominant, thicker layer to be $\sim 0.84\,\Omega$ cm.

5 Characterization of Dopants and Deep Level Defects

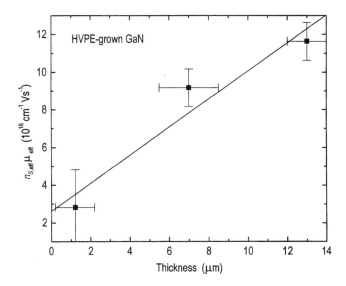

FIG. 3. Effective carrier density and mobility product vs sample thickness for a 13 μm n-type GaN sample thinned by mechanical polishing to a thickness of ~7 μm and ~1.2 μm. The solid line is a linear regression to the experimental data. The zero intercept with the "y-axis" indicates nonuniform distribution of electrons.

The presence of two distinct layers with a sharp interface for HVPE-grown GaN films on GaCl-pretreated sapphire was also observed in cross-sectional TEM micrographs (Götz et al., 1996f). The presence of a highly defective, ~300 nm thick layer near the GaN/sapphire interface was clearly resolved. This near interface layer consists mainly of stacking faults. Above this interface layer a uniform layer of characteristic GaN cells could be identified. It is possible that this structurally distinct interface layer is correlated with the highly conductive layer, the presence of which is predicted by the Hall-effect data near the GaN/sapphire interface. Under the assumption of $d_1 = 300$ nm, the resistivity of the near interface layer is estimated to be $\sim 7.2\,\mu\,\Omega$-cm.

IV. Deep Level Defects

1. Rectifying Devices and Capacitance Transient Methods

Capacitance transient techniques utilize devices that possess a voltage-modulable space-charge layer (SCL), such as a Schottky-barrier diode (with a metal-semiconductor rectifying contact) or a *p-n* junction diode. In these diodes the magnitude of the applied reverse bias voltage (V_r) controls the width of the spatial region beneath the junction. In this layer deep levels on the majority carrier half of the bandgap are unoccupied by majority carriers under steady-state conditions. Specifically, due to the band bending deep levels are unoccupied by majority carriers from the junction to the depth within the SCL at which the deep level intersects the quasi-Fermi level; this depth is called the crossover point.

In capacitance transient spectroscopy, the measurement sequence includes a pulse bias step (reduction of the magnitude of the reverse bias), which collapses the SCL for a short period of time (e.g., 1 ms) to fill deep levels with majority carriers (filling pulse). Immediately after the original reverse bias is reinstated, a metastable occupancy of deep levels exists within the SCL between the crossover points determined by the filling pulse and the reverse bias. For conventional DLTS the sample temperature is selected to provide enough thermal energy to excite the trapped carrier into the majority carrier band. Due to the applied electrical field these carriers are swept out of the SCL. As a consequence the differential capacitance of the rectifying device increases. This event is recorded as a capacitance transient that exhibits an exponential dependence on time, provided that the deep level concentration is small compared to the shallow doping concentration (Miller *et al.*, 1977). The emission rate of electrons (e_n) from a deep level depends exponentially on the activation energy for electron emission to the conduction band (ΔE_n) and the sample temperature, as follows:

$$e_n = (\sigma_n v_n N_{C,\text{eff}}/g_n) \cdot \exp[-\Delta E_n/kT] \qquad (7)$$

This expression is derived from the principal of detailed balance [Schockley–Read–Hall Statistics (Blakemore, 1962)]. In Eq. (7), σ_n is the capture cross section, v_n is the thermal velocity for electrons, $N_{C,\text{eff}}$ is the effective density of states of electrons, and g_n is the degeneracy factor of the deep level. A similar equation applies for the emission rate of holes (e_p) from deep levels to the valence band. Analysis of the temperature dependence of the emission rate yields the activation energy for electron (*n*-type) or hole (*p*-type) emission for the emitting deep level [Arrhenius analysis] (Götz and Johnson, 1994). However, the temperature dependence of the capture cross section generally has to be assumed, which introduces an uncertainty for the determination of the activation energy. Moreover, in many cases, the activation energy exhibits a pronounced dependence on the electrical field (reverse bias) applied to the diode during carrier emission (i.e., the Poole-Frenkel effect). The concentration of the deep levels can be estimated from the magnitude of the corresponding capacitance transient. The field dependence of the emission rate for deep levels can be utilized to determine whether a deep level is donor-like in n-type semiconductors or acceptor-like in p-type semiconductors. For this purpose capacitance transients are recorded for a set of filling pulses between V_r and 0 V. The individual capacitance transients of two neighboring pulse voltages are subtracted. Thus, a capacitance–transient signal is generated that results only from carrier emission from deep levels located between two crossover points defined by a pair of "correlated" filling pulses. Because the electrical field decreases monotonically from the junction position to the edge of the SCL, capacitance transients are obtained for carrier emission over a range of electrical field intensities. This technique is labeled double-correlation DLTS (DDLTS) and not only is used to investigate the field dependence of emission rates but also for spatial depth profiling (Lefevre and Schulz, 1977). For donors and acceptors in p-type and n-type semiconductors, respectively, the emission rates are typically not affected by the electrical field in the SCL due to the repulsive nature of the Coulomb interaction between donors and holes or acceptors and electrons, respectively. However, the emission rates for donors and acceptors in n-type and p-type semiconductors, respectively, are likely to exhibit a pronounced dependence on the electrical field.

For photoemission capacitance transient spectroscopy (ODLTS) the sample temperature is set low enough to freeze out any thermal emission processes associated with the deep levels under investigation. After trap filling, monochromatic light is used to excite majority carriers to the majority carrier band. The capacitance transients are recorded as a function of the photon energy (Chantre et al., 1981). A threshold voltage for carrier emission and the concentration of detectable deep levels can be estimated from the capacitance transients. Photoemission capacitance transient spectroscopy is especially useful for the investigation of GaN because for wide-bandgap semiconductors conventional thermal DLTS can detect only deep levels, which are energetically located within ~ 1 eV of either bandedge for practical measurement conditions.

2. Conventional (Thermal) Deep Level Transient Spectroscopy

Figure 4 shows a DLTS spectrum for an n-type GaN Schottky diode. The spectrum displays the DLTS signal plotted as a function of the sample temperature. The DLTS signal is the difference between the diode capacitance at two different delay times ($t_1 < t_2$) after the filling pulse. The delay times define the emission rate window e_0 ($=\ln(t_2/t_1)/(t_2 - t_1)$). The

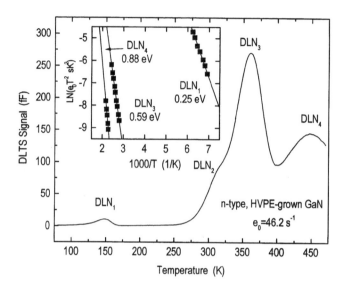

FIG. 4. The DLTS spectrum for n-type GaN grown by HVPE. The spectrum displayed for an emission rate of 46.2 s^{-1} reveals the presence of four discrete deep levels in the sensitivity range of the measurement. They are labeled DLN_1–DLN_4. Here DLN_2 appears only as a shoulder, whereas DLN_1, DLN_3, and DLN_4 introduce peaks into the DLTS spectrum. The inset demonstrates Arrhenius analyses for DLN_1, DLN_3, and DLN_4 that yield activation energies for electron emission to the conduction band. These activation energies are depicted in the inset.

key feature of the DLTS signal is that it goes through a maximum at that temperature for which the trap emission rate (e_n or e_p, Eq. (7)) equals e_0.

The DLTS spectrum displayed in Fig. 4 for $e_0 = 46.2$ s^{-1} contains the signature of four discrete deep levels, which appear at 148 K, ~318 K, 361 K, and 448 K. It was obtained from a 13 μm thick GaN sample grown by HVPE. The deep levels are labeled DLN_1–DLN_4. Three of the DLTS peaks are well resolved; the peak for DLN_2 appears only as a shoulder. The activation energies for electron emission to the conduction band for DLN_1, DLN_3, and DLN_4 were determined from Arrhenius analyses, which are shown in the inset of Fig. 4. The activation energies were obtained from the slope of the solid lines under the assumption of temperature-independent electron capture cross sections and are depicted in the inset. The concentrations and the electron capture cross sections of the deep levels DLN_1, DLN_3, and DLN_4 were also determined (Götz and Johnson, 1994); they are $\sim 1.7 \times 10^{14}$ cm^{-3} and $\sim 5 \times 10^{-15}$ cm^2 (DLN_1), 5.0×10^{15} cm^{-3} and $\sim 5 \times 10^{-16}$ cm^2 (DLN_3), and 3.3×10^{15} cm^{-3} and $\sim 1 \times 10^{-14}$ cm^2 (DLN_4).

The shallower two deep levels (DLN_1 and DLN_2) are consistently observed for as-grown, *n*-type GaN (Götz *et al.*, 1994; Hacke *et al.*, 1994; Hacke *et al.*, 1996). The activation energies for DLN_1 and DLN_2 were observed to range from 0.14–0.26 eV and from 0.44–0.62 eV, respectively, depending on measurement conditions and on assumption of the temperature dependence for the capture cross section. The chemical or structural nature of the defects that give rise to the electronic levels DLN_1 and DLN_2 are unknown. The DLN_3 was observed for *n*-type GaN grown by HVPE with an activation energy of ~ 0.67 eV (Hacke *et al.*, 1994). It appeared also in DLTS spectra of GaN grown by MOCVD after implantation of N ions and was subsequently removed by annealing (Haase *et al.*, 1996). The latter experiment indicates that DLN_3 may be associated with a native defect or a complex that involves a native defect. The defect level DLN_4 has not been reported in the literature.

Figure 5 shows a DLTS spectrum for a Schottky diode fabricated with AlGaN. The AlGaN film was grown by MOCVD on a highly conductive SiC substrate. The AlN composition was estimated by energy-dispersive x-ray analysis to be $\sim 12\%$. The DLTS spectrum in Fig. 5, which is

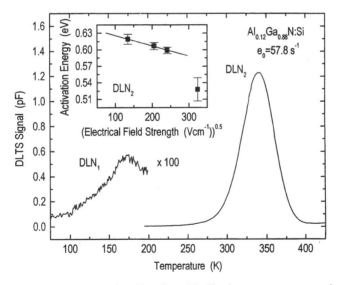

FIG. 5. The DLTS spectrum for $Al_{0.12}Ga_{0.88}N$. The low temperature portion of the spectrum is amplified by a factor of 100. The two peaks that are visible are labeled with the corresponding deep levels. The inset summarizes results from DDLTS for the same AlGaN sample. The activation energies for DLN_2 are shown as a function of the square root of the applied electrical field strength. For lower magnitudes of the field the activation energies follow a square-root dependence that is indicated by the solid straight line.

displayed for an emission rate window of $57.8 \, \text{s}^{-1}$, exhibits two peaks that relate to deep levels in the AlGaN material. The dominant peak is DLN_2, which appeared in the DLTS spectrum for the HVPE-grown GaN sample only as a shoulder, but is commonly observed for MOCVD-grown GaN; DLN_1 also appears in the DLTS spectrum, however, only as a very weak signal. Its concentration is estimated to be $\sim 10^{14} \, \text{cm}^{-3}$. The peak associated with DLN_1 and DLN_2 appears at 175 K and at 345 K, respectively. With respect to GaN their energy positions are shifted by ~ 25 K (Götz et al., 1996c). The activation energy for electron emission to the conduction band was determined for DLN_2 to be (0.62 ± 0.02) eV (with an assumed temperature–independent cross section). The concentration of DLN_2 was computed from the DLTS data to be $(4 \pm 1) \times 10^{16} \, \text{cm}^{-3}$.

The concentration of DLN_2 in the AlGaN sample was high enough to apply DDLTS. The reverse bias was set at -4 V and the filling pulse voltages varied between 0 and ~ 3 V in 1 V steps. From the DDLTS measurement the activation energies for electron emission of DLN_2 were determined in different, well defined spatial regions of the SCL and, thereby, under different electrical field strengths ranging from $1.1 \times 10^5 \, \text{V cm}^{-1}$ to $1.8 \times 10^5 \, \text{V cm}^{-1}$. The activation energies are plotted in the inset of Fig. 5 as a function of the square root of the electrical field strength. The dependence of the activation energies on the electrical field suggests that DLN_2 is a donor-like defect. Due to the coulombic attraction between a donor and the emitted electron, the emission rate exhibits an electrical field dependence (Poole-Frenkel effect), which in some cases follows a simple model suggested by Hartke (1968). As can be seen in the inset of Fig. 5, the activation energies for electrical field strengths up to $5.8 \times 10^4 \, \text{V cm}^{-1}$ follow the square-root dependence for electron emission over a Coulomb barrier. The zero-field activation energy for DLN_2 is thus estimated to be ~ 0.65 eV. For higher electrical fields, our data show a strong deviation from the square-root dependence. This behavior may suggest that tunneling through the Coulomb barrier becomes the dominant electron–emission process at higher magnitudes of the electrical fields.

3. OPTICAL DEEP LEVEL TRANSIENT SPECTROSCOPY

Deep levels with activation energies for electron or hole emission > 1 eV are usually not detectable with conventional DLTS due to practical limitations, for instance, the temperature range of the DLTS measurement; ODLTS is a sensitive tool to detect deep levels over the entire bandgap of wide–bandgap semiconductors.

In Figs. 6 and 7, the ODLTS results shown were obtained for n-type, Si-doped GaN from a Schottky diode and for p-type, Mg-doped GaN from

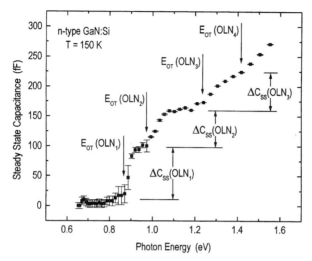

FIG. 6. Dependence of the steady-state photocapacitance ΔC_{ss} on photon energy for a Schottky diode on n-type GaN. The spectrum displays distinct steps, marked by arrows, due to photoemission from four distinct deep levels, which are labeled OLN_1–OLN_4. The magnitude of the steady-state photocapacitance for each of the deep levels is labeled $\Delta C_{ss}(OLN\#)$ ($\# = 1$–4).

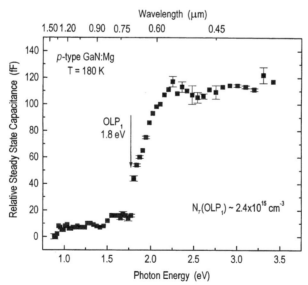

FIG. 7. Dependence of the steady-state photocapacitance on photon energy for a p-n^+ junction diode. The spectrum displays a dominant step marked by an arrow and the corresponding optical threshold energy due to photoemission from a distinct deep level labeled OLP_1. The magnitude of the step yields an estimate of the concentration $N_T(OLP_1)$.

a "one-sided" p-n^+ junction diode, respectively. The data are presented as relative steady-state capacitance vs energy of the excitation light ($\Delta C_{ss}(hv)$); $\Delta C_{ss}(hv)$ is the magnitude of the photocapacitance transient obtained by fitting one or two exponentials to the experimental capacitance transient, which was recorded for 4000 s. For each energy step, three transients were averaged. The photocapacitance signal increases monotonically with hv because $\Delta C_{ss}(hv)$ is proportional to the sum of charge-change contributions from optically active deep levels that reside within an energy interval that extends up to hv. Each discrete level introduces a step in the ΔC_{ss} spectrum, near its optical threshold energy E_{OT}.

For n-type, Si-doped GaN (Fig. 6) the energy of the excitation light was varied between 0.65 and 1.55 eV. The sample temperature was 150 K and the filling pulse and the reverse bias were 0 and -3 V, respectively. After switching to the reverse bias no capacitance change could be observed until the light was turned on. The spectrum displays four abrupt increases in slope, each of which can be associated with a distinct deep level. Each feature is identified by an arrow, with a label (i.e., OLN_1-OLN_4) and the approximate threshold energy E_{OT}. The first two steps at ~ 0.87 and ~ 0.97 eV are clearly resolved and followed by plateaus. The second two, at ~ 1.25 eV and ~ 1.45 eV, are less well resolved but still suggest the onset of photoemission from additional deep levels. With each step in the spectrum of Fig. 6 is associated a step height that is proportional to the density of the corresponding deep level; saturation was not achieved for OLN_4 within the chosen range of photon energies. The trap densities are estimated to be 9×10^{14} cm^{-3}, 5×10^{14} cm^{-3}, and 6×10^{14} cm^{-3} for OLN_1, OLN_2, and OLN_3, respectively. The origin of these deep levels is unknown. Speculatively, the comparable trap density may be indicative of a related origin for the several deep levels. It has also been speculated that one or more of these deep levels are involved in the recombination process that give rise to the "yellow luminescence" band commonly observed for n-type GaN (Glaser et al., 1995).

For the ODLTS measurement of p-type GaN (Fig. 7) a "one-sided" p-n^+ junction diode was employed. The sample temperature was kept at 180 K (± 0.1 K). The reverse bias was set at -4 V and a filling pulse of -1 V was chosen. Without light no capacitance transient was detected after switching from the filling pulse to the reverse bias. With no significant capacitance changes recorded for photon energies below $E_{gap}/2$ the energy range for the excitation light was extended to 3.6 eV. The spectrum for the p-type, Mg-doped GaN materials exhibits one dominant abrupt increase in slope at ~ 1.8 eV, which can be associated with a distinct deep level near midgap. Its concentration can be estimated from the step height to be 2.4×10^{15} cm^{-3}. There are several minor steps visible in the spectrum of Fig. 7, which may be associated with deep levels in Mg-doped, p-type GaN.

For the same p-n^+ junction diode a conventional, thermal DLTS measurement was conducted. Deep levels were detected in the lower half of the bandgap with activation energies for hole emission to the valence band of ~ 0.21, ~ 0.32, and ~ 0.47 eV (Götz et al., 1996a). The concentrations ranged from $\sim 4 \times 10^{14}$ to 9×10^{14} cm^{-3}. None of the deep-level defects observed with DLTS and ODLTS could significantly influence the hole concentration through compensation in p-type, Mg-doped GaN. However, it was reported from photocapacitance spectroscopy measurements of slightly Mg-doped, n-type GaN that several deep levels are present in the upper half of the bandgap with optical threshold energies for electron emission to the conduction band in the range between 1.0–1.8 eV (Yi and Wessels, 1996). These centers were not detectable after annealing at 850 °C, a temperature that is usually employed for acceptor activation (Götz et al., 1996g).

V. Conclusions

The characterization of dopants and deep-level defects by variable temperature Hall-effect measurements and capacitance transients techniques, respectively, have proven to be useful tools for the development of nitride materials for application in light emitting and electronic devices. The electronic properties of Si donors and Mg acceptors were successfully determined by variable temperature Hall-effect measurements. Silicon is a shallow donor and can be easily incorporated in concentrations up to $\sim 10^{19}$ cm^{-3}. Its thermal activation energy was measured to range from ~ 12 to ~ 16 meV depending on Si concentration and the concentration of the compensating acceptors. The relatively low hole concentrations for Mg-doped GaN were found to be due to the high activation energy for ionization of Mg acceptors, which was measured to range between ~ 160 to ~ 185 meV. Typically less than 1% of the Mg acceptors are ionized at room temperature in equilibrium. Variable temperature Hall-effect measurements have played a crucial role in understanding the interaction of hydrogen and acceptors in as-grown Mg-doped GaN. The Hall-effect data, which were obtained for the activation of the p-type conductivity by isochronal anneals in the temperature range between 500–850 °C were shown to be consistent with acceptor activation due to the dissociation of acceptor-hydrogen complexes.

Deep levels were detected in n- and p-type GaN only in concentrations below $\sim 10^{16}$ cm^{-3} by capacitance–transient spectroscopies. The DLTS has been successful in detecting deep levels down to concentrations as low as

$\sim 10^{13}$ cm^{-3}. The application of photoemission capacitance-transient spectroscopy has extended the sensitivity range of these techniques throughout the entire bandgap. It appears that deep levels do not play an important role as compensating centers to influence the electrical properties of nitride materials. However, their role as carrier traps or competing radiative or nonradiative recombination centers is still suspect. Some luminescence processes such as the commonly observed yellow luminescence may involve deep levels as constituents, but further studies are required to learn more about the role of deep levels and their relation with structural defects and their effect on GaN device performance.

Recently, a report appeared in the literature (Calleja et al., 1997) that relates the yellow luminescence bond to the presence of a deep level ~ 1 eV above the valence band edge of GaN, which was observed by ODLTS.

For the analysis of the Hall-effect data, some of the high doping effects can explicitly be taken into account. Similar to what has been shown for other semiconductors, the lowering of the activation energy of Mg acceptors in GaN is expected to depend linearly on the distance between ionized acceptors ($\propto N^{1/3}$). For Mg-doped GaN, the overlap of the acceptor potentials is likely to be one of the major effects for the lowering of the activation energy. The mean distance between two acceptors at a concentration of $\sim 10^{19}$ cm^{-3} is ~ 50 Å. At this distance there is sufficient overlap of the Coulomb potentials, although it is significantly greater than the Bohr radius for an acceptor in GaN which is only ~ 5 Å. The lowering of the activation energy δE_A due to the Coulomb interaction between closely spaced ionized acceptors (concentration N_A^-) is given by

$$\partial E_A(N_A^-) = f \frac{q^2}{4\pi\varepsilon_S}(N_A^-)^{1/3} \qquad (8)$$

where $\Delta E_{A,0}$ is the activation energy that would be measured for very low acceptor concentration, f is a geometric factor, q is the electronic charge, and ε_S is the dielectric constant. If Eq. (4) is modified according to

$$p = N_A \left[1 + \frac{g_p p}{N_{V,\text{eff}}} \exp\left(\frac{\Delta E_{A,0} - \partial E(N_A^-)}{kT}\right) \right]^{-1} - N_{\text{comp}} \qquad (9)$$

$\Delta E_{A,0}$ can be derived from the temperature dependence of the Hall-effect data. The analysis of a large set of samples yielded activation energies in the range between 202 and 214 meV independent of the concentration of the acceptors (Götz et al., 1998). $\Delta E_{A,0}$ is expected to be close to the true ionization energy of Mg acceptors in GaN (i.e., energy difference between the acceptor level and the valence band maximum).

Acknowledgments

The authors are pleased to thank J. Neugebauer and C. G. Van de Walle for helpful discussions and J. Walker for technical support. They also thank D. P. Bour (Xerox PARC), H. Liu (Hewlett-Packard Company), R. Molnar (Lincoln Laboratory), and M. D. Bremser (NCSU) for supplying the nitride materials. This work was partially supported by DARPA (agreement No. MDA972-95-3-008).

References

Ager III, J. W., Suski, T., Ruvimov, S., Krueger, J., Conti, G., Weber, E. R., Bremser, M. D., Davis, R., and Kuo, C. P. (1997). Intrinsic and thermal stress in gallium nitride epitaxial films. *Mat. Res. Soc. Symp. Proc.*, **449**, 775–780.

Akasaki, I. and Amano, H. (1996). Crystal growth of column-III nitride semiconductors and their electrical and optical properties. *J. Crystal Growth*, **163**, 86–92.

Amano, H., Kito, M., Hiramatsu, K., and Akasaki, I. (1989). P-type conduction in Mg-doped GaN treated with low-energy electron beam irradiation (LEEBI). *Jpn. J. Appl. Phys.*, **28**, L2112–L2114.

Blakemore, J. (1962). *Semiconductor Statistics*, p. 170. New York: Pergamon.

Calleja, E., Sánchez, F. J., Basak, D., Sánchez-Garciá, M. A., Muñoz, E., Izpura, I., Colle, F., Tijero, J. M. G., Sánchez-Rojas, J. L., Beaumont, B., Lorenzini, P., and Gibert, P. (1997). Yellow luminescence and related deep states in undoped GaN. *Phys. Rev. B*, **55**, 4689–4694.

Chantre, A., Vincent, G., and Bois, D. (1981). *Phys. Rev. B*, **23**, 5335.

Chung, B-C. and Gershenzon, M. (1992). The influence of oxygen on the electrical and optical properties of GaN crystals grown by metalorganic vapor phase epitaxy. *J. Appl. Phys.*, **72**, 651–659.

Glaser, E. R., Kennedy, T. A., Doverspike, K., Rowland, L. B., Gaskill, D. K., Freitas, Jr., J. A., Khan, M. Asif, Olson, D. T., Kuznia, J. N., and Wickenden, D. K. (1995). Optically-detected magnetic resonance of GaN films grown by organometallic chemical vapor deposition. *Phys. Rev. B*, **51**, 13326–13336.

Götz, W. and Johnson, N. M. (1994). Deep Level Transient Spectroscopy: A Case Study in GaAs. In *Characterization in Compound Semiconductor Processing*. G. E. McGuire and Y. E. Strausser, eds., pp. 109–123. Boston: Butterworth-Heinemann.

Götz, W., Johnson, N. M., Amano, H., and Akasaki, I. (1994). Deep level defects in n-type GaN. *Appl. Phys. Lett.*, **65**, 463–465.

Götz, W., Johnson, N. M., Street, R. A., Amano, H., and Akasaki, I. (1995). Photoemission capacitance transient spectroscopy of n-type GaN. *Appl. Phys. Lett.*, **66**, 1340–1342.

Götz, W., Johnson, N. M., and Bour, D. P. (1996a). Deep level defects in Mg-doped, p-type GaN grown by metalorganic chemical vapor deposition. *Appl. Phys. Lett.*, **68**, 3470–3472.

Götz, W., Johnson, N. M., Bour, D. P., Chen, C., Kuo, C., Liu, H., and Imler, W. (1996b). Shallow dopants and the role of hydrogen in epitaxial layers of gallium nitride. *Electrochem. Soc. Proc.* **96–11**, 87–99.

Götz, W., Johnson, N. M., Bremser, M. D., and Davis, R. D. (1996c). A donor like deep level defect in $Al_{0.12}Ga_{0.88}N$ characterized by capacitance transient spectroscopies. *Appl. Phys. Lett.*, **69**, 2379–2381.

Götz, W., Johnson, N. M., Chen, C., Kuo, C., and Imler, W. (1996d). Activation energies of Si donors in GaN. *Appl. Phys. Lett.*, **68**, 3144–3146.

Götz, W., Johnson, N. M., Bour, D. P., McCluskey, M. D., and Haller, E. E. (1996e). Local vibrational modes of the Mg-H acceptor complex in GaN. *Appl. Phys. Lett.*, **69**, 3725–3727.

Götz, W., Romano, L. T., Johnson, N. M., and Krusor, B. S. (1996d). Electronic and structural properties of GaN grown by hydride vapor phase epitaxy. *Appl. Phys. Lett.*, **69**, 242–244.

Götz, W., Walker, J., Romano, L. T., and Johnson, N. M. (1997). Thickness dependence of electronic properties of GaN epi-layers. *Mat. Res. Soc. Symp. Proc.* **449**, 525–528.

Götz, W., Johnson, N. M., Walker, J., Bour, D. P., and Street, R. A. (1996g). Activation of acceptors in Mg-doped GaN grown by metalorganic chemical vapor deposition. *Appl. Phys. Lett.*, **68**, 667–669.

Götz, W., Kern, R. S., Chen, C., Liu, H. Steigerwald, D. A., and Fletcher, R. M. (1998). Hall-effect characterization of III-V nitride semiconductors for high efficiency light emitting diode. Proceedings of the E-MRS 1998 Conference. Symposium L. Materials Science and Engineering B, in press.

Haase, D., Schmid, M., Dörnen, A., Härle, V., Bolay, H., Scholz, F., Burkhard, M., and Schweizer, H. (1996). DLTS and CV analysis of doped and N-implanted GaN. *Mat. Res. Soc. Symp. Proc.*, **423**, 531–537.

Hacke, P., Detchprohm, T., Hiramatsu, K., Sawaki, N., Tadatomo, K., and Miyake, K. (1994). Analysis of deep levels in n-type GaN by transient capacitance methods. *J. Appl. Phys.*, **76**, 304–309.

Hacke, P., Nakayama, H., Detchprohm, T., Hirmatsu, K., and Sawaki, N. (1996). Deep levels in the upper band-gap region of lightly Mg-doped GaN. *Appl. Phys. Lett.*, **68**, 1362–1364.

Hacke, P., and Okushi, H. (1997). Characterization of the dominant midgap levels in Si-doped GaN by optical-isothermal capacitance transient spectroscopy. *Appl. Phys. Lett.*, **71**, 524–526.

Hartke, J. L. (1968). The three-dimensional Poole-Frenkel effect. *J. Appl. Phys.*, **39**, 4871–4873.

Herring, C. (1960). Effect of random inhomogeneities on electrical and galvanomagnetic measurements. *J. Appl. Phys.*, **31**, 1939–1953.

Ilegems, M. and Montgomery, H. C. (1973). Electrical properties of n-type vapor-grown gallium nitride. *J. Phys. Chem. Solids*, **34**, pp. 885–895.

Lefevre, H. and Schulz, M. (1977). Double correlation technique (DDLTS) for the analysis of deep level profiles in semiconductors. *Appl. Phys.*, **12**, 45–53.

Miller, G. L., Lang, D. V., and Kimerling, L. C. (1977). Capacitance Transient Spectroscopy. In *Annual Review of Materials Science*, **7**, 377–448.

Molnar, R. J., Lei, T., and Moustakas, T. D. (1993). Electron transport mechanism in gallium nitride. *Appl. Phys. Lett.*, **62**, 72–74.

Molnar, R. J., Götz, W., Romano, L. T., and Johnson, N. M. (1996). Hydride vapor phase epitaxy of gallium nitride films for quasi-bulk substrates. *Electrochem. Soc. Proc.*, **96-11**, 212–226.

Nakamura, S., Mukai, T., and Senoh, M. (1994). Candela-class high-brightness InGaN/AlGaN double-heterostructure blue-light-emitting diodes. *Appl. Phys. Lett.*, **64**, 1687–1689.

Nakamura, K. S., Mukai, T., and Senoh, M. (1992). Si- and Ge-doped GaN films grown with GaN buffer layers. *Jpn. J. Appl. Phys.*, **31**, 2883–2888.

Nakamura, S., Iwasa, N., Senoh, M., and Mukai, T. (1992). Hole compensation mechanism of p-type GaN films. *Jpn. J. Appl. Phys.*, **31**, 1258–1266.

Neugebauer, J. and Van de Walle, C. G. (1995). Role of hydrogen in doping of GaN. *Appl. Phys. Lett.*, **68**, 1829–1831.

Neugebauer, J. and Van de Walle, C. G. (1996). Atomic geometry and electronic structure of native defects in GaN. *Phys. Rev. B*, **50**, 8067–8070.

Schubert, E. F. (1993). Doping in III-V Semiconductors. In *Cambridge Studies in Semiconductor Physics and Microelectronic Engineering: 1*, pp. 482–487. London: University Press.

Schubert, E. F. (1993). Doping in III-V Semiconductors. *Cambridge Studies in Semiconductor Physics and Microelectronic Engineering: 1*, pp. 34–53, London: University Press.

Tanaka, T., Watanabe, A., Amano, H., Kobayashi, Y., Akasaki, I., Yamazaki, S., and Koike, M. (1994). P-type conduction in Mg-doped GaN and $Al_{0.08}Ga_{0.92}N$ grown by metalorganic chemical vapor phase epitaxy. *Appl. Phys. Lett.*, **65**, 593–594.

Van der Pauw, L. J. (1958). A method of measuring specific resistivity and Hall effect of discs of arbitrary shape. *Philips Res. Repts.*, **13**, 1–9.

Van Vechten, J. A., Zook, J. D., Hornig, R. D., and Goldenberg, B. (1992). Defeating compensation in wide gap semiconductors by growing in H that is removed by low temperature de-ionizing radiation. *Jpn. J. Appl. Phys.*, **31**, 3662-3663.

Yi, G-C. and Wessels, B. C. (1996). Deep level defects in Mg-doped GaN. *Mat. Res. Soc. Symp. Proc.*, **423**, 525–530.

CHAPTER 6

Stress Effects on Optical Properties

Bernard Gil

CENTRE NATIONAL DE LA RECHERCHE SCIENTIFIQUE
GROUPE D'ETUDE DES SEMICONDUCTEURS
UNIVERSITÉ DE MONTPELLIER II
MONTPELLIER, FRANCE

I. INTRODUCTION 209
II. THE CRYSTALLINE STRUCTURES OF III-NITRIDES 213
III. EFFECTS OF STRAIN FIELDS ON THE ELECTRONIC STRUCTURE OF WURTZITE III
   NITRIDES 213
   1. *The Problem of the Determination of the Bandgap of GaN* 213
   2. *Influence of Strain Fields on the Band Structure of III Nitrides* 217
   3. *Deformation Potentials of Wurtzite GaN* 227
IV. EXCITON OSCILLATOR STRENGTHS AND LONGITUDINAL-TRANSVERSE SPLITTINGS . 234
   1. *Exchange Interaction and Strain-Induced Variations of the Oscillator
      Strengths for Γ_5 and Γ_1 Excitons* 234
   2. *Strain-Induced Variations of the Longitudinal-Transverse Splittings* 239
V. ORIGIN OF THE STRAIN 244
VI. PHONONS UNDER STRAIN FIELDS 254
VII. SHALLOW VS DEEP-LEVEL BEHAVIOR UNDER HYDROSTATIC PRESSURE 257
VIII. INFLUENCE OF STRAIN FIELDS ON OPTICAL PROPERTIES OF GaN-AlGaN
    QUANTUM WELLS 259
    1. *Introduction* 259
    2. *Dispersion Relations in the Valence Band of GaN* 260
    3. *Valence Band Levels and Excitons in GaN-AlGaN Quantum Wells* 262
IX. SELF-ORGANIZED QUANTUM BOXES 267
X. CONCLUSION 268
   REFERENCES 270

I. Introduction

This chapter treats strain effects on nitride semiconductor compounds. Before dealing with these materials it is necessary to recall some basic properties of strain and to review briefly some of the effects evidenced during the last decades of intense investigations of zinc blende semiconductors. This introduction is followed by what is, for the III-nitrides, a state-of-the-art of the physics of strain effects, and concomitantly a correlation with similar phenomena in more classical zinc blende semiconductors will also be presented.

The application of a homogeneous strain in a solid changes the equilibrium position of the atoms, *may* change the crystal symmetry and *will* change the electronic band structure *and* the phonon spectrum. When this homogeneous strain field is moderate, the modifications of the lattice parameter are small and the strain-induced variations of the electronic or vibrational states can be described as a perturbation of the quantum states in the strain-free crystal. If the strain is sufficiently strong, the crystal will undergo one (or several) phase transitions toward a more compact structure with significant implications on band structure and phonon modes (the best probes of chemical bonding).

In semiconductors, even for strain below the critical value for the onset of a phase transition, the energy gaps are altered and, in some cases, degeneracies are lifted, or the symmetry of the fundamental conduction or valence band can be changed (Pikus and Bir, 1974). It is very convenient to express the strain-induced modifications of the energy of each critical point using deformation potentials. These quantities represent the changes in energy per unit strain and are of the order of electronvolt. In the simple picture of a chemical bond for the hydrogen molecule, the splitting between the bonding and antibonding orbital increases with the increase of the interatomic overlap integral, when the hydrogen–hydrogen distance decreases. The corresponding deformation potential is negative if compression is taken to be negative. In real crystals, the situation is much more subtle:

First, there is a large number of deformation potentials, which are defined from joint considerations of group theory and quantum mechanics by extensive application of the theory of invariants for the lattice considered (Pikus and Bir, 1974).

Second, given a critical point, because, in short the deformation potentials depend on the wavefunction of the crystal for this wavenumber and energy (in other words of the whole band structure at this point), their signs and values may not be trivial. For example, in cubic semiconductors, under pressure, the direct bandgap at zone center Γ-Γ increases, and the Γ-X indirect one decreases (Camphausen *et al.*, 1971). This difference is due to a significant contribution of d states to wavefunction of the conduction band at X. Accurate computation of such deformation potentials is not easy, and it requires conceptualizing complicated models that rely on progressively more experimental determinations and with time-increasing degrees of sophistication.

Under strain field, it has been shown that the nature of the fundamental bandgap of several semiconductors, for example, GaAs, GaSb or their alloys with aluminum, could be switched from direct to indirect in reciprocal space, with, for instance, a significant collapse of the radiative recombination efficiency (Wolford and Bradley, 1985). It has also been shown by application of external strain fields that untrivial transport properties in these compounds could be related to the existence of localized states, deeply resonant with the continuum of the conduction band (Zhao et al., 1984; Wolford et al., 1984). The wavefunctions of these levels are spread along the Γ, X, and L extrema and their pressure coefficients in direct bandgap crystals are smaller than the one for a fundamental bandgap. When pushing them into the gap of the semiconductor, pressure can reveal them explicitly because the electron statistics are drastically modified or they can be revealed via detection of additional photoluminescence peaks, saturation of mobility, or modification of the coupling of LO phonons with the radiation field.

In the area of the heteroepitaxy of silicon-related, zinc blende III-V and II-VI low-dimensional semiconductors, strain fields have also been shown to be very important: most of the binary materials are lattice mismatched with the commercial substrates Si, GaAs, and InP extensively used in the recently developed modern electronics, as shown in Fig. 1. In contrast to earlier studies on strain effects on bulk crystals, in the case of heteroepitaxies, strain is no longer an external perturbation applied to a crystal: it is in the crystal, it influences its growth, and it may favor growth of novel structural phases (Froyen et al., 1988). Coherent epitaxy of a semiconductor layer lattice mismatched to the substrate cannot exceed a critical value, beyond which the amount of elastic energy in the deformed layer is sufficient to generate a plethora of lattice dislocations, that is, topological defects that have a disastrous influence on mobility and lifetime (Biefeld, 1989). In zinc blende crystals several approaches have been attempted in order to compute the critical thickness for coherent growth of an epilayer as a function of its lattice mismatch with the substrate. When reaching and slightly exceeding the critical thickness for coherent growth, a transition occurs between the two-dimensional growth mode and a stochastic three-dimensional one. Below this critical thickness, coherent two-dimensional epitaxy occurs and the material quality is satisfactory, but the band structure of the deposited layer differs from the band structure of the unstrained bulk due to the homomorphism of the in-plane lattice parameter of the epitaxy with the one of the substrate. There are also subsequent effects on the disperson relations, joint densities of states for the interband absorption processes, and light polarization effects (Mailhiot and Smith, 1990). The advantages of these effects have been maximized by designers in the area of strained-layer

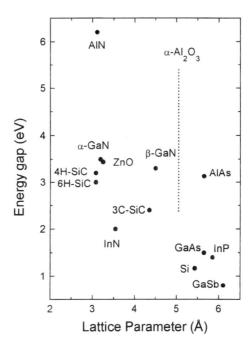

FIG. 1. Bandgap of several cubic and hexagonal semiconductors vs their lattice parameters. For wurtzitic compounds, the plot is given as a function of a.

devices such as quantum well lasers or optical modulators (Pearsall, 1990). It has also been found to be useful to have strained epilayers with strong built-in piezoelectric field if grown in directions other than [001] (Smith and Mailhiot, 1990).

It may be energetically favorable to have the strained layer grown as a self-organized island of material. This is caused by a balancing effect between the reduction of island elastic energy (the material on the top of the islands can relax elastically in contrast to the strained-layer case) and the increase of the surface energy (the area increases). Islanding is possible when the former energy dominates the latter (Ledentsov, 1996). These islands may have controlled sizes if growth occurs so that ripening susceptible to minimize the surface energy term occurs. Such islands are generally several monolayers high, with small lateral dimensions. When the island formation follows a Stranski-Krastanov growth mode, they lie upon a one-monolayer thick layer of the deposited material, often referred to as a "wetting" layer. Quantum dots are obtained when these islands are grown with a capping barrier material. There is a subsequent significant confinement energy of the carriers in these zero-dimensional semiconductors. InAs-GaAs is a prototype case (Leonard et al., 1993). It is interesting to note that there are *no* dislocations in these dots, and that injection lasers based on single sheets of quantum dots demonstrates lasing via the dot ground state, low threshold current density, ultrahigh material gain and ultrahigh temperature stability of the threshold current (Ledentsov, 1996).

II. The Crystalline Structures of III-Nitrides

There are three common crystal structures shared by the III-nitrides: the wurtzitic, zinc blende, and rocksalt structures. At ambient conditions the thermodynamically stable structures are wurtzite for bulk AlN, GaN, and InN. Nitrides can be forced to grow in the zinc blende phase by epitaxial growth of thin films on cubic substrates such as Si, MgO, and GaAs. The rocksalt structure can be induced at very high pressures. In GaN, 50 GPa hydrostatic pressure is required to produce the phase transition from the wurtzitic to the NaCl phase. This occurs at lower values for AlN (14 GPa) and InN (12 GPa). In terms of lattice constants, GaN has the following parameters: $a_w \sim 3.19$ Å and $c_w \sim 5.185$ Å for the wurtzite phase; $a_c \sim 4.5$ Å for the cubic stacking; and $a_r \sim 4.2$ Å for the rocksalt structure. For AlN, the corresponding values for the three phases are $a_w \sim 3.11$ Å and $c_w \sim 4.98$ Å; $a_c \sim 4.38$ Å; and $a_r \sim 4.05$ Å, respectively. For InN, the wurtzite parameters are $a_w \sim 3.54$ Å; $c_w \sim 5.70$ Å; the zinc blende crystal parameter is $a_c \sim 4.98$ Å (Popovici et al., 1998). In this chapter we will consider the wurtzite phases P6$_3$mc (C_{6v}^4) of the III-nitrides.

III. Effects of Strain Fields on the Electronic Structure of Wurtzite III Nitrides

1. THE PROBLEM OF THE DETERMINATION OF THE BANDGAP OF GaN

From simple arguments based on chemical bonding, and using the language of atomic physics, one can say that, at zone center, the Bloch wavefunction of the lowest conduction band of the usual semiconductors is essentially built from *s* atomic functions of the anion while *p* atomic functions of the anion dominate the wavefunction of the valence band. Transfer of these locutions to the language of group theory requires inclusion of actual crystal symmetry and the use of labels linked to irreducible representations of the crystal point group. In a cubic (T_d)

environment the threefold (Γ_5) and single (Γ_1) irreducible representations account for the symmetries of the spinless valence and conduction electron (Koster et al., 1963; Klingshirn, 1995). A general result of group theory is to predict that, under any lowering of cubic symmetry, the Γ_5 representation splits. *A wurtzite (C_{6v}) crystal field results in a discrimination of z-like valence states from x-like and y-like ones.* This produces a valence band splitting sketched in Fig. 2 (left-hand side). On the basis of eigenvectors of the angular momentum L, we usually write it (Hopfield, 1960) $\Delta_1 L_z^2$. Theoretical calculations (Lambrecht et al., 1996) show that the variations of Δ_1 are linear with the ratio c/a as shown in Fig. 3. The band-to-band transition between the Γ_1 conduction band and the Γ_1 valence band is allowed in π polarization ($\boldsymbol{E} // \boldsymbol{c}$) while the band-to-band transition between the Γ_1 conduction band and the Γ_5 valence band is allowed in σ polarization ($\boldsymbol{E} \perp \boldsymbol{c}$).

The double group, which takes into account the existence of a spin, is more adapted than the simple group to describe symmetry properties of electrons (Bassani and Parravicini, 1979). Among the relativistic corrections, we neglect the relativistic correction to the kinetic energy (known to give a smooth deformation of the band structure), and the angular momentum-dependent correction to the potential (the Darwin term). Neither of them gives any splitting, *which is not the case* for the spin-orbit term (H_{so} hereafter), and which couples the operators σ in spin space with the angular momentum $\boldsymbol{L}$ in ordinary space. We parametrize H_{so} and prescript it to be

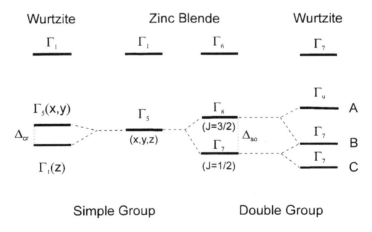

FIG. 2. Sketch of the symmetries of the conduction and valence band of the zinc blende (middle of the figure) and wurtzitic direct bandgap semiconductors. The left-hand side corresponds to the simple group; the right-hand side shows symmetries in the context of the double group.

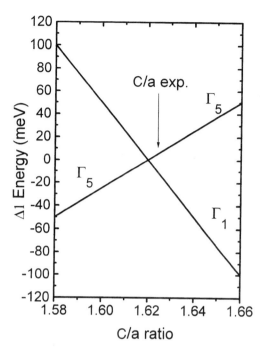

FIG. 3. Plot of the evolution of the crystal field splitting in GaN as a function of the C/a ratio. See Lambrecht et al. (1996).

invariant under symmetry operations in both spin and ordinary spaces. This gives: $H_{so} = 1/3\Delta_{so}(\mathbf{L}\cdot\sigma)$ and $H_{so} = \Delta_2 \mathbf{L}_z\sigma_z + \Delta_3(\mathbf{L}_x\sigma_x + \mathbf{L}_y\sigma_y)$ in cubic and wurtzite symmetry, respectively (Cho, 1976). The resulting effect is dramatic for the valence band and is also sketched in Fig. 2 (right-hand side) where the ordering of the valence sublevels is given for ZnS. The band-to-band transitions between the Γ_7 conduction band and the Γ_7 valence bands are allowed in σ and π polarization ($\mathbf{E}\perp\mathbf{c}$ and $\mathbf{E}//\mathbf{c}$), whereas the band-to-band transition between the Γ_7 conduction band and the Γ_9 valence band is allowed in σ polarization ($\mathbf{E}\perp\mathbf{c}$).

Figure 4 illustrates the observation of selection rules in the polarized reflectance experiments that enabled Dingle et al. (1971) to propose the first determination of the GaN bandgap, crystal field splitting and spin-orbit interaction. On the figure, the extinction of line A for the π polarization is well shown, as well as the dependence of the oscillator strengths of the other transitions with the polarization of the incident photon. Works performed on other samples (Monemar, 1974; Shan et al., 1995; Orton, 1996) showed relatively large differences in the energy positions of excitonic transitions,

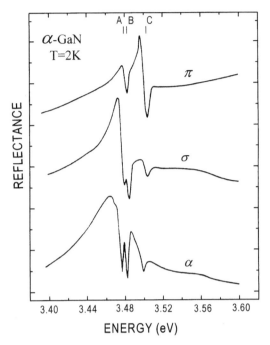

FIG. 4. The 2 K reflectance spectra taken on one of the various GaN samples studied by Dingle et al. (1971), for π, σ, and α polarization conditions.

which were reported using different spectroscopic techniques. This caused a significant amount of confusion until Gil et al. (1995) quantitatively interpreted this scattering in terms of residual built-in strain. What they did first was to retain data taken by reflectance or photoreflectance spectroscopy, to ensure that the energies people were measuring were due to free exciton, and thus were not the result of a radiative extrinsic recombination process. We note that they restricted their examination to samples grown on sapphire along the [0001] direction. They then plotted the energies of lines A, B, C as a function of the energy position of line A as shown in Fig. 5. The straight dotted lines which connect the different energies are nice evidence of a correlation between the energies of A, B, and C lines. Such a correlation results in different strain states in the different samples. We note that the samples used were grown during some 25 years of investigations by researchers using very different growth techniques (Gil et al., 1995). From this observation it was concluded that the bandgap, the crystal field splitting of the topmost valence band, and the spin-orbit interaction were not established in unstrained GaN. However, it was obvious that the consider-

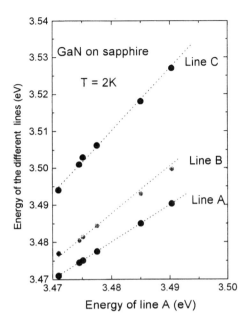

FIG. 5. Plot of the energies of A, B, and C lines measured near 2 K by reflectance spectroscopy, against energy of line A. The dashed lines indicate the correlation between these energies. The slopes are 1, 1.24, and 1.68. Details are extensively given in Gil et al. (1995).

able amount of theoretical and experimental work on the effects of strain fields on the electronic structure of semiconductors could be used in addressing this issue.

2. INFLUENCE OF STRAIN FIELDS ON THE BAND STRUCTURE OF III NITRIDES

a. *Basic Considerations and Strain Hamiltonians*

In the most general way, the time-reversal independent Hamiltonian $H_{wurtzite}$ for the valence band in wurtzite semiconductors is given as follows from straightforward applications of invariants theory (Cho, 1976):

$$\begin{aligned}H_{wurtzite} =& [a_1 + a_2 L_z^2 + 2a_3 L_z \sigma_z + 2a_4(L_x\sigma_x + L_y\sigma_y)S + \sqrt{2}b_1(L_x\sigma_y - L_y\sigma_x)]T \\ &+ 2\sqrt{2}e_1(XL_x + YL_y)\sqrt{2} + 2e_2(X\sigma_x + Y\sigma_y)L_z \\ &+ 2\sqrt{2}e_3(X\{L_x,L_z\} + Y\{L_y,L_z\}) + f_1[W(L_x^2 - L_y^2) + 2Z\{L_x,L_y\}] \\ &+ \sqrt{2}f_2[(L_x\sigma_x - L_y\sigma_y)W + (L_x\sigma_y - L_y\sigma_x)Z]\end{aligned} \quad (1)$$

TABLE I

SIMPLE EXAMPLES OF TIME-REVERSAL INDEPENDENT X IN WURTZITE SYMMETRY

Representation	X	Operators
Γ_1, A_1	S	$C_{6v}, z, k_z^2, (k_x^2 + k_y^2), e_{zz}, (e_{xx} + e_{yy})$
Γ_2	T	
Γ_5, E_1	(X, Y)	$(x, y), (k_x k_z, k_y k_z), (e_{xz}, e_{yz})$
Γ_6, E_2	(V, W)	$[(k_x^2 - k_y^2), 2k_x k_y], [(e_{xx} - e_{yy}), 2e_{xy}]$

In this equation, physical quantities S, T, X, Y, W, and Z are associated with representative variables having the $\Gamma_1(S)$, $\Gamma_2(T)$, $\Gamma_5(X, Y)$, and $\Gamma_6(W, Z)$ symmetries (Table I) and L_i and σ_j are the components of the orbital and spin hole angular momenta. Coefficients $a_1, a_2, a_3, a_4, b_1, e_1, e_2, e_3, f_1$, and f_2 linked to $S, T, \ldots$ correspond to crystal field splitting or spin-orbit interaction when $X = C_{6v}$, to the deformation potentials when X is the strain tensor. This equation will also be extremely useful to treat, *mutatis mutandis*, the valence band dispersion relations when X is K (k_x, k_y, k_z), the wavevector. It is also worth noticing that, as it will be shown further, the growth along the [0001] direction does not affect the wurtzite crystal symmetry, which means that the only nonvanishing components of the strain tensor have Γ_1 symmetry.

In this chapter, we will limit this Hamiltonian to most *spin-independent* deformation potentials. Because of the strongly ionic character, the only contribution for the nitrides (except for InN) to the spin orbit interaction at the topmost of the valence band is from the atomic N2p states (13.6 meV in the free atom). The contribution from the lower cation d states (Ga3d or In4d) is negative and much more significant for InN (Lambrecht et al., 1996). As a consequence, the spin orbit splitting is small in nitrides. Its strain-induced variations are second-order relativistic corrections, which can be reasonably ignored. For the convenience of any future discussion, we label the valence band deformation potentials following Pikus and Bir (1974). The strain Hamiltonian is

$$\mathbf{H}_{\text{strain}} = (C_1 + C_3 L_z^2)e_{zz} + (C_2 + C_4 L_z^2)e_\perp + C_5(L_-^2 e_+ + L_+^2 e_-) + C_6\{[L_z L_+]e_{+z} + [L_z L_-]e_{-z}\} \quad (2)$$

where

$$L_\pm = 1\sqrt{2}(L_x \pm iL_y), \quad 2[L_i L_j] = L_i L_j + L_j L_i \quad (3)$$

$$e_\perp = e_{xx} + e_{yy}, \quad e_\pm = e_{xx} \pm 2ie_{xy} - e_{yy}, \quad e_{\pm z} = e_{xz} \pm ie_{yz} \quad (4)$$

The total time-reversal independent Hamiltonian $\mathbf{H}_{\text{wurtzite}}$ for the valence band in wurtzite semiconductors is $\mathbf{H}_{\text{wurtzite}} = \mathbf{H}_{\text{strain}} + H_{\text{cr}} + H_{\text{so}}$.

The number of deformation potentials is larger than for a cubic crystal. Among the six valence band deformation potentials, C_1 and C_2 are *pure hydrostatic deformation potentials* of the zinc blende type, and the second series of deformation potentials, C_3 and C_4, account for the strain-induced modification of the crystal field splitting. The remaining two, C_5 and C_6, are *shear deformation potentials in the strictest sense*, the analogues of quantity labeled d in zinc blende crystals.

The strain Hamiltonian acts as follows on the conduction band (Pikus and Bir, 1974; Cho, 1976):

$$H_{cstrain} = D_1 e_{zz} + D_2 e_\perp \tag{5}$$

It is very convenient for clarity of presentation to introduce the following notations: $\delta_1 = (D_1 + C_1)e_{zz} + (D_2 + C_2)e_\perp = a_1 e_{zz} + a_2 e_\perp$, $\delta_2 = C_3 e_{zz} + C_4 e_\perp$. For the purpose of the calculation of energy transitions, from the p-like basis functions, which can be noted p_x, p_y, and p_z or even simpler: x, y, z, we define the eigenvectors of the angular momentum $|+\rangle$ (or p_+), $|-\rangle$ (or p_-) as the following linear combinations: $|+\rangle = |p_+\rangle = -(p_x + ip_y)/\sqrt{2}$, and $|-\rangle = |p_-\rangle = (p_x - ip_y)/\sqrt{2}$.

b. *Strain Hamiltonian for Samples Grown Along the $\langle 0001 \rangle$ Direction, on Substrates such as C Plane Sapphire*

The epitaxial relation between C-plane sapphire and C-plane GaN occurs as shown in Fig. 6 (Nakamura and Fasol, 1997). The two lattices are rotated 60° with respect to each other so that the strong lattice mismatch between hexagonal sapphire ($a = 5.19$ Å) and GaN ($a = 3.19$ Å) is minimized. There is thus a coherence of the two lattices along the sixfold axis. The in-plane thermal expansion coefficient of sapphire is $7.5 \times 10^{-6} \, K^{-1}$, the corresponding GaN one is $5.6 \times 10^{-6} \, K^{-1}$. The experimental investigations of the GaN layers are made at temperatures considerably lower than for the growth one, and the GaN layers experience a biaxial compression in the growth plane (Amano *et al.*, 1988).

The corresponding stress tensor has the following form:

$$\begin{array}{ccc} |X\rangle & |Y\rangle & |Z\rangle \end{array}$$
$$\begin{bmatrix} \zeta & 0 & 0 \\ 0 & \zeta & 0 \\ 0 & 0 & 0 \end{bmatrix} \tag{6}$$

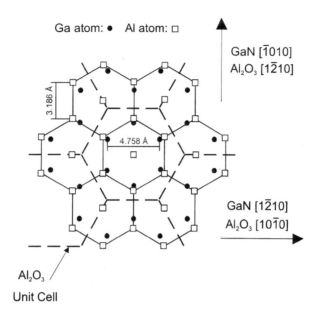

FIG. 6. Relative orientation of the GaN and sapphire lattices in the C plane. Note the 60 °C rotation of the crystallographic axis a of GaN with respect to the a axis of Al_2O_3.

where Z is parallel to c. The compliance tensor of the $S_{ij}s$ connects the component of the strain to the stress. In wurtzite crystals, this equation holds (Nye, 1957):

$$\begin{bmatrix} e_{xx} \\ e_{yy} \\ e_{zz} \\ 2e_{yz} \\ 2e_{xz} \\ 2e_{xy} \end{bmatrix} = \begin{bmatrix} S_{11} & S_{12} & S_{13} & 0 & 0 & 0 \\ S_{12} & S_{11} & S_{13} & 0 & 0 & 0 \\ S_{13} & S_{13} & S_{33} & 0 & 0 & 0 \\ 0 & 0 & 0 & S_{44} & 0 & 0 \\ 0 & 0 & 0 & 0 & S_{44} & 0 \\ 0 & 0 & 0 & 0 & 0 & 2(S_{11}-S_{12}) \end{bmatrix} \cdot \begin{bmatrix} \xi_{xx} \\ \xi_{yy} \\ \xi_{zz} \\ \xi_{yz} \\ \xi_{xz} \\ \xi_{xy} \end{bmatrix} \quad (7)$$

If we assume the applicability of elasticity theory, in case of growth on C-plane sapphire, the nonvanishing components of the strain tensor are

$$e_{xx} = e_{yy} = (S_{11} + S_{12})\xi \quad (8)$$

$$e_{zz} = 2S_{13}\xi \quad (9)$$

TABLE II

WAVEFUNCTIONS FOR THE VALENCE BAND — ↑ AND ↓ ARE THE
SPIN COMPONENTS IN THE VALENCE BAND

$|\Gamma_9\rangle$	$|\Gamma_7^1\rangle$	$|\Gamma_7^2\rangle$
$\|+\rangle\uparrow$	$\|-\rangle\uparrow$	$\|z\rangle\downarrow$
$\|-\rangle\downarrow$	$\|+\rangle\downarrow$	$\|z\rangle\uparrow$

In terms of symmetrized combinations along the irreducible representations of C_{6v}, the nonvanishing ones have the Γ_1 symmetry: $e\Gamma_1^1 = e_{zz}$, and $e\Gamma_1^2 = (e_{xx} + e_{yy}) = e_\perp$. The important point we wish to outline here is that, in contrast with the situation in the case of zinc blende crystals, the isotropic in-plane biaxial stress does not reduce the crystal symmetry.

Related to the $\{|\Phi\Gamma_9\rangle, |\Phi\Gamma_7^1\rangle, |\Phi\Gamma_7^2\rangle\}$ basis set, which wavefunctions are given in Table II, the strain-induced shifts of the valence band levels are eigenvalues of the following matrix:

$$\begin{bmatrix} |\Phi\Gamma_9\rangle & |\Phi\Gamma_7^1\rangle & |\Phi\Gamma_7^2\rangle \\ \langle E_0^v\rangle + \Delta_1 + \Delta_2 + C_1 e_{zz} + C_2 e_\perp + C_3 e_{zz} + C_4 e_\perp & 0 & 0 \\ 0 & \langle E_0^v\rangle + \Delta_1 - \Delta_2 + C_1 e_{zz} + C_2 e_\perp + C_3 e_{zz} + C_4 e_\perp & \sqrt{2}\Delta_3 \\ 0 & \sqrt{2}\Delta_3 & \langle E_0^v\rangle + C_1 e_{zz} + C_2 e_\perp \end{bmatrix} \tag{10}$$

The shifts of A, B, and C lines are, in this model, obtained by replacing the deformation potentials C_1 and C_2 by a_1 and a_2, respectively.

The eigenvalues are obtained analytically as

$$E_A = E_0 + \Delta_1 + \Delta_2 + \delta_1 + \delta_2 \tag{11}$$

$$E_{B,C} = E_0 + (\Delta_1 - \Delta_2)/2 + \delta_1 + \delta_2/2 \pm \frac{\sqrt{[\Delta_1 - \Delta_2 + \delta_2]^2 + 8\Delta_3^2}}{2} \tag{12}$$

It is very convenient to make a linear expansion of E_B and E_C a function of the stress (Gil et al., 1995; Sandomirskii, 1964):

$$E_{B,C} = E_{B,C}^0 + \delta_1 + \delta_2\left(1 \pm \frac{\Delta_1 - \Delta_2}{\sqrt{(\Delta_1 - \Delta_2)^2 + 8\Delta_3^2}}\right)\bigg/2 \tag{13}$$

or
$$E_{B,C} = E^0_{B,C} + \delta_1 + \delta_2(1 \pm \eta)/2 \qquad (14)$$

c. Determination of the Electronic Structure of Strain-Free GaN

In the dimensionless diagram where the energies of A, B, and C lines are plotted as a function of the shift of A, we have the following scale (Gil et al., 1995):

$$\delta_1 + \delta_2 = 1, \quad \delta_1 + \delta_2(1 + \eta)/2 = 1.24 \quad \text{and} \quad \delta_1 + \delta_2(1 - \eta)/2 = 1.68$$

Within the crudeness of the linear model and taking into account the experimental uncertainties, we find $\delta_1 \sim 1.86$, $\delta_2 \sim -0.86$, and $\eta \sim 0.44$. The interest of this linear model is that it gives a relation between δ_1 and δ_2, in other words a relation between four of the six deformation potentials. For nonlinear equations, we found the best fit to the data to be obtained using $\Delta_1 = 10 \pm 0.1$ meV, $\Delta_2 = 6.2 \pm 0.1$ meV, and $\Delta_3 = 5.5 \pm 0.1$ meV and strain-free values of A, B, and C lines at 3470.3 meV, 3476.6 meV and 3492.6 meV, respectively, at 2 K (Gil et al., 1995). Shromme (1998) suggests 3467 meV, 3469 meV, and 3485 meV. At this stage, the deformation potentials are unknown. Extrapolating the present calculation to mathematical situations where the A line is red-shifted with respect to its zero stress value, we predict the possibility of offsetting the crystal field splitting, and changing the ordering of the valence band levels (Gil et al., 1996). We also note that the extra diagonal coupling terms $\sqrt{2}\Delta_3$ are comparable or even sometimes larger than the energy splitting between the two diagonal terms, and we expect strong mixings and exchange of oscillator strengths for σ and π polarizations.

In σ polarization, where the electric field of the z-propagating incident photon is taken along the x axis, if we attribute value 0.5 to the strength of transition involving the $|\Phi\Gamma_9\rangle$ state, the strengths of the transitions involving the two other levels are **half** of the square of the modules of the contribution of $|\Phi\Gamma_7^1\rangle$ in the eigenvectors produced by resolution of the two level system in Eq. (10). In a cubic situation we would obtain one-half, one-sixth, and one-third, respectively.

Concerning now the π experimental configuration, when the electric field of the incident photon is collinear with z, the transition between conduction band and $|\Phi\Gamma_9\rangle$ is silent while the strengths of the other two processes are proportional to the square of the modules of the contribution of $|\Phi\Gamma_7^2\rangle$ in the eigenvectors produced by resolution of Eq. (10). The values obtained in cubic symmetry are 0, two-thirds, and one-third, respectively. Figure 7 reports the evolution of the band-to-band oscillator strengths for B and C

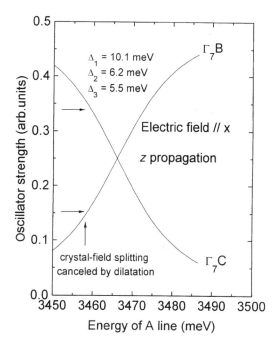

FIG. 7. Plot of the strain-induced evolution of the oscillator strengths of the B and C band-to-band transitions in σ polarization. The stress is expressed as a function of the position of line A. The strength of B (resp. C) increases (resp. decreases) when A is blue shifted with respect to the unstrained situation, that is, when the interband splittings increase and when the built-in compression increases.

lines when the energy of line A varies. This theoretical prediction is quantitatively observed as shown in Fig. 8 where several of the reflectance spectra that are plotted are taken for the σ polarizations on samples grown by metallorganic chemical vapor deposition (MOCVD) on C plane sapphire (Gil et al., 1997). When the energy of line A increases, we observe an increase of intensity of line B and a collapse of line C.

d. Optical Properties of GaN Films Deposited with Wurtzite Symmetry on 6H-SiC and Si

The in-plane thermal expansion coefficient (Madelung, 1991) of 6H-SiC is $4.2 \times 10^{-6}\,\mathrm{K}^{-1}$, the silicon one is $3.6 \times 10^{-6}\,\mathrm{K}^{-1}$. One expects layers grown on such substrates to experience a strong biaxial dilatation. Figure 9

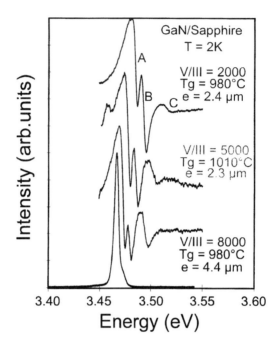

FIG. 8. Evolution of the lineshape of several reflectance spectra for GaN epilayers grown by MOVPE under several growth conditions. Note that the increase of B and the collapse of C are associated with the overall blue shift of the spectra when the biaxial compression increases. The spectra are taken with the permission of Gil and Briot (1997).

presents a typical 2 K reflectivity experiment taken for a sample grown on 6H SiC (Nelson et al., 1996). In contrast to the initial interpretation, we attribute to the line labeled B by Nelson et al. the actual signature of C. In addition, the well resolved, 4 meV splitted doublet A is nothing than a few A and B lines. This is consistent with our calculation as shown in Fig. 10. The remarkable strength of the line we associate to C, and which was observed at 3.489 eV, comparing with the strength of the low-energy doublet at 3.470–3.474 eV serves as well to reinforce the interpretation we address here. Edwards et al. (1997) reported extensive study of the optical properties of GaN on 6H SiC. In case of epitaxy on silicon, Chichibu et al. (1997) reported that transition energies at 3.455 eV, 3.440 eV and 3.463 eV were obtained at 10 K. These transitions are associated to lines A, B, and C, respectively, as shown by Fig. 10, and correspond to a situation for which reversal of the nature of the fundamental valence band of GaN has been produced by biaxial dilatation. It is interesting to note that there are some

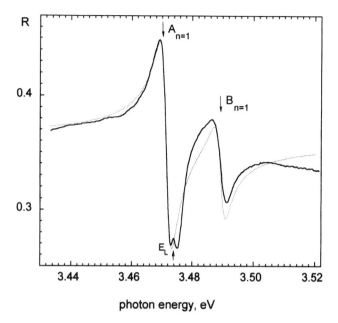

FIG. 9. Low-temperature reflectivity spectrum (full line) for GaN on 6H-SiC. The data are taken from Nelson et al. (1996) with their permission. The line shape fitting is given as dashed lines together with the initial identification of the transition, which has been revised, taking into account the biaxial dilatation in the epilayer.

gaps between the experimental data plotted in Fig. 10. We have to emphasize that the literature of the field is now extremely well disseminated (Buyanova et al., 1996; Volm et al., 1996; Tchounkeu et al., 1996; Kovalev et al., 1996; Shan et al., 1996; Chichibu et al., 1996; Shikanai et al., 1997). The experimental data reported on the figure have been restricted for the sake of clarity.

e. *Effects of Hydrostatic Pressure*

Under hydrostatic pressure, the stress tensor writes:

$$\begin{bmatrix} P & 0 & 0 \\ 0 & P & 0 \\ 0 & 0 & P \end{bmatrix} \quad (15)$$

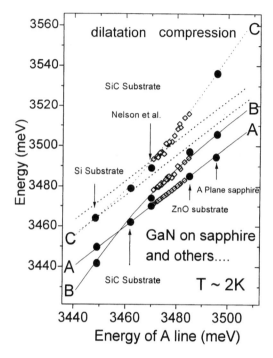

FIG. 10. Universal dependence of the energies of B and C lines against position of line A for GaN epilayers grown along the [0001] direction on C-plane oriented substrates and on A-plane oriented sapphire. Note the reversal of ordering of the valence band states for growth on silicon. See text for more details.

The components of the strain are now: $e_{xx} = e_{yy} = (S_{11} + S_{12} + S_{13})P$ and $e_{zz} = (S_{33} + 2S_{13})P$. We are dealing with equations similar to that of symmetrical built-in strain in the (0001) plane but with the strain-dependent coefficients changed. Studies of pressure dependence of the near bandgap absorption and of a broad deep photoluminescence band have appeared elsewhere for bulk GaN (Camphausen et al., 1971; Teisseyre et al., 1995; Perlin et al., 1992). Camphausen et al. (1971) reported 42 MeV/GPa. From photoluminescence data up to 5.5 GPa, Teisseyre et al. (1995) measured a larger value in zinc-doped GaN: 47 meV/GPa. Perlin et al. (1992), in a similar range of pressure, reported slight nonlinear dependence of the pressure-induced shift of the absorption coefficient at room temperature:

$$E_g(P) = E_g(0) + 47(P/\text{GPa}) - 0.36(P/\text{GPa})^2 \qquad (16)$$

Perlin et al. (1992) argued that this nonlinear behavior with pressure masked a linear dependence with lattice parameter. These values match very well with the linear dependence published in the theoretical paper by Gorczyca and Christensen (1993): 41 meV/GPa or with their more recent nonlinear value (Christensen and Gorczyca, 1994):

$$E_g(P) = E_g(0) + 39(P/\text{GPa}) - 0.32(P/\text{GPa})^2 \qquad (17)$$

Perlin et al. (1995) also measured the pressure shift of the yellow luminescence band and found the peak maximum to shift linearly with pressure up to 20 GPa, at a rate of 30 meV/GPa, then to be insensitive to the pressure up to 27 GPa.

More recently, the pressure dependence of the I_2 line was found to be 44 meV/GPa at 9 K and 47 ± 1 meV/GPa at 300 K from photoluminescence data on GaN epilayers on sapphire up to 7 GPa (Kim et al., 1995). This experiment reveals a slight nonlinear dependence with pressure and the pressure coefficient of the yellow band is estimated to be significantly lower: 34 ± 3 meV/GPa at 9 K and 33 ± 8 meV/GPa at 300 K. Shan et al. (1995) report, from experiments up to 6 GPa, a smaller value of 39 meV/GPa for the free exciton and a similar value for the yellow band, which, however, exhibits a larger nonlinear behavior than the near bandedge feature. The values of $\partial^2 E/\partial P^2$ they measured are -0.18 meV/GPa^2, -0.08 meV/GPa^2, and -2 meV/GPa^2 for the free exciton, donor-bound exciton and yellow band, respectively.

3. Deformation Potentials of Wurtzite GaN

a. a_1, a_2, C_3, C_4 in the Quasi-Cubic Model

These values have evolved over considerable time and are linked to an increase in collection of experimental information (especially with improved knowledge of GaN stiffness coefficients, which were only recently accurately measured). This evolution is illustrated in the discussion that follows. Experimental results taken from optical spectroscopy analysis establish several relations between some of these deformation potentials.

(i) The shift of B, C excitons vs the shift of the A line for biaxial (0001) stress. This gives the following relation (Gil et al., 1995):

$$a_1(2S_{13}) + 2a_2(S_{11} + S_{12}) \sim -2.2[C_3(2S_{13}) + 2C_4(S_{11} + S_{12})] \qquad (18)$$

(ii) Under hydrostatic pressure, the shift of the A, or free exciton line, is 42 meV/GPa, a value that is the average of the results of hydrostatic pressure measurements. We note that very careful experiments reported in recent papers regarding the effects of hydrostatic pressure on samples with needle-like optical transitions have not reported any differences between slopes of A, B, C excitons when resolved (Kim et al., 1995; Shan et al., 1995). This strongly suggests that δ_2 might be a vanishingly small quantity when the components of the strain are those produced by hydrostatic pressure (Tchounkeu, 1996). This gives the following relation: $C_4/C_3 \sim -0.5$, which is expected in the context of the quasi-cubic description and which states that a hexagonal lattice is very similar to a rhombohedrically deformed cubic lattice (Pikus and Bir, 1974). Then:

$$a_1(2S_{13}+S_{33})+2a_2(S_{11}+S_{12}+S_{13}) \sim 42\,\text{meV/GPa} = 42\,\text{meV/GPa} \quad (19)$$

*At this stage **one relation still misses** obtaining the absolute values of these for deformation potentials. It is required to **postulate an additional relation** such as a quasi-cubic or a cylindrical approximation, for performing quantitative computation.* Making the assumption $a_1 = a_2$ (cylindrical approach), we could obtain good agreement between computation and experiment for (001)-grown GaN layers using $a_1 = -8.16\,\text{eV}$, $C_3 = -2C_4 = -1.88\,\text{eV}$ (Tchounkeu, 1996), and taking for the elastic constants S_{ij} the values calculated from the old measurements of the stiffness coefficients: $C_{11} = 296\,\text{GPa}$, $C_{12} = 130\,\text{GPa}$, $C_{13} = 158\,\text{GPa}$, and $C_{33} = 267\,\text{GPa}$ (Savastenko and Sheleg, 1978).

Similar agreement between theory and experiment can be obtained if computing the **exciton** deformation potentials in the context of the quasi-cubic model introduced by Pikus and Bir to compare the valence-band physics in both true wurtzite environments and quasi-cubic ones (Pikus and Bir, 1974). This approximation states that $C_2 - C_1 = C_3$ and $C_3 = -2C_4$. The latter relation ($C_3 = -2C_4$) seems to be observed from hydrostatic pressure measurements in GaN as discussed by Tchounkeu et al. but not for wurtzite II-VIs as was shown in Table II of the same paper. Applying the quasi-cubic model to excitons in GaN requires postulating $D_1 = D_2$. Using for the elastic constants of GaN the values given here, this approximation leads to the following values: $a_1 = -6.85\,\text{eV}$, $a_2 = -8.84\,\text{eV}$, $C_3 = -2C_4 = -1.99\,\text{eV}$ (Gil and Alemu, 1997). In order to illustrate that getting these values is actually an issue, and to illustrate the difficulty one encounters in obtaining them experimentally and theoretically, we begin to compare them to the theoretical quasi-cubic data obtained by Suzuki and Uenoyama (1995) using a full-potential linearized augmented plane-wave (FLAPW) method with a local density approximation (LDA): $a_1 = -15.35\,\text{eV}$,

$a_2 = -12.32$ eV, $C_3 = -2C_4 = -3.03$ eV. The situation is more complicated when we observe that experimentalists no longer agree concerning experimental values of the C_{ij}s.

Recently, Polian et al. (1996), using Brillouin scattering experiments, determined within an experimental uncertainty of less than 20 GPa the following values: $C_{11} = 390$ GPa, $C_{12} = 145$ GPa, $C_{13} = 106$ GPa, and $C_{33} = 398$ GPa. If we use these values, the values we get for deformation potentials are $a_1 = -5.22$ eV, $a_2 = -10.8$ eV, $C_3 = -2C_4 = -5.58$ eV (Gil and Alemu, in press). These values are in fairly good agreement with the full set of data obtained by Shan et al. (1996), who fitted optical data using a model where the components of the strain tensor had been obtained from x-ray measurements (Figs. 11 and 12) rather than using elasticity theory. Shan et al. (1996) obtained: $a_1 = -6.5$ eV, $a_2 = -11.8$ eV, $C_3 = -2C_4 = -5.3$ eV. These two sets of values obtained in the context of the quasi-cubic model exhibit slight discrepancies that only reflect the experimental uncertainties. For instance, from their biaxial stress investigation, Shan et al. (1996) extract a hydrostatic pressure coefficient that is only slightly higher (4.96 meV/kbar) than

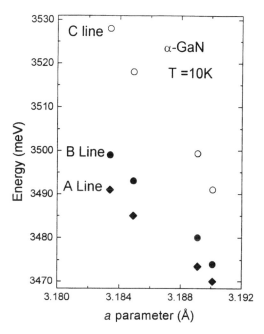

FIG. 11. The energies of A, B, and C lines obtained by reflectance spectroscopy are plotted against the a parameter measured by x-ray diffractometry. The data are taken from the work of Shan et al. (1996) with their permission.

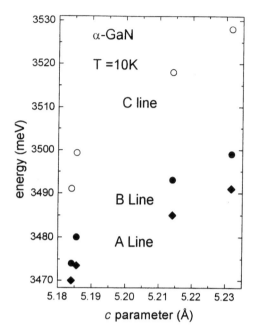

FIG. 12. The analogue of Fig. 11, but the energies of the transitions are plotted against value of the C parameter in the epilayer. See Shan et al. (1996) for more details.

their experimental one (3.90 meV/kbar) (Shan et al., 1995) and the average value we take (4.2 meV/kbar). Second, if using their deformation potentials we compute the numerical coefficient of the right-hand side of Eq. (18), we find it to be -2.26. This value is extremely close to the average value we take (-2.2) and is in the range of experimental uncertainties we discussed in a previous work (Tchounkeu et al., 1996). This, we believe, demonstrates the sensitivity of the deformation potentials to the ingredients used in the fitting procedure. A novel measurement of stiffness coefficients was reported (Yamaguchi et al., 1997). Measurements were also obtained from Brillouin scattering experiments and are close to the values of Polian et al. (1996). The novel data are: $C_{11} = 365$ GPa; $C_{12} = 135$ GPa; $C_{13} = 114$ GPa; and $C_{33} = 381$ GPa. This gives $a_1 = -5.32$ eV, $a_2 = -10.23$ eV, $C_3 = -2C_4 = -4.91$ eV. For the sake of completeness, we note that Chichibu et al. (1996) unfortunately took overly large values of the crystal field splitting (22 meV), but used these C_{ij}s and derived deformation potentials $C_3 = -2C_4 = -5.73$ eV. Similar work was done by Shikanai et al. (1997), who theoretically addressed the strain-induced variations of the free exciton binding energies, and proposed $C_3 = -2C_4 = -8.82$ eV.

b. Determination of C_5 and C_6

Determining the two remaining deformation potentials requires working on epilayers for which the built-in strain has reduced hexagonal symmetry. This is possible if epilayers grown on A plane sapphire are used (Alemu et al., 1998).

Figure 13 illustrates the orientation of the GaN epilayer with respect to the axes of the A plane sapphire substrate. The (0001) plane of the GaN film grown on the A plane sapphire substrate is parallel to the (11-20) face of the A plane sapphire. One side of the hexagonal GaN crystal is parallel to the C face of the A plane sapphire substrate. When cooling down the sample, the thin GaN epilayer experiences strong in-plane anisotropic deformation between its [11-20] and [1-100] directions because the thermal expansion coefficient of sapphire is strongly anisotropic ($\alpha_{//c} = 8.5 \times 10^{-6}\,\text{K}^{-1}$ and $\alpha_{\perp c} = 7.5 \times 10^{-6}\,\text{K}^{-1}$). As a consequence, the C_{6v} hexagonal symmetry of the GaN lattice is reduced to C_{2v} (Alemu et al., 1998). In addition, we expect and detect an in-plane anisotropy of the optical response as shown in Fig.

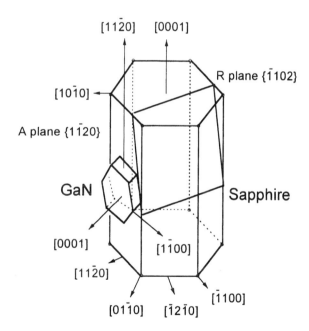

FIG. 13. Relative orientations of the crystallographic axes of GaN in case of heteroepitaxy on A plane sapphire. Note that the (0001) plane of sapphire is parallel to the (11-20) plane of the GaN epilayer.

14. If we now wish to address the deformation potential issue, we have to write the combined effects of the strain fields, crystal field splitting, and spin-orbit interaction on this sample. In order to obtain the three non-vanishing components of the strain in the GaN epilayer, it is convenient to assume that elasticity theory is still valid and to introduce an anisotropic stress tensor $(\sigma_1, \sigma_2, 0, 0, 0, 0)$, which represents the stress acting on the GaN due to the differences between its thermal expansion coefficients and the ones of the sapphire substrate. The symmetrized component ts of the strain tensor in the GaN epilayer is:

$$e_{xx} + e_{yy} = e_\perp = (S_{11} + S_{12})(\sigma_1 + \sigma_2) \quad (20)$$

$$e_{zz} = S_{13}(\sigma_1 + \sigma_2) \quad (21)$$

$$e_{xx} - e_{yy} = (S_{11} - S_{12})(\sigma_1 - \sigma_2) \quad (22)$$

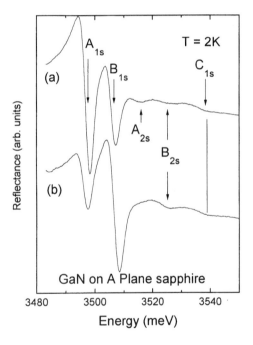

FIG. 14. Evolution of the intensity of the reflectance structures with in-plane orientation of the electric field: (a) [10-10] orientation; (b) [-12-10] orientation. Note the increase of A_{1s} is accompanied by a decrease of B_{1s} and makes it easier to detect A_{2s}. Arrows indicate the average positions of transverse excitonic polaritons.

The problem is here reduced to the resolution of the following 3 × 3 matrix in the basis of the three Γ_5 functions of C_{2v} given by the compatibility relations with the Γ_9 and Γ_7 representations of C_{6v} (Alemu et al., 1998).

$$\begin{array}{ccc} |\Gamma_5(X)\rangle & |\Gamma'_5(-iY)\rangle & |\Gamma''_5(Z)\rangle \end{array}$$

$$\begin{bmatrix} \Delta_1 + \delta_1 + \delta_2 - \delta_3 & \Delta_2 & \Delta_3 \\ \Delta_2 & \Delta_1 + \delta_1 + \delta_2 + \delta_3 & \Delta_3 \\ \Delta_3 & \Delta_3 & \delta_1 \end{bmatrix} \quad (23)$$

Here the orbital components of the wavefunctions are indicated between parentheses, and the strain-induced shifts are given as a function of the GaN deformation potentials $\delta_1 = (D_1 + C_1)e_{zz} + (D_2 + C_2)e_\perp = a_1 e_{zz} + a_2 e_\perp$, $\delta_2 = C_3 e_{zz} + C_4 e_\perp$, and $\delta_3 = C_5(e_{xx} - e_{yy})$. It is interesting to note that we have three unknown quantities C_5, σ_1, and σ_2 together with three equations [the eigenvalues of the problem in Eq. (23)], and three experimental values: the energies of A, B, and C lines. Fitting the sum of these three energies, the trace of matrix equation 23, which is only a function of Δ_1, δ_1, δ_2, gives $\sigma_1 + \sigma_2 = -2.438$ GPa. Next, if we express the wavefunctions of the resolved problem according to first-order perturbation theory, we get expressions such as:

$$\Psi_A = N[|\Gamma_5(X)\rangle + \Delta_2/(2\delta_3)|\Gamma'_5(-iY)\rangle + \Delta_3/(\Delta_1 + \delta_2 - \delta_3)|\Gamma''_5(Z)\rangle] \quad (24)$$

with

$$N^{-1} = \sqrt{1 + [\Delta_2/(2\delta_3)]^2 + [\Delta_3/(\Delta_1 + \delta_2 - \delta_3)]^2} \quad (25)$$

and similar equations for B and C. The oscillator strengths for line A in $X[10\text{-}10]$, $Y[-12\text{-}10]$, $Z[0001]$ polarization are given by the square of the contributions of the three Γ_5 eigenvectors. From Fig. 14, we note that the oscillator strengths vary by a factor of two when the polarization is changed; this gives $\Delta_2^2/\delta_3^2 \approx 8$. The fit to the data requires using $C_5 = -2.4$ eV. In terms of the oscillator strengths, the agreement between theory and experiment is illustrated in Fig. 15 where the evolution of the oscillator strengths are plotted relative to the angle with the [10-10] direction. For the sake of completeness, the strength we compute for line C is also given, but magnified by a factor of 8. Values of σ_1 and σ_2 are -1.11 GPa and -1.32 GPa, respectively. Last, the deformations along the [11-20] and [1-100] directions

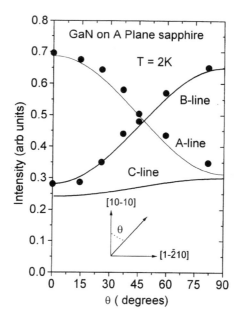

FIG. 15. Evolution of the oscillator strength of A and B lines as a function of the polarization of the photon (circles). Full lines represent the results of our calculation. Theoretical values for C line are magnified by 8.

are $\varepsilon_{[11\text{-}20]}/\varepsilon_{[1\text{-}100]} \approx 8.5/7.5$, the ratio between the sapphire thermal expansion coefficients. According to the quasi-cubic model.

$$4C_5 - \sqrt{2}C_6 = -C_3 \tag{26}$$

This gives $C_6 = -7.22 \text{ eV}$.

IV. Exciton Oscillator Strengths and Longitudinal-Transverse Splittings

1. EXCHANGE INTERACTION AND STRAIN-INDUCED VARIATIONS OF THE OSCILLATOR STRENGTHS FOR Γ_5 AND Γ_1 EXCITONS

Following Gil and Briot (1997), the exciton Hamiltonian Ξ_{exc} is written

$$\Xi_{\text{exc}} = H_{c=0} + H_{c\text{strain}} + H_{v=0} + H_{v\text{strain}} + H_{\text{exc}} \tag{27}$$

where $H_{v=0}(H_{c=0})$ are the strain-free valence (conduction) band Hamiltonians and $H_{v\text{strain}}(H_{c\text{strain}})$ are their respective, which account for the strain-related effects on the evolution of band extrema. The last operator in the right-hand section of Eq. (27) is

$$H_{exc} = R^* + 1/2\gamma\sigma_h, \sigma_c \qquad (28)$$

where R^* is the exciton binding energy and the last term is the crystalline exchange interaction. Operators σ_h and σ_c operate on valence hole and conduction electron spin functions. We restrict our attention to excitons in epilayers grown on (0001) substrates. The twelvefold exciton transforms are according to $3\Gamma_5 + 2\Gamma_1 + 2\Gamma_2 + \Gamma_6$. The construction of such exciton states from Γ_7^c, Γ_9^v, Γ_7^v states (we add superscripts to avoid confusion) is sketched in Fig. 16 and can be found in Table III. The twofold Γ_6 exciton is forbidden, as are also the two Γ_2 ones. The three Γ_5 excitons are created using an adapted photon, σ-polarized (electric field $\perp[0001]$) and the two Γ_1 excitons are created for π-polarized photons (electric field $//[0001]$).

Eigenenergies of Γ_5 excitons are obtained from resolution of the following matrix (Gil and Briot, 1997)

$$\begin{array}{ccc} |\Gamma_5\rangle & |\Gamma_5'\rangle & |\Gamma_5''\rangle \end{array}$$

$$\begin{bmatrix} \Delta_1 + \Delta_2 + \delta_1 + \delta_2 - 1/2\gamma & -\gamma & 0 \\ -\gamma & \Delta_1 - \Delta_2 + \delta_1 + \delta_2 - 1/2\gamma & \sqrt{2}\Delta_3 \\ 0 & \sqrt{2}\Delta_3 & \delta_1 + 1/2\gamma \end{bmatrix} \qquad (29)$$

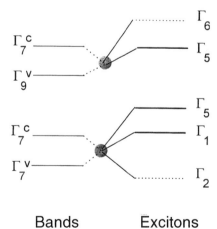

FIG. 16. Construction of the exciton states from the Bloch conduction and valence band states in C_{6v} symmetry.

TABLE III

Basis Functions for the $\Gamma = 0$ Excitons in Wurtzite Symmetry — Spin Components of the Missing Valence Electron Are $\uparrow$ and $\downarrow$, and α and β Represent the Spin Components of the Conduction Electron

Exciton State	Basis Functions
Γ_1	$(p_-\uparrow\beta + p_+\downarrow\alpha)/\sqrt{2}$
Γ'_1	$p_z(\uparrow\alpha + \downarrow\beta)/\sqrt{2}$
Γ_2	$(p_-\uparrow\beta - p_+\downarrow\alpha)/\sqrt{2}$
Γ'_2	$p_z(\uparrow\alpha - \downarrow\beta)/\sqrt{2}$
Γ_6	$\{p_+\uparrow\beta, p_-\downarrow\alpha\}$
Γ_5	$\{p_+\uparrow\alpha, p_-\downarrow\beta\}$
Γ'_5	$\{p_+\downarrow\beta, p_-\uparrow\alpha\}$
Γ''_5	$\{p_z\downarrow\alpha, p_z\uparrow\beta\}$

The exciton eigenvectors are written

$$\Psi_{\text{exc}\Gamma_5} = v|\Gamma_5\rangle + \bar{\omega}|\Gamma'_5\rangle + \omega|\Gamma''_5\rangle \tag{30}$$

and their oscillator strengths (σ-polarization) are

$$|\Omega_{\text{exc}\Gamma_5}|^2 = (v^2 + \bar{\omega}^2)/2 \tag{31}$$

Similar treatment performed for Γ_1 excitons gives

$$\begin{array}{cc} |\Gamma_1\rangle & |\Gamma'_1\rangle \end{array}$$
$$\begin{bmatrix} \Delta_1 - \Delta_2 + \delta_1 + \delta_2 + 1/2\gamma & \sqrt{2}\Delta_3 \\ \sqrt{2}\Delta_3 & \delta_1 - 3/2\gamma \end{bmatrix} \tag{32}$$

The oscillator strengths of these excitons allowed in π polarization are obtained by the squares of the contribution of $|\Gamma'_1\rangle$ in the two eigenvectors.

γ was not measured in GaN at the time this chapter was written (July 1997). Its computation is formally extremely tricky, even impossible, without making numerical approximations. During the past forty years we have learned from the physics of excitons in crystals that the influence of the electron-hole exchange interaction on the energy spectrum of the Wannier exciton is large in crystals that exhibit small dielectric constants or with bands having flat dispersion relations in reciprocal space. We have calculated γ, extending to the GaN case the two-body calculation developed by Röhner (1971) for a few II-VI compounds. The pertinent parameter of the model is the product p of the exciton radius with an *ad hoc* radius k_m of an effective Brillouin zone (BZ). This k_m is introduced to replace the tedious integration in k space through the whole *actual* BZ by an integration over a spherical space having the volume of this BZ. It has been shown by Röhner (1971) that k_m can be connected to the lattice parameter a using a well-adapted scaling argument. For GaN, taking $a = 3.19$ Å gives $k_m = 0.873$ Å^{-1}. To calculate p, we need the exciton Bohr radius in GaN. Recent determination of exciton masses (Suzuki and Uenoyama, 1995; Suzuki *et al.*, 1995; Drechsler *et al.*, 1995; Knap *et al.*, 1996; Merz *et al.*, 1996) suggests $a_B = 34 \pm 2$ Å. This gives $p = 30 \pm 2$, which corresponds to an exchange-influenced 1s exciton binding energy of about 85% of $R^*[1/\rho_1^2 = 0.851 \pm 0.006$ in the model by Köhner (1971)], and gives a GaN result that matches very well the trend in wurtzite II-VIs as shown in Fig. 17. The exciton exchange energy 2γ is thus estimated to be some 15% of the binding energy. Further, using a Rydberg energy of 25 meV (Alemu *et al.*, 1998), the

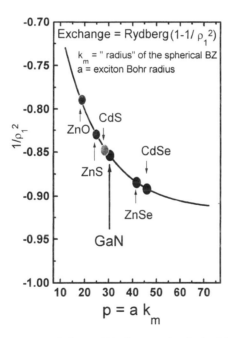

FIG. 17. Plot of the exchange-influenced binding energies obtained from Röhner (1971) for wurtzite II-VIs after a two-body calculation. The long arrow indicates the GaN value.

magnitude of the corresponding γ we get (~ 2 meV), is far from being negligible and might have drastic consequences on both: (i) energy spectrum; and (ii) oscilator strength in GaN epilayers under biaxial tension when the crystal field splitting is offset by the strain field. Results seen in Fig. 18 display the evolution of Γ_5 (full lines) and Γ_1 (thin lines) excitons as a function of biaxial stress. On the figure, positive values of the stress correspond to biaxial compression. We note that our calculation predicts that anticrossing of Γ_5 excitons may occur with biaxial tension. To the best of our knowledge, no quantitative information has been reported, to date, concerning the Γ_5-Γ_1 splittings in GaN, although Julier et al. (1998) measured $\gamma = 0.6 \pm 0.1$ meV in GaN on A-plane sapphire.

We now compare the oscillator strengths for band-to-band and excitonic processes, and we study their evolution with strain. The results presented in Fig. 19 concern σ polarization. Inclusion of excitonic effect in σ-polarization reinforces the light-crystal coupling for C excitons having Γ_5 symmetry and simultaneously reduces the oscillator strength of $\Gamma_5 B$ and $\Gamma_5 A$. Although $\Gamma_5 A$ and $\Gamma_5 B$ states strongly interact for biaxial tension of about 7.5 kbar, the corresponding exchange of oscillator strengths remains small. For

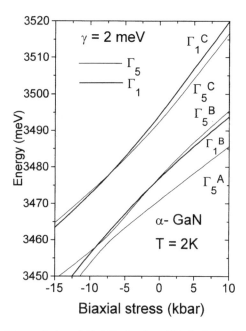

FIG. 18. Plot of the evolution of Γ_5 (thin lines) and Γ_1 (bold lines) excitons as a function of the cylindrical biaxial stress in the C plane. Used with the permission of Gil and Briot (1997).

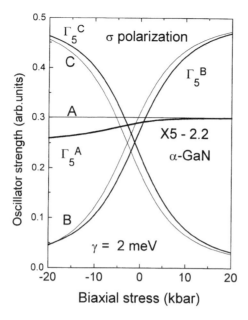

FIG. 19. Comparison between the stress-induced variations of the oscillator strengths for the band-to-band process (thin lines) and the Γ_5 exciton ones (bold lines) in σ polarization.

π-polarization, the two optically active levels also exchange their oscillator strengths as shown in Fig. 20. At this stage, we would like to emphasize the fact that we have no accurate information concerning eventual variations of the binding energy with strain. For this reason the assumption that γ is independent of the strain and has the same value for A, B, and C excitons has been made. This modeling could be improved in light of novel experimental information. However, it is obvious that such variations will not drastically change the results on strain-induced variations of the oscillator strengths, which are dominantly ruled by the variation of the band-to-band matrix elements.

2. STRAIN-INDUCED VARIATIONS OF THE LONGITUDINAL-TRANSVERSE SPLITTINGS

When performing reflectance experiments, we examine properties of exciton polaritons, which are freely propagating elementary excitations of the crystal. In the simplest model, the coupling of the electromagnetic field

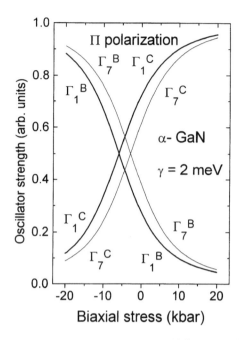

FIG. 20. The oscillator strengths for Γ_1 excitons (bold lines) and for the band-to-band process (thin lines) in P polarization as a function of the biaxial stress in the (0001) plane.

represented by a photon having an energy $E = hkc/2\pi$ with the exciton $E = E_T + h^2k^2/8\pi^2 m$ gives two propagating modes (Hopfield, 1958) as shown in Fig. 21. Fitting the line shape of reflectivity experiments such as those from Fig. 8, the GaN dielectric constant is modeled phenomenologically using a set of three oscillators

$$\varepsilon(k, \omega) = \varepsilon_b + \sum_{j=1}^{3} \frac{4\pi\alpha_{0j}\omega_{0j}^2}{\omega_{0j}^2 + (\hbar k^2/m_j^*)\omega_{0j} - \omega^2 - i\Gamma_j\omega} \quad (33)$$

where ε_b is the background dielectric constant, ω_{0j} is the transverse frequency related to the exciton j with an effective mass m_j^*, $A_j = 4\pi\alpha_{0j}$ is the polarizability of the exciton resonance at $\omega = 0$ and $\mathbf{k} = \mathbf{0}$, and Γ_j is the damping parameter used to account for the interactions of the excitons with the phonons and extrinsic defects. The energy of transverse excitons are obtained by solving equation $\varepsilon(k, \omega) = k^2c^2/\omega^2$, while the energies of the longitudinal excitons are obtained making $\varepsilon(k, \omega) = 0$ in Eq. (33). In the

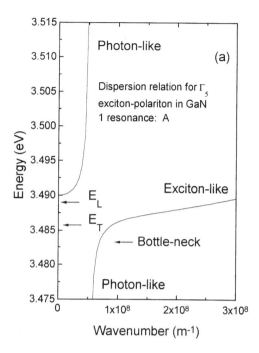

FIG. 21. Dispersion relations for a simple exciton polariton picture. Under nonresonant excitation conditions, one has an accumulation in the bottleneck region and in the lower part of the upper branch.

reflectivity spectrum of Fig. 22, dashed lines were fitted using $\varepsilon_b = 8.75$ and the following values for the ω_{0j} quantities, 3487 meV, 3496 meV and 3520 meV, respectively. The corresponding A_js are: 5×0^{-3}, 3×10^{-3}, and 2×10^{-3}. The broadening parameters Γ_j included in the fitting are 6, 6, and 16 meV, respectively. For the convenience of this fit, we used the values of the excitonic translation mass m_j obtained when previously fitting the reflectance of another GaN epilayer, 1.3, 0.5, and 1 in units of the free electron mass. The longitudinal-transverse splittings can be extracted from the reflectance lineshape fitting. For each exciton, the longitudinal transverse splitting is given by: $\omega_{LT} \approx 2\pi\alpha\omega/\varepsilon_b$. They are found to be 1 meV, 0.6 meV, and 0.4 meV for A, B, and C, respectively. It is now well established after experimental results taken on CdS (Hein and Wiesner, 1973) and GaAs (Sell *et al.*, 1973), for instance, that under nonresonant excitation conditions, free exciton photoluminescence has two peaks, which in order to be interpreted require inclusion of the notion of exciton polariton. Due to the *k*-selection rule for optical matrix element, excitons having exactly the

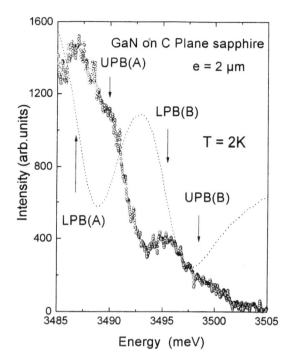

FIG. 22. Reflectance (dashed line) and photoluminescence (open circles) spectra of a GaN epilayer grown on C plane sapphire. The photoluminescence spectrum was taken using the 325-nm radiation of an He-Cd laser, at an excitation density of 1 W-cm^{-2}.

photon k vector should radiate. In the pure bulk material, the coupling of the electromagnetic field with the exciton relaxes the k selection rule and transforms the fluorescence mechanism into a transport of the coupled excitation from the depth it is created toward the surface where part of it is reflected and part of it is transmitted.

Under nonresonant excitation conditions, an electron-hole pair created in the continuum of states will relax toward lower energies within the two polariton branches. In particular, it will thermalize efficiently via most probably acoustic phonon emission within the low polariton band, down to the knee of the polariton dispersion, just below the transverse frequency, due to the high density of states in this region (Benoit à la Guillaume et al., 1970). Below the transverse frequency, when the wavefunction becomes more and more photon-like, the decrease of both the scattering matrix element and of the density of final states, combined with the large increase of the group velocity $d\omega/dk$ causes the radiative lifetime to dominate over the thermal relaxation (Andreani, 1995) (in this region, the escape of the photon out of the crystal becomes very efficient). This gives a "polariton relaxation bottleneck" (Toyozawa, 1959) and, at the end, the polariton exhibits a pronounced distribution peak just below the exciton energy. These peaks correspond to photoluminescence maxima at 3486.6 meV and 3495.4 meV (Gil et al., 1997b) and are labeled LPB(A) and LPB(B) in Fig. 22. There is also a significant population of the upper polariton branch, which gives the high-energy contribution peaks labeled UPB(A) and UPB(B) in Fig. 22. The dips in the photoluminescence correspond to the energy of the longitudinal exciton when $k \to 0$. The magnitudes of these luminescence peaks are ruled by the whole dynamical equilibrium of the population of polaritons in *real* and *reciprocal* spaces (Weisbuch and Ulbrich, 1979). The energy distribution of the polariton is larger within the LPB than in the UPB branch.

In the case of our sample, the situation is more complex than in Fig. 21: there is an additional oscillator B which bends the exciton photon-like branch of A (Fig. 23). Then, the distribution of polaritons in the UPB(A), LPB(B), and UPB(B) scan comparable energy ranges, which are significant-

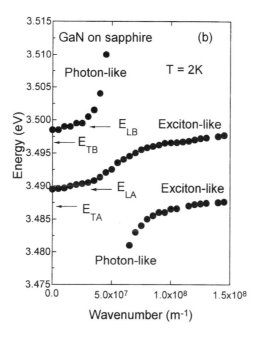

FIG. 23. Dispersion relations for the exciton polaritons calculated from the reflectance data of Fig. 22.

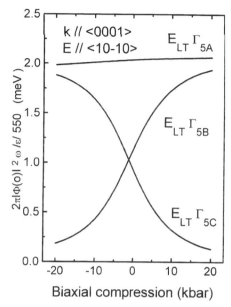

FIG. 24. Stress-induced variations of the longitudinal-transverse splittings for Γ_5 exciton polaritons.

ly smaller than in the LPB(A) branch. This is consistent with the result of a PL lineshape fitting using Lorentzian functions. Comparing the splittings between energies of dips in PL bands at 3489.4 meV and 3497.8 meV and values of transverse excitons, we get 2.4 meV and 1.8 meV. The two approaches give slightly different values, which are in the range of the experimental uncertainty. We note that a 2-meV value of the longitudinal transverse splitting for A exciton has been measured in a GaN homoepiaxial epilayer (Baranowski et al., 1997). Observation of these LT splittings requires very good samples that work at low excitation densities (1 ~ W/cm^{-2}). It is also important to remark that, to have some chance of measuring the LT splitting for Γ_5 B exciton polaritons, strongly biaxially compressed layers are welcome, as shown in Fig. 24.

V. Origin of the Strain

Hiramatsu et al. (1993) suggested that the lattice mismatch strain between GaN and sapphire should be relieved after several nanometers of growth, according to the critical thickness theory. The primary relief mechanism should be formation of dislocations at the film/substrate interface. The situation is not so simple, because GaN epilayers are in general obtained by sophisticated heteroepitaxial processes that vary, depending on substrate, growth method, and even, growers. There are several methods to obtain high-quality GaN epilayers on sapphire substrates, namely, using a low-temperature grown GaN buffer layer, or an AlN buffer layer, or even no buffer layer. All these approaches lead to different structural interface morphologies that may influence the strain relaxation. Then, when cooling down the sample, the difference between the thermal expansion coefficients of the substrate and the GaN will produce a residual biaxial compression that is dependent upon the entire sample structure.

Figure 25 represents the variations of two reflectance spectra taken on a 2-μm thick GaN epilayer grown on a C plane sapphire substrate with a 250 Å GaN buffer layer. The 240 K spectrum has been blue-shifted some 46 MeV so that the signature of the A exciton is energy matched with the energy of this transition at 2 K (at 77 K the shift required to match the two

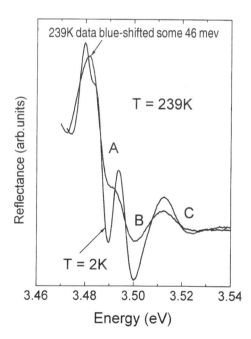

FIG. 25. 2 K and 239 K reflectance spectra typical of GaN on sapphire. The 239-K data have been blue-shifted some 46 meV.

spectra is 4 meV). Obviously, at first sight, and within the experimental uncertainty, the splittings are temperature-independent.

In Fig. 26 the energies of line B and of line C, obtained for this sample from lineshape fittings of the reflectance spectra taken between liquid helium and room temperature, are plotted as a function of the position of line A. From the least mean square fits of these dependences, we find a **slight** decrease of the splittings B-A and C-A when the temperature increases (i.e., when the energy of A decreases) from 2 K up to room temperature. This is expected from the simplest model based on thermal expansion coefficients. A similar result was obtained in a sample grown using an AlN buffer layer. In this range of temperature, our finding is consistent with results reported by Li et al. (1997) in Fig. 2 of their paper. These authors, however, observed a sudden increase of the residual compression if the sample temperature was increased beyond 300 K up to 475 K. These two findings are not inconsistent as shown by Buyanova et al. (1996), who studied the temperature dependence of the bandgap of several GaN epilayers deposited on sapphire and 6H SiC, and compared their findings with the behavior of bulk GaN. What they found is offered in Fig. 27. Surprisingly, the GaN homoepitaxial layers

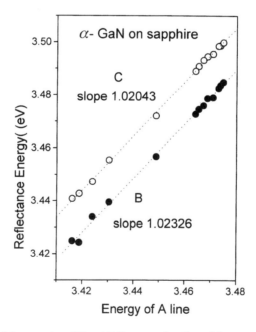

FIG. 26. Plot of the energies of C and B lines as a function of the energy of line A, which decreases when the temperature increases.

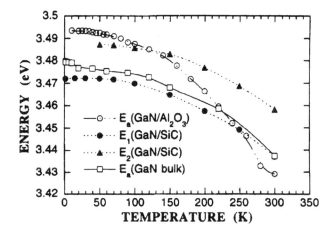

FIG. 27. Temperature dependence of the bandgap for various GaN samples: bulk GaN and several heteroepitaxies on 6H-SiC and sapphire. Data taken from Buyanova et al. (1996) with their permission.

exhibit the same behavior as epilayers deposited on 6H SiC, while the heteroepitaxy on sapphire exhibits a stronger red shift when increasing the sample temperature. The behavior of the latter sample is very suggestive of a reduction of the built-in strain when the temperature is raised. It is obvious from these investigations, which confirm x-ray diffractometry experiments (Leszczinski et al., 1994), that the relaxation of the stress with temperature is extremely sample dependent.

From a more mechanical point of view it is important to keep in mind that bending of the substrate/epilayer system can be the subject of variation with temperature. Kisielowski et al. (1996) report that the strain state of epilayers could be described by the coexistence of both biaxial strain (due to post-growth cooling of the sample) and purely hydrostatic strain due to the presence of point defects. This explains very well the departure of some reflectance experiments (Buyanova et al., 1996) from the universal plot of Gil et al. (1995), on the one hand, but also some shifts of the phonon modes from the bulk values, which shifts do not follow the trend expected in the case of pure biaxial compression (see Section VI of this chapter). To examine how large the influence of point defects may be, at the scale of optical properties, Briot grew a series of samples on C plane sapphire substrates, all of which were bought from the same commercial company and had extremely close x-ray diffraction patterns. For all samples the deposition of a 250 Å GaN buffer layer at 550 °C followed a 10 min nitridation of the substrate according to the method described by Briot et al. (1996). Then he proceeded to deposition of the high-temperature GaN layer. He varied the V/III molar ratio. The experiment was reproduced for two growth temperatures. We found that the residual stress increases when the V/III molar ratio is decreased as shown in Fig. 28 where there are plotted (vs this growth parameter) the energies of the A exciton measured by reflectance at 2 K on these samples. This increasing of the residual stress was accompanied by an increase of the electron concentrations, which were found to scale four orders of magnitude, between 10^{16} cm^{-3} and 10^{20} cm^{-3}. This result is interesting: homoepitaxial layers are generally obtained in a crystalline state with **exact** in-plane lattice matching with the GaN substrate, but with slightly lower c constant (0.02 ~ %) (Baranowski and Porowski, 1996). Because in bulk crystals the number of structural defects (probably nitrogen vacancies) and the density of residual electrons (mainly due to auto-doping) and larger than in epilayers, we address here one of the physical phenomena that rule the residual stress in the GaN epilayer: the structural quality of the layer itself.

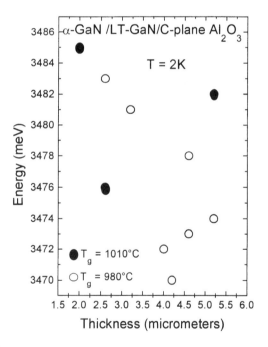

FIG. 28. Evolution of the bandgap of GaN on sapphire as a function of the thickness of the epilayer. The layer thickness has an influence on the stress relaxation, but the scatter of the data reveals the contribution of other effects.

We have also varied the growth time and the growth temperature. The growth temperature has poor influence (in the narrow range of values giving high-quality GaN). In terms of thickness effects, a trend is only roughly found if the V/III molar ratio is kept constant: the strains decrease when the layer thickness increases (Fig. 29). We have noticed that the residual compression is higher if we use an aluminum nitride buffer layer than if a GaN buffer layer is used; this probably is the case because we introduce in the sample a heavily strained region at the interface between the GaN epilayer and the AlN buffer layer. Rieger *et al.* (1996) have studied the influence of the thickness of this AlN layer on the residual strain and found the biaxial compression to decrease when the thickness of this buffer layer increases.

The low-temperature grown buffer layers are very thin compared to the substrate and epilayer. However, they drastically influence the growth; they are not only supposed to reduce residual strain, but also to act as nucleation layers because they offer a high density of nucleation sites and, as compared to a direct heteroepitaxy on the sapphire substrate, the heteroepitaxy of the GaN on the AlN surface is facilitated by a lower interface energy. These two

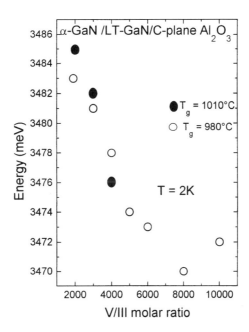

FIG. 29. Evolution of the bandgap of GaN on sapphire as a function of the V/III molar ratio, which is found to have a dominating influence.

mechanisms promote an initial two-dimensional GaN growth. To understand the dependence of the structural properties of the buffer layers and to correlate the residual strain to these structural properties, Le Vaillant et al. (1998) made a very careful study of the x-ray diffraction patterns of AlN buffer layers deposited on C plane sapphire. They applied the Warren-Averbach analysis (Warren, 1959), a method based on the precise line shape analysis of the x-ray diffraction peaks, which permits extraction of both grain size and intragrain strain distributions from XRD experiments. These x-ray diffraction data have been recorded in standard Bragg-Brentano (θ-2θ), asymmetrical (ω-20) and (ω-fixed 2θ) geometry. As a result of a low-temperature deposition process followed by thermal annealing, the AlN buffer layers as shown in AFM images are made of small grains of various sizes depending on annealing times; the rms-roughness values range from 0.25–0.9 nm. The 2-min annealed sample shows the higher grain-size discrepancy. For an annealing time ranging between 2–5 min, a coalescence of the grains begins, smoothing the surface. An investigation of both grain size and local strain is also needed through the bulk of the AlN buffer layer: in a crystallite, the local strain depends on the volume to the surface ratio and is inhomogeneously relaxed through the bulk. AlN (0002) and (0004) diffraction peaks have been systematically recorded, using the Bragg-Brentano geometry and fitted using adjustable pseudo-Voigt functions.

The frequency distribution function of the crystal column length is presented in Fig. 30 for all annealing times. In this figure, two characteristic column-length values can be pointed out. They represent the intragrain distributions as summed over the entire volume of the layer. The as-grown sample appears to be almost exclusively composed of small crystallized columns, each about 6 nm long. Then, with annealing time, an increasing proportion of larger columns centered around the value of 30 nm occurs. The column-length distribution function corresponding to an annealing time of 2 min consists of two sets of column size of 6 and 30 nm long. The maxima of the frequency of apparition are sharply centered around these main column-size values and well separated. Then, with a longer annealing time the column-length distribution function is smoothed and the proportion of column lengths reaching the thickness of the layer (~ 500 Å) is increased. Nevertheless, their relative proportion never exceeds 1%. The variation of the interplanar spacing d of the crystallite lattices around its value in a perfect crystal is taken into account in the Fourier analysis of the XRD by a distortion parameter. This distortion parameter provides pertinent information about the strain in the crystallites, averaged over all the columns of equal length.

In Fig. 31 the relative distortion of column length is reported. The smaller the grain size, the higher the corresponding strain. The annealing process

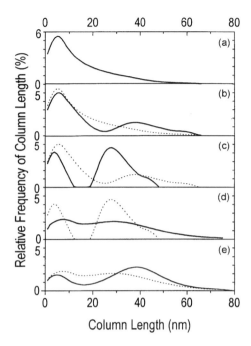

FIG. 30. Frequency distribution function of column lengths in the crystal given by Warren-Averbach method for samples: as-grown (a); annealed at 1070 °C for 0.5 min (b); 2 min (c); 5 min (d); and 15 min (e). To better appreciate the evolution, the frequency distribution function of column lengths of the previous annealing time are also represented by dashed lines.

leads to an increase of the distortion, which is maximum for a thermal annealing time of 2 min. This holds mainly for small crystallite lengths. Then, after relaxation, it can be observed that the strain increases again. The highest strain values are also observed for the 2-min anncaled sample.

Correlatively, the rocking curves presented in Fig. 32 show that the higher the annealing time, the sharper the angular distribution of the c axis of AlN buffer layers. The cube of the mean column length is linearly dependent on annealing time. This can be attributed to the Ostwald ripening. Generally, recrystallization phenomena are classified into three stages: nucleation; primary recrystallization by atomic migration through a matrice; and secondary recrystallization by Ostwald ripening, resulting in the growth of bigger grains to the detriment of smaller ones. The distribution function of column sizes (see Fig. 30), showing an important concentration of columns of about 30 nm is the result of such a phenomenon. The surface of the 2-min annealed sample has the highest number of low-energy nucleation sites. The

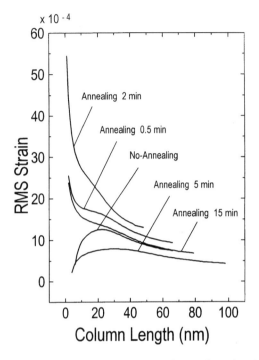

FIG. 31. Root mean square (rms) strain vs column length.

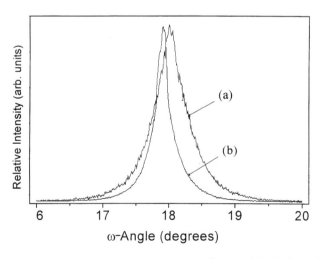

FIG. 32. (0002)-AlN rocking curves: (a) as-grown; (b) annealed 15 min at 1070 °C.

chemisorption of GaN molecules is favored on such a surface, because of the great number of available nucleation sites and the small possible diffusion lengths. Hence, a pseudo two-dimensional growth process of the GaN is favored on such a surface. Figure 33 shows the correlation between the optical properties and the annealing time of the AlN buffer layer in case of deposition on C plane sapphire.

Correlation of biaxial strains, bound exciton energies, and defect microstructures in GaN films grown on 6H SiC (0001) substrate has been addressed by Perry et al. (1996). They found that GaN films grown on vicinal SiC were compressively strained compared to those grown on axis SiC. To understand these results they examined the defect microstructures of GaN films grown on an AlN buffer on 6H SiC(0001) substrates. The primary effects in the AlN buffer layers are on the off-axis SiC substrates and their associated domain boundaries; these effects include a marked amount of strain and a high density of threading dislocations in these AlN buffer layers. A reduction of the density of inversion domain boundaries due to the reduction in the density of SiC steps is observed in the AlN films

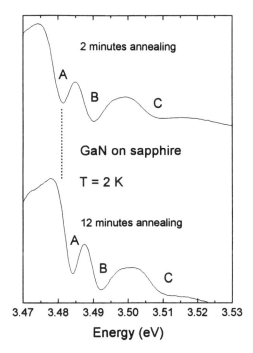

FIG. 33. 2 K reflectance spectra in GaN epilayers on sapphire showing that the strain decreases with increasing time of the AlN buffer layer.

grown on on-axis SiC. In that case the GaN/AlN interface is epitaxial, free of domains, threading dislocations, and low angle grain boundaries. The TEM contrast at distances of 50–80 Å of the GaN interface revealed traces of rounded peaks and grooves typical for films that are under compression. Perry et al. (1996) conclude from their high resolution transmission electron microscopy (HRTEM) studies that the stress due to lattice mismatch is not fully relieved at the GaN/AlN interface for films grown on on-axis SiC. This results in a compressive stress that counteracts the tensile stress due to lattice mismatch. Thin films (<0.6 μm) grown on on-axis SiC are in compression while those on vicinal SiC are in tension. As thickness increases (>1 μm) this compressive stress lessens and at greater thicknesses (for thicknesses >4 μm) cracking may occur to relieve thermal mismatch stress.

Ponce (1998) has performed detailed investigations of the spatial variation of the photoluminescence and of the energy and intensity of Raman active phonons. He found that the residual strain is spatially inhomogeneous.

Gfrorer et al. (1996) reported the possibility of obtaining AlGaN with biaxial elastic tension if growing it on a GaN buffer layer.

It is obvious that extension of such elaborate investigations based on a combination of careful investigations of optical properties together with subtle investigations of x-ray diffraction using electron microscopy will help to elucidate the origin and nature of residual strains in III-nitrides.

VI. Phonons Under Strain Fields

In this section we will use the notations of Raman spectroscopy (which differ from the notations used for band structure). The relationship between the symmetries are given in Table I. A group theory analysis of $q = 0$ lattice vibrations predicts 6 optical modes, which decompose as follows into the following representations of the C_{6v} point group: $\Gamma_{opt} = A_1 + E_1 + 2E_2 + 2B_1$. The polar modes A_1 and E_1 — both infrared and Raman active — are polarized along the z optical axis and in the basal (x, y) plane, respectively $(x, y, z$ represent the crystal principal axes). Among the nonpolar lattice vibrations, the E_2 modes are Raman active and the B_1 modes are silent. The anisotropy in short-range atomic forces is thus responsible for the A_1-E_1 splitting while the long-range Coulomb field is at the origin of the longitudinal-transverse (LO-TO) splitting of polar modes. This results in direction-dependent frequencies and polarization properties when approaching the limiting $q = 0$ value along or perpendicularly to the optical axis. The effect of stress on $q = 0$ phonons in wurtzite structure was addressed long ago in piezospectroscopic studies of CdS (Briggs and Ramdas, 1976). In the approximation of a linear response the phenomenological Hamiltonian can be expressed as $H_\varepsilon = V_{ij} e_{ij}$ where the e_{ij} ($i, j = x, y, z$) are the components of the strain tensor. The changes in frequency of $q = 0$ phonons under stress are obtained in first-order perturbation theory in terms of deformation potential constants by rearranging this contracted tensorial product in order to form combinations of V_{ij} transforming according to the A_1, E_1, and E_2 irreducible representations of the crystal point group. Under hydrostatic or biaxial (0001) oriented stress, the shifts of the phonon frequencies can be expressed in the following way:

$$\Delta v_J = 2a'_J \sigma_{xx} + b'_J \sigma_{zz}, \qquad J = A_1, E_1, \text{ or } E_2 \qquad (34)$$

The σ_{ii} are the diagonal components of the stress tensor, with $\sigma_{xx} = \sigma_{zz}$ for the hydrostatic pressure and $\sigma_{zz} = 0$ for biaxial (0001) stress. The coefficients a'_J and b'_J are expressed in terms of elastic constants and deformation potential constants as

$$a'_J = a_J(S_{11} + S_{12}) + b_J S_{13} \quad \text{and} \quad b'_J = 2a_J S_{13} + b_J S_{33} \qquad (35)$$

where a_J represents the change in frequency of phonon J per unit relative change of length along z, and b_J is its counterpart along x.

The hydrostatic pressure dependence of $q = 0$ phonon frequencies in GaN have been measured by Raman spectroscopy on bulk samples by Perlin et al. (1992) and Wetzel et al. (1996). Concerning the phonon shifts under biaxial stress, they were measured again by Raman spectroscopy by various groups (Rieger et al., 1996; Demangeot et al., 1996; Kozawa et al., 1995). Demangeot et al. (1996) reported on a series of layers grown under different V/III ratio conditions, in which stress calibration was carried out through reflectance measurements as shown in the section V of this chapter. Values of the biaxial stress coefficient $K_J^B = dv_J / d\sigma$, where $\sigma = -\sigma_{xx}$ ($\sigma > 0$ for compression) were estimated from Raman data on the strained layers. A first relation between the a_J and b_J, obtained from Eqs. (34) and (35), can thus be established

$$2a_J(S_{11} + S_{12}) + 2b_J S_{13} = -K_J^B \qquad (36)$$

One more relation is needed for the determination of the $\{a_J, b_J\}$ set of deformation potential constants. It can be obtained from the hydrostatic pressure Raman measurements carried out by Perlin et al. (1992) on bulk GaN crystals because, according to Eqs. (34) and (35)

$$2(S_{11} + S_{12} + S_{13})a_J + (2S_{13} + S_{33})b_J = -K_J^H \qquad (37)$$

Figure 34 shows the nonanalyzed Raman spectra of the heterostructures in the frequency range of the highest E_2 mode. A shift of the E_2 mode frequency is clearly observed with respect to the sapphire line at 578 cm^{-1}, corresponding to mode hardening with decreasing V/III ratio. To a lesser extent, the same trend is noted for the $A_1(LO)$ mode. Phonon frequency as a function of biaxial stress in GaN is deduced from our Raman measurements for the E_2 and $A_1(LO)$ modes and is displayed in Fig. 35. Least square fit gives for the slopes the following values $K_{E_2}^B = 2.9 \pm 0.3$ cm^{-1}/GPa and $K_{A_1(LO)}^B = 0.8 \pm 0.4$ cm^{-1}/GPa, taking, for elastic compliance constant values, the results of a calculation obtained from the elastic stiffness coefficients $C_{11} = 296$ GPa, $C_{12} = 130$ GPa, $C_{13} = 158$ GPa, $C_{33} = 267$ GPa (Demangeot et al., 1996). In spite of the large imprecision in the $K_{A_1(LO)}^B$ pressure coefficient, due to rather weak Raman efficiency of the mode, its lower value undoubtedly reflects the weak influence of the in-plane biaxial stress to the $A_1(LO)$ compared to the E_2 mode.

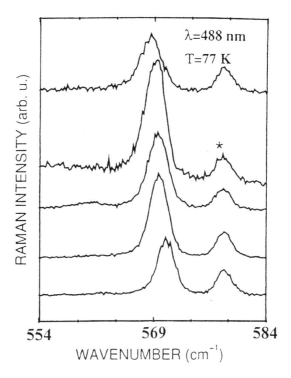

FIG. 34. Unanalyzed Raman spectra taken at 77 K for GaN layers with different biaxial stress. Details are given in Demangeot et al. (1996).

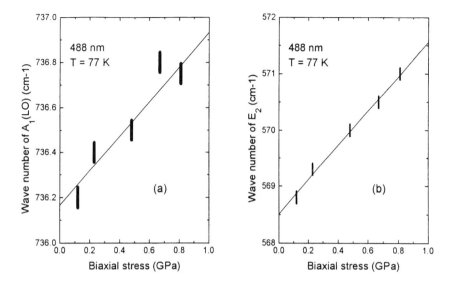

FIG. 35. Least square fits of mode wavenumbers vs the internal stress: (a) A_1 mode; (b) E_2 mode.

The hydrostatic pressure coefficients of the modes were taken from the measurements of Perlin et al. (1992) on a heavily doped sample. They failed to observe the $A_1(LO)$ mode in backscattering geometry along [0001] at atmospheric pressure but measured it at $850\,\text{cm}^{-1}$ at 32.2 GPa. We adopt the pressure coefficient of their $A_1(TO)$ mode, in fact the phonon-like component of the $A_1(LO)$ plasmon coupled mode. For $K_{E_2}^H = 4.17\,\text{cm}^{-1}/\text{GPa}$ and $K_{A_1(LO)}^H = 4.06\,\text{cm}^{-1}/\text{GPa}$, we obtain the two sets of deformation potential constants below $a_{E_2} = -818 \pm 14\,\text{cm}^{-1}$, $b_{E_2} = -797 \pm 60\,\text{cm}^{-1}$, $a_{A_1(LO)} = -685 \pm 38\,\text{cm}^{-1}$, $a_{A_1(LO)} = -997 \pm 70\,\text{cm}^{-1}$. As expected, they are two orders of magnitude smaller than those describing strain effects on the electronic band extrema.

VII. Shallow vs Deep-Level Behavior Under Hydrostatic Pressure

Hydrostatic pressure has been used to evidence resonances in the conduction band of GaN. First, Perlin et al. (1995) noticed that electron concentrations in undoped bulk GaN crystals ranged between $10^{17}\,\text{cm}^{-3}$ to $10^{20}\,\text{cm}^{-3}$. They proposed interpreting the nature of the residual donor in bulk GaN in terms of nitrogen vacancy. They considered that *ab initio* calculations had shown that V_N introduces a resonant level some 0.8 eV above the minimum of the conduction band of GaN (Neugebauer and Van de Walle, 1994; Bogulawski et al., 1994, 1995). They argued that in the neutral charge state, the one electron that should occupy this resonance autoionizes to the bottom of the conduction band and becomes bound by the Coulomb tail of the vacancy potential, forming a shallow level. Following arguments developed in Section I of this chapter, from the spreading of the wavefunction over the whole Brillouin zone, this resonance is expected to shift slower than the bottom of the fundamental conduction band when the pressure increases. A pressure-induced crossover is expected to occur between the resonant level and the shallow one, making the resonance a genuine gap state with efficient cross section for electron trapping with noticeable consequences for the density of free conduction electrons in the conduction band. Perlin et al. (1995) have evidenced the decrease of IR absorption due to the freeze out of the free carriers, which above 20 GPa results in trapping to the pressure-induced genuine gap state. They have reinforced this observation by Raman experiments, via the appearance of the A_1(LO) phonon above 20 GPa, when the screening by dense plasma is canceled. Wetzel et al. (1996) later studied in more detail the screening of the LO phonon with pressure in these samples. From their Raman measurements, which showed as expected that the intensities of E_1 and E_2 modes are not sensitive to the electron density, they could plot the evolution of the intensity of A_1 normalized to the integrated intensity of E_2, and thus calibrate the dependence with electron concentration in their samples. They concluded that this resonant level lies some 0.4 ± 0.1 eV above the conduction band at atmospheric pressure and found it to be some 126^{+20}_{-5} meV below the conduction band at 27 GPa.

Wetzel et al. (1997) extended these experiments to silicon-doped and oxygen-doped GaN grown by hydride vapor phase epitaxy (HVPE). What they found from Raman studies up to 40 GPa is that the ground state of oxygen undergoes a transition from a hydrogenic level (dilute doping) or a degenerate level (high doping) below a critical pressure of 20 GPa to a strongly localized gap state above this value. They could not assign the strongly localized state to either a DO or DX state. Repeating this experiment on silicon-doped samples, but up to 25 GPa only, they did not observe free electron freeze out and concluded that silicon forms a purely hydrogenic state for $P \leqslant 25$ GPa. They concluded that Si has to be excluded as the source for high n doping in all other samples studied, namely, bulk GaN crystals. In the opinion of the author it is most interesting to assume that when 1 GPa corresponds to 2.1% Al, silicon should be a good hydrogenic donor for at least $0 < x < 0.56$. This is interesting because gallium precursors always contain trace amounts of silicon, which in the range of high-growth temperatures used in metallorganic vapor phase epitaxy (MOVPE) *want* to incorporate into the nitrides (Briot, private communication). A second conclusion is that oxygen is expected to induce a strongly localized state for $x \geqslant 0.4$.

VIII. Influence of Strain Fields on Optical Properties of GaN-AlGaN Quantum Wells

1. INTRODUCTION

Quantum wells are realized by embedding, in a direction that we shall take as z here, a thin slice of semiconductor material between two thick pieces of higher bandgap material. Then, the energy of the carriers is quantified along the growth direction and, for bound states, the continuum of k_z allowed values is restricted to a few numbers. For these quantum states, at zone center, the envelope part of the total wavefunction of a given carrier can be computed using basic quantum mechanics arguments (Bastard, 1988; Weisbuch, 1991) when the distribution of gap difference between the conduction and valence band is known. Inversely combined comparison of the calculation with experiment can be used to determine this distribution. If the carriers present are photo-injected the resolution of the Shrödinger equation occurs simultaneously and self-consistently with the resolution of a Poisson equation (Boring et al., 1993). The situation may be slightly complicated if we have to include built-in strain. However, taking into account intervalence band couplings is much more subtle, and brings to the calculation of envelope functions, complexities that were established earlier when, after understanding the donor impurity problem in bulk zinc blende crystal, researchers focused on understanding acceptor impurity states. In other words, instead of resolving a second-order differential equation, one has to solve a system of coupled equations that witness the complexity of the valence band. Then the envelope function of the carriers is multicomponent, and is expanded along the different Bloch states of the valence band. This complexity cannot be bypassed in the zinc blende crystal for computing the joint density of states (Andreani et al., 1987), or for computing confined light-hole states in thin quantum wells (Gil et al., 1991; Arnaud et al., 1992). It was shown in particular that the coupling between the light-hole and spin-orbit split-off state drastically influences the optical properties of such quantum wells. There has been little attention to date regarding III-nitride systems with wurtzite symmetry. The pioneers in this area are Suzuki and Uenoyama (1995, 1996, 1997), Uenoyama and Suzuki (1995), Suzuki et al.

(1995), Chuang and Chang (1996), Chang (1997), Sirenko et al. (1996a, 1996b, 1997), and Jeon et al. (1997). All of them have shown how interesting the physics of such quantum wells is, as it contains all the ingredients required to make the understanding of the problem difficult:

> The built-in strain may modify the hierarchy of the valence band, with subsequent consequences on laser thresholds for instance.
> The energy proximity of the valence bands requires work in the context of many-band envelope calculations.
> The effective mass is poorly known for the valence band of the III-nitrides.
> If GaN has been intensively studied, it is not the case for AlGaN, and to the best of our knowledge the crystal field splitting has never been determined experimentally for this alloy.
> Piezoelectric fields may have a drastic influence.
> Only a few experimental papers have been published on samples grown in similar conditions. These papers reflect the experimental trends and help to understand physical phenomena (Morkoç, 1997).

2. Dispersion Relations in the Valence Band of GaN

In the case of a layer grown along the [0001] direction with in-plane deformation (compression or extension), the total valence Hamiltonian should take into account the band dispersions and strain energies. The different operators are

$$\lambda = \frac{\hbar^2}{2m_0}[A_1 k_z^2 + A_2(k_x^2 + k_y^2)] + C_1 e_{zz} + C_2(e_{xx} + e_{yy})$$

$$\theta = \frac{\hbar^2}{2m_0}[A_3 k_z^2 + A_4(k_x^2 + k_y^2)] + C_3 e_{zz} + C_4(e_{xx} + e_{yy})$$

$$F = \Delta_1 + \Delta_2 + \lambda + \theta$$

$$G = \Delta_1 - \Delta_2 + \lambda + \theta$$

$$K = \frac{\hbar^2}{2m_0} A_5(k_x + ik_y)^2$$

$$H = \frac{\hbar^2}{2m_0} A_6(k_x + ik_y)k_z$$

$$\Delta = \sqrt{2}\Delta_3$$

The valence matrix writes then

$$\begin{bmatrix} F & -K^* & -H^* & 0 & 0 & 0 \\ -K & G & H & 0 & 0 & \Delta \\ -H & H^* & \lambda & 0 & \Delta & 0 \\ 0 & 0 & 0 & F & -K & H \\ 0 & 0 & \Delta & -K^* & G & -H^* \\ 0 & \Delta & 0 & H^* & H & \lambda \end{bmatrix}$$

in the base

$$\begin{bmatrix} |11\uparrow\rangle = -(X\uparrow + iY\uparrow)/\sqrt{2} \\ |1\bar{1}\uparrow\rangle = (X\uparrow - iY\uparrow)/\sqrt{2} \\ |10\uparrow\rangle = Z\uparrow \\ |1\bar{1}\downarrow\rangle = (X\downarrow - iY\downarrow)/\sqrt{2} \\ |11\downarrow\rangle = -(X\downarrow + iY\downarrow)/\sqrt{2} \\ |10\downarrow\rangle = Z\downarrow \end{bmatrix}$$

it can be simplified to a (3 × 3) block diagonal matrix in a smart base (Chuang and Chang, 1996) with Γ_{9v}, Γ_{7v}^1, and Γ_{7v}^2 paired states, which are roughly speaking the analogues of the heavy hole, light hole, and split-off in GaAs, the electron symmetry being Γ_{7c}. We note that the coupling term Δ is a constant for wurtzite quantum wells while it varies like k_z^2 for quantum wells based on zinc blende materials (Gil et al., 1991). This comes from the choice of the basis functions in terms of irreducible representation rather than in terms of the z-component $|j, m_j\rangle$ of the total hole angular momentum. To perform a numerical calculation (Bigenwald et al., 1998) we have, as in Chuang and Chang (1996), used the valence mass parameters of Suzuki et al. (1995) but we took $\Delta_1 = 10$ meV, $\Delta_2 = 6.2$ meV, and $\Delta_3 = 5.5$ meV.

Figure 36 gives the hole effective masses for a strained GaN crystal lattice-matched to $Al_{0.2}Ga_{0.8}N$ as a function of the in-plane (on the left) and on-axis (on the right) wavevector. It can be seen that far from zone center, for every k direction, the particles in Γ_{7v}^1 and Γ_{9v} bands are very heavy $(m > m_0)$, whereas the Γ_{7v}^2 state remains light $(m \sim 0.2 m_0)$. The large modifications of the hole masses correspond to strong anticrossings between bands due to the limitation of the Hamiltonian to terms of k up to second order. Figure 37 shows, for the same crystal, the dipolar matrix elements vs

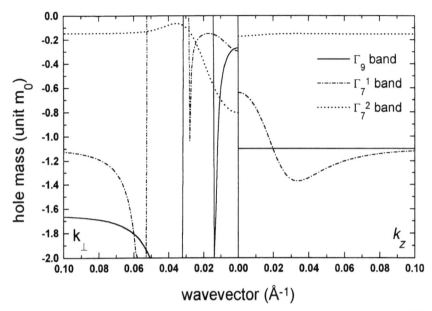

FIG. 36. Hole masses as a function of the wavevector [z direction (on-axis) on the right, in-plane on the left] for a compressively strained GaN crystal ($\varepsilon_{zz} = 0.5\%$). The sharp variations exhibited for z direction and Γ_9 and Γ_7^1 states correspond to the anticrossing between the bands.

the wavevector (in-plane and on-axis) for the three possible transitions (Γ_{7c}-Γ_{9v}, Γ_{7c}-Γ_{7v}^1, and Γ_{7c}-Γ_{7v}^2 written, respectively, A, B, and C) the photon being π-polarized. We clearly evidence that, except for the A transition, both π and σ photon polarization can lead to optical absorption at the zone center. Away from the Γ position, the phenomena become clearly cut: at large k_z, C is allowed while A and B are forbidden, and in-plane and z wavevector directions play opposite roles for the selection rules of the electron-hole transitions.

3. VALENCE BAND LEVELS AND EXCITONS IN GaN-AlGaN QUANTUM WELLS

We have next considered the quantum well built from a GaN well layer lattice matched to barrier $Al_{0.2}Ga_{0.8}N$ layers. The parameter values for the alloy are linearly interpolated from those of the bulks and the effects of built-in piezoelectric fields have been ignored. Note that the crystal field

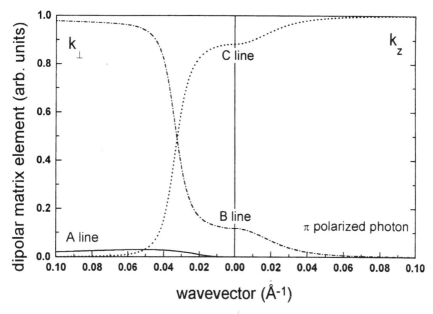

FIG. 37. Dipolar matrix elements $(\langle\chi_e|\vec{\varepsilon}\cdot\vec{r}|\chi_h\rangle)^2$ for the different transitions in a compressively strained GaN ($\varepsilon_{zz} = 0.5\%$) crystal as a function of the wavevector. The incident photon is π polarized.

splitting of AlGaN, which was interpolated between the GaN value and the AlN one (-58.5 meV) (Suzuki et al., 1995), is negative, giving a reversal of the ordering of Γ_9 and Γ_7^1. Figure 38 gives the evolution of the 2D valence band levels when the width of the GaN layer is changed. We remark that the calculation predicts the fundamental valence band state to be either a Γ_{9v}-like hole or a Γ_{7v}^1 state.

The exciton binding energies and oscillator strengths are computed with the formalism previously developed by Bigenwald and Gil (1994) with a 2 parameters trial function. Due to the anisotropy of the structure, the dielectric constant was globalized (Shikanai et al., 1997) and the masses were weighted with the probability densities. In Fig. 39 are represented the exciton binding energies of the fundamental transitions as a function of the GaN layer width. The meaning of this calculation collapses for well widths larger than three exciton Bohr radius. The hierarchy in binding energies reflects essentially two points. The envelope functions of the first Γ_{9v}-like and first Γ_{7v}^1-like states with analogous masses are confined similarly as the electron in the well and the higher binding energy corresponds to $L \sim 15$ Å when the dispersions of both electron and hole functions are minimal. The

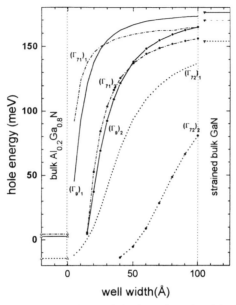

FIG. 38. Well thickness dependence of the 2D valence band levels in GaN-$Al_{0.2}Ga_{0.8}N$ quantum wells.

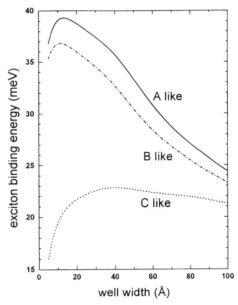

FIG. 39. Exciton binding energies of the fundamental transitions in a strained $Al_{0.2}Ga_{0.8}N$/GaN single quantum well as a function of the well width.

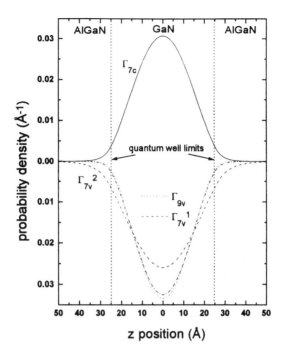

FIG. 40. Probability densities of the conduction and valence fundamental states in a strained 50-Å wide $Al_{0.2}Ga_{0.8}N/GaN$ single quantum well.

rather light Γ_{7v}^2 hole state leaks so much out of the well (for $L < 100$ Å) that the electron-hole pair has a small binding energy and is almost constant for $20 < L < 100$ Å.

These points are confirmed on Fig. 40 where are drawn the probability densities of the electron and valence states as a function of the z coordinate in a 50 Å wide quantum well. From the exciton parameters, we have simulated optical density spectrum of a 50 Å wide $Al_{0.2}Ga_{0.8}N/GaN$ single quantum well for π and σ polarized photons in Figs. 41 and 42. The model relies on the fractional dimensional formalism as explained by Christol *et al.* (1994). For this width, the modes are well pronounced: two lines (A and B), optically active in σ polarization, are forbidden or little permitted for π photons while the contrary occurs for the C line, so that major peaks have similar intensities. A small peak appears for B excitons in π polarization because the well-confined associate hole state belongs mainly to GaN and is active in σ and π polarizations at small k. Straightforward comparison of the spectra reveal similar shapes.

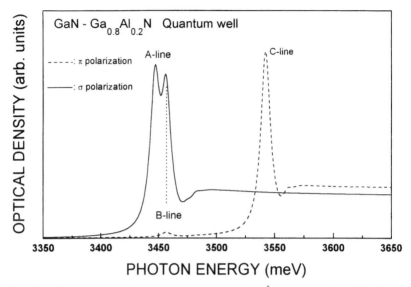

FIG. 41. Simulated optical density spectrum of a 50-Å wide $Al_{0.2}Ga_{0.8}N/GaN$ single quantum well. The solid curve corresponds to π polarized photons while the dashed curve is for σ polarized light. The broadening parameter Γ is set to 5 meV. The bandgaps are given their 300 K values.

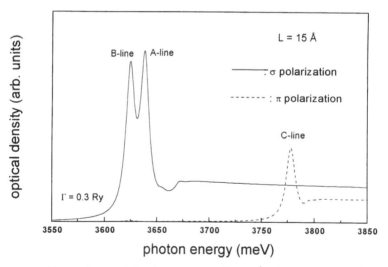

FIG. 42. Simulated optical density spectrum of a 15-Å wide $Al_{0.2}Ga_{0.8}N/GaN$ single quantum well. The solid curve corresponds to π polarized photons while the dashed curve is for σ polarized light. The broadening parameter Γ is set to 5 meV. The bandgaps are given their 300 K values. Note that the fundamental transition involves the Γ_7^1 hole Bloch state.

IX. Self-Organized Quantum Boxes

Combined effects of bulk elastic deformation energies and surface ones may favor growth of material following the Stranski-Krastanov (1939) mode as detailed in Section I of this chapter. Such a growth mode has been observed for InAs-GaAs (Leonard et al., 1993; Moison et al., 1994), SiGe (Eaglesham and Cerullo, 1990), and CdSe-ZnSe (Xin et al., 1996). Gallium nitride dot structures were reported first by Dmitiev et al. (1996). They were obtained by direct growth on 6H-SiC substrates. Tanaka et al. (1994) later demonstrated the realization of boxes of GaN embedded between two $Al_{0.2}Ga_{0.8}N$ cladding layers. This was achieved by MOCVD, using the Si faces of on-axis 6 HSiC substrates on which had been deposited an AlN buffer layer. The GaN dot widths and heights could be reduced down to 40 and 6 nms, respectively. The optical properties of such boxes are shown in Fig. 43, where one can clearly distinguish, for the 40 × 6 boxes, an 80-MeV blue shift of the PL with respect to the value expected for bulk GaN. The photoluminescence is broad, due to the combined effects of size distributions, piezoelectric fields, but comparable in width to the PL of larger boxes (120 nm × 100 nm), which do not produce significant confinement of the carriers. More recently, Daudin et al. (forthcoming, 1998) took advantage of the SK growth mode to deposit by MBE 8.5 nm × 8.5 nm GaN/AlN

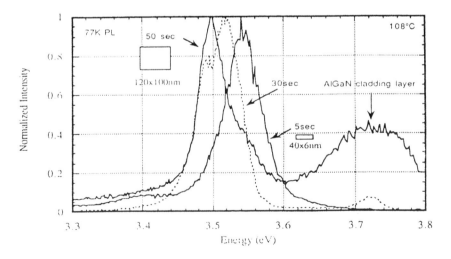

FIG. 43. 77 K photoluminescence spectra of GaN quantum dots confined in AlGaN layers. Figure taken from Tanaka et al. (1994) with their permission.

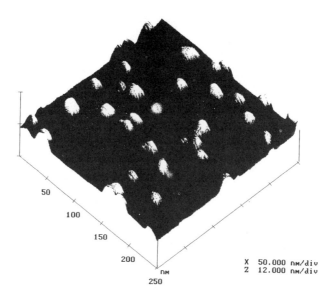

FIG. 44. AFM image of GaN quantum dots, deposited at 720 °C by Daudin et al. (1998) in MBE. [Daudin et al. (1998) and courtesy of Y. Samson, F. Widmann, B. Daudin, and G. Feuillet.]

superlattices with superior interfaces morphologies on nitrided C plane sapphire, using an AlN buffer layer. They also demonstrated the possibility of GaN quantum dots embedded between two AlN cladding layers. Figure 44 illustrates the AFM image of such GaN quantum dots, deposited at 720 °C. Changing the growth conditions, they could tune the sizes of these boxes. Room temperature cathodoluminescence experiments evidenced quantum confinement up to 5.5 eV (Daudin et al., private communication). To conclude, they could also realize quantum dot superlattices as shown in Fig. 45a, and 45b.

X. Conclusion

Residual strain-fields are always present in III-nitrides and in their related heterostructures. They can produce a reversal of the ordering of the valence band in bulk crystals under biaxial tension. In addition, the biaxial compression favors the observation of photoluminescence in LPB and UPB branches of the B exciton polariton. The oscillator strengths of the optical

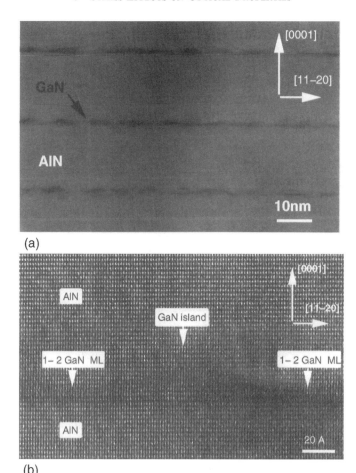

FIG. 45. GaN-AlN quantum dot superlattices realized by Daudin et al. (1998) in MBE. (a) General TEM picture, (b) HRTEM details of the morphology of an individual quantum dot. (Repeated with the permission of Daudin et al.; material contained in a private communication.)

transitions at zone center are extremely dependent on the biaxial strain. Growing a sample on A plane sapphire produces an anisotropic strain field and an anisotropy of the in-plane optical response. These strain fields are linked to growth conditions, buffer layers, substrates, and microscopic details of the interface morphologies. Though nitrides are super hard materials, and the built-in strain is rather weak, they substantially modify phonon frequencies. The screening of the LO phonon by the electron sea can be canceled using an external perturbation such as a hydrostatic pressure. A metal-to-semiconductor transition can be induced near 20 GPa. This method is extremely useful in understanding the doping properties of native defects or impurities such as silicon or oxygen in GaN, and extrapolating their effects to AlGaN. The growth of a thin GaN layer may be ruled by a Stranski-Krastanov mode giving quantum boxes. These quantum boxes, which are similar in size to the quantum boxes so important in lasing systems based on InGaN confining layers, have a bright future. Although piezoelectric fields are not understood to date in nitrides, they are extant and should create interesting effects.

Acknowledgments

My colleague and friend O. Briot has grown most of the GaN samples I used. I am grateful to Jin Joo Song, A. Hoffmann, B. Monemar, H. Morkoç, F. Ponce, and S. Nakamura for many stimulating discussions. I want to thank my colleagues A. Alemu, R. Aulombard, P. Bigenwald, F. Binet, R. Bisaro, J. Campo-Ruiz, P. Christol, B. Daudin, F. Demangeot, J. Duboz, G. Feuillet, G. Fishman, J. Frandon, M. Julier, J. Lascaray, M. Leroux, P. Lefebvre, Y. Le-Vaillant, K. P. O'Donnell, M. Renucci, J. L. Rouviere, S. Ruffenach-Clur, and M. Tchounkeu for their help and for being in the same boat. G. Fishman performed a sanity check of the first draft manuscript.

References

Alemu, A., Gil, B., Julier, M., and Nakamura, S. (1998). *Phys. Rev. B*, **57**, 3761.
Amano, H., Hiramatsu, K., and Akasaki, I. (1988) *J. Jour. Applied Physics*, **27**, L1384.
Andreani, L. C. (1995). In *Confined Electrons and Photons: New Physics and Applications*. Elias Burstein and Claude Weisbuch, eds., Nato ASI series, series B, **340**, p. 57.
Andreani, L. C., Pasquarello, A., and Bassani, F. (1987). *Phys. Rev. B*, **36**, 5887.
Arnaud, G., Boring, P., Gil, B., Garcia, J. C., Landesman, J. P., and Leroux, M. (1992). *Phys. Rev. B* (RC), **46**, 1886.
Baranowski, J. M., Lilienthal-Weber, Z., Korona, K., Pakula, K., Stepniewski, R., Wysmolek, A., Grzegory, I., Nowak, G., Porowski, S., Monemar, B., and Bergman, P. (1997). *Materials Research Society Symposium Proc.* **449**, 393.
Baranowski, J. and Porowski, S. (1996). In *Proc. IEEE International Conference on Semiconducting and Semiinsulating Materials*, Toulouse, France, 1996, p. 77.
Bassani, F. and Pastori Parravicini, G. (1979). In *Electronic States and Optical Transitions in Solids*. New York: Pergamon Press.
Bastard, G. (1988). In *Wave Mechanics Applied to Semiconductor Heterostructures*. Les Editions de Physique, Les Ulis, France.
Benoit à la Guillaume, C., Bonnot, A., and Debever, J. M. (1970). *Phys. Rev. Lett.*, **24**, 1235.

Biefeld, M. (1989). Compound Semiconductor Strained-Layer Superlattices. Switzerland: Trans. Tech. Publications.
Bigenwald, P. and Gil, B. (1994). *Solid State Commun.*, **91**, 33.
Bigenwald, P., Christol, P., Konczewicz, L., Testud, P., and Gil, B. (1998). Presented at the EMRS Meeting, Strasbourg, France, June 1997. *Materials Science and Engineering B*, **50**, 208.
Bogulawski, P., Briggs, E., White, T. A., Wensell, M. G., and Bernholc, J. (1994). *Materials Research Society Symposium Proc.*, **339**, 693.
Bogulawski, P., Briggs, E., and Bernholc, J. (1995). *Phys. Rev. B*, **51**, 17255.
Boring, P., Gil, B., and Moore, K. J. (1993). *Phys. Rev. Lett.*, **71**, 1875.
Briggs, R. J. and Ramdas, A. K. (1976). *Phys. Rev. B*, **13**, 5518.
Briot, O., Gil, B., Tchounkeu, M., Demangeot, F., Frandon, J., and Renucci, M. (1996). *Journal of Electronic Materials*, **26**, 294.
Briot, O. Private communication.
Buyanova, I. A., Bergman, J. P., Monemar, B., Amano, H., and Akasaki, I. (1996). *Appl. Phys. Lett.*, **69**, 1255.
Buyanova, I. A., Bergman, J. P., Li, W., Monemar, B., Amano, H., and Akasaki, I. (1996). *Materials Research Society Symposium Proc.*, **423**, 675.
Camphausen, D. L., Neville Connel, G. A., and Paul, W. (1971). *Phys. Rev. Lett.*, **26**, 184.
Chichibu, S., Azuhata, T., Sota, T., Amano, H., and Akasaki, I. (1997). *Appl. Phys. Lett.*, **71**.
Chichibu, S., Shikanai, A., Azuhata, T., Sota, T., Kuramata, A., Horino, K., and Nakamura, S. (1966). *Appl. Phys. Lett.*, **68**, 3766.
Cho, K. (1976). *Phys. Rev. B*, **14**, 4463.
Christensen, N. E. and Gorczyca, I. (1994). *Phys. Rev. B*, **50**, 4397.
Christol, P., Lefebvre, P., and Mathieu, H. (1994). *IEEE J. Quantum Electron*, **30**, 2287.
Chuang, S. L. (1996). *IEEE Journal of Quantum Electronics*, **32**, 1791; (1997). *Semicond. Sci. Technol.*, **12**, 252.
Chuang, S. L. and Chang, C. S. (1996). *Appl. Phys. Lett.*, **68**, 1657; (1996). *Phys. Rev. B*, **54**, 2491.
Daudin, B., Widmann, F., Feuillet, G., Adelmann, C., Samson, Y., Arlery, M., and Rouviere, J. L. (1998). Presented at the EMRS meeting, Strasbourg, France, June 1997. *Materials Science and Engineering B*, **50**, 8.
Daudin, B., Feuillet, G., Fishman, G., and Rouviere, J. L. Private communication.
Demangeot, F., Frandon, J., Renucci, M., Briot, O., Gil, B., and Aulombard, R. L. (1996). *Solid State Communications*, **100**, 207.
Dingle, R., Sell, D. D., Stokowski, S. E., and Illegems, M. (1971). *Phys. Rev. B*, **4**, 1211.
Dmitriev, V., Irvine, K., Zubrilov, A., Tsvetskov, D., Nikolaev, V., Jacobson, M., Nelson, D., and Sitnikova, A. (1996). *Mater. Res. Soc. Symp. Proc.*, **395**, 295.
Drechsler, M., Hoffmann, D. M., Meyer, B. K., Detchprohm, T., Amano, H., and Akasaki, I. (1995). *Jpn. J. Appl. Phys.*, **34**, L1178.
Eaglesham, D. J. and Cerullo, M. (1990). *Phys. Rev. Lett.*, **64**, 1963.
Edwards, N. V., Yoo, S. D., Bremser, M. D., Weeks, Jr., T. W., Nam, O. H., Iiu, H., Stall, R. A., Orton, M. N., Perkins, N. R., Kuech, T. F., and Aspnes, D. E. (1997). *Appl. Phys. Lett.*, **70**, 2001.
Froyen, S., Wei, S., and Zunger, A. (1988). *Phys. Rev. B*, **38**, 10124.
Gfrorer, G., Schlusener, T., Harle, V., Scholz, F., and Hangleiter, A. (1996). *Materials Research Society Symposium Proc.*, **449**, 429.
Gil, B., Briot, O., and Aulombard, R. L. (1995). *Phys. Rev. B*, **52**, R17 028.
Gil, B., Hamdani, F., and Morkoç, H. (1996). *Phys. Rev. B*, **54**, 7678.
Gil, B. and Briot, O. (1997). *Phys. Rev. B*, **55**, 2530.
Gil, B. and Alemu, A. (1997) *Phys. Rev. B*, **56**, 12446.

Gil, B., Clur, S., and Briot, O. (1997). *Solid State Communications*, **104**, 267.
Gil, B., Lefebvre, P., Boring, P., Moore, K. J., Duggan, G., and Woodbridge, K. (1991). *Phys. Rev. B* (RC), **44**, 1942.
Gorczyca, I. and Christensen, N. E. (1993). *Physica B*, **185**, 410.
Heim, U. and Wiesner, P. (1973). *Phys. Rev. Lett.*, **30**, 1205.
Hiramatsu, K., Detchprohm, T., and Akasaki, I. (1993). *Japanese Journal of Appl. Physics*, **32**, 1528.
Hopfield, J. J. (1960). *J. Phys. Chem. Solids*, **15**, 97.
Hopfield, J. J. (1958). *Phys. Rev.*, **112**, 1555.
Jeon, J. B., Lee, B. C., Sirenko, Yu. M., and Littlejohn, M. A. (1997). *J. Appl. Phys.*, **82**, 386.
Julier, M., Campo, J., Gil, B., Lascoray, J. P., and Nakamura, S. (1998). *Phys. Rev. B.*, **57**, R6791.
Kim, S., Herman, I. P., Tuchman, J. A., Doverspike, K., Rowland, L. B., and Gaskill, D. K. (1995). *Appl. Phys. Lett.*, **67**, 380.
Kisielovski, C., Krüger, J., Ruminov, S., Suski, T., Ager, III, J. W., Jones, E., Lilienthal-Weber, Z., Rubin, M., Weber, E. R., Brenner, M. D., and Davis, R. F. (1996). *Phys. Rev. B*, **54**, 17745.
Klingshirn, C. F. (1995). In *Semiconductor Optics*. Berlin, Heidelberg, New York: Springer-Verlag.
Knap, W., Alause, H., Bluet, J. M., Camassel, J., Young, J., Asif Khan, M., Chen, Q., Huant, S., and Shur, M. (1996). *Solid State Communications*, **99**, 195.
Koster, J. G. F., Dimmock, J. O., Wheeler, R. G., and Statz, H. (1963). In *Properties of the Thirty-Two Point Groups*. Cambridge: MIT Press.
Kovalev, D., Averboukh, B., Volm, D., Meyer, B. K., Amano, H., and Akasaki, I. (1996). *Phys. Rev. B*, **54**, 2518.
Kozawa, T., Kachi, T., Kano, H., Nagase, H., Koide, N., and Manabe, K. (1995). *J. Appl. Phys.*, **77**, 4389.
Lambrecht, W. R. L., Kim, K., Rashkeev, S., and Segall, B. (1996). *Materials Research Soc. Symposium Proc.*, **395**, F. A. Ponce, R. D. Dupuis, S. Nakamura, and J. A. Edmond, eds., 455.
Ledentsov, N. N. (1996). In *Proc. 23rd Int. Conf. Physics of Semiconductors*, Berlin, 1996, M. Scheffler and R. Zimmermann, eds., Singapore: World Scientific, p. 19.
Leonard, D., Krishnamurthy, M., Reaves, C. M., Denbaars, S. P., Petroff, P. M. (1993). *Appl. Phys. Lett.*, **63**, 3203.
Leszczinski, M., Suski, T., Teisseyre, H., Perlin, P., Grzegory, I., Jun, J., Porowski, S., and Moustakas, T. D. (1994). *Appl. Phys. Lett.*, **76**, 4909.
Le Vaillant, Y.-M., Bisaro, R., Oliver, J., Durand, O., Duboz, J.-Y., Ruffenach-Clur, S., Briot, O., Gil, B., and Aulombard, R. L. (1998). *Proc. 1997 EMRS Spring Meeting*. To appear in *Materials Science and Engineering B*.
Li, C. F., Huang, Y. S., Malikova, L., and Pollak, F. H. (1997). *Phys. Rev. B*, **55**, R9251.
Madelung, O. (1991). *Data in Science and Technology — Semiconductors: Group IV Elements and III-V Compounds*. Berlin, Heidelberg: Springer-Verlag.
Mailhiot, C. and Smith, D. L. (1990). In *Critical Reviews in Solid State and Materials Science*. Vol. **16**, pp. 131–160.
Manasreh, O. (1996). *Phys. Rev. B*, **53**, 16425.
Merz, C., Kunzer, M., and Kaufmann, U. (1996). *Semicond. Science Technol.*, **11**, 712.
Mohammad, S. N. and Morkoç, H. (1995). *Progress in Quantum Electronics*. Marek Osinski, ed., Amsterdam: Elsevier Science Publishers.
Moison, J. M., Houzay, F., Barthe, F., Leprince, L., André, E., and Vatel, O. (1994). *Appl. Phys. Lett.*, **64**, 196.

Monemar, B. (1974). *Phys. Rev. B*, **10**, 676.
See, for instance, Morkoç, H. (1997). In *Widebandgap Nitrides and Devices*. Berlin: Springer-Verlag.
Nakamura, S. and Fasol, G. (1997). *The Blue Laser Diode*. Berlin: Springer-Verlag.
Nelson, D. K., Melnik, Yu. V., Selkin, A. V., Yacobson, M. A., Dmitriev, V. A., Irvine, K., and Carter, Jr. C. H. (1996). *Fiz. Tverd. Tela*, **38**, 651 [*Sov. Phys. Solid State*, **38**, 455 (1996)].
Neugebauer, J. and Van de Walle, C. G. (1994). *Phys. Rev. B*, **50**, R8067.
Nye, J. F. (1957). *Physical Properties of Crystals*. Oxford: The Clarendon Press.
Orton, J. W. (1996). *Semicond. Sci. Technol.*, **11**, 1026.
Pearsall, T. P. (1990). *Semiconductors and Semimetals*. Vol. **32**, Strained-Layer Superlattices. Boston: Academic Press.
Perlin, P., Suski, T., Teisseyre, H., Leczcynski, M., Grzegory, I., Jun, J., Porowski, S., Bogulawski, P., Bernholc, J., Chervin, J. C., Polian, A., and Moustakas, T. (1995). *Phys. Rev. Lett.*, **75**, 296.
Perlin, P., Gorczyca, I., Christensen, N. E., Grzegory, I., Teisseyre, H., and Suski, T. (1992). *Phys. Rev. B*, **45**, 13307.
Perlin, P., Jauberthie-Carillon, C., Itie, J. P., San Miguel, A., Grzegory, I., and Polian, A. (1992). *Phys. Rev. B*, **45**, 83.
Perry, W. G., Zheleva, T., Bremser, M. D., Davis, R. F., Shan, W., and Song, J. J. (1996). *Journal of Electronic Materials*, **26**, 224.
Pikus, G. E. and Bir, L. G. (1974). In *Symmetry and Strained-Induced Effects in Semiconductors*. New York: Wiley.
Polian, A., Grimsditch, M., and Grzegory, I. (1996). *J. Appl. Phys.*, **79**, 3343.
Ponce, F. (1998). In: *Group III Nitriole Semiconductor Compounds: Physics and Application*, B. Gil, ed., Oxford: Clarendon Press, p. 123.
Popovici, G., Morkog, H., and Mohammad, S. N. (1998). In: *Group III Nitriole Semiconductor Compounds: Physics and Application*, B. Gil, ed., Oxford: Clarendon Press, pp. 19–69.
Rieger, W., Metzger, T., Angerer, H., Dimitrinov, R., Ambacher, O., and Stutzmann, M. (1996). *Appl. Phys. Lett.*, **68**, 970.
Rohner, P. G. (1971). *Phys. Rev. B*, **3**, 433.
Sandomirskii, V. B. (1964). *Fiz. Tverd. Tela*, **6**, 324 [*Sov. Phys. Sol. St.*, **6**, 261 (1964)].
Savastenko, V. A. and Sheleg, A. U. (1978). *Phys. Status Solidi*, **A48**, K135.
Sell, D. D., Stokovski, S. E., Dingle, R., and DiLorenzo, J. V. (1973). *Phys. Rev. B*, **7**, 4568.
Shan, W., Schmidt, T. J., Yang, X. H., Hwang, S. J., and Song, J. J. (1995). *Appl. Phys. Lett.*, **66**, 987.
Shan, W., Hauenstein, R. J., Fischer, A. J., Song, J. J., Perry, W. G., Bremser, M. D., Davis, R. F., and Goldenberg, B. (1996). *Phys. Rev. B*, **54**, 13460.
Shan, W., Schmidt, T. J., Hauenstein, R. J., Song, J. J., and Goldenberg, B. *Appl. Phys. Lett.*, **66**, 3492.
Shikanai, A., Azuhata, T., Sota, T., Chichibu, S., Kuramata, A., Horino, K., and Nakamura, S. (1997). *J. Appl. Phys.*, **81**, 417.
Sirenko, Yu. M., Jeon, J. B., Kim, K. W., Littlejohn, M. A., and Stroscio, M. A. (1996a). *Phys. Rev. B*, **53**, 1997.
Sirenko, Yu. M., Jeon, J. B., Kim, K. W., Littlejohn, M. A., and Stroscio, M. A. (1996). *Appl. Phys. Lett.*, **69**, 2504.
Sirenko, Yu, M., Jeon, J. B., Lee, B. C., Kim, K. W., Littlejohn, M. A., Stroscio, M. A., and Iafrate, G. J. (1997). *Phys. Rev. B*, **55**, 4360.
Skromme, B. (1998). *Materials Sci. Engineering B.*, **50**, 117.
Smith, D. L. and Mailhiot, C. (1990). *Review of Modern Physics*, **62**, 173.
Stranski, I. N. and Krastanov, Von L. (1939). *Akad. Wiss. Lit. Mainz Math-Natur. Kl. lib.*, **146**, 797.

Suzuki, M. and Uenoyama, T. (1995a). *Jpn. J. Appl. Phys.*, **35**, L953.
Suzuki, M. and Uenoyama, T. (1995b). *Jpn. J. Appl. Phys.*, **34**, 3442.
Suzuki, M. and Uenoyama, T. (1996). *Jpn. J. Appl. Phys.*, **35**, 543; (1996). **35**, 1420, *Appl. Phys. Lett.*, **69**, 3378; (1996). *J. Appl. Phys.*, **80**, 6868; (1997). *Solid State Electronics*, **41**, 271.
Suzuki, M., Uenoyama, T., and Yanase, A. (1995). *Phys. Rev. B*, **52**, 8132.
Tanaka, S., Hirayama, H., Iwai, S., and Aoyagi, Y. (1994). *Materials Research Society Symposium Proc.*, **449**, 135.
Tchounkeu, M., Briot, O., Gil, B., and Aulombard, R. L. (1996). *J. Appl. Phys.*, **80**, 5352.
Teisseyre, H., Perlin, P., Suski, T., Grzegory, I., Jun, J., and Porowski, S. (1995). *J. Phys. Chem. Solids*, **56**, 353.
Toyozawa, Y. (1959). *Prog. Theor. Phys. Suppl.*, **12**, 111.
Uenoyama, T. and Suzuki, M. (1995). *Appl. Phys. Lett.*, **67**, 2527.
Volm, D., Oettingen, K., Streibl, T., Kovalev, D., Ben-Chorin, M., Diener, J., Meyer, B. K., Majewski, J., Eckey, L., Hoffmann, A., Amano, H., Akasaki, I., Hiramatsu, K., and Detchprohm, T. (1996). *Phys. Rev. B*, **53**, 16543.
Warren, B. E. (1959) *Progress in Metal Physics*. London: Pergamon Press, Vol. **8**, p. 147, and references therein.
Weisbuch, C. and Ulbrich, R. (1979). *J. of Luminescence*, **18/19**, 27.
Weisbuch, C. and Vinter, B. (1991). *Quantum Semiconductor Structures*. Boston: Academic Press.
Wetzel, C., Walukiewicz, W., Haller, E. E., Ager, III, J., Grzegory, I., Porowski, S., and Suski, T. (1996). *Phys. Rev. B*, **53**, 1322.
Wetzel, C., Suski, T., Ager, III, J. W., Webber, E. R., Haller, E. E., Fischer, S., Meyer, B. K., Molnar, R. J., and Perlin, P. (1997). *Phys. Rev. Lett.*, **78**, 3923.
Wolford, D. J. and Bradley, J. A. (1985). *Solid State Communications*, **53**, 1069.
Wolford, D. J., Bradley, J. A., Fry, K., and Thompson, J. (1984). *Proc. 17th Int. Conf. Physics of Semiconductors*, San Francisco, 1984. Berlin: Springer-Verlag, pp. 627–630.
Xin, S. H., Wang, P. D., Yin, Aie, Kim, C., Dobrolowska, M., Merz, J. L., Furdyna, J. K. (1996). *Appl. Phys. Lett.*, **69**, 3884.
Yamaguchi, M., Yagi, T., Azuhata, T., Sota, T., Suzuki, K., Chichibu, S., and Nakamura, S. (1997). *J. Phys. Cond. Matter*, **9**, 241.
Zhao, X., Li, G., Han, H., Wang, Z., Tang, R., and Che, R. (1984). *Chinese Physics Letters*, **1**, 19.

CHAPTER 7

Strain in GaN Thin Films and Heterostructures

Christian Kisielowski

DEPARTMENT OF MATERIALS SCIENCE AND MINERAL ENGINEERING
UNIVERSITY OF CALIFORNIA
BERKELEY, CALIFORNIA

I. THIN-FILM GROWTH AT LOW TEMPERATURES 275
   1. *Introduction* 275
   2. *Origin of Stresses* 277
   3. *Growth Modes* 283
II. STRESS/STRAIN RELATIONS 286
   1. *Elastic Constants of GaN* 286
   2. *Experimental Stress/Strain Calibrations* 289
III. CONTROL OF HYDROSTATIC AND BIAXIAL STRESS AND STRAIN COMPONENTS . . . 292
   1. *Stress-Controlling Parameters* 292
   2. *Lattice Constants* 297
   3. *Engineering of Materials Properties by Active Utilization of Stress and Strain* 299
IV. STRAINED AlN/InN/GaN HETEROSTRUCTURES 304
   1. *Al and In Distribution in $Al_xGa_{1-x}N/In_xGa_{1-x}N/GaN$ Quantum Well Structures* 304
   2. *Strain-Induced Transition from a Two-Dimensional to a Three-Dimensional Growth Mode* 308
V. PERSPECTIVES 313
   REFERENCES 315

I. Thin-Film Growth at Low Temperatures

1. INTRODUCTION

In recent years gallium nitride and the related aluminum and indium nitrides have attracted substantial scientific and industrial interest. This development was triggered by reports on *p*-type doping of the wide bandgap material by Amano *et al.* (1989) that allowed for the fabrication of *p/n* junctions, light emitting diodes, and laser structures with emission wavelengths that ranged from the visible spectrum to the ultraviolet (Nakamura and Fasol, 1997). From that point, a III-nitride technology developed in an unprecedented short period of time. This rapid progress is based on experience with thin-film growth commonly gained through trial-and-error processes. Consequently, many of the physical processes that control the performance of GaN thin films and heterostructures are "handled" in some way even though they are not yet fully understood.

Stress and strain in GaN thin films is one of these issues. It was recognized early that large stresses originate from the growth on lattice mismatched substrates, sapphire or SiC, with thermal expansion coefficients that differ from that of GaN (Table I). Film cracking was reported and investigated (Hiramatsu et al., 1993). It could be avoided if the film thickness was kept below 4 μm. Amano et al. (1986) introduced the growth of AlN buffer layers in order to improve the structural quality of the GaN main layers. Dislocation annihilation in the buffer-layer region was stimulated by this growth process. Similar results were obtained by growing a GaN buffer layer at temperatures of around 770 K (Nakamura, 1991). However, it was not until recently that the large impact of buffer-layer growth on the stresses in the films was uncovered (Kisielowski et al., 1996a; Reiger et al., 1996). Moreover, it was pointed out that the growth of buffer layers contributes with *biaxial* strain to the overall stress in the thin films that also can be altered by the growth of main layers of different lattice constants (Kisielowski et al., 1996a).

Lattice parameters are usually changed by a variation of native defect concentrations or by the incorporation of dopants and impurities. Such

TABLE I

SELECTED MATERIALS, PROPERTIES OF GaN, InN, AlN, SAPPHIRE, AND SiC[a,b,c]

Material	c-Lattice constant (nm)	a-Lattice constant (nm)	c-Thermal exp. coeff. (K^{-1})	a-Thermal exp. coeff. (K^{-1})	a-Lattice mismatch (%)
GaN	0.5185	0.3188	$3.17\ 10^{-6}$	$5.59\ 10^{-6}$	0.0
AlN	0.4982	0.3112	$5.3\ 10^{-6}$	$4.2\ 10^{-6}$	2.4
InN	0.572	0.3542	$\sim 3\ 10^{-6}$	$\sim 4\ 10^{-6}$	-12
Sapphire	1.299	0.4758	$8.5\ 10^{-6}$	$7.5\ 10^{-6}$	-14
6H-SiC	1.511	0.308	$4.68\ 10^{-6}$	$4.2\ 10^{-6}$	2.3

[a]Data are taken from Morkoç et al. (1994), Leszczynski et al. (1994), and Kisielowski et al. (1996a).
[b]Growth on SiC is not addressed in this paper.
[c]Linear thermal expansion coefficients at room temperature are listed.

point defects introduce *hydrostatic* strain components. Thus, hydrostatic and biaxial strain components coexist in the films and they are physically of different origin. The biaxial strain comes from growth on lattice-mismatched substrates having different thermal expansion coefficients. The hydrostatic strain, however, originates from the incorporation of point defects. It was shown that a balance of these strains can be exploited to strain-engineer desired-film properties (Kisielowski *et al.*, 1996b; Klockenbrink *et al.*, 1997; Fujii *et al.* 1997).

This chapter reports on our current understanding of stress- and strain-related phenomena in GaN thin films and heterostructures. It focuses on their impact on thin-film growth if any physical property of thin films is affected. Some examples will be given. They concern the surface morphology of the GaN films, native defect formation, the incorporation of dopants, and photoluminescence.

Most of the investigated epitaxial films and heterostructures were grown by molecular beam epitaxy (MBE) utilizing a hollow anode nitrogen source (Anders *et al.*, 1996; Kim *et al.*, 1998). Moreover, many other samples grown in different reactors either by MBE or by metal organic chemical vapor deposition (MOCVD) were also analyzed. This guarantees that the MBE results presented here are typical for GaN thin films. Differences between films grown by MBE or by MOCVD are seen in terms of characteristic numerical values.

2. ORIGIN OF STRESSES

Many of the materials' properties relate to growth temperatures that are low compared with the melting point $T_m \sim 2800\,K$ of GaN (Van Vechten, 1973). Typically, MBE and MOCVD growth proceeds around 1000 K and 1300 K, respectively. Thus, the thin films are grown at only 36% and 46% of T_m even though absolute growth temperatures are high. In fact, the MBE growth of GaN compares well with the growth of low temperature (LT-) GaAs, if measured in fractions of the melting point temperature ($T_{growth,LT\text{-}GaAs} = 0.3\,T_m$). LT-GaAs is known to deviate from a stoichiometric composition (Luysberg *et al.*, 1998; Smith *et al.*, 1988). Early lattice parameter measurements revealed stoichiometric fluctuations in GaN, also (Lagerstedt *et al.*, 1979). It is a main topic of this review to uncover how stress and strain influence the stoichiometry of GaN thin films.

Plastic materials' properties are specifically different at low temperatures. Little is known about the plasticity of GaN around $0.4\,T_m$. Some guidance can be obtained from Fig. 1 (Specht, private communication). It shows the brittle to ductile transition temperature of different materials as well

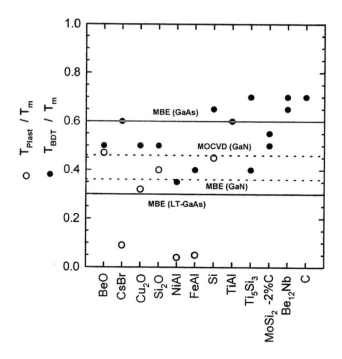

FIG. 1. Brittle to ductile transition temperature and onset of dislocation motion in selected materials systems. Values are given in fractions of the melting point temperature. Typical growth temperatures of GaN and of GaAs are indicated. For details see text.

as onset temperatures for dislocation glide. Values are given in fractions of the melting point temperature. The brittle to ductile transition typically occurs around 0.4–0.6 T_m. Thus, it is likely that GaN films are grown in the brittle region if MBE is employed. On the other hand, MOCVD growth proceeds close to typical brittle to ductile transition temperatures. Dislocation glide may be observed at considerably lower temperatures, in particular, if large stresses are involved. The relative MBE growth temperature of GaN coincides with that of LT-GaAs. Therefore, it is expected that stoichiometric variations are significant in GaN. Growth processes are limited kinetically in such materials. They proceed far away from a thermodynamic equilibrium.

The stresses in GaN thin films originate from different sources. Throughout this chapter we refer to GaN epitaxy on sapphire c-plane substrates because it is the most commonly used substrate material. A growth cycle consists of three steps: the surface nitridation, the buffer layer, and the main layer growth. MOCVD and MBE growth are performed this way even though temperatures differ (Klockenbrink et al., 1997; Nakamura and Fasol, 1997). Figure 2 schematizes those steps of a growth cycle that cause strain. First, it is the lattice mismatch of -14% between GaN and sapphire (Table I) that introduces a tensile strain into the buffer layers. This layer relaxes the lattice mismatch to some unknown extent. It was shown recently that only a fraction of the total lattice mismatch contributes to a residual stress because misfit dislocations are present at the sapphire/buffer layer interface (Ning et al., 1996; Ruvimov et al., 1997). Heating to the growth temperature increases tensile strain. The main layer deposition itself may introduce strain if its lattice parameter differs from that of the buffer layer at the growth temperature. Finally, cooling adds a compressive strain. Thus, the resulting stresses may change in a complex manner during a growth cycle as indicated in the inset of Fig. 2.

The stresses that originate from thermal strain were estimated to be as large as 1.5 GPa if an epitaxial GaN layer on sapphire is cooled from an MBE growth temperature of 1000 K to room temperature or vice versa (Fig. 3) (Kisielowski et al., 1996b). One concludes from the calculation that stresses in MOCVD-grown GaN must be larger than those in MBE-grown

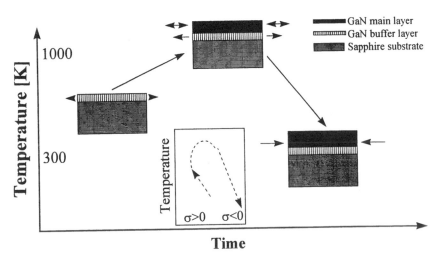

FIG. 2. A schematic representation of the strains that are caused during a GaN growth cycle. Arrows indicate the amount and the sign of the resulting stresses if GaN is deposited on sapphire. Buffer layer deposition, heating, main-layer growth and the post-growth cooling to room temperature cause stress and strain. The inset shows a schematic stress-temperature diagram. Details of the curve are unknown. Current research accesses this information.

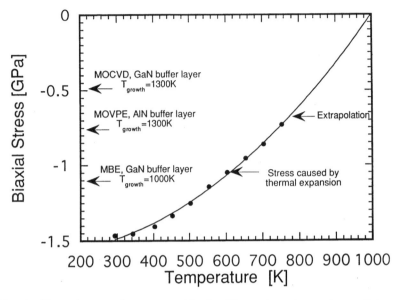

FIG. 3. Thermal stresses that are caused by the different thermal expansion coefficients of GaN and sapphire. They are extrapolated to an MBE growth temperature of 1000 K (solid line). Three largest stress values from measurements of about 250 MBE- and 100 MOCVD-grown samples are indicated. They are representative. An extrapolation to an MOCVD-growth temperature of 1300 K is not meaningful.

samples. This is simply because of the higher main-layer deposition temperature of 1300 K in the MOCVD process. Experiments have surprisingly shown the opposite. The largest stress values are indicated in Fig. 3. They were obtained by measuring stresses in approximately 300 samples obtained from different sources. Therefore, they are representative. Even the largest stresses in MOCVD grown samples do not reach the values that are commonly present in MBE grown samples. Also, it seems that the deposition of an AlN buffer layer in an MOCVD growth process may increase the stress compared with the deposition of a GaN buffer layer. However, none of the measured stresses can account for the expected thermal stresses that were estimated assuming an elastic response of the materials. It is for these reasons that plasticity has to be considered.

The thermal stresses in the epitaxial layers are biaxial by nature. Any biaxial stress, however, can be deconvoluted into a hydrostatic — and two pure shear stresses (Kelly and Groves, 1970)

$$\begin{bmatrix} \sigma_y & 0 & 0 \\ 0 & \sigma_y & 0 \\ 0 & 0 & \sigma_z \end{bmatrix} = \begin{bmatrix} \sigma_a & 0 & 0 \\ 0 & \sigma_a & 0 \\ 0 & 0 & 0 \end{bmatrix}$$

$$= 1/3 \begin{bmatrix} \sigma_p & 0 & 0 \\ 0 & \sigma_p & 0 \\ 0 & 0 & \sigma_p \end{bmatrix} + 1/3 \begin{bmatrix} 0 & 0 & 0 \\ 0 & \sigma_a & 0 \\ 0 & 0 & -\sigma_a \end{bmatrix} + 1/3 \begin{bmatrix} \sigma_a & 0 & 0 \\ 0 & 0 & 0 \\ 0 & 0 & -\sigma_a \end{bmatrix} \quad (1)$$

$$\quad\quad\quad\text{hydrostatic} \quad\quad\quad\quad\quad \text{pure shear} \quad\quad\quad\quad \text{pure shear}$$

where $\sigma_a = \sigma_x = \sigma_y$ is the biaxial stress in the c-plane of the materials and $1/3 \, \sigma_p = 2/3 \, \sigma_a$ is the invariant hydrostatic stress. A hydrostatic stress can be altered by the introduction of point defects because they change the crystal volume (Kisielowski et al., 1996a)

$$\Delta V/V = 3\varepsilon = ((r_s/r_h)^3 - 1)CN^{-1} := 3bC \quad (2)$$

and, thereby, the stresses. In Eq. (2) ε is a hydrostatic strain, r_h, r_s are the radii of the host and the solute atoms and C, N are the concentration of solute atoms and lattice sites (8.8×10^{22} cm^{-3}), respectively. Hence, b is the factor that describes the change of the lattice constant. This argument can be inverted to understand why the application of a biaxial stress may alter the materials stoichiometry or the incorporation of dopants and impurities.

The pure shear stresses act in the xz and yz planes and form an angle of 45° with the normal of the (basal) c-plane. They cause dislocation glide and plastic relaxation (Hirth and Lothe, 1982). Nothing is known about dislocation motion in GaN at about $0.4 \, T_m$. However, dislocation motion at low temperature was investigated in silicon (Alexander et al., 1987). Typically, the dislocation velocities are thermally activated and depend sensitively on the applied resolved shear stress $\tau = S*\sigma$ (S = Schmidt factor)

$$v = v_0(\tau/\tau_0)^m \exp(-Q/kT) \quad (3)$$

where v_0, m, and τ_0 are constants and Q is the activation energy for dislocation motion. In silicon the factor m equals 2 at deformation temperatures as low as $0.4 \, T_m$. This gives rise to a quadratic dependence of v on the resolved shear stresses. $Q = 2.1$ eV was recently estimated for GaN (Sugiura, 1997). This value is close to the activation energy of dislocation motion in silicon (Alexander et al., 1987). Therefore, dislocations may glide well below the brittle to ductile transition if the applied shear stresses are large.

Uniaxial, dynamic deformation experiments of Si were performed under a hydrostatic confinement at temperatures as low as $0.3\,T_m$ (Castaing et al., 1981). They are a close analogy to what happens if an epitaxial film of GaN on sapphire is heated or cooled because the biaxial stresses decompose into a hydrostatic pressure and a uniaxial shear. Stress/strain relations of dynamic deformation experiments were modeled (Suezawa et al., 1979). Figure 4 schematizes some of the results. It is seen that plastic flow reduces the elastic stresses. The upper yield stress σ_{uy} can reach in silicon values as large as 1–3 GPa if a deformation is done at $0.34\,T_m$. The crystals crack if σ_{uy} becomes too large. A further comparison of Si with GaN is useful in conjunction with film cracking because hardness H and fracture toughness K_c of the materials are similar (Si: $H = 9$ GPa, $K_c = 0.7$ MPa m$^{1/2}$; GaN: $H = 12$ GPa, $K_c = 0.8$ MPa m$^{1/2}$) (Drory et al., 1996). Typically, σ_{uy} drops largely to 10–30 MPa at a deformation temperature of $0.64\,T_m$. More generally, the upper yield stress decreases with increasing temperature and an increasing initial dislocation density but it increases with an increasing strain rate (Suezawa et al., 1979). Thus, it can be concluded from Figs. 3 and 4 that it is not the elasticity that solely determines the stresses in the thin

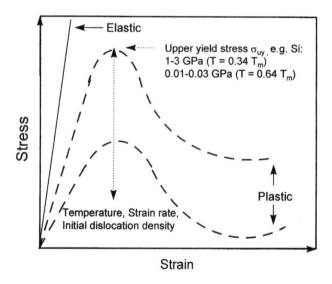

FIG. 4. Elastic and plastic response of materials that are subject to an external strain. A plastic response reaches a maximum stress value (σ_{uy}) in contrast to elastic materials behavior. Plastic flow reduces the initial slope of the curves. Values for upper yield stresses in Si are given. Yield stresses decrease with increasing temperature and increasing initial dislocation densities. They increase with an increasing strain rate.

films. Instead, plastic relaxation must occur. Consequently, stresses must be large in the MBE-grown films because plastic flow or even yield is more difficult to obtain at a low deposition temperature.

A large dislocation density in and close to the buffer region is reported in many papers (see, e.g., Hwang, 1997). The dislocation network consists of a dense array of basal dislocations. It differs significantly from the threading dislocation network in the main layers (see, e.g., Kisielowski et al., 1997a). The example shown in Fig. 5 depicts the different dislocation arrangements in the main layer and in the buffer layer of a GaN film grown on sapphire. The specific dislocation arrangement in the buffer-layer region suggests that plastic relaxation occurs preferentially there. Yield and flow stresses are expected to be smallest in the buffer-layer region because the dislocation density is largest in this area. Additionally, they depend on the material chosen for the buffer-layer growth. An AlN buffer layer should be "harder" than a GaN one because of the larger Al-N bond strength. Consequently, stresses in films grown on AlN buffer layers should exceed those in films deposited on GaN buffers. This is observed for MOCVD-grown samples and shown in Fig. 3.

Finally, stresses come from the growth of GaN/InN/AlN quantum well structures because of a lattice mismatch of -12% and 2.4% to GaN, respectively (Table I).

3. Growth Modes

Both MOCVD and MBE growth proceed at growth temperatures of roughly 1300 K and 1000 K, respectively. Film decomposition limits the MBE growth to higher temperatures (Newman et al., 1993). Figure 6 depicts the surface morphology of GaN films grown at different temperatures. The films exhibit surface features of a characteristic size that increase with T_{growth}. The TEM reveals that the surface features mark oriented grains in the films. Stress influences their size, shape, and grain coalescence (Fujii et al., 1997). Thin GaN layers grown around 770 K are the buffer layers that are utilized in the MBE (Klockenbrink et al., 1997) and the MOCVD (Wu et al., 1996) growth process. A Volmer-Weber three-dimensional (3D) growth mode (Volmer and Weber, 1926) dominates at growth temperatures of about 1000 K and below. In contrast, a Stranski-Krastanov 2D/3D growth mode (Stranski and Von Krastanov, 1939) is observed at 1300 K (Fujii et al., 1997).

Figure 7 is an estimation of a surface diffusion coefficient from the size of the grains and the growth rate of the crystals. The estimated activation energy is stress dependent and an average value of 2.6 ± 0.6 eV can be

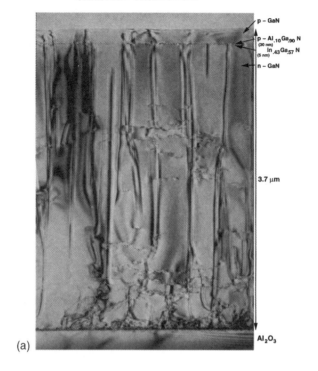

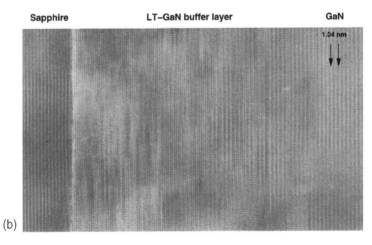

FIG. 5. (a) Cross-sectional TEM image of a GaN diode structure. A network of threading dislocations dominates in the main layer. (b) Lattice image of the buffer layer region. The layer exhibits a dense network of stacking faults and partial dislocations. The dislocation density decreases with increasing distance from the sapphire/LT-GaN interface. This dislocation arrangement is typical for buffer layers.

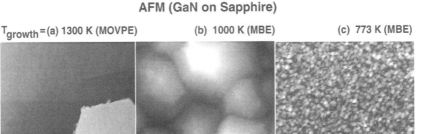

FIG. 6. Surface morphology of GaN thin films grown at different temperatures. Measurements by atomic force microscopy (AFM).

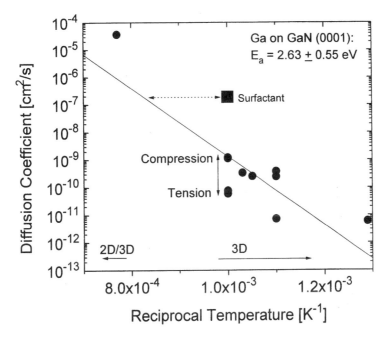

FIG. 7. Estimation of a surface diffusion coefficient on GaN (0001). It is argued that Ga ad-atoms limit the surface diffusion that is stress dependent and can be greatly enhanced by using the Bi surfactant. Both 3D and 2D/3D growth is observed in the indicated temperature ranges. It does not depend on the growth method.

extracted that agrees well with measurements of the Ga surface diffusion rate on cubic GaN {111} surfaces (Brandt et al., 1996). Therefore, it was concluded that the diffusion of Ga ad-atoms on {0001} surfaces limits the size of the grains in the hexagonal films (Fujii et al., 1997).

Surface diffusion can be stimulated by the use of surfactants. The grain size in MBE-grown films could be increased drastically by using Bi as a surfactant (Fig. 7) (Klockenbrink et al., 1997). The experiment supports the interpretation that surface diffusion is the grain-size limiting factor and not, for example, the buffer-layer structure that also must influence grain sizes. The grains in MBE-grown films are almost as large as in MOCVD-grown layers if a surfactant is used. Nevertheless, the films were grown at a considerably lower temperature (Fig. 7).

II. Stress/Strain Relations

1. ELASTIC CONSTANTS OF GaN

Principally all necessary stress-strain relations for uniaxial, biaxial, and hydrostatic stresses can be obtained if the elastic constants of GaN are known within reasonable limits. However, experimental measurements as well as theoretical calculations scatter significantly as shown in Table II.

The elastic constants of a 0.3-mm thick GaN crystal were measured to an accuracy of less than 1% (Schwarz et al., 1997). It is seen from Table II that only some part of the other anisotropic constants agree with these values. For many purposes it is convenient to use isotropic elastic constants. Table II lists a consistent set of isotropic elastic constants that was calculated in the Voigt average from the c_{ij}. This way, the anisotropic elastic constants can be compared with estimations of the isotropic moduli and the Poisson ratio v (Kisielowski et al., 1996a; Schwarz et al., 1997). It is seen from Table II that only the isotropic values from Kisielowski et al. (1996a) and Schwarz et al. (1997) agree within experimental errors. Thus, it is reliable to use a bulk modulus of 200 GPa and a Poisson ratio between 0.23 and 0.29 in combination with the anisotropic constants from Schwarz et al. (1997).

Suitable stress-strain relations for thin films with varying lattice constants that are subject to a biaxial stress are (Kisielowski et al., 1996a)

$$\varepsilon_c = (1 + bC)(1 - 2v\sigma_a/E) - 1 \tag{4a}$$

$$\varepsilon_a = (1 + bC)(1 + (1 - v)\sigma_a/E) - 1 \tag{4b}$$

TABLE II
Anisotropic Elastic Constants of GaN[a]

c_{11}	c_{12}	c_{13}	c_{33}	c_{44}	Poisson ratio	G, Shear modulus	E, Young modulus	B, Bulk modulus	Reference
296	130	158	267	24	0.37	54	148	194	Savastenco and Sheleg (1978)
396	144	64	476	91	0.24	128	317	201	Kim et al. (1994)
369	94.2	66.7	397	118	0.20	135	323	177	Azuhata et al. (1996)
390	145	106	398	105	0.26	121	305	210	Polian et al. (1996)
377	160	114	209	81.4	0.29	93	240	193	Schwarz et al. (1997)
					0.23 ± 0.06	118	290	200 ± 20	Kisielowski et al. (1996a)
					0.18 ± 0.02	123	290	151	Perry et al. (1997)

[a]Isotropic elastic constants represent the Voigt average over all orientations. Consistent sets of data are given. Values are in units of GPa.

In Eq. (4) the change of the lattice parameters a and c is measured in terms of the strains $\varepsilon_c = (c - c_0)/c_0$ and $\varepsilon_a = (a - a_0)/a_0$. a_0 and c_0 are the "ideal" lattice constants of GaN that are listed in Table I and are currently debated (Skromme et al., 1997). A finite biaxial stress σ_a creates a tetragonal distortion of pseudomorphically grown films. $\varepsilon_{c,a}$ change linearly with σ_a. The slope of the line is determined solely by the elastic constants. A substitution of host atoms with atoms of different sizes causes parallel off-sets of the linear relations. Experimentally, however, it is not unique to determine elastic constants by measuring ε_c and ε_a, only, because a simple change of the lattice constant cannot be distinguished from an elastic tetragonal distortion. Instead, some knowledge of the elastic constants must be explored (Kisielowski et al., 1996a). This confusion contributes to the scatter of elastic constants if the problem is not recognized (Perry et al., 1997). It is the amount to which GaN can be stressed and strained that makes the material special. Stress values that exceed 1 GPa were reported (Fig. 3). Strain can be altered by the occupation of lattice sites with point defects. The resulting lattice expansion/contraction can be estimated from the different tetrahedral covalent radii of host and solute atoms. Table III lists values.

The substitution of O on N-sites, for example, will hardly affect the GaN lattice constant. In fact, it was recently reported that the lattice constant of oxinitrides equals that of GaN (Sudhir et al., 1997). Bismuth is the other extreme case. The atom is more than twice the size of nitrogen. This makes it difficult to solve any substantial amount of Bi on the nitrogen sublattice. Therefore, Bi is a good choice for a surfactant (Klockenbrink et al., 1997). The huge difference between r_{Ga} and r_N causes unusually large changes of the lattice constant if native defects such as Ga interstitials (Ga_i) or antisite

TABLE III

TETRAHEDRAL COVALENT RADII r OF SELECTED ATOMS[a]

Element	Ga	N	C	Si	Ge	O	Be
r (nm)	0.124	0.070	0.077	0.117	0.122	0.066	0.107
Element	Mg	Ca	Al	In	As	Sb	Bi
r (nm)	0.140	0.197	0.126	0.144	0.118	0.136	0.146

[a] Modified from L. Pauling, *The Nature of the Chemical Bond*, Ithaca, 1960.

defects (N_{Ga} or Ga_N) are formed. Doping with Si, Ge, or O donor atoms, however, would change the lattice constants to a much smaller extent. Native defect formation under n-doping conditions can be investigated best if Si or Ge is the dopant because substantial changes of the lattice constants can hardly be caused by these elements. It can be estimated from Eq. (4) that impurity concentrations of 10^{20} cm^{-3} may expand or contract the lattice by values larger than 0.1% (Kisielowski *et al.*, 1996a). In contrast, it takes a large biaxial stress of 1 GPa to strain the lattice constants by the same amount. Certainly, such estimations are crude because they neglect lattice-relaxation effects and ionicity. However, they can be guidelines.

Local compositional modulations that are caused by local stress fluctuations are the subject of intensive studies (Milluchick *et al.*, 1997). In GaN thin-film growth the effect makes it possible to tailor the incorporation of dopants and the film stoichiometry through the biaxial stress if it is controlled properly (Kim *et al.*, 1998).

2. Experimental Stress/Strain Calibrations

The presence of strain in GaN thin films affects the position of near bandedge photoluminescence lines because the resulting stresses alter the bandgap energy. The application of hydrostatic stresses increases the bandgap energy by 40 meV/GPa (Kisielowski *et al.*, 1996a). Amano *et al.* (1988) pointed out that biaxial stresses shift the bandedge luminescence by a smaller amount. In principle there are two options to investigate the stresses in the films by photoluminescence.

The first one consists of an analysis of free excitons (Kaufmann *et al.*, 1996; Orton, 1996; Meyer, 1997). However, it is not always possible to observe the separation of the A, B, and C excitons. Also, the transitions broaden with an increasing temperature of the PL measurement. The 40 meV broad transitions are typically observed at room temperature.

Alternatively, the stresses can be determined simply by measuring the line position of donor-bound excitons. In this case, an uncertainty exists as to the identity of the donors. Oxygen, silicon, gallium interstitials (Ga_i), and nitrogen vacancy (N_v) are discussed to have closely spaced energy levels (Perlin *et al.*, 1995; Goetz *et al.*, 1996; Meyer, 1997). Their energy separation may be as small as 5 meV. They can be confused if the lines are shifted by a biaxial stress. Therefore, Raman spectroscopy was utilized to confirm stress values that were determined by photoluminescence (Kisielowski *et al.*, 1996a).

Figure 8 depicts the effect of a biaxial stress on PL spectra. A parallel

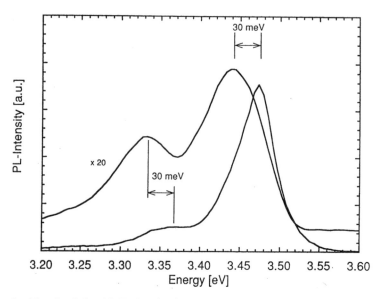

FIG. 8. Near bandedge 4.2 K photoluminescence spectra of MBE-grown GaN thin films deposited on sapphire c-plane. A displacement of the lines by 30 meV is seen. Stress is the origin of the line displacements.

displacement of the lines is seen that amounts to 30 meV. This displacement is caused by stress.

Figure 9 depicts results of stress and strain measurements. Lattice constants (strains) were measured by x-ray diffraction. A "zero" stress calibration was obtained by homoepitaxial growth of GaN on bulk GaN (Gassmann et al., 1996). However, neither homoepitaxially grown films exhibit luminescence lines at well-defined energy positions. Shifts of about 3 meV are common (Gassmann et al., 1996; Ponce et al., 1996; Pakula et al., 1997). It is likely that the related stresses stem from the growth of GaN films with lattice constants that differ from the bulk substrates. This effect is well known from the growth of nonstoichiometric LT-GaAs on GaAs bulk substrates (Liu et al., 1995; Luysberg et al., 1998). It is seen from Fig. 9 that ε_c changes linearly with the shift of the E_2 Raman mode and the donor-bound exciton transition. This meets the expectations from Eq. 4. However, part of the available data cannot be used for the modeling because of their large deviation from the linear regression. These deviations were taken as an indication that GaN thin-film growth can be largely nonstoichiometric (Kisielowski et al., 1996a).

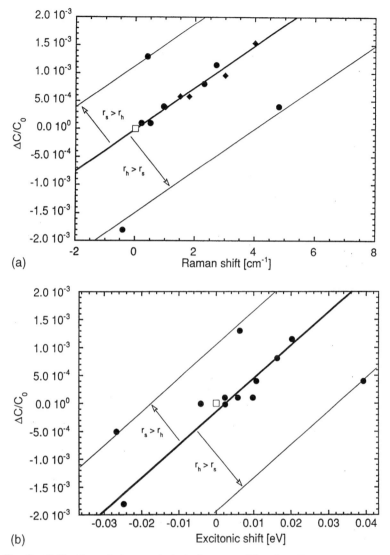

FIG. 9. Calibration of stress and strain by x-ray diffraction, Raman spectroscopy (E_2 mode), and photoluminescence. (a) A linear dependence of the excitonic shift is utilized to measure the stress that changes the c-lattice constants $\Delta c/c_0$. The slope of the linear regression is determined by elastic constants. (b) A linear dependence of the Raman shift (E_2 mode) is utilized to measure the stress that changes the c-lattice constants $\Delta c/c_0$. Significant deviations of some data from the linear regression indicate that compositional changes alter the lattice constants and introduce the parallel line shifts. r_h and r_s are radii of host and solute atoms, respectively.

A consistent description of the PL-, the Raman-, and the lattice constant measurements was achieved by using B = 200 GPa and $v = 0.23$ (Table II) to obtain (Kisielowski et al., 1996a)

$$\omega = \omega_0 - 4.2\,\text{cm}^{-1}\,\text{GPa}^{-1} * \sigma_a \qquad (5a)$$

$$E_{pl} = E_0 - 0.027\,\text{eV}\,\text{GPa}^{-1} * \sigma_a \qquad (5b)$$

where $\omega_0 = 566.2\,\text{cm}^{-1}$ and $E_0 = 3.467\,\text{meV}$. The elastic constants used in Eq. (5) determine largely the calibration factors $4.2\,\text{cm}^{-1}\,\text{GPa}^{-1}$ and $0.027\,\text{eV}\,\text{GPa}^{-1}$ for the biaxial stresses (Kisielowski et al., 1996a).

III. Control of Hydrostatic and Biaxial Stress and Strain Components

1. STRESS-CONTROLLING PARAMETERS

Stress in the GaN thin films can be manipulated by four parameters: the buffer-layer thickness (Fig. 10), the buffer-layer growth temperature (Fig. 11), the V/III flux ratio (Fig. 12), and doping (Fig. 13) (Kisielowski et al., 1996b; Kim et al., 1998; Krüger et al., 1998). The buffer-layer thickness and its growth-temperature affect the biaxial stresses. Compositional changes and doping alter the lattice constants of the main layers and, thereby, hyedrostatic stress components.

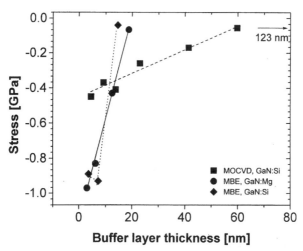

FIG. 10. Dependence of stresses in GaN thin films on the buffer-layer thickness. The MOCVD samples were grown at 1300 K, MBE samples at 1000 K.

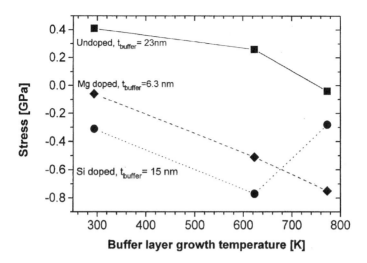

FIG. 11. Dependence of stresses on the buffer-layer growth temperature. The MBE main layers were grown at 1000 K. The buffer-layer thicknesses are indicated.

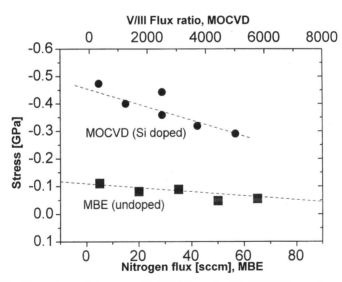

FIG. 12. Dependence of stresses on the V/III flux ratio. The MOCVD samples were grown at 1300 K, MBE samples at 1000 K. In case of MBE growth the Ga source temperature was kept constant at 1210 K and the N flux was varied as indicated. Buffer layer thicknesses are MOCVD 5 nm; MBE 20 nm.

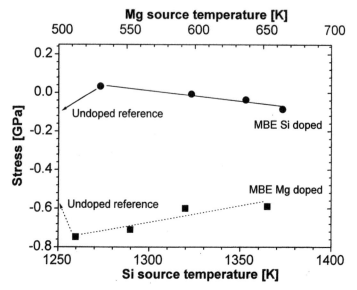

FIG. 13. Dependence of stresses on p- and n-doping. The MBE main layers were grown at 1000 K. Source temperatures are indicated. The dopant concentration ranges from $\sim 10^{17}$ to 10^{19} cm^{-3} in both cases. The undoped reference samples reveal an abrupt stress variation even at low doping levels.

Figure 10 shows that an increasing buffer-layer thickness reduces the compressive stresses in MBE, as well as in MOCVD-grown samples. However, absolute values differ substantially. The stresses in MOCVD-grown crystals rarely exceed -0.5 GPa. They reach twice this value in MBE-grown material even though the GaN buffer layers were grown at similarly low temperatures of about 770 K. Stresses reduce rapidly in MBE-grown films if the buffer-layer thickness increases. It takes only a 20-nm thick buffer layer to relax them. In contrast, the buffer layers need to be 60 nm thick to achieve a similar stress relaxation if MOCVD growth is employed. The large stress gradients that come with the MBE process make growth control difficult.

The deposition temperature of buffer layers has the largest impact on stresses. This is depicted in Fig. 11 for MBE-grown samples. It is seen that tensile stresses can reach 0.4 GPa. The compression in the films may be as large as -0.9 GPa. The stresses increase linearly with the buffer-layer growth temperature in the GaN:Mg films. A temperature dependence with a slope identical to that measured for GaN:Mg occurs for undoped GaN and for GaN:Si only in a limited temperature interval. The slopes change drastically if the stresses tend to exceed extreme values of about -0.9 and

+0.4 eGPa. This puts an upper and a lower limit to the stresses that can be present in the thin films. We interpret these extreme stress values to yield stress for compression and tension, respectively (see Fig. 4). Alternatively, the films may crack (which did not happen to our MBE films grown on GaN buffer layers). It seems that cracking occurs more often if AlN buffer layers are deposited (Hiramatsu et al., 1993; Rieger et al., 1996).

The materials stoichiometry adds to the stresses in MBE- and MOCVD-grown films. This is shown in Fig. 12. The V/III flux ratio was varied in these experiments. A flux ratio variation was achieved in the MBE-growth process by keeping the Ga source temperature at a constant value of 1210 K while increasing the nitrogen flux. The data are intentionally scaled such that relative flux changes are identical for the different growth processes. It is seen that the stresses decrease with an increasing V/III flux ratio if all other growth parameters are kept constant. Obviously, the stresses can be altered by growing the main layers of a different stoichiometry. This fact implies that the main layers are stressed during deposition.

Doping of the main layers with impurity atoms affects the stresses in a rather unexpected manner. In particular, the substitution of Ga by Si donors at a doping level of 10^{17}–10^{18} cm^{-3} should not affect the GaN lattice constant or the stresses (Table III, Eq. 4). A similar argument holds for a low Mg doping concentration. Figure 13 contrasts this expectation. Low doping levels are already sufficient to change the stresses by 0.1 to 0.2 GPa as well as the lattice constants (see Section III.3, for a comparison). The different size of the doping atoms cannot account for the measured changes. Thus, doping must alter native defect concentrations to an amount that causes lattice constant variations around 10^{-3}. This introduces the quoted stresses. A dependence of the defect formation energy on the Fermi level was predicted by theory (Neugebauer and Van de Walle, 1994; Boguslawski et al., 1995). However, the calculated defect concentrations are too small to account for the observed lattice expansion/contraction. Therefore, the defect formation is kinetically limited, too. Furthermore, stresses of opposite sign are introduced if the doping level increases from about mid-10^{17} to mid-10^{19} cm^{-3} as shown in Fig. 13. Similar dependence of the stresses on Si doping was reported for GaN:Si grown by MOCVD (Ruvimov et al., 1996; Suski et al., 1996).

Figure 2 illustrated the origin of the stresses. Tensile stresses can be introduced during buffer-layer growth and during heating if the material is deposited on sapphire (Table I). In our experiments, the buffer-layer deposition temperature varies between room temperature and 1000 K. The buffer-layer thickness and the growth temperature at which the main layer is deposited determine how much of the stress can be relaxed. The buffer layer is usually prestrained at the growth temperature because of its lattice mismatch to the substrate and the thermal stresses. Therefore, its lattice constant will *a priori not match* the main-layer lattice constant. Consequently, an additional compressive or tensile stress can be introduced during main-layer growth. The lattice constant of this prestrained buffer layer and the V/III flux ratio during main-layer growth fix the stoichiometry of the growing film. Additionally, it can be altered by doping. Finally, the post-growth cooling generates large compressive thermal stresses.

It can be deduced from Fig. 2 that stress-free growth can be accomplished if a buffer layer is grown at room temperature and the lattice constant of the deposited main layer matches that of the buffer layer at the growth temperature. In the examples depicted in Fig. 11 this was achieved for GaN:Mg. Typically, an increasing buffer-layer growth temperature increases the compressive stresses introduced by the post-growth cooling. They simply cannot be balanced by the smaller tensile stresses that are generated during heating across a narrower temperature interval.

Stress-temperature cycles were well investigated for metal films deposited on silicon (Nix, 1989). The similarity of these experiments with GaN thin-film growth is striking. A goal of both experiments is to access the strain-temperature curve that is schematized in the inset of Fig. 2. Yield stresses and plastic flow determine the shape of these curves to a large extent. Time dependencies are not taken into account in this simple picture. The results depicted in Fig. 11 provide the first experimental data to quantify these dependencies. In the experiments with GaN:Mg the material reacts primarily in an elastic manner. Some plastic flow must be involved because the stresses increase linearly with temperature while the thermal expansion coefficients do not (Fig. 3). On the other hand, yield stresses were reached in the experiments with GaN:Si and with undoped GaN (Fig. 11). In the case of GaN:Si a substantial amount of the large compressive stress relaxes. In the undoped GaN the tensile stress becomes too large if the buffer layer is grown at room temperature. Therefore, stresses can change abruptly if critical values are exceeded. Consequently, the same linear dependence of the stress on the buffer-layer growth temperature is restricted to limited temperature intervals. Initial differences between the differently doped materials were introduced by growing buffer layers of different thicknesses and by the doping. It is our understanding that the substantial plastic relaxation occurs in the buffer-layer region. Therefore, the dislocation arrangement is different there (compare Fig. 5).

Generally, the role of buffer-layer growth in the GaN thin-film technology is seen in providing a suitable nucleation layer for the growth of the main layers. In contrast, our results stress the important role of the buffer-layer growth in controlling stresses and, thereby, affecting stoichiometry. In addition, its presence confines plastic relaxation to a well-defined and *extended* region close to the film substrate interface. Thus, it is not only the presence of misfit dislocations at the substrate/film interface that controls the amount of stress relaxation. Instead, it is the dislocation network in the whole buffer-layer region that relaxes the very large stresses in this system with a lattice mismatch of 14% and the different thermal expansion coefficients. This mechanism differs substantially from what has been reported for other semiconductor systems (van der Merve, 1963; Matthews and Blakeslee, 1975; People, 1986).

2. LATTICE CONSTANTS

The nonstoichiometry of the GaN growth causes variations of the lattice parameter up to 1%. This variation compares well with what has been measured in LT-GaAs (Liu *et al.*, 1995). There, the formation of As_{Ga} antisite defects causes a lattice expansion and the doping with Be a contraction (Specht, 1997). In GaN the dominant native defects that contribute to the stresses and change the materials lattice constant could not yet be identified spectroscopically.

The buffer-layer thickness, its growth temperature, the V/III flux ratio, and the doping are again the key parameters that control the lattice constants and the strain. Figure 14 depicts the dependence of the *c*-lattice constant on the buffer-layer thickness (Kisielowski *et al.*, 1996b; Kim *et al.*, 1998, Krüger *et al.*, 1998). Both MBE and MOCVD growth are compared. In all investigated cases a "largest" lattice constant can be measured. However, the maxima occur at different buffer-layer thicknesses in the MBE and the MOCVD growth. Figure 14 compares directly with the stress variations in these samples shown in Fig. 10. The raw data are not corrected for a tetragonal distortion. It is estimated from Eq. (4) that a compressive stress of one GPa would change the *c*-lattice parameter by 0.1% (Kisielowski *et al.*, 1996a). This is only 10% of the total lattice parameter variation that can be caused by the nonstoichiometric growth. In fact, a correction would further decrease the already small lattice constants measured in films with thin buffer layers (Fig. 14).

Similarly, the lattice parameters can be altered by changing the V/III flux ratio for a given buffer-layer thickness. The lattice parameter measurements from Fig. 15 compare with the stress measurements shown in Fig. 12. Commonly, the minimum at low V/III flux ratios is explored to grow material with low intrinsic carrier concentration and large mobilities (Hwang *et al.*, 1997). Lattice constants of around 0.5180 nm are typical values measured there. They are close to the indicated "ideal" *c*-value of GaN. It was argued that this is the stoichiometric point for GaN thin-film

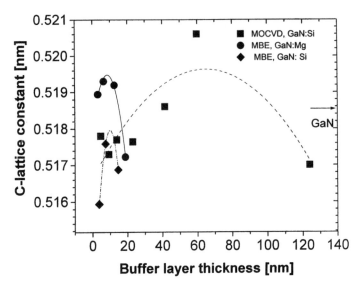

FIG. 14. Dependence of c-lattice constants in GaN thin films on the buffer layer thickness. The MOCVD samples were grown at 1300 K; MBE samples at 1000 K. The plot compares with Fig. 10. The lattice constant of "ideal" GaN is indicated.

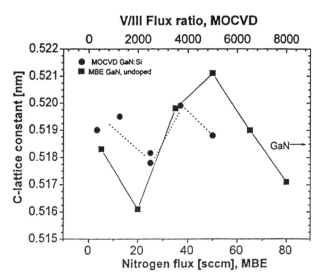

FIG. 15. Dependence of c-lattice constants on the V/III flux ratio. The MOCVD samples were grown at 1300 K, MBE samples at 1000 K. In case of MBE growth the Ga source temperature was kept constant at 1210 K and the N flux was varied as indicated. The lattice constant of "ideal" GaN is indicated.

growth (Hwang et al., 1997). However, similar lattice constants can be grown in films with very different film compositions. Thus, it is not a unique procedure to judge film stoichiometry by considering only lattice constants.

The significance of these measurements relates to the empirical finding that buffer layers of about 20 nm have to be grown by MOCVD in order to obtain films with a low intrinsic carrier concentration and a high mobility (Hwang et al., 1997; Nakamura and Fasol, 1997). Our investigatioens explain the need to grow a buffer layer of a particular thickness. Stress and strain are propagated across the buffer layer into the main film. They fix the chemical composition of the films for a given V/III flux ratio during growth. The resultant stoichiometry determines the mobility and the native dopant concentration. Consequently, materials properties can be altered by correctly engineering the stresses in the films through a design of the buffer layers.

3. ENGINEERING OF MATERIALS PROPERTIES BY ACTIVE UTILIZATION OF STRESS AND STRAIN

GaN thin-film growth must be optimized to obtain desired characteristics such as good mobility, good optical properties, or smooth surfaces. It is not yet possible to grow one film that performs well in all respects. Usually, experience with GaN thin-film growth allows this goal to be reached. In the preceding paragraphs it was shown that stress and strain are key parameters to understanding the low-temperature growth process. They can be properly controlled by the dependencies described in Sections III.1 and III.2. An active utilization of stress and strain makes GaN growth predictive. This process was named strain engineering (Kisielowski et al., 1996a). Some examples follow.

Figure 16 shows PL spectra of samples stressed to different amounts (Fujii et al., 1997). The large compressive and tensile stresses broaden the luminescence lines. In contrast, a narrow PL transition can be observed if the films are unstressed. Thus, correct design of the stresses allows growth of material with excellent optical properties. Figure 17 is a photoluminescence (PL) spectrum from a film that is best in this respect (Krueger et al., 1997). A narrow donor-bound transition of only 1.2 meV width is seen. Many phonon replica are visible and the yellow luminescence band is almost absent. These properties are often taken as an indication that very good material was grown. In fact, a 1.2 meV narrow donor-bound transition is currently the best reported value for GaN heteroepitaxy no matter whether MOCVD or MBE growth was employed. Narrower transitions (~ 0.7 meV) were observed only if GaN was deposited homoepitaxially on

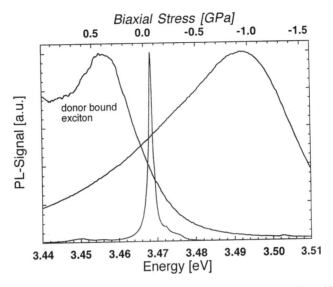

FIG. 16. Near bandedge 4.2 K photoluminescence spectra of GaN thin films. All samples were grown on sapphire c-plane. Stresses can be engineered to range from 0.4 to −1 GPa.

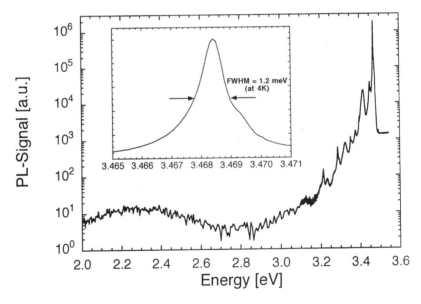

FIG. 17. 4.2 K photoluminescence (PL) spectrum of GaN thin films that was engineered to be stress-free. A line width of 1.2 meV is unusually narrow for heteroepitaxially grown films.

bulk GaN crystals (Pakula *et al.*, 1997). Thus, films with outstanding optical properties can be grown by correctly engineering strain and stress.

However, there is a drawback to such films. Figure 18 depicts the surface morphology of three films measured by atomic force microscopy (AFM) (Fujii *et al.*, 1997). The purpose of this example is twofold. First, Fig. 18(b) and (c) depict the surfaces of crystals that are unstressed. They suffer from a large surface roughness. Such films may exhibit outstanding luminescence spectra as shown in Fig. 17. Nevertheless, they are of little use for electronic applications. Second, the example depicts the equivalence to alter stress by either growing buffer layers of a different thickness or by varying the V/III flux ratio during main layer growth. The film in Fig. 18(a) was deposited on a 19-nm thick buffer layer with a nitrogen flow of 35 sccm. It is compressed and its surface is reasonably flat. The compression can be reduced if a thicker buffer layer is grown (b) or if the nitrogen flux ratio during main-layer growth is increased (c). The stoichiometry of both films differs even though their surface morphology is the same.

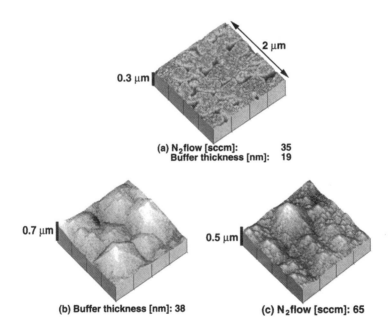

FIG. 18. Surface morphology of GaN thin films. AFM measurements. The equivalent impact of hydrostatic and biaxial stresses on the morphology of the films is demonstrated. An initial surface morphology (a) can be altered by either changing the film composition (c) or the buffer layer thickness (b) to obtain stress free (rough) films.

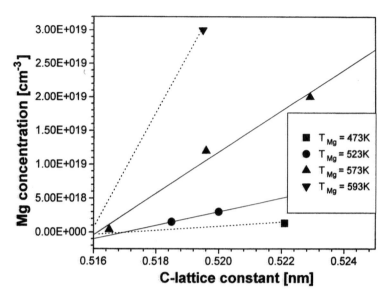

FIG. 19. Dependence of the Mg concentration in GaN thin films on the c-lattice constant. [Mg] was measured by secondary ion mass spectroscopy (SIMS). The source temperature of the Mg cells is indicated. All films were grown at 1000 K. Full lines represent a linear regression. Dotted lines indicate expected dependencies. A variation of the Mg concentration by two orders of magnitude is measured in films grown with $T_{Mg} = 593$ K. Buffer layers and V/III flux ratios were altered to change the lattice constants. The lattice constant of "ideal" GaN is indicated.

Unexpectedly, the design of a proper lattice constant is crucial for the incorporation of doping atoms. This is demonstrated in Fig. 19 for p-doping of GaN with Mg (Kriegseis *et al.*, to be published). The concentration of dissolved atoms increases with an increasing c-lattice constant in spite of a constant Mg source temperature. The Mg concentration in the layers grown with a constant source temperature of $T_{Mg} = 573$ K increases by two orders if the c-lattice parameter expands from 0.516 to 0.522 nm. The different lattice constants were obtained by a proper buffer-layer growth and by a V/III flux ratio variation. It is tempting to compare the observed dependencies with experiments performed on SiC (Larkin *et al.*, 1994). There, it was shown that a dopant incorporation could be altered to a similar extent by changing the C/Si flux ratio during growth. The process was explained by site competition. Unfortunately, it was not reported how much the procedure changed the stresses and the strains of the epitaxial SiC films and how much the Fermi energy was affected.

Figure 20 reveals further useful information. The dependence of the

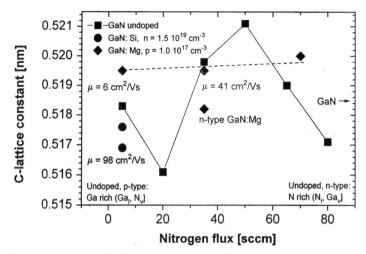

FIG. 20. Dependence of c-lattice constants on the V/III flux ratio for selected MBE-grown samples (1000 K). The growth is Ga rich at low fluxes and under p-doping conditions. The growth is N rich at large fluxes and under n-doping conditions. The variation of lattice constants is simplified under heavy p-doping conditions. The growth of p- and n-type films with good mobility proceeds at different fluxes. The Mg-doped n-type films tend to have smaller lattice constants.

c-lattice parameter on stoichiometry is analyzed. Undoped films are compared with p- and with n-type material. The lattice parameter of undoped GaN changes in a complex manner. This cannot be explained because of unknown native defect concentrations. However, the growth is Ga-rich at low nitrogen fluxes. It can lead to the formation of Ga droplets. Gallium interstitials (Ga_i) and nitrogen vacancies (N_v) are primary defects in Ga-rich material. In contrast, the growth is nitrogen-rich at large V/III fluxes. Thus, nitrogen interstitials (N_i) and gallium vacancies (Ga_v) should be formed there.

The complex variation of the lattice parameter simplifies substantially under defined doping conditions. This is demonstrated for p-type doping in Fig. 20. Theory predicts that p-type crystals are Ga rich while n-type crystals will grow N rich (Neugebauer and Van de Walle, 1994; Boguslawski et al., 1995). The lattice constants of nitrogen-rich crystals should be small because of the smaller size of nitrogen compared with gallium and vice versa. In fact, it is seen from Fig. 20 that this expectation is met. In addition, the lattice constants of Mg-doped n-type crystals tend to be smaller than those of p-type GaN:Mg. The experiments suggest strongly a dependence of native defect formation energies on the Fermi level. Even in the case of undoped material the lattice expands in the extreme Ga-rich regime and contracts if nitrogen-rich growth conditions are approached.

Heavily doped p- and n-type material with mobility comparable to those of MOCVD-grown samples can be grown by MBE at low nitrogen fluxes (Fig. 20). One would expect to meet the requirements for optimal growth with similar fluxes (Nakamura and Fasol, 1997; Hwang et al., 1997). However, an unusually large p-type mobility of 41 cm^2/V s could be achieved by growing with a larger V/III flux ratio. Thus, the conditions for "optimal" doping need to be adjusted for every particular dopant at a desired concentration. Again, stress and strain are the key parameters in this process.

The n- and p-type conductivities of 230 and 0.7 Ω^{-1} cm^{-1} obtained by MBE growth are very competitive with MOCVD material if it is heavily doped with Si and Mg, respectively. In addition, the large compressive stress (up to -0.9 GPa) that can come with heavy Si doping can be reduced to moderate values (≥ -0.3 GPa) by strain engineering. This minimizes the common risk of film fracture in heavily doped GaN:Si.

IV. Strained AlN/InN/GaN Heterostructures

1. Al AND In DISTRIBUTION IN $Al_xGa_{1-x}N/In_xGa_{1-x}N$/GaN QUANTUM WELL STRUCTURES

The growth of $Al_xGa_{1-x}N/In_xGa_{1-x}N$/GaN quantum well structures allows for bandgap engineering between 2 and 6 eV. $In_xGa_{1-x}N$ quantum wells are utilized to access the visible spectrum (Nakamura and Fasol, 1997). $Al_xGa_{1-x}N$ structures are explored to produce high-power devices (Kahn et al., 1993). A combination of $Al_xGa_{1-x}N/In_xGa_{1-x}N$/GaN structures is often grown into light-emitting devices (Fig. 5) even though it is often not clear what might be the purpose of adding a particular $Al_xGa_{1-x}N$ layer to the $In_xGa_{1-x}N$/GaN quantum wells. It seems to be experience that informs choice.

The InN and AlN exhibit a lattice mismatch of -12% and $+2.4\%$ to GaN, respectively (Table I). Stresses can be compensated by an alternating growth of $Al_xGa_{1-x}N$ and $In_xGa_{1-x}N$ layers because of the inverted signs of the lattice mismatch. A mismatch of -12% is unusually large. Consequently, the growth of InN on GaN is difficult to accomplish. Theory predicts a miscibility gap in the quasi-binary InN-GaN system (Ho and Stringfellow, 1997). Experiments confirmed this expectation (Singh et al., 1997). Nevertheless, high-quality $In_xGa_{1-x}N$ quantum well structures could be grown with an indium concentration x_{In} as large as 0.5 (Nakamura and

Fasol, 1997). In spite of the large lattice mismatch, the growth is usually pseudomorph. Misfit dislocations are rarely observed. There are anomalies that occur at an atomic scale. Phase separation (Fig. 21), the formation of alloy clusters (Fig. 22), and indium segregation into top layers (Fig. 23) (Kisielowski *et al.*, 1997a, b) were investigated by quantitative high-resolution electron microscopy (HREM).

Many of the observed phenomena can be influenced today by crystal growth details. For example, the formation of indium-rich plate-like precipitates (Fig. 21) was only observed in one particular sample. It does not commonly occur. Also, the segregation of In into top layers can be controlled to some extent. This is depicted in Fig. 24, which compares the atomic scale roughness of GaN/In$_x$Ga$_{1-x}$N and In$_x$Ga$_{1-x}$N/GaN interfaces. Generally, In$_x$Ga$_{1-x}$N/GaN interfaces are broader. A substantial fluctuation of the total well width may occur. Moreover, it is seen from Fig. 24 that in the case of $x_{In} = 0.3$ different results were obtained on samples from different suppliers. In fact, in one of the samples the widths of both interfaces are almost equal. Therefore, it was suggested that during thin-film growth In atoms may accumulate on top of the growing crystal (Kisielowski *et al.*, 1997).

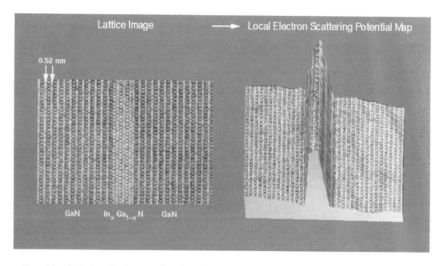

FIG. 21. Left: Lattice image of a plate-like precipitate in GaN thin films. MOCVD growth. The sample surface is etched to be atomically flat. Such precipitates are sometimes observable in the tail of quantum well structures. The precipitate is In rich. It probably consists of InN. Right: Reconstruction of the electron scattering potential by quantification of the information from the lattice image (QUANTITEM). Local pattern differences are exploited in the reconstruction process.

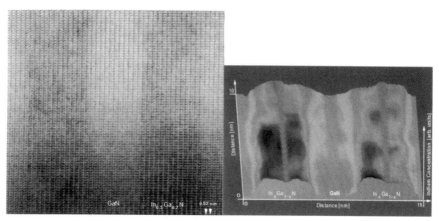

FIG. 22. Indium clusters in the $In_xGa_{1-x}N/GaN$ quantum well structure. The sample surface is etched to be atomically flat. Local pattern differences in the lattice image (left) are exploited to determine the local In concentration (right). The composition map reveals the presence of alloy clusters.

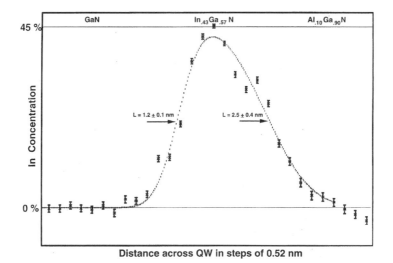

FIG. 23. Asymmetric composition profile across the $In_xGa_{1-x}N$ quantum well MOCVD growth. Growth direction: from left to right. The interfacial widths and the total well width can be measured by averaging the local In concentration in atomic columns parallel to a quantum well to obtain a compositional profile. $Erf(x/L)$ are fitted to the data to extract the interfacial width L.

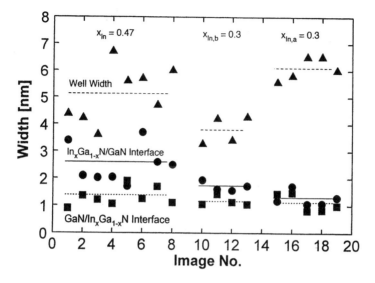

FIG. 24. Local variations of well width and interfacial width in MOCVD grown samples with different In content. $In_xGa_{1-x}N/GaN$ and $GaN/In_xGa_{1-x}N$ interfaces are of different widths. Results depend on details of the growth process. They vary in samples with nominally $x_{In} = 0.3$ dependent on the source of the crystals. The growth of symmetrical $GaN/In_xGa_{1-x}N/GaN$ structures becomes increasingly difficult as x_{In} increases.

In contrast, the formation of alloy clusters was observed in all investigated samples from the concentration range $0.2 < x_{In} < 0.5$. Their size is roughly 1–2 nm (Kisielowski et al., 1997a). They contain an uncertain number of indium atoms. The impact of this inhomogeneous indium distribution in the quantum wells on the optical performance of devices is currently debated (Chichibu et al., 1996). Much smaller indium clusters of 2–3 atoms were reported to be present in InGaAs (Zheng et al., 1994). It was argued that their formation is driven by strain.

The incorporation of Al into GaN does not suffer from cluster formation within the well. Here, additional complications arise at the $GaN/Al_xGa_{1-x}N/GaN$ interfaces (Kisielowski and Yang, 1998). This is depicted in Fig. 25. Lattice images exhibit a bright contrast at both interfaces. A map of the mean inner potential of the sample reveals an additional contribution to the scattering of electons at these places. The contrast variation may be attributed to an Al accumulation at interfaces. A dot-like structure is only seen in these interfacial areas. It is not certain if this interpretation is unique. Other contributions to electron scattering such as the presence of large electrostatic fields may cause a similar effect.

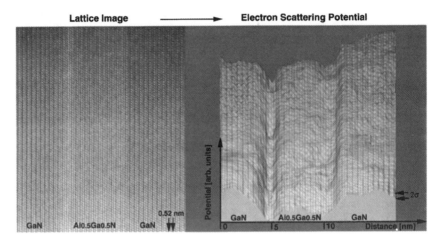

FIG. 25. Lattice image of the $Al_xGa_{1-x}N/GaN$ quantum well structure (left). Right: Map of the electron scattering potential. The sample surface is etched to be atomically flat. An unusual large change of the scattering potential at the AlN/GaN interfaces is present. It exhibits a dot-like structure.

2. Strain-Induced Transition from a Two-Dimensional to a Three-Dimensional Growth Mode

A particularity of the GaN/AlN/InN system is the rare observation of misfit dislocations. Instead, it is common that stress relaxation occurs via a transition from a two-dimensional (2D) to a three-dimensional (3D) growth mode. The phenomenon can be investigated best if multiquantum well structures are grown. In the case of $In_xGa_{1-x}N$ such a transition was introduced by growing 10 nominal identical $In_xGa_{1-x}N/GaN$ stacks as shown in Fig. 26. Similarly, 3D growth can be initiated in the $Al_xGa_{1-x}N/GaN$ system by growing wells with an increasing Al content x_{Al} (Fig. 27). These experiments serve two purposes.

First, the quantum wells are marker layers reflecting the growth history. It is seen from both figures that the growth was planar in its initial stage. It takes about 5 $In_xGa_{1-x}N/GaN$ stacks with a nominal indium concentration of 0.3 to initiate 3D growth on {0001} and {01-11} surfaces. The growth proceeds almost exclusively on {0001} planes from this point on. As a result the final surface is rough (Fig. 26, AFM image). Thus, the growth rate of GaN {0001} planes is significantly larger than that of {01-11} surfaces. Typically, the 3D growth is initiated at positions where threading dislocations are present.

7 Strain in GaN Thin Films and Heterostructures

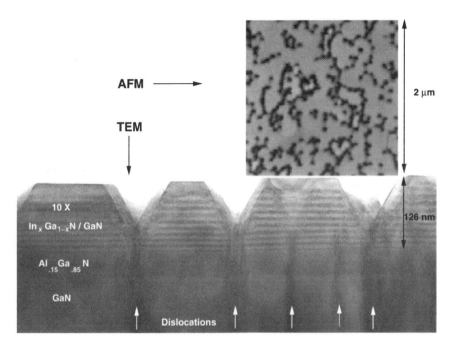

FIG. 26. Cross-sectional transmission electron micrograph of the $In_xGa_{1-x}N/GaN$ multi-quantum well structure. $In_xGa_{1-x}N$ quantum wells were used as marker layers to reveal the flatness of growing surfaces. A transition from a 2D to a 3D growth on (0001) and on (01-11) planes is shown. The samples surface is rough (AFM image, top).

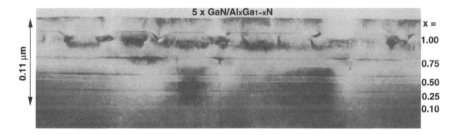

FIG. 27. Cross-sectional transmission electron micrograph of the $Al_xGa_{1-x}N/GaN$ multi-quantum well structure. x was increased as indicated. The spacing of the wells and layers vary. A transition from a 2D to a 3D growth occurs in the well with $x_{Al} = 0.75$. A rough surface can be flattened by overgrowth with GaN.

In $Al_xGa_{1-x}N/GaN$ the growth becomes 3D in the well with $x_{Al} = 0.75$ (Fig. 27). The {01-11} planes are also formed as shown in Fig. 28. From Fig. 28 it is also seen that extended defects are present close to the positions where the growth becomes 3D. The extended defects are stacking faults in this case. Unlike the $In_xGa_{1-x}N/GaN$ quantum well structure, the resulting holes can be filled by overgrowth with GaN. As a result, the final GaN c-surface grew in a 2D manner. Thus, the structure offers the possibility of alteration between 3D and 2D growth. Buried quantum dots can be grown this way even though their spacing and size are not regular at this point (Widmann et al., 1997).

Figure 29 shows calculated values for the critical layer thicknesses that are required to form misfit dislocations in the InN/GaN/AlN system. Different models are used (van der Merve, 1963; Matthews and Blakeslee, 1975; People, 1986). The transition from 2D to 3D growth is indicated. For InGaN it occurs above any calculated value for critical thickness. Yet, no misfit dislocations are formed. In contrast, the transition occurs in AlGaN without being even close to any of the critical thickness values.

Moreover, the reproducibility of growth can be investigated if quantum wells are deposited successively. Figure 30 shows the width and the spacing

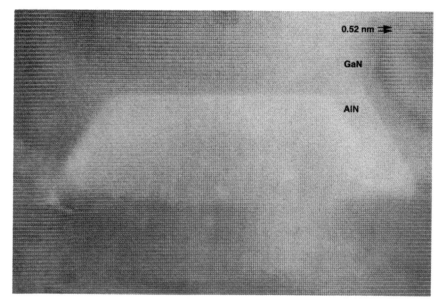

FIG. 28. Lattice image of a AlN/GaN quantum well. Three-dimensional growth is initiated at extended defects.

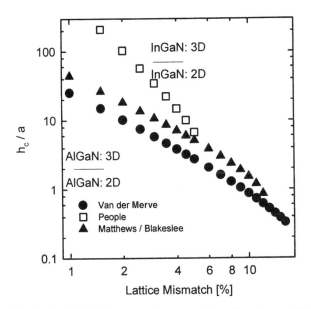

FIG. 29. Calculated critical thicknesses and measured transitions from a 2D to a 2D/3D growth mode in the $Al_xGa_{1-x}N/GaN$ and the $In_xGa_{1-x}N/GaN$ system.

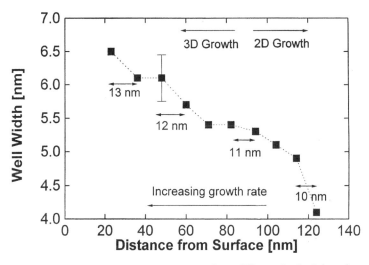

FIG. 30. Width of the $In_xGa_{1-x}N$ quantum wells at different depths below the surface. Both the well width and their layer spacing change continuously with the deposition of each additional layer.

of the 10 $In_xGa_{1-x}N/GaN$ stacks that were shown in Fig. 26. Both parameters increase with each additional deposited layer. Consequently, the growth rates must increase. An increasing growth rate is expected if the growth is 3D because the growing surface area reduces for geometrical reasons. However, even the first 5 stacks exhibit the same tendency. As a result of the changing growth rates the optical emission from a multiquantum well structure degenerates to a broadband compared with the emission from a single quantum well (Krueger et al., 1997). Similarly, an irregular spacing of wells and layers in the $Al_xGa_{1-x}N/GaN$ stacks is evident from Fig. 27. Thus, it is concluded that growth rates can be affected if a multiquantum well structure is deposited even if the growth is 2D. A stress dependence of the diffusing Ga ad-atom (Fujii et al., 1997) can account for these observations.

Largely different growth rates of {0001} and {01-11} planes were observed in bulk GaN (Liliental-Weber, 1997). They were found to depend on the materials polarity. A model was put forward that links the formation of pinholes to growth velocities that differ from plane to plane (Liliental-Weber et al., 1997). In chemical etching experiments a pronounced formation of {01-11} surfaces was reported (Kisielowski et al., 1997). Thus, {01-11} surfaces exhibit a distinguished low surface energy. Therefore, it is not surprising that they are also formed in the transition from 2D to 3D growth.

It is remarkable that the 2D growth transforms into a Volmer-Weber growth after deposition of a well-defined number of quantum well stacks. Its occurrence requires the presence of nucleation sites and a driving force. The sites observed here are located in the vicinity of extended defects. The stress resulting from the pseudomorphic growth of the strained layers must be the driving force because strain energy builds up with the deposition of each additional layer. Strain fields fluctuate around dislocations and other extended defects (Hirth and Lothe, 1982). Therefore, they are suitable sites to make 2D growth unstable. On the other hand, extended impurity clusters may serve as nucleation sites, too, if the stresses become large.

In conclusion, there are three crucial areas that involve stress and strain. First, it is the transition from 2D to 3D growth, which can be driven by stress. Second, the pseudomorphic growth of AlN/GaN/InN quantum wells introduces stress and strain that affects growth rates. As a result, the deposition of identical multiquantum well structures has to be tailored by adjusting growth parameters for each individual well of the stack. At present, this process is not yet well controlled. The p- and n-doping adds an additional dimension to the problem. Doping alters the lattice constant (Fig. 20) or, worse, a particular lattice constant may have to be grown in order to achieve a substantial incorporation of dopants (Fig. 19). Third, the large strain in $In_xGa_{1-x}N/GaN$ quantum well structures favors cluster formation. Finally, it is noted that the intended In or Al content in a layer might deviate from its real concentration in the wells or barriers.

V. Perspectives

This paper addresses the long-standing issue of how to better understand GaN thin-film growth better. The presented results form a basis that can make the thin-film growth predictive. Important aspects could be clarified but many questions remain open for discussions and investigations.

There is the most important issue of buffer-layer growth. It is certain now that stresses and strains in the main layers are controlled through the buffer-layer growth. In addition, the V/III flux ratio and doping determine the stoichiometry of materials. A hydrostatic strain, determined by point defects, interacts with the hydrostatic component of the biaxial strain coming from growth on lattice mismatched substrates and from thermal strain. The resulting stresses reach 1 GPa, even if growth only on c-plane sapphire is considered. They can be designed to be compressive or tensile. The formation of point defects is found to depend on the Fermi energy. This adds complexity to the overall system.

Both MBE- and MOCVD-thin-film growth proceeds at low temperatures. The shear components of biaxial stresses enforce dislocation glide below or close to the brittle-to-ductile transition of these materials systems. It is argued that plastic flow or even yield stresses can be reached. Here, σ_{uy} was estimated to be -0.9 GPa at MBE growth temperatures and at most -0.5 GPa at 1300 K for GaN buffer layers. The plastic relaxation is mostly confined to the buffer-layer region. However, such a model simplifies the complex relaxation phenomena that may occur or may be introduced intentionally. In particular, the stress-strain and the temperature-stress curves that are schematized in Figs. 2 and 4 are unknown. There is urgent need for this information. Some insight was gained by growing buffer layers at different temperatures. The presented model explains why stresses can be larger in MBE-grown samples than in MOCVD material even though the temperature differences in an MBE-growth cycle are smaller.

The large dependence of the doping incorporation on the lattice constants opens exciting possibilities. It explains why p-doping is restricted to such a narrow window of growth parameters. The result provides guidelines for how to increase the dopant concentration in GaN thin films. Details of the process are not yet well understood. There are similarities with the "site competition" model that was proposed for the growth of heavily doped SiC. However, stresses and Fermi energies must be considered, too.

Implicitly, the stress and strain were always utilized to grow high-quality GaN thin films. Typically, a trial-and-error process is employed to make progress in short time periods. It seems unlikely that predictive GaN thin-film growth can be developed faster at this point. However, limits of the trial-and-error approach may be reached if it comes to the growth and the optimization of complex thin-film structures. In this case it will be very beneficial to access models that predict growth procedures. The presented results form a basis for the development of models.

Principally, MBE and MOCVD growth can be controlled through the same set of parameters. There is no doubt that MOCVD growth of thin films is of larger industrial importance than MBE growth even if growth rates equalized. Two main factors make the MBE process more difficult. First, it is the smaller size of the grains in the films that comes with the lower growth temperature. Second, it is the larger sensitivity of the MBE process to small changes of growth parameters.

On the other hand, there are potential advantages of MBE. The larger stresses in the process and the lower growth temperature may be utilized to increase doping levels or the incorporation of indium in quantum well structures through active utilization of stress and strain. Some examples of how to strain engineer desired film properties were given in this chapter. In particular, strain engineering can be applied to avoid film cracking. It is now feasible to increase the grain size in films grown at low temperatures by the use of surfactants. Thus, MBE might become a very valuable tool for the growth of specialized structures that cannot be created by a MOCVD process.

ACKNOWLEDGMENTS

GaN growth was supported by the Office of Energy Research, Office of Basic Energy Sciences, Division of Advanced Energy Projects (BES-AEP) and by the Laboratory Technology Transfer Program (ER-LTT) of the U.S. Department of Energy under Contract No. DE-AC03-76SF00098. Characterization of the material was supported by BMDO administered by ONR under Contract No. N00014-96-1-0901. R. Klockenbrink, Y. Kim, and M. Leung performed MBE growth. J. Krueger (PL), D. Corlatan (PL), J. W. Ager III (Raman), G. S. Sudhir (electrical measurements, x-rays), Y. Peyrot (x-ray), H. Fujii, and Y. Kim (AFM) characterized the thin films. The TEM investigations were supported through the U.S. Department of Energy under Contract No. DE-AC03-76SF00098 and by Z. Liliental-Weber. The use of facilities in the National Center for Electron Microscopy is greatly appreciated. The MOCVD samples were provided by B. McDermott, S.

Nakamura, A. Khan, J. Yang, and C.-P. Kuo. P. Specht and P. Pirouz contributed with fruitful discussions. In particular, the author acknowledges the continuing interest and support of the GaN growth and characterization effort at LBNL and UCB by M. Rubin and E. R. Weber.

REFERENCES

Alexander, H., Kisielowski-Kemmerich, C., and Swalski, A. (1987). *Phys. Stat. Sol.* (a) **104**, 183.
Amano, H., Kito, M., Hiramatsu, K., and Akasaki, I. (1989). *Jpn. J. Appl. Phys.*, **28**, L2112.
Amano, H., Sawaki, N., Akasaki, I., and Toyoda, Y. (1986). *Appl. Phys. Lett.*, **48**, 353.
Amano, H., Hiramatsu, K., and Akasaki, I. (1998). *Jpn. J. Appl. Phys.*, **27**, L1384.
Anders, A., Newman, N., Rubin, M., Dickinson, M., Jones, E., Phatak, P., and Gassmann, A. (1996). *Rev. Sci. Instrum.*, **67**, 905.
Azuhata, T., Sota, T., and Suzuki, K. (1996). *J. Phys. Condens. Matter*, **8**, 3111.
Boguslawski, P., Briggs, E. L., and Bernholc, J. (1995). *Phys. Rev. B*, **51**, 17225.
Brandt, O., Yang, H., and Ploog, K. H. (1996). *Phys. Rev. B*, **54**, 4432.
Castaing, J., Veyssiere, P., Kubin, L. P., and Rabier, J. (1981). *Phil. Mag.* A, **44**, 1407.
Chichibu, S., Asuhata, T., Sota, T., and Nakamura, S. (1996). *Appl. Phys. Lett.*, **69**, 4188.
Drory, M. D., Ager, J. W. III, Suski, T., Grzegory, I., and Porowski, S. (1996). *Appl. Phys. Lett.*, **69**, 4044.
Fujii, H., Kisielowski, C., Krueger, J., Leung, M. S. H., Klockenbrink, R., Weber, E. R., and Rubin, M. (1997). *Mat. Res. Soc. Symp.*, **449**, 227.
Gassmann, A., Suski, T., Newman, N., Kisielowski, C., Jones, E., Weber, E. R., Liliental-Weber, Z., Rubin, M. D., Helawa, H. I., Grzegory, I., Bockowski, M., Jun, J., and Porowski, S. (1996). *Appl. Phys. Lett.*, **80**, 2195.
Goetz, W., Johnson, N. M., Chen, C., Liu, H., Kuo, C., and Imler, W. (1996). *Appl. Phys. Lett.*, **68**, 3144.
Hiramatsu, K., Detchprohm, T., and Akasaki, I. (1993). *Jpn. J. Appl. Phys.*, **32**, 1528.
Hirth, J. P. and Lothe, J. (1982). *Theory of Dislocations.* New York: John Wiley & Sons.
Ho, I. H. and Stringfellow, G. B. (1997). *Mat. Res. Soc. Symp.* F. A. Ponce, T. D. Moustakas, I. Akasaki, and B. A. Monemar, eds., **449**, 871.
Hwang, C.-Y., Schurman, M. J., Mayo, W. E., Lu, Y.-C., Stall, R. A., and Salagaj, T. (1997). *J. Electron. Mat.*, **26**, 243.
Kahn, M. A., Bhattarai, A., Kuznia, J. N., and Olsen, D. T. (1993). *Appl. Phys. Lett.*, **63**, 1214.
Kaufmann, U., Kunzer, M., Merz, C., Akasaki, I., and Amano, H. (1996). *Mat. Res. Soc. Symp.* (F. A. Ponce, R. D. Dupuis, S. Nakamura, and J. A. Edmond, eds.), **395**, 633.
Kelly, A. and Groves, G. W. (1970). *Crystallography and Crystal Defects.* Reading: Addison-Wesley.
Kim, K., Lambrecht, W. R. L., and Segall, B. (1994). *Phys. Rev. B*, **50**, 1502.
Kim, Y., Klockenbrink, R., Kisielowski, C., Krüger, J., Corlatan, D., Sudhir, G. S., Peyrot, Y., Cho, Y., Rubin, M., and Weber, E. R. (1998). *Mat. Res. Soc. Symp.*, **482**, 217.
Kisielowski, C., Krueger, J., Ruvimov, S., Suski, T., Ager, J. W. III, Jones, E., Liliental-Weber, Z., Rubin, M., Weber, E. R., Bremser, M. D., and Davis, R. F. (1996a). *Phys. Rev. BII*, **54**, 17745.
Kisielowski, C., Krueger, J., Leung, M., Klockenbrink, R., Fujii, H., Suski, T., Sudhir, G. S., Rubin, M., and Weber, E. R. (1996b). In *Proc. 23rd Internat. Conf. on the Physics of*

Semiconductors (ICPS-23), Berlin, M. Scheffler and R. Zimmerman, eds., Singapore: World Scientific, p. 513.

Kisielowski, C., Liliental-Weber, Z., and Nakamura, S. (1997a). *Jpn. J. Appl. Phys.*, **36**, 6932.

Kisielowski, C. (1997b). *Microscopy and Microanalyses '97*. G. W. Bailey, R. V. W. Dimlich, K. B. Alexander, J. J. McCarthy, and T. P. Pretlow, eds., Berlin: Springer, p. 935.

Kisielowski, C., Schmidt, O., and Yang, J. (1998). *Mat. Res. Symp. Proc.*, **482**, 369.

Klockenbrink, R., Kim, Y., Leung, M. S. H., Kisielowski, C., Krueger, J., Sudhir, G. S., Rubin, M., and Weber, E. R. (1997). *Mat. Res. Soc. Proc.* **468**, 75.

Kriegseis, W., Meyer, B. K., Kisielowski, C., Klockenbrink, R., Kim, Y., and Weber, E. R. (to be published).

Krueger, J., Kisielowski, C., Klockenbrink, R., Sudhir, G. S., Kim, Y., Rubin, M., and Weber, E. R. (1997). *Mat. Res. Soc. Proc.*, **468**, 299.

Krüger, J., Sudhir, G. S., Corlatan, D., Cho, Y., Kim, Y., Klockenbrink, R., Ruvimov, S., Liliental-Weber, Z., Kisielowski, C., Weber, E. R., McDermott, B., Pittman, R., and Gertner, E. R. (1998). *Mat. Res. Soc. Proc.*, **482**, 447.

Lagerstedt, O. and Monemar, B. (1979). *Phys. Rev. B*, **19**, 3064.

Larkin, D. J., Neudeck, P. G., Powell, J. A., and Matus, G. (1994). *Appl. Phys. Lett.*, **65**, 1659.

Leszczynski, M., Suski, T., Perlin, P., Teisseyre, H., Grzegory, I., Bockowski, M., Jun, J., Porowski, S., and Major, J. (1995). *J. Phys. D: Appl. Phys.*, **28**, A149.

Liliental-Weber, Z. (1997). *Gallium Nitride*. J. I. Pankove and T. D. Moustakas, eds., Boston: Academic Press.

Liliental-Weber, Z., Chen, Y., Ruvimov, S., and Washburn, J. (1997). *Phys. Rev. Lett.*, **79**, 2835.

Liu, X., Prasad, A., Nishio, J., Weber, E. R., Liliental-Weber, Z., and Walukiewicz, W. (1995). *Appl. Phys. Lett.*, **67**, 279.

Luysberg, M., Sohn, H., Prasad, A., Specht, P., Liliental-Weber, Z., and Weber, E. R. (1998). *J. Appl. Phys.*, **83**, 561.

Matthews, J. W. and Blakeslee, A. E. (1975). *J. Cryst. Growth*, **29**, 273.

Meyer, B. K. (1997). *Mat. Res. Soc. Symp.* (F. A. Ponce, T. D. Moustakas, I. Akasaki, and B. A. Monemar, eds.), **449**, 497.

Milluchick, J. M., Twesten, R. D., Lee, S. R., Follstaedt, D. M., Jones, E. D., Ahrenkiel, S. P., Zhang, Y., Cheong, H. M., and Mascarenhas, A. (1997). *MRS Bulletin*, **22**, 38.

Morkoç, H., Strite, S., Gao, G. B., Lin, M. E., Sverdlov, B., and Burns, M. (1994). *J. Appl. Phys.*, **76**, 1363.

Nakamura, S. and Fasol, G. (1997). *The Blue Laser Diode*. Berlin: Springer.

Nakamura, S. (1991). *Jpn. J. Appl. Phys.*, **30**, L1705.

Neugebauer, J. and Van de Walle, C. (1994). *Phys. Rev. B*, **50**, 8067.

Newman, N., Ross, J., and Rubin, M. (1993). *Appl. Phys. Lett.*, **62**, 1242.

Ning, X. J., Chien, R. F., Pirouz, P., Yang, J. W., and Khan, M. A. (1996). *J. Mat. Res.*, **11**, 580.

Nix, W. D. (1989). *Metallurgical Transactions* A, **20A**, 2217.

Orton, J. W. (1996). *Semicond. Sci. Technol.*, **11**, 1026.

Pakula, K., Wysmonek, A., Korona, K. P., Baranowski, J. M., Stepniewski, R., Grzegory, I., Bcokowski, M., Jun, J., Krukowski, S., Wrobelewski, M., and Porowski, S. (1997). *Solid State Communications*, **97**, 919.

People, R. (1986). *IEEE J. Quantum Electronics*, **22**, 1696.

Perlin, P., Suski, T., Teisseyre, H., Leszczynski, M., Grzegory, I., Jun, J., Perowski, S., Boguslawski, P., Bernholc, J., Chervin, J. C., Polian, A., and Moustakas, T. D. (1995). *Phys. Rev. Lett.*, **75**, 296.

Perry, W. G., Zheleva, T., Bremster, M. D., Davis, R. F., Shan, W., and Song, J. J. (1997). *J. Electron. Mat.*, **26**, 224.

Polian, A., Drimsdich, M., and Grzegory, I. (1996). *J. Appl. Phys.*, **79**, 3343.

Ponce, F. A., Bour, D. P., Goetz, W., Johnson, N. M., Helava, H. I., Grzegory, I., Jun, J., and Porowski, S. (1996). *Appl. Phys. Lett.*, **68**, 917.
Rieger, W., Metzger, T., Angerer, H., Dimitrov, R., Ambacher, O., and Stutzmann, M. (1996). *Appl. Phys. Lett.*, **68**, 970.
Ruvimov, S., Liliental-Weber, Z., Washburn, J., Koike, M., Amano, H., and Akasaki, I. (1997). *Mat. Res. Soc. Proc.*, **468**, 287.
Ruvimov, S., Liliental-Weber, Z., Suski, T., Ager, J. W. III, Washburn, J., Krueger, J., Kisielowski, C., Weber, E. R., Amano, H., and Akasaki, I. (1996). *Appl. Phys. Lett.*, **69**, 990.
Savastenco, V. A. and Sheleg, A. U. (1978). *Phys. Stat. Sol.* (a), **48**, K13.
Schwarz, R. B., Khachaturyan, K., and Weber, E. R. (1997). *Appl. Phys. Lett.*, **70**, 1122.
Singh, R., Doppalapudi, D., Moustakas, T. D., and Romano, L. T. (1997). *Appl. Phys. Lett.*, **70**, 1089.
Skromme, B. J., Zhao, H., Wang, D., Kong, H. S., Leonard, M. T., Bulman, G. E., and Molnar, R. J. (1997). *Appl. Phys. Lett.*, **71**, 829.
Smith, F. W., Calava, A. R., Chen, Chang-Lee, Manfra, M. J., and Mahoney, L. J. (1988). *IEEE Electron Device Letters*, **9**, 77.
Specht, P., Private communication.
Specht, P., Jeong, S., Sohn, H., Luysberg, M., Prasad, A., Gebauer, J., Krause-Rehberg, R., and Weber, E. R. (1997). *Proc. 19th International Conference on Defects in Semiconductors*, Aveiro, Portugal.
Stranski, I. N. and Von Krastanov, A. (1939). *Akad. Wiss. Mainz L. Math-Nat.* K1 IIb **146**, 797.
Sudhir, G. S., Fujii, H., Wang, W. S., Kisielowski, C., Newman, N., Dieker, C., Liliental-Weber, Z., Rubin, M., and Weber, E. R. (1998). *J. Electron. Mat.*, **27**, 215.
Suezawa, M., Sumino, K., and Yonenaga, I. (1979). *Phys. Stat. Sol.* (a), **51**, 217.
Sugiura, L. (1997). *J. Appl. Phys.*, **81**, 1633.
Suski, T., Ager, J. W. III, Liliental-Weber, Z., Krueger, J., Ruvimov, S., Kisielowski, C., Weber, E. R., Kuo, C.-P., Grzegory, I., and Porowski, S. (1996). *Proc. ICPS*, Berlin 1996, M. Scheffler and R. Zimmermann, eds., Singapore: World Scientific, p. 2917.
van der Merve, J. H. (1963). *J. Appl. Phys.*, **34**, 123.
Van Vechten, J. A. (1973). *Phys. Rev. B*, **7**, 1479.
Volmer, M. and Weber, A. (1926). *Z. Physik. Chem.*, **119**, 277.
Widmann, F., Daudin, B., Feuillet, G., Samson, Y., Arlery, M., and Rouviere, J. L. (1997). *MRS Internet J.*, **2**, 20.
Wu, H., Fini, P., Keller, S., Tarsa, E. J., Heying, B., Mishra, U. K., DenBaars, S. P., and Speck, J. S. (1996). *Jpn. J. Appl. Phys.*, **35**, L1648.
Zheng, J. F., Walker, J. D., Salmeron, M. B., and Weber, E. R. (1994). *Phys. Rev. Lett.*, **72**, 2414.

CHAPTER 8

Nonlinear Optical Properties of Gallium Nitride

Joseph A. Miragliotta and Dennis K. Wickenden

JOHNS HOPKINS UNIVERSITY
THE APPLIED PHYSICS LABORATORY
LAUREL, MARYLAND

I. INTRODUCTION	319
II. BACKGROUND	322
III. SECOND-ORDER NONLINEAR OPTICAL PHENOMENA	325
1. Second-Harmonic Generation	326
2. The Electro-Optic Effect (LEO)	338
IV. THIRD-ORDER NONLINEAR OPTICAL PHENOMENA	341
1. Third-Harmonic Generation (THG)	342
2. Electric Field-Induced Second-Harmonic Generation (EFISH)	346
3. Two-Photon Absorption (TPA)	351
4. Degenerate Four-Wave Mixing	355
V. POTENTIAL DEVICES	361
VI. CONCLUSIONS	366
REFERENCES	366

I. Introduction

Since the prediction and observation of nonlinear optical phenomena in semiconductors nearly 35 years ago, experimental and theoretical studies have examined the intrinsic nonlinearities of various III-V and II-VI compound systems (Lax *et al.*, 1962; Miller *et al.*, 1963; Butcher and McLean, 1963, 1964; Kelley, 1963; Soref and Moos, 1964). In part, the scientific activity afforded semiconductor nonlinear optics is motivated by the pursuit of new electro-optical devices for applications in areas such as telecommunications and optical computing. In fact, technologically important nonlinear optical effects such as frequency mixing, electro-optical modulation, and phase conjugation have already been demonstrated in various types of fabricated semiconductor structures (Tien *et al.*, 1970; Anderson and Boyd, 1971; Ito and Inaba, 1978; Miller *et al.*, 1986; Cada *et al.*, 1988; Fuji *et al.*, 1990; Heyman *et al.*, 1994; Takahashi *et al.*, 1994;

Chenault et al., 1994; Blanc et al., 1995; Suzuki and Iizuka, 1997). Parallel to these applied interests, numerous studies over the past three decades have shown that optical nonlinearities in semiconductors are versatile probes of phenomena such as phase transitions, interfacial electronic structure, and surface adsorption (Shank et al., 1983; Akhmanov et al., 1985; Lanzafame et al., 1994; Guyot-Sionnest et al., 1990; Reider et al., 1991; Harris et al., 1987).

Of current interest to the optical community are the continuing improvements in the growth and processing of wide-bandgap III-V semiconductors, materials which are well-suited for devices in the infrared to ultraviolet (UV) range of the spectrum. The advances in the deposition of gallium nitride (GaN) epitaxial films provides investigators with an optical waveguiding material with both a large optical bandwidth and high damage threshold (Strite and Morkoç, 1992; Strite et al., 1993). An added benefit is the ability to alloy GaN continuously with either aluminum nitride or indium nitride for the production of semiconductor devices with a bandgap that is tunable from ~ 2 to 6 eV.

Most activity regarding GaN has been directed towards the development of visible and near-ultraviolet LEDs and laser diodes (Denbaars, 1997; Mohammed and Morkoç, 1996; Mohs et al., 1997; Nakamura, 1996; Nakamura et al., 1996), but it has been known since the late 1960s that the symmetry of the GaN crystal structure assures the existence of an intrinsic second-order nonlinearity in the bulk region of the material (Phillips and Van Vechten, 1989; Levine, 1969). Shown in Table I is a listing of predicted values for the second-order susceptibility $\chi^{(2)}$ in GaN and other nonlinear

TABLE I

SECOND-ORDER NONLINEARITIES IN VARIOUS MATERIALS[a]

Material	Point group	d_{ijk} (pm/V)
Quartz	$32 = D_3$	$d_{11} = 0.40$
		$d_{14} = 0.01$
BaTiO$_3$	4 mm = C_{4v}	$d_{15} = -17.17$
		$d_{31} = -18.01$
		$d_{33} = -6.70$
LiNbO$_3$	3 m = C_{3v}	$d_{22} = 3.10$
		$d_{31} = 5.86$
		$d_{33} = 41.05$
KDP	42 m = D_{2d}	$d_{14} = 0.503$
		$d_{36} = 0.503$
GaN	6 mm = C_{6v}	$d_{33} = -26.40$ (Levine, 1969)

[a]Singh (1971).

crystals. As will be discussed in Section II, the degree of coupling between an optical field and the nonlinear material is linearly proportional to the magnitude of the nonlinear susceptibility. The results in Table I indicate that the early predictions of $\chi^{(2)}$ in GaN were comparable to or greater than many conventional nonlinear crystals such as KDP or $LiNbO_3$. Although the potential for efficient nonlinear optical coupling in GaN has been recognized for nearly 30 years, recent improvements in material quality have resulted in an increase in the characterization of the optical nonlinearities.

In this chapter, a review of the past and ongoing characterization of the nonlinear optical properties of GaN along with a discussion pertaining to their potential integration into optical devices will be presented. Section II provides an introduction to the general theory of nonlinear interactions in matter, which was introduced in the early 1960s (Armstrong *et al.*, 1962; Bloembergern and Pershan, 1962; Pershan, 1963; Bloembergen, 1965). This approach is used routinely in modeling the second- and third-order nonlinear optical responses in semiconductors and other materials, and issues pertinent to this review are presented. Sections III and IV are devoted to second- and third-order nonlinear effects in GaN, respectively, which include, when possible, a comparison between experimental and results. These two sections will outline the results from experimental investigations that use nonlinear optical techniques such as second-harmonic (SHG) and third-harmonic generation (THG) for the characterization of the second- and third-order susceptibilities in GaN.

Section V introduces the reader to the optical waveguide structures that are used to enhance nonlinear optical processes relative to the corresponding response in the bulk material. Despite the intrinsic nonlinearities in GaN, there are two characteristics of thin films that limit their use for optical frequency conversion and phase modulation devices. First, the transfer of energy between the incident and generated nonlinear optical fields is dependent on their propagation distances in the material. The inability to grow high-quality GaN samples conveniently to a thickness greater than $\sim 10 \mu m$ produces a system with inherently low conversion efficiencies. The second limitation, which will be discussed in more detail in Section III.1, is associated with frequency mixing applications such as SHG. In these processes, the generated nonlinear field in the material is actually composed of two waves at the upconverted frequency. Although these two waves have the same frequency, they may have different phase velocities. The phase velocity difference creates an interference that oscillates between constructive and destructive as the waves propagate through the nonlinear sample. When the phase velocity of the two waves is equal, the resultant upconverted field will increase linearly with propagation distance in the material; for example, the ability to "phase-match" in material such as KDP or $LiNbO_3$ results in conversion efficiencies that are greater than 25% when phase matching is maintained for propagation distances on the order of a few centimeters. However, the dispersive properties of the refractive index do not allow phase matching in bulk GaN (Ejder, 1971), which limits efficient upconversion to very short propagation distances. In nonphase matchable materials, this distance is termed the "coherence length," which is on the order of a few microns in GaN for optical frequencies between the infrared to near ultraviolet. Fortunately, it is possible to overcome the phase mismatch limitation in bulk GaN with the use of optical waveguide structures, which allow one to tailor the dispersive properties of the refractive index and produce phase matching under very select conditions (Sasaki *et al.*, 1984; Khanarian *et al.*, 1990; Sugihara *et al.*, 1991; Stegeman, 1992; Clays *et al.*, 1993; Noordman *et al.*, 1995). Waveguiding structures are routinely used to enhance nonlinear conversion efficiencies in many organic and inorganic systems (Malouin *et al.*, 1998; Nguyen, 1997; Phelps *et al.*, 1997; Tessler *et al.*, 1997; Diaz-Garcia *et al.*, 1997; Gase and Karthe, 1997), and their incorporation for potential GaN-based nonlinear devices should be straightforward. As is demonstrated in Section V, the use of optically engineered structures should augment the conversion efficiencies in GaN by orders of magnitude over those in the corresponding bulk material.

II. Background

In the typical treatment of the electromagnetic response of nonlinear material, the induced dipole moment per unit volume, that is, electric polarization, that arises from the material response to an incident optical field of amplitude $E(t)$ is represented as a power-series expansion of the field (Boyd, 1992)

$$\begin{aligned} P(t) &= \chi^{(1)}E(t) + \chi^{(2)}E^2(t) + \chi^{(3)}E^3(t) + \cdots \\ &= P^{(1)}(t) + P^{(2)}(t) + P^{(3)}(t) + \cdots \\ &= P^{(1)} + P^{nl} \end{aligned} \quad (1)$$

In the first term, $\chi^{(1)}$ is the optical susceptibility that describes phenomena that are linearly dependent on the amplitude of the incident optical field (Born and Wolf, 1980). The higher-order terms $\chi^{(2)}$ and $\chi^{(3)}$ are the second- and third-order nonlinear optical susceptibilities, respectively, which describe induced polarizations that exhibit quadratic and cubic field dependencies (P^{nl}). The more common nonlinear optical effects are listed in Table II, where the bold terms represent phenomena that have been observed or theoretically predicted in GaN.

TABLE II
Various Second- and Third-Order Nonlinear Optical Phenomena

Second-order phenomena	Third-order phenomena
Sum-frequency generation	Stimulated Raman scattering
Difference-frequency generation	Optical-field induced birefringence
Second-harmonic generation	**Two-photon absorption**
Parametric amplification	Self-focusing
Optical rectification	Phase conjugation
Optical field-induced magnetization	**Third-harmonic generation**
Linear electro-optic effect	**Electric-field induced SHG**
Magneto-optic effect	**Degenerate four-wave mixing**

For purposes of modeling the second and third-order nonlinear response to an electromagnetic source, it is convenient to express the total time-dependent incident field as consisting of more than one frequency component:

$$E(t) = \sum_n (E_n e^{-i\omega_n t} + \text{c.c}) \tag{2}$$

where the summation extends over the positive and negative frequencies of the individual field components, and c.c is the conjugate of the complex field amplitudes. In second-order processes, the incident field is composed of two distinct frequency components, ω_1, and ω_2, which results in an incident field that is given by

$$E(t) = E_1 e^{-i\omega_1 t} + E_2 e^{-i\omega_2 t} + \text{c.c} \tag{3}$$

Substitution of Eq. (3) into the second-order term in Eq. (1) generates a nonlinear polarization source that contains many distinct frequency components due to the various couplings between the ω_1 and ω_2 fields. The positive frequency terms in the nonlinear polarization are composed of a frequency and time dependent component, the former given by

$$\begin{aligned}
P^{(2)}(2\omega_1) &= \chi^{(2)}(2\omega_1; \omega_1, \omega_1)E_1^2 \\
P^{(2)}(2\omega_2) &= \chi^{(2)}(2\omega_2; \omega_2, \omega_2)E_2^2 \\
P^{(2)}(\omega_1 + \omega_2) &= 2\chi^{(2)}(\omega_1 + \omega_2; \omega_1, \omega_2)E_1 E_2 \\
P^{(2)}(\omega_1 - \omega_2) &= 2\chi^{(2)}(\omega_1 - \omega_2; \omega_1, -\omega_2)E_1 E_2^* \\
P^{(2)}(0) &= 2\chi^{(2)}(0; \omega_1, -\omega_1)E_1 E_1^* + 2\chi^{(2)}(0; \omega_2, -\omega_2)E_2 E_2^*
\end{aligned} \tag{4}$$

The terms from top to bottom represent SHG ($2\omega_1, 2\omega_2$), sum-frequency generation ($\omega_1 + \omega_2$), difference-frequency generation ($\omega_1 - \omega_2$), and optical rectification (0), respectively. The total expression for the nonlinear polarization contains negative frequency components due to the negative frequency field terms in Eq. (2). However, these components are simply the complex conjugates of the expressions in Eq. (4) and provide no further insight into the nature of the nonlinearity. Each term in Eq. (4) acts as a generation source for a new electric field that oscillates at the frequency of the respective nonlinear polarization.

In an analogous treatment, the various terms in the total third-order nonlinear polarization are obtained by substituting an incident source with three frequency components into Eq. (1). This procedure has been performed in a number of nonlinear optical textbooks (Boyd, 1992; Shen, 1984; Yariv, 1975) where the resultant frequency components in the third-order nonlinear polarization account for processes such as THG, degenerate four-wave mixing (DWFM), and stimulated Raman scattering.

Although the electric field and polarization terms in Eqs. (1)–(4) have been written as scalars, they are, in general, vector quantities. Therefore, the second- and third-order susceptibilities in the power-series field expansion must be third- and fourth-rank tensors, respectively, so as to preserve the vector nature of the nonlinear polarization (Shen, 1984; Yariv, 1975). In the case of wurtzite GaN, the coupling between the incident field and the nonlinear material is determined by the orientation of the electric field vector with respect to the nonvanishing elements in the nonlinear susceptibility tensors.

Finally, it is important to keep in mind that the observable in an optical measurement is typically an optical intensity, not a polarization source such as described by Eq. (4). The signal intensity is proportional to the square of the nonlinear polarization source:

$$I^{nl} \alpha P^{nl} \cdot P^{nl*} \tag{5}$$

where substitution of the nonlinear polarization terms in Eq. (1) into Eq. (5) produces a nonlinear intensity that exhibits quadratic and cubic dependencies on the incident optical intensity. Additionally, the nonlinear signal intensity scales as the square of the nonlinear susceptibilities, $\chi^{(2)}$ and $\chi^{(3)}$. As is demonstrated in the sections on SHG and THG, this relationship provides a means of determining the magnitude of the second and third-order nonlinearities in GaN.

In Sections III and IV, discussions will focus on the various nonlinear optical effects that arise from the second- and third-order polarization terms in Eq. (1), respectively. A detailed discussion of nonlinear optics in solids is beyond the scope of this chapter; the interested reader is referred to a number of excellent reviews on this subject matter (Bloembergen and Pershan, 1962; Bloembergen, 1965; Boyd, 1992; Shen, 1984). In Section III, second-order phenomena in GaN such as SHG and the linear electro-optic effect (LEO) are presented with particular emphasis on effects that have potential device applications. Similarly, Section IV discusses the various third-order nonlinear optical effects such as THG and DFWM.

III. Second-Order Nonlinear Optical Phenomena

Second-order nonlinear effects are represented by the nonlinear susceptibility tensor $\chi^{(2)}$, which describes the coupling interaction between two electric fields ω_1 and ω_2, and the illuminated material. The nonvanishing, independent tensor elements of $\chi^{(2)}$ in hexagonal GaN are: $\chi^{(2)}_{xzx}$, $\chi^{(2)}_{zxx}$, $\chi^{(2)}_{xxz}$, and $\chi^{(2)}_{zzz}$ (Boyd, 1992; Shen, 1984; Levenson and Kano, 1988). It is common practice to list the Cartesian components in the subscript index as $x = 1$, $y = 2$, and $z = 3$. Also, many authors refer to the second-order nonlinear coefficient as

$$d_{ijk} = \frac{\chi^{(2)}_{ijk}}{2}$$

In this chapter, references to the magnitude of the second-order nonlinear susceptibility will be in MKS units of picometer/volt (pm/V).

For an ideal hexagonal wurtzite structure, the nonzero tensor elements are not independent and are related by the following expressions (Robinson, 1968)

$$\chi^{(2)}_{xzx} = \chi^{(2)}_{zxx} = \chi^{(2)}_{xxz}$$
$$\chi^{(2)}_{zxx} = \frac{-\chi^{(2)}_{zzz}}{2} \tag{6}$$

Early attempts to predict the magnitude of $\chi^{(2)}$ in GaN were performed independently by Levine (1969) and Phillips and Van Vechten (1969). Both investigators derived simple and tractable theories that related the nonlinearity of the GaN crystal to the excess charge in the bonding region of the covalent bond between the Ga and N atoms. Physically, the nonlinear polarizability of an individual bond in the lattice arises when a large electric field induces a spatial perturbation in the linear susceptibility. In sp^3-bonded crystals such as GaN, the magnitude of the field-induced perturbation is dependent on the average energy gap between the bonding and antibonding levels, which is the sum of both the covalent (homopolar) and ionic (heteropolar) contributions of the bond (Phillips and Van Vechten, 1969; Phillips, 1969; Van Vechten, 1969a, 1969b). An analytical expression for the microscopic nonlinear polarizability is obtained from a differentiation of the linear polarizability with respect to the applied field along the bond axis. In determining the macroscopic nonlinearity, the crystal is treated as being constructed out of n identical, noninteracting symmetric bonds. Therefore, summing all individual microscopic polarizabilities induced on the bonds generates the macroscopic nonlinear susceptibility. One limitation of this simple approach is the inability to provide insight regarding the wavelength dependence of the susceptibility. However, the attractive aspect of the bond-charge model is the ability to determine the critical parameters of the nonlinear polarizability from the linear optical properties. In Levine's calculations for GaN, an estimate of these parameters allowed a determination of the heteropolar and homopolar nonlinearities, which were 21 and -48 pm/V, respectively, giving a total of -27 pm/V for $d_{33} = \chi^{(2)}_{zzz}/2$.

To date, three of the second-order effects listed in Table II have been applied as nonlinear probes of GaN: SHG, LEO, and sum-frequency generation (SFG). In the next two sections, a discussion of experimental investigations of $\chi^{(2)}$ in GaN using SHG and LEO is presented, with a comparison to recent theoretical predictions for the nonlinearities.

1. SECOND-HARMONIC GENERATION

For the specific case of SHG from a GaN crystal, where the incident fields have the same frequency ($\omega_1 = \omega_2 = \omega$), the second-order nonlinear polarization in the metal nitride film has the following form (Miragliotta et al., 1993):

$$P^{(2)}_{x'}(2\omega) = 2\chi^{(2)}_{x'z'x'} E_{z'}(\omega) E_{x'}(\omega)$$
$$P^{(2)}_{y'}(2\omega) = 2\chi^{(2)}_{x'z'x'} E_{z'}(\omega) E_{y'}(\omega) \qquad (7)$$
$$P^{(2)}_{z'}(2\omega) = \chi^{(2)}_{z'x'x'} (E^2_{x'}(\omega) + E^2_{y'}(\omega)) + \chi^{(2)}_{z'z'z'} E^2_{z'}(\omega)$$

where E_i are the ith components of the optical field in the nonlinear media, and the frequency notation has been omitted in the nonlinear susceptibility. Note that it has not been assumed that $\chi^{(2)}_{xzx} = \chi^{(2)}_{zxx}$ as this will only occur when the hexagonal structure is ideally wurtzitic. The primed notation on the expressions in Eq. (1) represents the source terms in the coordinate frame of the crystal (Goldstein, 1981). When the crystal and laboratory frame do not coincide, an appropriate transformation must be made between the two axes as was shown by Miragliotta et al. (1993).

The first reported experimental investigation of $\chi^{(2)}$ in GaN was performed by Catalano et al. (1977) on a 290 μm thick platelet deposited on a sapphire substrate. The investigators used a fixed wavelength laser at $\lambda_{in} = 1.064$ μm to generate a detectable SHG signal at 532 nm, examining the dependence of the nonlinear response on the incident angle of the IR source. An illustration of a typical SHG transmission measurement from a thin film is shown in Fig. 1. An optical field at frequency ω is incident at a film/air interface and transmitted into the film. The transmitted linear field induces a nonlinear polarization wave at 2ω that propagates with a wavevector $k_s(2\omega) = 2k_2(\omega)$, where $k_2(\omega)$ is the wavevector of the linear field. The nonlinear source polarization radiates a "forced" SHG wave at 2ω with a wavevector equal to $k_s(2\omega)$. At the surface of the nonlinear material, the forced wave produces a "free" SHG wave with wavevector $k_f(2\omega)$. Note that the propagation direction of the two waves is not equal in the illustration,

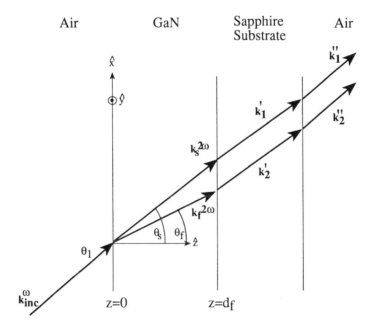

FIG. 1. Schematic of SHG in a thin GaN film on a sapphire substrate. Only the harmonic fields are shown in the film and sapphire substrate. The unit vectors $\hat{x}$ and $\hat{z}$ define the plane of incidence. The unit vector $\hat{y}$ points out of the page and is along the GaN surface.

because of differences in the refractive index at ω and 2ω. The resultant 2ω field in the film is the sum of the two SHG waves. The difference in their respective phase velocities produces an interference between the two waves as they propagate through the nonlinear material. When the propagation distance of the two waves is varied by changing their transmitted angle, the degree of interference between the two waves at the back (and front) of the film is also varied. Additionally, if the phase velocity mismatch is varied by changing the refractive index difference at ω and 2ω, the interference in the film will also vary. These two independent effects will manifest themselves as an oscillatory behavior in both the 2ω transmitted and reflected field from the film.

The transmitted SHG signal from the Catalano et al. (1977) measurement is shown in Fig. 2. The notable characteristic from this study is the oscillatory behavior in the SHG intensity as the angle of incidence is varied approximately every $\sim 10°$. This is a result of the inability to match the phase velocity of the forced ($k_s(2\omega)$) and free ($k_f(2\omega)$) SHG waves in the GaN film. As was just discussed, the variation in the incident angle produces a corresponding change in the propagation distance, resulting in a variation in the degree of interference of these two waves at the back surface of the film. Interestingly, the transmitted SHG intensity from the film will continue to increase with increasing film thickness as long as the thickness does not exceed the coherence length of the material (Jerphagnon and Kurtz, 1970)

$$I_c = \frac{\lambda_{\text{inc}}}{4|n_{2\omega} - n_\omega|} \qquad (8)$$

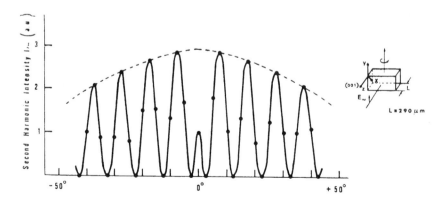

FIG. 2. The SHG intensity vs inclination of a GaN platelet to the incident laser beam. Rotation axis, normal to the beam, is parallel to the crystal axis. (Reprinted with permission from Catalano et al., 1977.)

where I_c is the coherence length, n_ω and $n_{2\omega}$ are the indices of refraction at the fundamental and harmonic wavelengths, respectively, and λ_{inc} is the incident wavelength. When the propagation distance extends beyond this length, typically in the range of 1–10 μm in GaN, the phase difference between the two fields leads to destructive interference and a decrease in the total SHG field in the film.

The investigators compared the results in Fig. 2 with an expression for the transmitted SHG intensity given by (Shen, 1984; Catalano, 1977)

$$I_{\text{SHG}} = \frac{2\mu_0^{3/2}\varepsilon_0^{1/2}\omega^2|\chi_{xzx}^{(2)}|^2 I_\omega^2 L^2}{An_{2\omega}n_\omega^2}\left[\frac{\sin(\Delta kL/2)}{\Delta kL/2}\right]^2 \quad (9)$$

where μ_0, ε_0, L, Δk, and A are the magnetic permeability, dielectric constant, film thickness, wavevector mismatch, and beam area, respectively. The sinusoidal term in brackets produces an SHG intensity that oscillates when either the propagation length or wavevector (phase velocity) difference is varied. Using previously determined values for this parameter, the authors estimated the value of the $\chi_{xzx}^{(2)}$ component to be 0.15 pm/V, with a coherence length of 4.5 μm. It is noted that one concern with this initial study was the appearance of an SHG signal at normal incidence, which is not expected because the coupling between the incident beam and the nonlinear susceptibility should have been zero. One possible explanation for this result is a misorientation between the crystallographic and laboratory frame of reference. A similar result was observed in a later SHG study by Miragliotta *et al.* (1993).

A more detailed study of the second-order nonlinearity was performed by Ishidate *et al.* (1980), where the magnitude of the three nonzero tensor elements of $\chi^{(2)}$ was evaluated from a wedge-shaped, vapor-phase deposited GaN film. Similar to the Catalano study, the investigators examined the transmitted SHG intensity and observed an oscillatory behavior in the signal, Fig. 3, as the GaN wedge was translated across the incident laser source ($\lambda_{\text{in}} = 1.064$ μm). The sample translation increased the propagation distance, which resulted in a variation in the phase velocity mismatch between the forced and free fields at 2ω. An important contribution from this study was the ability to select carefully the polarization direction of the incident fundamental source and upconverted field, allowing a detailed examination of the coupling of the incident field to the three nonzero components of the nonlinear susceptibility tensor. The investigators were able to determine the magnitude of $\chi_{xzx}^{(2)}$ and $\chi_{zxx}^{(2)}$ relative to $\chi_{zzz}^{(2)}$ in separate SHG transmission measurements, which was ~ -0.5 for both tensor element ratios. These ratios are consistent with the geometrically predicted

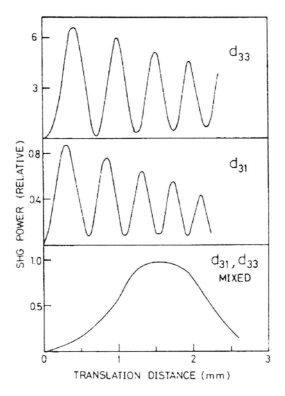

FIG. 3. Dependence of SHG power on the translation distance. The translation distance corresponds to a sample thickness of zero (reprinted with permission from Ishidate et al., 1980).

expectations (Levine, 1969; Robinson, 1968), and confirms that the values of $\chi^{(2)}_{xzx}$ and $\chi^{(2)}_{zxx}$ are nearly equal in wurtzite crystals. In addition to the ratio determinations, the absolute magnitude of $\chi^{(2)}_{zzz}$ was found to be 88 times the value of $\chi^{(2)}$ in quartz.

Interestingly, the next nonlinear characterization of GaN was not reported until 1993, which is a comment on the dramatic improvement in the materials quality of GaN during the late 1980s and early 1990s (Strite and Morkoç, 1992). In the SHG investigation by Miragliotta et al. (1993), the authors evaluated the $\chi^{(2)}_{xzx}$ and $\chi^{(2)}_{zxx}$ tensor elements of $\chi^{(2)}$ for a series of six films ranging in thickness from 0.74–5.31 μm. The authors noted that the crystalline orientation of the film, that is, the c-axis normal to the surface, prevented any appreciable coupling between the transmitted field at ω and the $\chi^{(2)}_{zzz}$ tensor element. This study was similar to the previous measurement by Catalano et al. (1977), where the transmitted SHG intensity at 532 nm was monitored as a function of the angle of incidence of the fundamental input beam. The authors utilized the coupling relationship between the ω field and the nonlinear susceptibility in Eq. (7) and limited the SHG response for a given measurement to either $\chi^{(2)}_{xzx}$ or $\chi^{(2)}_{zxx}$. An example of the SHG dependence on angle, fundamental polarization, and film thickness is shown in Fig. 4, where the ω field polarization provided efficient coupling to only $\chi^{(2)}_{zxx}$. The results clearly show the effects of the small coherence length in GaN as demonstrated by the decrease in the SHG intensity as the film increases from 0.75 to 5 μm. SHG measurements were also performed where the fundamental field coupling was limited to $\chi^{(2)}_{xzx}$ (not shown). The magnitudes of $\chi^{(2)}_{zxx}$ and $\chi^{(2)}_{xzx}$, shown in Table III, were obtained by curve fitting the SHG intensity scans to calculated electromagnetic expressions for

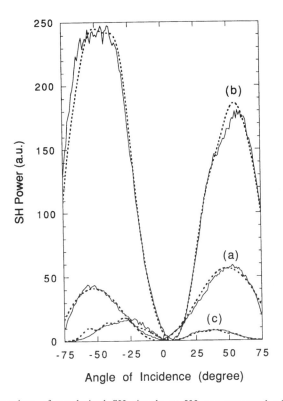

FIG. 4. Comparison of p-polarized SH signals at 532 nm generated with p-polarized incident radiation at 1.064 μm for (a) 0.74 μm; (b) 2.25 μm; and (c) 5.31 μm GaN film. The solid line is the experimental data and the dashed curve is the curve fitting analysis. (Reprinted with permission from Miragliotta et al., 1993.)

TABLE III
NONLINEAR COEFFICIENTS IN GaN FOR VARIOUS FILM THICKNESSES[a]

Film thickness (μm)	$\chi^{(3)}_{xzx}$(pm/V)	$\chi^{(3)}_{zxx}$(pm/V)	$\chi^{(3)}_{zzz}$(pm/V)
0.74	4.41	4.22	−8.82
1.09	4.53	4.53	−9.06
1.30	4.67	4.53	−9.35
2.25	4.16	4.33	−8.33
3.00	4.48	4.41	−8.96
5.31	4.80	4.80	−9.59

[a] Miragliotta et al. (1993).

the 2ω response. In addition to the relative magnitudes, the values $\chi^{(2)}_{zxx}$ and $\chi^{(2)}_{xzx}$ were found to have unity ratio, which was consistent with the wurtzitic relationship in Eq. (6) and the results of Ishidate et al. (1980).

An important contribution of this SHG study was the demonstration of the sensitivity of the 2ω response to the orientation of the crystal symmetry with respect to the laboratory frame of reference. The SHG intensity was capable of detecting small deviations in the relative angle between the c-axis direction of the GaN film and the surface normal of the sample, as was illustrated in Fig. 5. The peak intensity in the two SH transmission profiles indicated that the magnitude of the nonlinearity for these respective films was comparable; however, the asymmetric profile observed for the 1.30 μm film and the signal minimum at a measured angle away from normal incidence was a result of a 2.7° misorientation between the optical axis of the film and the surface normal of the sample.

The series of fixed-frequency SHG measurements were performed under conditions where both the incident and upconverted wavelengths were well below the absorption edge of the GaN semiconducting material. Recently, Chen et al. reported calculations on the nonresonant magnitude of $\chi^{(2)}_{xzx}$ in GaN using a Kohn-Sham local density approximation that incorporates local-field corrections (Chen et al., 1995; Levine and Allen, 1991). The results from the calculations determined values of −3.45 and 5.76 pm/V for $d_{31} = \chi^{(2)}_{xzx}/2$ and $d_{33} = \chi^{(2)}_{zzz}/2$, respectively, which are very close to the experimental values of these coefficients.

In a series of tunable SHG measurements, Miragliotta et al. (1994a) continued the characterization of $\chi^{(2)}$ in GaN by including a range of 2ω photon energies that examined the 2ω response from the visible wavelength region up to the fundamental absorption edge of the semiconductor. In their first tunable investigation, the authors examined the qualitative behavior of the transmitted SHG signal from a GaN sample using a tunable, p-polarized

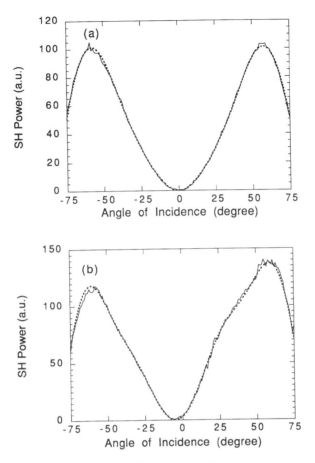

FIG. 5. Comparison of p-polarized SH signals at 532 nm generated with p-polarized incident radiation at 1.064 μm for (a) 1.09 μm; and (b) 1.30 μm films. The solid line is the experimental data and the dashed curve is the curve fitting analysis. (Reprinted with permission from Miragliotta et al., 1993.)

incident source. As is shown in Fig. 6, they observed an SHG signal that displayed an oscillatory behavior as the 2ω photon energy was tuned from ~500 nm to the fundamental absorption edge. Unlike the oscillations observed in the previous fixed frequency SHG measurements, this oscillatory behavior was attributed to the variation in the phase velocity mismatch which resulted from the wavelength dependence of the refractive index at ω and 2ω. The period associated with maximum interference between the second-harmonic waves, that is, the energy separation between successive

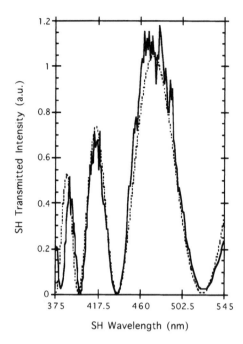

FIG. 6. Transmitted SHG signal from a 5.3 μm GaN film on sapphire. The solid line is the experimental data and the dashed curve is the curve fit analysis. (Reprinted with permission from Miragliotta et al., 1994a.)

SHG minima, was observed to increase with decreasing photon energy due to the increase in the nonlinear optical coherence length of the GaN film [see Eq. (8)]. The authors noted that the p-polarized SHG response was attributed to the coupling of the fundamental beam to the $\chi^{(2)}_{xzx}$ and $\chi^{(2)}_{zzz}$ tensor elements. This complication negated any attempt to determine the magnitude of these elements from the single SHG spectrum.

In a later quantitative study, Miragliotta et al. (1994b) looked more closely at the absorption edge region in order to determine the magnitude and dispersion of $\chi^{(2)}$. Similar to their previous tunable SHG measurement, the authors examined the transmitted SHG signal from a series of samples while limiting the tunable range of the incident source to within the fundamental absorption edge region. As is shown in Fig. 7, the SHG intensity exhibited oscillations below the absorption edge region, which was consistent with the results shown in Fig. 6. However, the phase mismatch oscillations terminated for 2ω photon energies $> E_g$ due to the strong absorption of the SHG field in the film. The investigators determined the

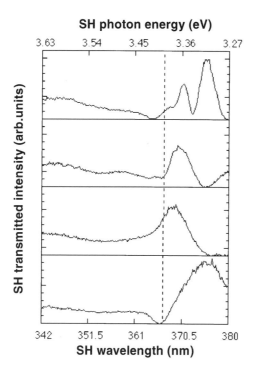

FIG. 7. Transmitted SH intensity from (a) 6.1 μm; (b) 2.6 μm; (c) 1.3 μm; and (d) 0.85 μm GaN film. Incident field was s-polarized and generated SH field was p-polarized. (Reprinted with permission from Miragliotta et al., 1994b.)

magnitude of the second-order nonlinearity of GaN near the absorption edge by comparing the SHG intensity in Fig. 7 with calculated electromagnetic expressions for the transmitted SHG intensity from a thin nonlinear (Miragliotta et al., 1994b) given by

$$I_{2\omega} = 2\pi T_{13}^2 [f_1 \exp(i\phi) + f_2]^* [f_1 \exp(i\phi) + f_2] |\chi_{zxx}^{(2)}|^2 \quad (10)$$

where

$$f_1 = \frac{\dfrac{(Q_4 \cos\theta_t - Q_3 n_t) R_{12} \exp(i\phi)}{(n_{2\omega} \cos\theta_t + n_t \cos\theta_{2\omega})} - \dfrac{(Q_1 n_1 - Q_2 \cos\theta_t)}{(n_1 \cos\theta_{2\omega} + n_{2\omega} \cos\theta_1)}}{1 + R_{12} R_{23} \exp(2i\phi)}$$

$$f_2 = \frac{Q_4 \cos\theta_{2\omega} + Q_3 n_{2\omega}}{n_{2\omega} \cos\theta_t + n_t \cos\theta_{2\omega}}$$

(11)

with

$$Q_{1(3)} = \frac{-E_y^2(0\,(d))n_\omega^2 \sin\theta_\omega \cos\theta_\omega}{n_{2\omega}^2(n_\omega^2 - n_{2\omega}^2)}$$

$$Q_{2(4)} = \frac{-E_y^2(0\,(d))n_\omega^2 \sin\theta_\omega}{n_\omega^2 - n_{2\omega}^2}$$

(12)

In these expressions, n_1, $n_{2\omega}$, and n_t are the refractive indices at 2ω of the incident media, the GaN film, and sapphire substrate, respectively, with n_ω the refractive index of the GaN film at the fundamental frequency; R_{12}, R_{23}, T_{13}, and T_{23} are the Fresnel coefficients for 2ω reflection and transmission across the sapphire/air and GaN/sapphire interfaces, respectively; $E_y(\omega, 0\,(d))$ is the fundamental s-polarized field at the front (back) surface of the sample, and ϕ is the optical phase of the SHG field at the GaN/sapphire interface, which is linearly dependent on the film thickness d. Using previously determined values for the refractive index and the film thickness, the spectrum of $|\chi_{zxx}^{(2)}|$, shown in Fig. 8, was obtained from the SHG

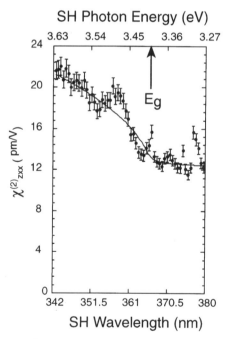

FIG. 8. Spectrum of $|\chi_{zxx}^{(2)}|$ as determined from the SH transmission spectrum in Fig. 7. The solid line is a rough fit to the data. E_g is the bandgap energy. (Reprinted with permission from Miragliotta et al., 1994b.)

spectrum. The experimentally derived result exhibited an enhancement for SHG photon energies above the absorption edge, reaching a maximum value of ~ 21 pm/V. The authors noted that the spectrum of $|\chi^{(2)}_{zxx}|$ increased 70% as the SHG wavelength is varied from 380–342 nm, while the linear dielectric function varies only 14% in the same region. The variational differences were attributed to the photon energy dependence of the linear (E_ω^{-3}) vs nonlinear (E_ω^{-5}) response in the GaN film (Moss *et al.*, 1990, 1991; Sipe and Ghahramani, 1993; Ghahramani *et al.*, 1991).

There has been one wavelength dependent theoretical calculation regarding $\chi^{(2)}$ in wurtzite GaN by Hughes *et al.* (1997) who performed a first principles electronic structure calculation using the full-potential linearized augmented plane-wave method within the local density approximation. Details of this approach are beyond the scope of this chapter, but can be found in Hughes and Sipe (1996). The results for the calculated wavelength dependence of $d_{31} = \chi^{(2)}_{xzx}/2$ are shown in Fig. 9, along with a comparison of the experimental results of this tensor component as measured by Miragliotta *et al.* (1994b). Although the experimental values are approximately twice that of the calculated magnitude, there is reasonable agreement between theory and experiment with respect to the dispersive nature of the susceptibility tensor.

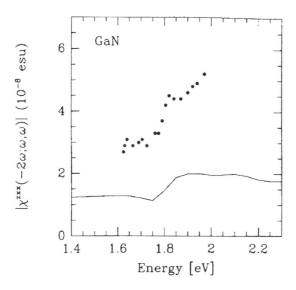

FIG. 9. The absolute value of $|\chi^{(2)}_{zxx}|$ (solid line) plotted with the experimental data from Miragliotta *et al.* (solid circles).

2. The Electro-Optic Effect (LEO)

The LEO, like SHG, is a second-order nonlinear phenomenon that can occur only in noncentrosymmetric material. The nonlinear response results from a dc electric field-induced variation in the refractive index of a nonlinear material, which results in a corresponding variation in the intensity or phase of an optical field that is transmitted through or reflected from the electrified sample. Although this nonlinear optical effect has not received as much attention in the nitride community as has SHG, the potential for electro-optic waveguide modulators with near ultraviolet to mid-infrared bandwidth holds much promise for GaN.

The magnitude of the LEO response is linearly dependent on the amplitude of the applied dc electric field, and is described in terms of a nonlinear polarization given by (Boyd, 1992)

$$P^{(2)}(\omega) = \chi^{(2)}(\omega; \omega, 0)E(\omega)E_{dc} \tag{13}$$

where $E(\omega)$ and E_{dc} are the incident optical and dc electric fields, respectively. A considerable advantage afforded the electro-optic effect relative to frequency conversion is the removal of the phase-matching requirement between the fundamental and the generated fields because the frequency of these two optical waves are identical.

Because the electro-optic effect is a nonlinear field-induced change in the refractive index, it is more convenient to describe this nonlinear effect by the following expression (Boyd, 1992; Shen, 1984; Kaminow, 1974):

$$\Delta\left(\frac{1}{n^2}\right)_{ij} = \sum_k r_{ijk} E_k \tag{14}$$

In this expression, $\Delta(1/n^2)_{ij}$ is the ijth component of the inverse refractive index variation, E_k is the kth component of the dc electric field, and the electro-optic tensor r_{ijk} is related to $\chi^{(2)}_{ijk}$ by the following expression (Salah and Teich, 1991):

$$r_{ijk} = \frac{2\chi^{(2)}_{ijk}}{n^4} \tag{15}$$

The electro-optic coefficient has the same symmetry properties as $\chi^{(2)}$, that is, it vanishes in centrosymmetric media. Also, the relationship between r_{ijk} and $\chi^{(2)}_{ijk}$ assures that the nonvanishing elements in these two tensors for a given noncentrosymmetric material are identical.

To date, the only experimental investigation of the electro-optic coefficient in GaN was performed by Long et al. (1995) for an MOCVD-grown thin-layer film on a c-axis sapphire substrate. The measurement involved an examination of the nonvanishing LEO coefficients in GaN, $r_{31} \equiv r_{311}$ and $r_{33} \equiv r_{333}$ at an optical wavelength of 632.8 nm (1.96 eV). The LEO effect typically generates very small variations in the refractive index ($<10^{-3}$); therefore the authors used an interferometric approach to examine the dc field induced changes. An applied dc electric field was directed normal to the surface of the sample (c-axis of the crystal) by placing the GaN/sapphire sample between two indium tin oxide (ITO) electrodes. The optical beam was transmitted through the GaN/sapphire while the sample was biased with the ITO electrodes. Under this arrangement, the field-induced index variations are given by (Long et al., 1995):

$$n_x = n_y \approx n_o - n_o^3 r_{31} \frac{E_{dc}}{2}$$
$$n_z \approx n_e - n_e^3 r_{33} \frac{E_{dc}}{2}$$
(16)

where n_o and n_e are the values for the ordinary and extraordinary index of refraction, and the dc field is applied along the c-axis of the sample. For an s-polarized input beam, the phase shift in an optical wave that results from the index variations in Eq. (16) is

$$\phi_s = -\frac{\pi}{2} n_o^3 r_{31} E_{dc} d_1 \left\{ \text{sqrt}\left(1 - \sin^2 \frac{\alpha}{n_o}\right)\right\}^{-1}$$
(17)

where d_1 is the thickness of the film and α is the external angle of incidence. For p-polarization, the phase shift is given by

$$\phi_p = -\frac{\pi}{2} n_o^3 E_{dc} d_1 \left\{ r_{33} \sin^2 \frac{\alpha}{n_o} + r_{31}\left[1 - \sin^2 \frac{\alpha}{n_o}\right]\right\} \left\{\text{sqrt}\left(1 - \sin^2 \frac{\alpha}{n_o}\right)\right\}^{-1}$$
(18)

The results for a p-polarized examination of the LEO coefficients are shown in Fig. 10. In this result, the transmitted optical probe was examined as a function of the modulation frequency of the applied electric field. Interestingly, the field-induced phase shift also shows contributions from acoustic resonances that arise from the piezoelectric effect in the GaN sample. The authors were able to evaluate the magnitude of the LEO

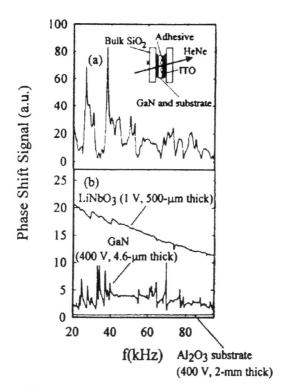

FIG. 10. Electric-field induced optical phase shift of a transmitted 632.8 nm beam vs modulation frequency from a GaN/sapphire/bulk silica system. The modulation voltage was 400 V. The electro-optic response from the GaN sample is compared to a LiNbO$_3$ standard. (Reprinted with permission from Long et al., 1995.)

coefficient by examining the phase shift at modulation frequencies where the piezoelectric contribution was minimal (~ 45 kHz). Under these conditions, the values for $\chi^{(2)}_{zzz}$ and $\chi^{(2)}_{zxx}$ were determined to be 30 ± 5.5 pm/V and 9 ± 1.7 V, respectively. These results are in close agreement with the estimates from Eq. (15), which allow a rough estimate of r_{ijk} from the second-order nonlinearity $\chi^{(2)}$.

Recently, Hughes et al. (1997) performed ab initio calculations of the linear electro-optic susceptibility for photon energies up to the bandgap of GaN. The results were based on a first-principles band structure calculation using the full-potential linearized augmented plane-wave (FLAPW) within the local density approximation. Their results predict a value of 7 pm/V and 6 pm/V for $\chi^{(2)}_{zzz}$ and $\chi^{(2)}_{zxx}$, respectively, at a photon energy of 1.96 eV, which are close to the experimentally determined values of Long et al.

(1995). The authors note that the clamped-lattice approximation in their calculations prevents a strict comparison to experimental values because the latter measurements of the LEO susceptibility measures the total LEO effect.

IV. Third-Order Nonlinear Optical Phenomena

In recent years, technological interest in third-order nonlinear effects in wide-bandgap zinc blende II-VI and III-V semiconductors has received considerable interest due to the large predicted magnitudes of the third-order susceptibilities in these materials (Moss *et al.*, 1990, 1991; Sipe and Ghahramani, 1993; Shaw *et al.*, 1992; Huang and Ching, 1993; Ching and Huang, 1993; Khurgin, 1994; Obeidat and Khurgin, 1995). Third-order nonlinear phenomena involve the interaction of four electromagnetic waves, and are therefore referred to as four-wave mixing (FWM) processes. These interactions are described by a third-order nonlinear susceptibility $\chi^{(3)}$ ($\omega_4 = \omega_1 + \omega_2 + \omega_3$), and occur in media with or without inversion symmetry. Although the third-order susceptibility is typically many orders of magnitude smaller than $\chi^{(2)}$, third-order effects are readily observed with tunable, high-intensity lasers that can resonantly enhance the magnitude of the susceptibility. Typically, the bandwidth of a resonance as observed by a $\chi^{(3)}$ process is narrower than a corresponding $\chi^{(2)}$ phenomenon; as such, it has been recognized for many years that third-order nonlinear phenomena are more spectroscopically versatile than $\chi^{(2)}$ phenomena because any one of the four electromagnetic waves in the interaction can resonantly enhance the third-order susceptibility. Over the past three decades, nonlinear techniques such as coherent anti-Stokes scattering, THG, and electric field-induced SHG (EFISH) have been routinely utilized in fundamental studies of both organic and inorganic materials (Shen, 1984; Levenson and Kano, 1988; Puccetti, 1995; Elsley, 1981).

There has been very little activity regarding theoretical predictions of the third-order nonlinear susceptibility in the wide-bandgap nitrides. Nayak *et al.* (1997) have calculated the nonresonant value of $\chi^{(3)}$ in the cubic zinc blende structure of the metal nitride family, including GaN, using an approach similar to the $\chi^{(2)}$ chemical bond calculations of Levine (1969) and Phillips and Van Vechten (1969). This apprach generates tractable expressions for the nonlinear susceptibility that rely on parameters that define the bond aspect of the crystal, for example, bond length, ionicity, and average energy bandgap. The calculations, however, are nondispersive and cannot provide information on the wavelength dependence of the nonlinearity. The authors chose a basis function set consisting of a linear combination of sp^3 hybrid orbitals for construction of both the valence and conduction bands in the semiconductor. The Bloch-like functions that were constructed from the basis sets were used to calculate the appropriate matrix elements for the $\chi^{(3)}_{xxxx}$ and $\chi^{(3)}_{xxyy}$ elements in the third-order susceptibility tensor. A number of approximations were made in the calculations, namely, the matrix elements were heavily weighted to reflect the contribution from intrasite (95%) rather than nearest-neighbor hybrids (5%), and effects associated with bonding/antibonding states in the valence band were not considered. Using published values for the GaN chemical bond parameters in the expression for $\chi^{(3)}$, the investigators calculated values of 2.24×10^{-21} and 7.7×10^{-22} m^2/V^2 for $\chi^{(3)}_{xxxx}$ and $\chi^{(3)}_{xxyy}$, respectively. Unfortunately, the nondispersive nature of the calculation negates any useful comparison between experiment and theory when the former measurements involve electronic or vibrational resonances.

In the following sections, discussions pertaining to the third-order nonlinear phenomena that have been observed in GaN, namely, THG in Section IV.1, EFISH in Section IV.2, two-photon absorption in Section IV.3, and degenerate FWM (DFWM) in Section VI.4.

1. THIRD-HARMONIC GENERATION (THG)

An electron energy schematic illustration of THG is shown in Fig. 11. In this nonlinear process, three photons of identical frequency ω are mixed to generate a new photon at frequency 3ω. This nonlinear process, like SHG, is parametric because there is no net transfer of energy to the material from the optical fields during the interaction. Analogous to second-order phenomena, the coupling between the incident laser source and the fourth-rank tensor $\chi^{(3)}_{ijkl}$ is dependent on the crystalline structure of the material and the propagation direction and polarization of fundamental light source. In hexagonal GaN, there are 21 nonvanishing elements in the susceptibility THG, where the permutation of the *jkl* indices has no physical significance, the nonvanishing, independent tensor elements are $\chi^{(3)}_{zzzz}$, $\chi^{(3)}_{xxxx}$, $\chi^{(3)}_{xxzz}$, and $\chi^{(3)}_{zzxx}$ (Shen, 1984). Under conditions of normal incidence at a GaN sample with the optical axis parallel to the surface normal, coupling between the incident fundamental beam and the third-order nonlinearity can only occur through the $\chi^{(3)}_{xxxx}$ element. The third-harmonic polarization induced in the GaN sample under normal incidence is (Miragliotta and Wickenden, 1994)

$$P_y^{(3)} = \chi^{(3)}_{xxxx}(3\omega; \omega, \omega, \omega)E_y^3(\omega) \qquad (19)$$

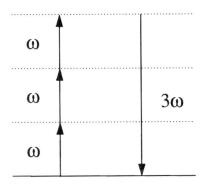

FIG. 11. Electron energy schematic illustration of third-harmonic generation. The three incident photons (ω) couple to the third-order susceptibility to generate a new photon of frequency 3ω. The solid line is the electronic ground state while the dashed lines represent virtual transitions in the material.

where the fundamental field in the film, $E_y(\omega)$, is directed along the y-axis of the crystal, and the xy-plane of the crystal is aligned with the surface plane GaN.

Miragliotta and Wickenden (1994) have examined $\chi^{(3)}_{xxxx}$ in GaN using a tunable laser source, which generated a third-harmonic signal near the fundamental absorption edge of the semiconductor. Figure 12 shows the THG transmitted intensity from a 6.08 μm thick film in the 3ω photon

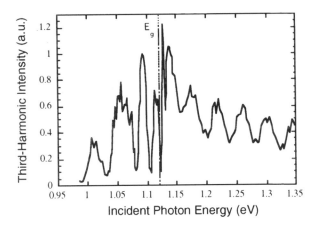

FIG. 12. The THG signal from a 6.08 μm GaN film on sapphire as a function of the incident photon energy. (Reprinted with permission from Miragliotta and Wickenden, 1994.)

energy range of 3.00–4.05 eV. For 3ω energies below E_g, the transmitted signal exhibited oscillations that were attributed to wavelength-dependent variations in phase velocity mismatch between the free and forced third-harmonic waves. Similar to the SHG response, the phase mismatch oscillations terminated for 3ω photon energies $>E_g$ due to the strong absorption of the THG field in the film.

Similar to the SHG analysis in Section III.1, the dispersion and absolute magnitude of $\chi^{(3)}_{xxxx}$ can be determined from the THG intensity, which is related to the nonlinear susceptibility by (Miragliotta and Wickenden, 1994; Burns and Bloembergen, 1971; Shelton and Shen, 1972)

$$I_{\text{THG}} = \frac{cf_1^* f_1 |\chi^{(3)}_{xxxx}|^2}{8\pi} \tag{20}$$

where

$$f_1 = \frac{4\pi f_2 T_{31}}{n_\omega^2 - n_{3\omega}^2} \tag{21}$$

$$f_2 = \frac{f_3 \exp(i\phi_f)(1 + R_{23})}{1 + R_{12} R_{23} \exp(2i\phi_f)} + f_4 \tag{22}$$

$$f_3 = \frac{E_y^3(\omega, d)\left[1 - \dfrac{n_\omega}{n_T}\right] R_{12} \exp(i\phi_f)}{\left[1 + \dfrac{n_{3\omega}}{n_T}\right]} - \frac{E_y^3(\omega, 0)[1 + n_\omega]}{[1 + n_{3\omega}]} \tag{23}$$

and

$$f_4 = \frac{E_y^3(\omega, d)[n_\omega + n_{3\omega}]}{[n_T + n_{3\omega}]} \tag{24}$$

In these expressions, $n_{\omega(3\omega)}$ is the refractive index of the GaN film at the fundamental (third-harmonic) frequency and n_T is the refractive index of the sapphire substrate at 3ω; R_{12}, R_{23}, and T_{31} are the Fresnel coefficients that account for 3ω reflection at the air/GaN and GaN/sapphire interfaces and transmission across the sapphire/air interface; $E_y(w, 0(d))$ is the fundamental field at the front (back) surface of the film and ϕ_f is the optical phase of the third harmonic field at the GaN/sapphire interface. Figure 13 shows the wavelength dependence of $|\chi^{(3)}_{xxxx}|$ as obtained from the spectrum in Fig. 12.

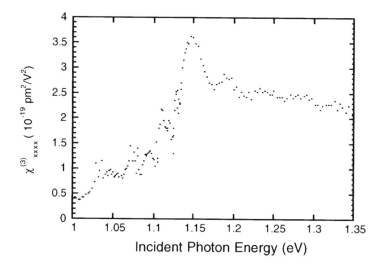

FIG. 13. The magnitude of $\chi^{(3)}_{xxxx}$ in GaN as a function of the photon energy of the incident laser source. (Reprinted with permission from Miragliotta and Wickenden, 1994.)

These results show that $\chi^{(3)}_{xxxx}$ exhibits a resonant enhancement for 3ω photon energies near the absorption edge, peaking at $3.8 \times 10^{-19}\,\mathrm{m^2/V^2}$ with a monotonic decrease in value with increasing energies. The absorption edge resonance is much more noticeable in the $\chi^{(3)}_{xxxx}$ spectrum as compared to the same resonance in the spectrum of $\chi^{(2)}_{zxx}$ in Fig. 8, which is due to the $[E_{3\omega} - E_{\mathrm{gap}}]^{-3}$ and $[E_{2\omega} - E_{\mathrm{gap}}]^{-2}$ dependence for the third- and second-order susceptibilities, respectively.

The magnitude of $\chi^{(3)}_{xxxx}$ at photon energies just below the absorption edge is approximately $3 \times 10^{-20}\,\mathrm{m^2/V^2}$, which is nearly an order of magnitude of the theoretical prediction of Nayak *et al.* (1997) for the nonresonant value of this tensor element in GaN (Malouin *et al.*, 1998). This is most likely due to the proximity of the experimental measurements to the absorption edge of GaN, where the resonant contribution to $\chi^{(3)}_{xxxx}$ negates any reasonable comparison to theory. Unfortunately, it is not possible at this time to compare the experimental results directly with a wurtzite band structure calculation of $\chi^{(3)}_{xxxx}$. Miragliotta and Wickenden (1994) compared the results of $\chi^{(3)}_{xxxx}$ in GaN to previous band structure calculations by Ghahramani *et al.* (1991) of $\chi^{(3)}$ in zinc blende ZnSe, shown in Fig. 14. ZnSe was chosen for the comparison because it has a direct valence to conduction band transitions at the fundamental absorption edge. Also, the density of states in ZnSe and GaN have upper valence band states that are dominated by the *p*-orbitals of the anion (Se or N) and a lower conduction band region

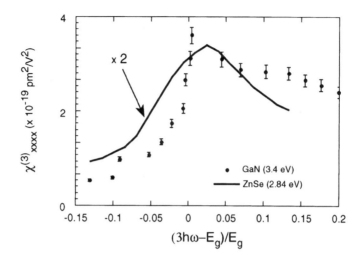

FIG. 14. Comparison of the experimental value of $\chi^{(3)}_{xxxx}$ in GaN with the theoretical prediction of this nonlinearity in zinc blende ZnSe. The calculated spectra (solid line), reprinted with permission from Ghahramani et al. (1991) are scaled by a factor of 2 with respect to the GaN data. The $\chi^{(3)}_{xxxx}$ values are plotted against the normalized (E_g) difference between the 3ω photon and absorption edge energies (energy values in parenthesis). (Reprinted with permission from Miragliotta and Wickenden, 1994.)

dominated by the s-orbitals of the cation (Zn or Ga). From the comparison in Fig. 14, $\chi^{(3)}_{xxxx}$ is resonantly enhanced at the absorption edge in both materials, with little contribution from higher interband transitions at this energy. The dispersion in GaN is similar to the wavelength dependence in ZnSe for photon energies prior to and including the absorption edge. The E_1 critical point resonance tends to influence the dispersion in the E_0 region more strongly in ZnSe because this optical resonance is much closer to the absorption edge in ZnSe ($E_1 \sim 4.8\,\text{eV}$) than in the GaN ($E_1 \sim 7\,\text{eV}$). The $\chi^{(3)}_{xxxx}$ spectrum in GaN does not decrease as rapidly above the absorption edge due to the weak overlap of these two resonances; $\chi^{(3)}_{xxxx}$ is approximately twice as large in GaN as compared to the predicted values in ZnSe despite the lower bandgap energy of the latter, which is most likely due to the higher ionicity in the II-VI vs III-V wide-bandgap semiconductors.

2. ELECTRIC FIELD-INDUCED SECOND-HARMONIC GENERATION (EFISH)

The ability of an applied dc electric field to perturb the optical properties of semiconductors has been utilized in many forms of nonlinear optical spectroscopy (Aspnes and Rowe, 1972; Aspnes and Studna, 1973; Aspnes, 1973). One type of dc electric field-dependent nonlinearity that occurs in semiconductors is EFISH (Lee et al., 1967; Levine and Bethea, 1976; Singer and Garito, 1981). Although this nonlinear process generates a second-harmonic field, it is a third-order effect because it involves the coupling of two optical fields of frequency ω with one dc electric field. Similar to other FWM processes, the EFISH response is described by the third-order susceptibility, $\chi^{(3)}(2\omega;\omega,\omega,0)$, which means that the generation of 2ω radiation can occur in either centro- or noncentro-symmetric media.

In an EFISH measurement from a noncentrosymmetric media like GaN, the 2ω polarization response from an electrified sample contains a second- and third-order nonlinear contribution. In an EFISH experimental study by Miragliotta and Wickenden (1996) the dc electric field was directed along the optical axis of the GaN sample with the x-y plane of the crystal rotated about the optical axis until the plane of incidence coincided with the x-z plane of the crystal. Using s-polarized incident radiation, the nonlinear polarization was induced along the optical axis (p-polarized) and contained a single $\chi^{(2)}$ and $\chi^{(3)}$ tensor element:

$$P_z^{nl}(z, 2\omega) = \chi_{eff}^{nl}(z)E_y^2(z, \omega) \tag{25}$$

where the effective nonlinearity $\chi_{eff}^{nl}(z)$ is given by (Miragliotta and Wickenden, 1996)

$$\chi_{eff}^{nl}(z) = \chi_{zxx}^{(2)}(2\omega;\omega,\omega) + 3\chi_{zxxz}^{(3)}(2\omega;\omega,\omega,0)E_s(z) \tag{26}$$

In these two expressions, $E_y(z, \omega)$ is the y component of the incident field and $E_s(z)$ is the dc electric field directed along the optical axis. The reflected SH intensity that is radiated from this nonlinear polarization contains a cross term, which is dependent on the phase difference ψ between the second- and third-order susceptibility

$$I_{2\omega} \propto |\chi_{zxx}^{(2)}|^2 + |3\chi_{zxxz}^{(3)}E_s(z)|^2 + 6E_s(z)|\chi_{zxx}^{(2)}\chi_{zxxz}^{(3)}|\cos\psi \tag{27}$$

The EFISH study by Miragliotta and Wickenden (1996) was performed at an electrified GaN interface, where the SHG photon energy was tuned through the fundamental absorption edge. The GaN film was placed in an electrochemical solution, which produced a barrier at the interface via an electron transfer from the n-type sample to the electrolyte solution (Reineke and Memming, 1992). The depletion region in the GaN was characterized by a dc electric field that decreased linearly with distance away from the surface. Control and modulation of the dc electric field was accomplished by varying the potential difference between the bulk regions of the solution and the GaN, where the field in the sample was expressed by (Nussbaum, 1981)

$$E_s(z) = \frac{qN_d}{\varepsilon_0 \varepsilon_s}(z - z_d) \qquad (28)$$

with the depletion layer depth given by

$$z_d = \left[\frac{2\varepsilon_0 \varepsilon_s}{qN_d}\left(V_{\text{bias}} - V_{\text{fb}} - \frac{kT}{q}\right)\right]^{1/2} \qquad (29)$$

In these two expressions, z is the distance into the GaN sample, z_d is the depletion layer thickness, V_{bias} is the applied potential, V_{fb} is the flat band potential, N_d is the dopant concentration, and ε_s is the dielectric constant of GaN.

Wavelength dependent EFISH results are shown in Figs. 15 and 16, where

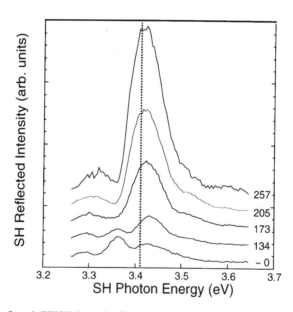

FIG. 15. Reflected EFISH intensity from an electrolyte/GaN interface as a function of the 2ω photon energy. The dashed line represents the position of the absorption edge in the unbiased sample. The number to the right of each spectrum is the value of the dc field at the GaN surface in kV/cm. (Reprinted with permission from Miragliotta and Wickenden, 1996.)

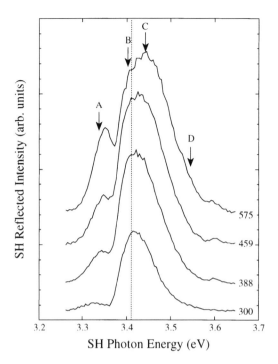

FIG. 16. Reflected EFISH intensity from an electrolyte/GaN interface as a function of the SHG photon energy. The number to the right of each spectrum is the value of the dc field at the GaN surface in kV/cm. (Reprinted with permission from Miragliotta and Wickenden, 1996.)

the dashed vertical line in both graphs denotes the position of the fundamental absorption edge, $E_g \sim 3.41$ eV. In the flatband spectrum, shown in Fig. 15 (0 V/cm), the two weak peaks observed at SHG photon energies just below the absorption edge position were attributed to the wavelength dependence of the phase mismatch between the forced and free SHG fields in the film. According to Eq. (27), the zero dc field spectrum is dependent only on the intrinsic second-order susceptibility of the GaN, and as such it exhibited a wavelength dependence that was consistent with the previous SHG measurements from unbiased samples in Section III.1. The application of a reverse bias dc field produces significant changes in both the magnitude and dispersion of the SHG response near the absorption edge. At a surface dc field of 134 kV/cm, the EFISH response is characterized by a single resonance at 3.43 eV with a linewidth value of ~ 50 meV. The resonance increases in amplitude as the surface field is increased to 257 kV/cm with a slight decrease in energy position (3.41 eV) and a broadening in the

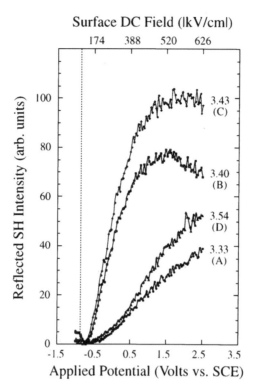

FIG. 17. A series of fixed frequency EFISH reflectance scans from a GaN/electrolyte interface as a function of the applied potential. The SH photon energy of each scan (A to D) is given at the right of each scan in eV. (Reprinted with permission from Miragliotta and Wickenden, 1996.)

linewidth. As was noted by the authors, the sharp spectral response in the 2ω response due to the $\chi^{(3)}$ nonlinearity provides EFISH with a spectroscopic capability that is far superior to that of the intrinsic $\chi^{(2)}$ response shown in Section III.1.

In addition to the wavelength dependent results, a series of fixed frequency EFISH measurements, shown in Fig. 17, were performed to determine the magnitude of $\chi^{(3)}_{zxxz}(2\omega;\omega,\omega,0)$ near the fundamental absorption edge. Details of the data analysis are given in Miragliotta and Wickenden (1996); however, Eq. (27) shows that the magnitude of $\chi^{(3)}_{zxxz}(2\omega;\omega,\omega,0)$ and phase shift ψ can be evaluated from the EFISH signal when the magnitude of the dc electric field and $\chi^{(2)}_{zxx}(2\omega;\omega,\omega)$ are established prior to the measurement. The dc field is evaluated from Eq. (28) while the value for

$\chi^{(2)}_{zxx}(2\omega;\omega,\omega)$ is obtained from the SHG investigation in Section III.1 (Fig. 8). In modeling the field dependent SHG signal in Fig. 17 at 1.72 eV, a value of $5.3 \times 10^{-19}\,\text{m}^2/\text{V}^2$ was determined for $\chi^{(3)}_{zxxz}(2\omega;\omega,\omega,0)$ with a relative phase difference of 0.55 radians. This result is consistent with the previous absorption edge measurement of $\chi^{(3)}_{xxxx}(3\omega;\omega,\omega,\omega) = 3.8 \times 10^{-19}\,\text{m}^2/\text{V}^2$ as determined by THG.

3. TWO-PHOTON ABSORPTION (TPA)

Illustrated by the electron energy diagram in Fig. 18, TPA is a nonlinear process that involves an optical excitation in a material system from the ground to an excited state by the simultaneous absorption of two incident photons. Unlike the linear absorption case, the TPA cross section σ is linearly proportional to the incident source intensity

$$\sigma = \beta(\omega)I(\omega) \qquad (30)$$

where $\beta(\omega)$ is the TPA coefficient in units of cm/GW. Because the transition rate R from the ground state to the excited state is proportional to the product of the absorption cross section and incident intensity, the TPA rate is

$$R = \frac{\beta(\omega)I^2}{\hbar\omega} \qquad (31)$$

In most materials, attenuation of the incident beam by the $\beta(\omega)$ term is negligible until very high intensities are achieved in the illuminated materials

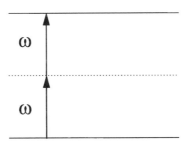

FIG. 18. Electron energy diagram of two-photon absorption. The two incident photons (ω) are simultaneously absorbed by the nonlinear media at an energy of 2ω. The process is nonparametric because there is a net transfer of energy to the material.

(Mizrahi et al., 1989; Aitchison et al., 1990; Hutchings and Van Stryland, 1992; Murayama and Nakayama, 1995). However, the potential utilization of GaN for nonlinear optical waveguide applications will likely require the use of sub-bandgap intensity sources in excess of 1–10 MW/cm² in order to achieve high conversion efficiencies. In this regime, the nonlinear absorption can be significant, particularly at photon energies in the midgap region where the linear absorption coefficient is very small.

The first investigation of $\beta(\omega)$ in GaN was performed by Miragliotta and Wickenden (1995) using a transient photoconductive approach that was previously demonstrated in TPA studies of GaAs and other semiconducting material (Yee, 1969; Laughton et al., 1994). In the photoconductivity approach, the relationship between the absorption coefficients and the photogenerated carrier density is given by

$$\Delta n = \Delta p = \frac{\{\alpha(\omega)I_\omega^n + 0.5\beta(\omega)I_\omega^2\}\tau_p}{\hbar\omega} \tag{32}$$

In Eq. (32), $\alpha(\omega)$ is the linear absorption coefficient, τ_p is the laser pulse width, $\hbar\omega$ is the incident photon energy, and the exponent n in the one-photon absorption term ranges between 0.5 and 1.0. The sublinear exponent is an empirical result from previous studies of sub-bandgap photoconductivity, where single photon-induced conductivity is attributed to the excitation of carriers from defect or trap states in the bandgap region into the conduction band (Schmidlin, 1977; Rose, 1963).

In the electrochemical approach to measuring the photocarrier density, it is necessary to measure the transient photocurrent across the semiconductor sample. As is shown by the photoconductivity results from a GaN/electrolyte interface in Fig. 19, the intensity dependence of the peak photocurrent was measured for a range of photon energies that scanned both above and below the midgap energy, $E_{g/2} = 1.71$ eV. The log plot of the peak photocurrent exhibits a marked increase as the incident photon energy is increased above the midgap energy of the semiconductor. In this energy regime, the photocurrent displays both one- and two-photon contributions as evidenced by the presence of two slopes in the data.

The relationship between the photocarrier concentration in Eq. (32) and the peak photocurrent is derived by utilizing an equivalent RC circuit model for the GaN depletion layer/electrolyte system, as has been developed by Wilson et al. (1985) for photocurrent investigations at a CdSe/electrolyte interface. The peak photocurrent is expressed by

$$i_p = \frac{Q_{sc}}{R_s C_{sc}} \tag{33}$$

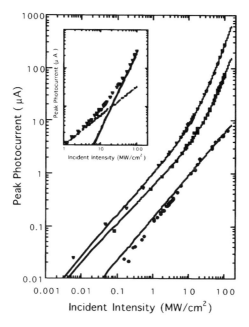

FIG. 19. Peak values of the transient photocurrent vs the incident laser intensity for 1.63 eV (●), 1.77 eV (■), and 1.90 eV (▼) photon energy inputs. The inset shows the deconvolution of the total peak photocurrent using the 1.90 eV incident beam into one- and two-photon contributions (dashed and solid line, respectively). (Reprinted with permission from Miragliotta and Wickenden, 1995.)

where R_s represents the series resistance of the cell (electrode, electrolyte, and external measuring resistor) and Q_{sc} was the initial photoinduced charge in the depletion layer. For the sub-bandgap excitation, the peak photocurrent due to the initial charge separation in the depletion layer is given by

$$i_p = \frac{e\{\alpha(\omega)I_\omega^n + 0.5\beta(\omega)I_\omega^2\}\tau_p V}{R_s C_{sc} \hbar\omega} \quad (34)$$

where V is the irradiated volume in the depletion layer, and e is the electronic charge. After the appropriate curve fitting of the data to Eq. (34), the value for the exponent n for the sub-bandgap contribution was found to range from 0.71–0.75 as the incident photon energy was varied between 1.96 and 1.63 eV. The intensity dependence of the 1.90 eV photocurrent data can also be deconvoluted into one- and two-photon contributions, as is shown in the inset of Fig. 19. The plot illustrates that the crossover between one-photon absorption and two-photon dominance occurs at an incident intensity of 18.5 MW/cm^2.

The dispersion and magnitude of the TPA coefficient were obtained from the curve fitting results of the data in Fig. 19, which are shown in Fig. 20 at a fixed incident intensity of 50 MW/cm^2. As the photon energy of the incident beam crosses the midgap energy, the TPA coefficient exhibits a sharp increase in magnitude. Above $E_{g/2}$, the TPA coefficient increases to a value of ~ 1.5 cm/GW at $\hbar\omega = 1.94$ eV. The data in Fig. 20 were found to be consistent to a previous model of $\beta(\omega)$, which was derived from second-order perturbation theory using a simple Kane parabolic band structure model for valence and conduction bands (Hutchings and Van Stryland, 1992). In this model, $\beta(\omega)$ has the following form:

$$\beta(\omega) = K_{\text{pb}} \frac{\sqrt{E_p}}{n_0^2 E_{\text{gap}}^3} F_2\left(\frac{\hbar\omega}{E_{\text{gap}}}\right) \qquad (35)$$

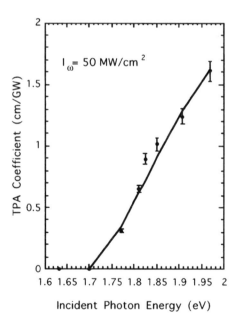

FIG. 20. Two-photon absorption coefficient in GaN vs the photon energy of the incident light source. Solid line through the data is a fit to the data using Eq. (35) in the text. (Reprinted with permission from Miragliotta and Wickenden, 1995.)

where the function F_2 is given by

$$F_2(x) = \frac{(2x-1)^{3/2}}{(2x)^5} \tag{36}$$

In these expressions, K_{pb} is a material independent constant equal to 1940 cm/GW (eV), E_p is related to the interband momentum matrix element, and n_0 is the linear refractive index. The solid line in Fig. 20 is the fit of the experimental data to Eq. (35). The only parameter in the model that was varied in the curve fit analysis, E_p, was found to be 41 eV. This value was nearly twice that of the theoretical predictions; however, the model does not include any effects such as a two-photon absorption due to defects or other lattice imperfections.

4. Degenerate Four-Wave Mixing

When all four waves in the FWM interaction are at the same frequency, the mixing process is referred to as degenerate FWM (DFWM) (Malouin et al., 1998; Yariv and Pepper, 1977; Prasad and Williams, 1991). Although the response arises from an inherently weak third-order nonlinearity, DFWM has advantages over incoherent optical probes such as photoluminescence and linear absorption in that multibeam mixing has the ability to perform both high spectral resolution and time-resolved measurements (Akimoto et al., 1997; Rodenberger et al., 1995; Kaminski et al., 1997; Prasad, 1992). Typically, the calculations that describe DFWM can be very complicated; however, the DFWM process can be physically described by the simple schematic shown in Fig. 21. In this illustration, three input waves, k_1, $-k_1$, and k_{in}, coherently mix in the nonlinear media via a third-order nonlinearity to generate a fourth wave, k_s, at the same frequency of the three incident fields. In the most commonly utilized DFWM configuration, two of the three input waves (k_1, k_{in}) coherently mix in the nonlinear media to generate a static, spatially periodic grating in the material. The grating diffracts the third input wave ($-k_1$) into two separate beams: one which is counterpropagating to the input wave k_{in}, and another which is directed toward $k_i - 2k_1$. Typically, the counterpropagating beam is examined in most experimental setups because it is phase matched to the input beam at all frequencies. The generation process involves the interaction of three input waves; it is customary to relate the phenomena to a third-order nonlinear polarization (Shen, 1984)

$$\mathbf{P}_s^{nl}(\mathbf{k}_s, \omega) = \chi^{(3)}(\omega; \omega, \omega, -\omega) : \mathbf{E}_l(\mathbf{k}_l, \omega)\mathbf{E}_l(-\mathbf{k}_l, \omega)\mathbf{E}_i^*(\mathbf{k}_i, -\omega) \tag{37}$$

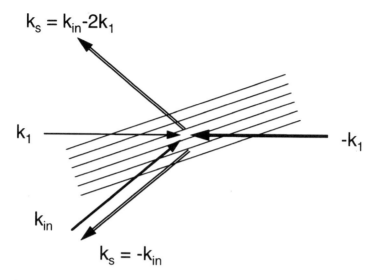

FIG. 21. The DFWM resulting from scattering of an incident wave $(-k_1)$ by the static grating (dashed parallel lines) formed by the other two incident waves, k_1 and k_{in}. The scattered wave $k_s = -k_{in}$ is always phase-matched for DFWM, while the $k_s = k_{in} - 2k_1$ wavevector may not satisfy this condition.

where $\mathbf{P}_s^{nl}$ is the polarization source that generates the diffracted field, $\mathbf{E}_l(\mathbf{k}_l)$, $\mathbf{E}_l(-\mathbf{k}_l)$, and $\mathbf{E}_i^*(\mathbf{k}_i)$ are the three input fields, and $\chi^{(3)}(\omega; \omega, \omega, -\omega)$ is the third-order susceptibility. Similar to the discussion regarding LEO in Section III.2, the nonlinearity and incident waves in Eq. (37) are at the same frequency, which allows the nonlinearity to be described as a variation in the refractive index

$$\Delta n(\omega) = n_2 I_0(\omega) \tag{38}$$

In Eq. (38) I_0 is the incident intensity of the pump beam and n_2 is the nonlinear refractive index, in units of cm^2/GW, given by (Boyd, 1992)

$$n_2 = \frac{3\pi \chi^{(3)}}{n_0} \tag{39}$$

where n_0 is the unperturbed refractive index. Therefore, the DFWM process shown in Fig. 21 is interpreted as a diffraction process that arises from an optically induced refractive index grating. The diffracted wave from the index grating is particularly attractive for spectroscopic and temporal investigations, because the intensity of the diffracted beam is dependent on the magnitude of the refractive index grating. The index grating is strongly influenced by the spectral and transient behavior of electronic and vibrational excitations in the material, because the magnitude of $\chi^{(3)}$ is triply resonant under these conditions.

Several DFWM studies have been performed on epitaxial GaN in an attempt to study the spectroscopic and temporal behavior of electronic excitations either near or above the fundamental absorption edge. The first DFWM investigation was performed by Taheri *et al.* (1996) as a probe of the transient behavior of photoexcited carriers in the conduction band. The authors used a sub-bandgap, fixed frequency pump and probe source at 532 nm (13 ps pulsewidth) to generate and examine, respectively, the temporal behavior of an optically induced index grating in a thin epitaxial GaN film. A plot of the diffracted signal intensity vs probe delay is shown in Fig. 22, where the time delay is with respect to the temporal position of the pump beam. The primary contribution to the initial change in the refractive index was attributed to free carrier generation involving either band-to-band or band-to-trap excitations. The nanosecond relaxation of the index grating, however, was believed to be dominated by the free carrier recombination to trap states in the gap region, because the observed lifetime was too short for band-to-band recombination. The authors related the efficiency of refractive

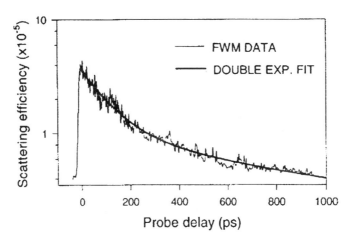

FIG. 22. Scattering efficiency as a function of the probe beam arrival delay. Solid line is a double exponential fit to the data. (Reprinted with permission from Taheri *et al.*, 1996.)

index diffraction grating to the magnitude of the index change by the following expression

$$\eta(t) = \sin^2(k\Delta\, n(t)L/2)$$
$$\approx (k\Delta n(t)L/2)^2 \tag{40}$$

where η is the diffraction efficiency, Δn is defined by Eq. (37), k is the wavevector of the probe beam, and L is the thickness of the sample. Using Eqs. (38) and (39), a peak value of $1 \times 10^{-3}\,\text{cm}^2/\text{GW}$ was determined for n_2, which is orders of magnitude larger than that observed in similar nonresonant DFWM studies of wide-bandgap semiconductors such as ZnSe [$6.7 \times 0^{-5}\,\text{cm}^2/\text{GW}$ (Sheik-Bahae et al., 1990)]. The authors postulated that the large value for n_2 at sub-bandgap excitation energies was due to single photon carrier generation from trap states in the energy gap region to the empty states in the conduction band. The hypothesis of a one-photon absorption was supported by the observation of a scattering efficiency that exhibited an $I^{3.1}$ dependence on the incident intensity rather than the expected I^4 dependence for two-photon excitations.

Two DFWM measurements have extended the work of Taheri et al. (1996) to include temporal and spectral examinations at the absorption edge region of wurtzite GaN. Pau et al. (1997) and Zimmermann et al. (1997) independently performed DFWM measurements to investigate the temporal dephasing characteristics of excitonic resonances near the fundamental absorption edge. Both studies utilized broadband (10 MeV), tunable fs laser sources to examine the DFWM response from the A, B, and C excitons as opposed to the conventional approaches of incoherent linear spectroscopy. In the latter techniques, including linear absorption and photoluminescence, the observed optical signals are typically influenced by inhomogeneous broadening of the resonances; however, the optical configuration used in the DFWM studies eliminated the effects of inhomogeneous broadening. In the work by Pau et al. (1997), the authors initially examined the decay time and linewidth of the A exciton resonance as a function of exciton density, which is shown in Fig. 23, and compared the time-dependent DFWM signal to the various exciton dephasing mechanisms. In this measurement, the photon energy of the incident pump and probe sources was carefully chosen for strong interaction with the A exciton, with little or no interaction with the higher energy B exciton. The DFWM linewidth exhibited a linear increase with exciton density in addition to a decay time of $\sim 3\,\text{ps}$ at low exciton density. These characteristics indicated that the dominant dephasing mechanism in the GaN sample was exciton-exciton scattering, where the degree of carrier dephasing from free-carrier/exciton scattering was negligible because the carrier concentration and mobility in the top layer of the GaN

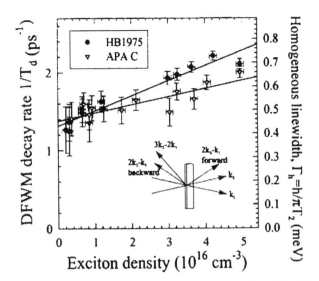

FIG. 23. The DFWM signal decay rate and the homogeneous linewidth of the A exciton for different laser intensities in two different samples. The solid line is a fit to the data. The inset shows the experimental setup. (Reprinted with permission from Pau et al., 1997.)

film were too low. Interestingly, the DFWM signal from the GaN/sapphire interface was found to decay more rapidly than the corresponding response from the GaN/air interface. This was attributed to the poor quality of the latter interface, which provided rapid decay mechanisms for the carrier population.

Pau et al. (1997) also used the transient DFWM response to detect quantum beating between the A and B excitons under conditions where the photon energy of the pump laser was chosen to excite both excitonic transitions. The results shown in Fig. 24 show the transient decay of the DFWM signal from a GaN/air interface at an excitation energy of 3.494 eV, which was of sufficient intensity to generate an average exciton density of 10^{16} cm^{-3}. At this photon energy, where excitation of the A and B exciton transitions can occur, the DFWM temporal decay exhibits an oscillatory behavior that is due to the quantum beating between the two exciton transitions. The beating period of ~500 fs corresponds to an exciton energy difference of 8.2 MeV, which is very close to the value of 8 MeV as determined by photoluminescence on the same sample. However, the accuracy of the temporal DFWM measurement is clearly superior to that of the spectral luminescence result because the latter is limited by the resolution of the source and detector. Finally, the authors also noted that

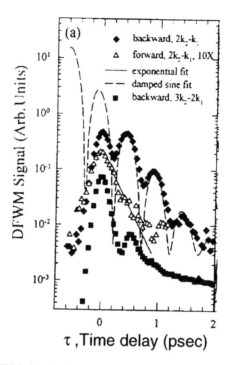

FIG. 24. The DFWM signal taken in the forward and backward directions. The dashed curve shows the fit of the data to a damped sine and the solid curve shows the fit of the data to an exponential. Also shown is the backward fifth-order signal from the GaN sample. (Reprinted with permission from Pau et al., 1997.)

at high exciton densities and short delay times, the DFWM signature indicated the presence of higher-order nonlinear processes.

The study by Zimmermann et al. (1997) was very similar to the Pau et al. (1997) investigation. Using similar experimental operating conditions, the investigators observed the energy splitting of ~ 8 MeV between the A and B exciton. In addition to the excitonic study, the authors were able to selectively couple into the biexciton by using an appropriately orientated optical polarization of the incident pump and probe source. The results of the optical polarization study are shown in Fig. 25, where the biexciton excitation is clearly evident in the cross-polarized excitation scheme. Specifically, the biexciton energy was determined by tuning the excitation source to the low-energy tail of the A exciton resonance and measuring the power spectrum of the DFWM. The solid and dashed line spectra represent the DFWM response for cross-polarized and circularly polarized excitation,

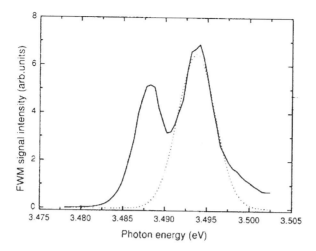

FIG. 25. Spectrum of the DFWM signal at an excitation energy of 3.494 eV. The solid line is for cross-polarized linear polarization of the pump/probe excitation while the dashed line is for cocircular pump/probe excitation. (Reprinted with permission from Zimmermann et al., 1997.)

respectively. The disappearance of the peak in the latter configuration is consistent with polarization selection rules for biexciton transitions, and confirms the peaks origin as biexcitonic. The energy separation of 5.7 MeV was also consistent with previous photoluminescence data (Kawakami et al., 1996).

V. Potential Devices

The ability to incorporate the relatively large nonlinearities in GaN into practical devices will rely on the ability to enhance the conversion efficiency of the device relative to the bulk material. Fortunately, GaN can be deposited on a substrate in the form of an optical waveguide with high transparency throughout the infrared and near ultraviolet regions. One way to increase the nonlinear conversion efficiencies in the GaN is to inject the incident light into the sample and propagate the incident beam along the surface length of the film. Optical waveguiding in films and fibers has received considerable attention in past years, in part due to the interest in optical communications. The structure of optical waveguides is ideal for nonlinear interactions because the thin film or fiber structures provide beam confinement over propagation distances that can be on the order of centimeters (Laurell and Arvidsson, 1988; Tien et al., 1990; Harada et al., 1991).

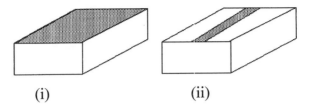

FIG. 26. Schematic illustration of the (i) planar and (ii) channel waveguide structure on a dielectric substrate. The shaded region is the confining thin film deposit.

The two most popular planar structures for integrated optical waveguides are shown in Fig. 26. These structures consist of thin film slabs or channels that confine the guided optical wave in either two or one dimensions, respectively, with the waveguide dimensions on the order of the wavelength of the field. The channel guide is the preferred structure because guided waves can actually propagate without beam diffraction for distances that are limited only by the attenuation effects of scattering and absorption. The ability to confine optical fields of moderate input powers can lead to surprisingly large intensities over long (~ 1 cm) propagation lengths in channel guides. For example, if one watt of power is injected into a 1 μm × 1 μm rectangular GaN channel waveguide, this leads to an intensity of 100 MW/cm^2. This intensity level is more than sufficient to generate a detectable response from the nonlinearities in most narrow and wide bandgap compound semiconductors.

Another characteristic of a waveguide structure is related to its ability to modify the dispersive properties of a confined electromagnetic wave relative to a plane wave propagation in the same material. The confined wave has a refractive index that is dependent on the geometric properties of the structure, for example, film thickness. In the nonlinear process of SHG, the dispersive properties of the waveguide can, in fact, be tailored so that the refractive index at the fundamental and upconverted frequencies is equal, which creates phase-matching in the waveguide (Conwell, 1973; Haus and Reider, 1987; Regener and Sohler, 1988; Bratz et al., 1990).

To illustrate the SHG enhancing characteristics of a waveguide structure, it is necessary to examine the expression for conversion efficiency in a wurtzite semiconductor crystal structure when the c-axis of the crystal is parallel to the surface normal. Assuming that the guided fundamental and SHG waves are propagating along the surface, the conversion efficiency is given by (Blanc et al., 1995)

$$\eta = \frac{P^{2\omega}}{P^{\omega}} = R\left[\frac{2\omega^2 \mu_0^{3/2} \varepsilon_0^{1/2} N_{\text{eff}}^3 d_{33}^2 L^2 P(\omega, 0) \sin^2(\beta^{mp} L/2)}{n_{2\omega}^2 n_\omega^4 (\beta^{mp} L/2)^2}\right] \quad (41)$$

where L is the propagation length, $P(\omega, 0)$ is the fundmental power at the film surface, n_ω and $n_{2\omega}$ are the mean values of the ordinary and extraordinary indices of refraction, d_{33} is the nonlinear coefficient, and N_{eff} is the mean value of the effective indices of the waveguide modes. β^{mp} is the phase mismatch between the two waveguide modes and is given by $\beta_{2\omega}^m - \beta_\omega^p$, where $\beta_{2\omega}^m$ and β_ω^p are the wavevectors of the waveguide modes at 2ω (m) and ω (p), respectively. A critical condition regarding the ω and 2ω fields in waveguide is that their respective modes must be different, that is, $m \neq p$, when normal dispersion exists for the refractive index of the bulk material (Tien and Ulrich, 1970; Stegeman and Stolen, 1989). It is important to recognize that the degree of difference in the spatial profiles of the two modes may alter the conversion efficiency significantly if there is poor overlap. To account for the spatial differences in the overlap between the m and p modes, the factor R in Eq. (41) represents the normalized overlap integral between the fundamental and SHG modes, which ranges between 0 (no overlap) and 1 (complete overlap).

To date, there have been no reported examples of phase-matched upconversion in GaN; however a demonstration of enhanced SHG in an AlN/glass waveguide structure was recently reported by Blanc *et al.* (1995). The AlN and GaN share the same wurtzite structure with similar linear and nonlinear optical properties. The investigation by Blanc *et al.* was able to demonstrate that the SHG response from a 1.342 μm AlN waveguide was enhanced by nearly a factor of 400 relative to the bulk material when the fundamental and SHG waves were in TM_0 and TM_2 modes, respectively. The effective index of the two modes for this AlN/glass structure vs wavelength is shown in Fig. 27, where the two indices are found to be equal

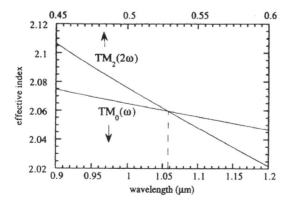

FIG. 27. The effective index of refraction of the $TM_0(\omega)$ and the $TM_2(2\omega)$ modes as a function of wavelength for a 1.342 μm AlN film on sapphire. The crossing point gives the phase-matching wavelength. (Reprinted with permission from Blanc *et al.*, 1995.)

at a fundamental wavelength of ~1.05 μm. The SHG signal from this structure, shown in Fig. 28, is enhanced at the phase-matching condition, with a dramatic decrease in the intensity on either side of the optimum wavelength. The authors noted that the value of the conversion efficiency for their structure was approximately 5×10^{-5}, which is lower than the theoretical prediction in Eq. (41). The lower value was attributed to the poor quality of the AlN film, which limited the propagation length to ~250 μm. It is important to note that the extension of this AlN study to potential GaN/sapphire waveguide structures is straightforward. The differences in the refractive indices and second-order nonlinearities of these two materials are not significant. The superior quality of state-of-the-art GaN films should allow for achievable propagation lengths on the order of 1–5 mm, which would provide orders of magnitude enhancements over the AlN/glass waveguide. For these distances, the level of conversion for SHG would produce SHG powers of ~1 mW when using a focused cw diode laser system of 100 mW power. This is comparable to SHG production in polymer and ferroelectric inorganic thin films ($LiNbO_3$), where demonstrated SHG efficiencies in the visible region of the spectrum are considered suitable for practical applications (Regener and Sohler, 1988; Shuto et al., 1997).

It is also possible to enhance the SHG efficiency in GaN by creating multilayer structures that are composed of alternating layers of GaN and a transparent linear dielectric material. A theoretical examination of SHG

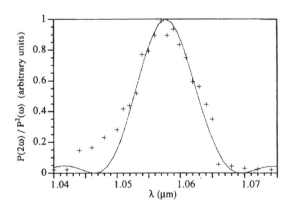

FIG. 28. Experimental tuning curve (filled circles) showing the dependence of the normalized SHG power vs the fundamental wavelength in an AlN/sapphire waveguide. The solid line is the calculated response assuming a 250 μm interaction length. (Reprinted with permission from Blanc et al., 1995.)

8 NONLINEAR OPTICAL PROPERTIES OF GALLIUM NITRIDE

from multilayer linear/nonlinear film composites has been examined by Bethune (1989, 1991) for both isotropic and anisotropic materials. The alternating nonlinear/linear layers in the structure provide "quasi" phase matching between the SHG fields in alternating layers. For a given nonlinear/linear film pair, the thickness of the linear layer is chosen to compensate for any phase shift between the fundamental and SHG fields that would lead to destructive interference between 2ω waves in the next nonlinear film. This structure eliminates the conversion efficiency problems associated with the short coherence length in a single GaN layer. Miragliotta *et al.* (1993) proposed the multilayer structure that is shown in Fig. 29. The device is composed of alternating GaN and sapphire layers, where the thickness of each GaN film was 2.2 μm. This is close to an optimum value for generating SHG in one film when the fundamental wavelength is 1.064 μm. The correct phase-matching conditions are achieved by varying the thickness of the sapphire layer until a maximum in the SHG

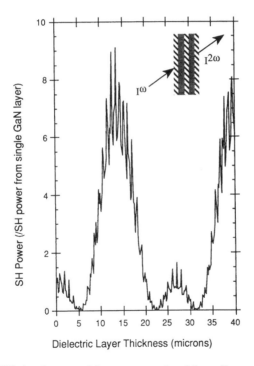

FIG. 29. The SHG signal generated from a proposed multilayer film structure (normalized to a single GaN layer) which is shown in the inset. In the device, the hatched regions are 2.2 μm GaN films (nonlinear) and the shaded regions are sapphire films (linear).

transmitted signal is obtained. The result in Fig. 29 shows the SHG signal from a three linear/nonlinear layer composite as the film thickness of the sapphire layers was varied from 0 to 40 μm. At an incident angle of 45°, the transmitted SHG power is an oscillatory function of the thickness of the dielectric layer. The SH power for this structure reached a maximum enhancement factor of ~ 9 over that of a single nonlinear film when the sapphire layer thickness was 13.5 μm. Clearly, the addition of n layers in the composite will enhance the SHG production by n^2 relative to a single GaN layer.

VI. Conclusions

The characterization of the second- and third-order nonlinearities in GaN has accelerated over the past decade due to the rapid improvement in the quality of epitaxially grown material. The results that have been presented in this chapter from recent nonlinear optical studies suggest that the optical nonlinearities are of sufficient magnitude for realistic device applications, for example, frequency conversion and phase modulation. The intrinsic waveguide nature of the film deposit/substrate suggests that waveguiding or multilayer structures are likely candidates for optical device platforms where propagation lengths on the order of a few millimeters (high conversion efficiencies) are possible. In addition to the potential device development, it is evident that second- and third-order phenomena such as SHG and DFWM are sensitive to the spectral and temporal electronic characteristics in GaN; as such, the use of nonlinear optical probes for analytical studies of GaN films or related structures is certainly warranted for future investigations.

REFERENCES

Aitchison, J. S., Oliver, M. K., Kapon, E., Colas, E., and Smith, P. W. E. (1990). *Appl. Phys. Lett.*, **56**, 1305.
Akhmanov, S. A., Koroteev, N. I., Paitian, G. A., Shumay, I. L., and Galjautdinov, M. F. (1985). *Opt. Comm.*, **47**, 202.
Akimoto, R., Ando, K., Sasaki, F., Kobayashi, S., and Tani, T. (1997). *J. Lumin.*, **72–74**, 309–11.
Anderson, D. B. and Boyd, J. T. (1971). *Appl. Phys. Lett.*, **19**, 266.
Armstrong, J. A., Bloembergen, N., Ducuing, J., and Pershan, P. S. (1962). *Phys. Rev.*, **127**, 1918.
Aspnes, D. E. and Rowe, J. E. (1972). *Phys. Rev. B*, **5**, 4022.
Aspnes, D. E. and Studna, A. A. (1973). *Phys. Rev. B*, **7**, 4605.
Aspnes, D. E. (1973). *Surf. Sci.*, **37**, 418.
Bethune, D. S. (1989). *J. Opt. Soc. Am.*, **B6**, 910.

Bethune, D. S. (1991). *J. Opt. Soc. Am.*, **B8**, 367.
Blanc, D., Bouchoux, A. M., Plumereau, C., Cachard, A., and Roux, J. F. (1995). *Appl. Phys. Lett.*, **66**, 659.
Bloembergen, N. and Pershan, P. S. (1962). *Phys. Rev.*, **128**, 606.
Bloembergen, N. (1965). *Nonlinear Optics*. New York: Benjamin-Cummings.
Born, M. and Wolf, E. (1980). *Principles of Optics*, 6th ed., New York: Pergamon.
Boyd, R. (1992). *Nonlinear Optics*. New York: Academic Press.
Bratz, A., Felderhof, B. U., and Marowsky, G. (1990). *Appl. Phys. B*, **50**, 393.
Burns, W. K. and Bloembergen, N. (1971). *Phys. Rev. B*, **4**, 3437.
Butcher, P. N. and McLean, T. P. (1963). *Proc. Phys. Soc.* (London), **81**, 219; (1964). **83**, 579.
Cada, M., Keyworth, B. P., Glinski, J. M., SpringThorpe, A. J., and Mandeville, P. (1988). *J. Opt. Soc. Am. B*, **5**, 462.
Catalano, I. M., Cingolani, A., Lugara, M., and Minafra, A. (1977). *Optics Comm.*, **23**, 419.
Chen, J., Levine, Z. H., and Wilkins, J. W. (1995). *Appl. Phys. Lett.*, **66**, 1129.
Chenault, D. B., Chipman, R. A., and Lu, S.-Y. (1994). *Appl. Opt.*, **33**, 7382.
Ching, W. Y. and Huang, M.-Z. (1993). *Phys. Rev. B*, **47**, 9479.
Clays, K., Armstrong, N. J., and Penner, T. L. (1993). *J. Opt. Soc. Am.*, **B10**, 886.
Conwell, E. M. (1973). *IEEE J. Quantum Electron.*, **9**, 867.
Denbaars, S. P. (1997). *Proc. IEEE*, **85**, 1740.
Diaz-Garcia, M. A., Hide, F., Schwartz, B. J., McGehee, M. D., Andersson, M. R., and Heeger, A. J. (1997). *Appl. Phys. Lett.*, **70**, 3191.
Eesley, G. I. (1981). *Coherent Raman Spectroscopy*. New York: Pergamon Press.
Ejder, E. (1971). *Phys. Status Solidi A*, **6**, 442.
Fujii, K., Shimizu, A., Bergquist, J., and Sawada, T. (1990). *Phys. Rev. Lett.*, **65**, 1808.
Gase, T. and Karthe, W. (1997). *Opt. Comm.*, **133**, 549.
Ghahramani, E., Moss, D. J., and Sipe, J. E. (1991). *Phys. Rev. B*, **43**, 9700.
Goldstein, H. (1981). *Classical Mechanics*, 2nd ed., Reading, MA: Addison-Wesley, Chap. 4.
Guyot-Sionnest, P., Dumas, P., Chabal, Y. J., and Higashi, G. S. (1990). *Phys. Rev. Lett.*, **64**, 2156.
Harada, A., Okazaki, Y., Kamiyama, K., and Umegaki, S. (1991). *Appl. Phys. Lett.*, **59**, 1535.
Harris, A. L., Chidsey, C. E. D., Levinos, N. J., and Loiacono, D. N. (1987). *Chem. Phys. Lett.*, **141**, 350.
Haus, H. A. and Reider, G. A. (1987). *Appl. Opt.*, **26**, 4576.
Heyman, J. N., Craig, K., Galdrikian, B., Sherwin, M. S., Campman, K., Hopkins, P. F., Fafard, S., and Gossard, A. C. (1994). *Phys. Rev. Lett.*, **72**, 2183.
Hillrichs, G., Graf, D., Marowsky, G., Roders, O., Schnegg, A., and Wagner, P. (1994). *J. Electrochem. Soc.*, **141**, 3145.
Huang, M.-Z. and Ching, W. Y. (1993). *Phys. Rev. B*, **47**, 9464.
Hughes, J. L. P., Wang, Y., and Sipe, J. E. (1997). *Phys. Rev. B*, **55**, 13630.
Hughes, J. L. P. and Sipe, J. E. (1996). *Phys. Rev. B*, **53**, 10751.
Hutchings, D. C. and Van Stryland, E. W. (1992). *J. Opt. Soc. Am.*, **B9**, 2065.
Ishidate, T., Inoue, K., and Aoki, M. (1980). *Jpn. J. Appl. Phys.*, **19**, 1641.
Ito, H. and Inaba, H. (1978). *Opt. Lett.*, **2**, 139.
Jerphagnon, J. and Kurtz, S. K. (1970). *J. Appl. Phys.*, **41**, 1667.
Kaminow, I. P. (1974). *An Introduction to Electrooptic Devices*. New York: Academic Press.
Kaminski, C. F., Hughes, I. G., and Ewart, P. (1997). *J. Chem. Phys.*, **106**, 5324.
Kawakami, Y., Peng, Z. G., Narukawa, Y., Funita, Sz., Fujita, Sg., and Nakamura, S. (1996). *Appl. Phys. Lett.*, **69**, 1414.
Kelley, P. L. (1963). *J. Phys. Chem. Solids*, **24**, 607, 1113 (1963); **24**, 1113.
Khanarian, G., Norwood, R. A., Haas, D., Feuer, B., and Karim, D. (1990). *Appl. Phys. Lett.*, **57**, 977.

Khurgin, J. B. (1994). *J. Opt. Soc. Am.* **B11**, 624.
Lanzafame, J. M., Palese, S., Wang, D., Miller, R. J. D., and Muenter, A. A. (1994). *J. Phys. Chem.*, **98**, 11020.
Laughton, F. R., Marsh, J. H., Barrow, D. A., and Portnoi, E. L. (1994). *IEEE J. Quantum Electron.*, **30**, 838.
Laurell, F. and Arvidsson, G. (1988). *J. Opt. Soc. Am.*, **B5**, 292.
Lax, B. J., Mavroides, G. and Edwards, D. F. (1962). *Phys. Rev. Lett.*, **8**, 166.
Lee, C. H., Chang, R. K., and Bloembergen, N. (1967). *Phys. Rev. Lett.*, **18**, 167.
Levenson, M. D. and Kano, S. (1988). *Introduction Nonlinear Laser Spectroscopy*. New York: Academic Press.
Levine, B. F. (1969). *Phys. Rev. Lett.* **22**, 787.
Levine, B. F. (1973). *Phys. Rev. B*, **7**, 2600.
Levine, Z. H. and Allan, D. C. (1991). *Phys. Rev. B*, **44**, 12781.
Levine, B. F. and Bethea, C. G. (1976). *J. Chem. Phys.*, **65**, 2429.
Long, X.-C., Myers, R. A., Brueck, S. R. J., Ramer, R., Zheng, K., and Hersee, S. D. (1995). *Appl. Phys. Lett.*, **67**, 1349.
Malouin, C., Villeneuve, A., Vitrant, G., Cottin, P., and Lessard, R. A. (1998). *J. Opt. Soc. Am.*, **B15**, 826.
Miller, R. C., Kleinman, D. A., and Savage, A. (1963). *Phys. Rev. Lett.*, **11**, 146.
Miller, D. A. B., Weiner, J. S., and Chemla, D. S. (1986). *IEEE J. Quantum Electron.*, **22**, 1816.
Miragliotta, J., Wickenden, D. K., Kistenmacher, T. J., and Bryden, W. A. (1993). *J. Opt. Soc. Am.*, **B10**, 1447.
Miragliotta, J., Bryden, W. A., Kistenmacher, T. J., and Wickenden, D. K. (1994a). *Inst. Phys. Conf. Ser.*, **137**, Chap. 6, 537.
Miragliotta, J., Bryden, W. A., Kistenmacher, T. J., and Wickenden, D. K. (1994b). In *Diamond, SiC, and Nitride Wide Bandgap Semiconductors*, C. H. Carter, Jr., G. Gildenblat, S. Nakamura, and R. J. Nemanich, eds., Pittsburgh, PA: *Mater. Res. Soc. Symposia Proc.* **339**, p. 465.
Miragliotta, J. and Wickenden, D. K. (1994). *Phys. Rev. B*, **41**, 1401.
Miragliotta, J. and Wickenden, D. K. (1996). *Phys. Rev. B*, **53**, 1388.
Miragliotta, J. and Wickenden, D. K. (1995). *Appl. Phys. Lett.*, **69**, 2095.
Mizrahi, V., DeLong, K. W., Stegeman, G. I., Saifi, M. A., and Andrejco, M. J. (1989). *Opt. Lett.*, **14**, 1140.
Mohammed, S. N. and Morkoç, H. (1996). *Progress in Quantum Electronics*, **20**, 361.
Mohs, G., Aoki, T., Nagai, T. M., Shimano, R., Kuwata-Gonokami, M., and Nakamura, S. (1997). *Solid State Communications*, **104**, 643.
Moss, D. J., Ghahramani, E., Sipe, J. E., and van Driel, H. M. (1990). *Phys. Rev. B*, **41**, 1542.
Moss, D. J., Ghahramani, E., and Sipe, J. E. (1991). *Phys. Stat. Sol.* (b), **164**, 587.
Murayama, M. and Nakayama, T. (1995). *Phys. Rev. B*, **52**, 4986.
Nakamura, S. (1996). *Advanced Materials*, **8**, 689.
Nakamura, S., Senoh, M., Nagahama, S., Iwasa, N., Yamada, T., Matsushita, T., Kiyoku, H., and Sugimoto, Y. (1996). *Japanese Journal of Applied Physics*, Part 2 (Letters), **35**, L74.
Nayak, S. K., Sahu, T., Mohanty, S. P., and Misra, P. K. (1997). *Semicond. Sci. Technol.*, **12**, 544.
Nguyen, H. (1997). *Microwave and Optical Technology Letters*, **16**, 283.
Noordman, O. F. J., van Hulst, N. F., and Bolger, B. (1995). *J. Opt. Soc. Am.*, **B12**, 2398.
Nussbaum, A. (1981). In *Semiconductors and Semimetals*, R. Willardson and A. C. Beer, eds., New York: Academic Press, Vol. **15**, p. 39.
Obeidat, A. and Khurgin, J. B. (1995). *J. Opt. Soc. Am.* **B12**, 1222.
Pau, S., Kuhl, J., Scholz, F., Haerle, V., Khan, M. A., and Sun, C. J. (1997). *Phys. Rev. B*, **56**, R12718.

Pershan, P. S. (1963). *Phys. Rev.*, **130**, 919.
Phelps, C. W., Barry, T. S., Rode, D. L., and Krchnavek, R. R. (1997). *J. Lightwave Technol.*, **15**, 1900.
Phillips, J. C. (1969). *Bonding in Crystals, Molecules, and Polymers*. Chicago: University of Chicago Press.
Philips, J. C. and Van Vechten, J. A. (1969). *Phys. Rev.*, **183**, 709.
Prasad, P. N. and Williams, D. J. (1991). *Introduction to Nonlinear Optical Effects in Molecules and Polymers*. New York: John Wiley and Sons.
Prasad, P. N. (1992). Nonlinear Optical Effects in Organic Materials, In *Contemporary Nonlinear Optics*, G. P. Agrawal and R. W. Boyd, eds., New York: Academic Press, p. 265.
Puccetti, G. (1995). *J. Chem. Phys.*, **102**, 6463.
Regener, R. and Sohler, W. (1988). *J. Opt. Soc. Am.*, **B5**, 267.
Reider, G. A., Hofer, U., and Heinz, T. F. (1991). *J. Chem. Phys.*, **94**, 4080.
Reineke, R. and Memming, R. (1992). *J. Phys. Chem.*, **96**, 1310.
Robinson, F. N. H. (1968). *Phys. Lett.*, **26A**, 435.
Rodenberger, D. C., Heflin, J. R., and Garito, A. F. (1995). *Phys. Rev. A*, **51**, 3234.
Rose, A. (1963). *Concepts in Photoconductivity and Allied Problems*. New York: Interscience Publishers.
Salah, B. E. A. and Teich, M. C. (1991). *Fundamentals of Photonics*. New York: John Wiley and Sons, p. 746.
Sasaki, K., Kinoshita, T., and Karasawa, N. (1984). *Appl. Phys. Lett.*, **45**, 333.
Schmidlin, F. W. (1977). *Phys. Rev. B*, **16**, 2362.
Shank, C. V., Yen, R., and Hirlimann, C. (1983). *Phys. Rev. Lett.*, **51**, 900.
Shaw, M. J., Ninno, D., Adderley, B. M., and Jaros, M. (1992). *Phys. Rev. B*, **45**, 11031.
Sheik-Bahae, M., Hutchings, D. C., Hagan, D. J., and Van Stryland, E. W. (1990). *IEEE J. Quantum Electron.*, **27**, 1296.
Shelton, J. W. and Shen, Y. R. (1972). *Phys. Rev. A*, **5**, 1867.
Shen, Y. R. (1984). *Principles of Nonlinear Optics*, 1st ed. New York: John Wiley and Sons.
Shuto, Y., Watanabe, T., Tomaru, S., Yokohama, I., Hikita, M., and Amano, M. (1997). *IEEE J. Quantum Electron.*, **33**, 349.
Singer, K. D. and Garito, A. F. (1981). *J. Chem. Phys.*, **75**, 3572.
Singh, S. (1971). In *Handbook of Lasers*. Cleveland, Ohio: Chemical Rubber Company.
Sipe, J. E. and Ghahramani, E. (1993). *Phys. Rev. B*, **48**, 11705.
Soref, R. A. and Moos, H. W. (1964). *J. Appl. Phys.*, **35**, 2152.
Stegeman, G. I. (1992). Nonlinear Guided Waves, In *Contemporary Nonlinear Optics*. G. P. Agrawal and R. W. Boyd, eds., New York: Academic Press, p. 1.
Stegeman, G. I. and Stolen, R. H. (1989). *J. Opt. Soc. Am.*, **B6**, 652.
Strite, S. and Morkoç, H. (1992). *J. Vac. Sci. Tech. B*, **10**, 1283.
Strite, S., Lin, M. E., and Morkoç, H. (1993). *Thin Solid Films*, **213**, 197.
Sugihara, O., Kunioka, S., Nonaka, Y., Aizawa, R., Koike, Y., Kinoshita, T., and Sasaki, K. (1991). *J. Appl. Phys.*, **70**, 7249.
Suzuki, N. and Iizuka, N. (1997). *Jpn. J. Appl. Phys.*, **36**, L1006.
Taheri, B., Hays, J., and Song, J. J. (1996). *Appl. Phys. Lett.*, **68**, 587.
Takahashi, H., Ohashi, M., Kondo, T., Ogasawara, N., Shiraki, Y., and Ito, R. (1994). *Jpn. J. Appl. Phys.*, **33**, L1456.
Tessler, N., Denton, G. J., and Friend, R. H. (1997). *Synth. Met.*, **84**, 475.
Tien, P. K., Ulrich, R., and Martin, R. J. (1970). *Appl. Phys. Lett.*, **17**, 447.
Tien, P. K., Ulrich, R., and Martin, R. J. (1990). *J. Opt. Soc. Am.*, **B7**, 768.
Tien, P. K. and Ulrich, R. (1970). *J. Opt. Soc. Am.*, **60**, 1325.
Van Vechten, J. A. (1969a). *Phys. Rev.*, **182**, 891.

Van Vechten, J. A. (1969b). *Phys. Rev.*, **187**, 1007.
Wilson, R. H., Sakata, T., Kawai, T., and Hashimoto, K. (1985). *J. Electrochem. Soc.*, **132**, 1082.
Yariv, A. (1975). *Quantum Electronics*. New York: John Wiley and Sons.
Yariv, A. and Pepper, D. M. (1977). *Opt. Lett.*, **1**, 16.
Yee, J. H. (1969). *Appl. Phys. Lett.*, **14**, 231.
Zimmermann, R., Euteneuer, A., Mobius, J., Weber, D., Hofmann, M. R., Ruhle, W. W., Gobel, E. O., Meyer, B. K., Amano, H., and Akasaki, I. (1997). *Phys. Rev. B*, **56**, 12722.

CHAPTER 9

Magnetic Resonance Investigations on Group III-Nitrides

B. K. Meyer

I. PHYSICS INSTITUTE
JUSTUS LIEBIG UNIVERSITY GIESSEN
GIESSEN, GERMANY

I. INTRODUCTION	371
II. MAGNETIC RESONANCE — THE BASIS OF IDENTIFICATION	373
1. *Electron Paramagnetic Resonance (EPR)*	373
2. *Optically Detected Magnetic Resonance (ODMR)*	375
3. *Electrically Detected Magnetic Resonance (EDMR)*	377
III. SHALLOW DONORS IN CUBIC AND HEXAGONAL GaN (EPR RESULTS)	377
1. *The Conduction Band/Shallow Donor Spin Resonance*	377
2. *The Conduction Band/Shallow Donor Resonance in AlGaN*	382
3. *Overhauser Shift and Nuclear Double Resonance*	383
IV. SHALLOW AND DEEP DONORS IN GaN (ODMR RESULTS)	385
1. *ODMR and Nuclear Double Resonance Investigations*	385
2. *Magneto-Optical and Infrared Absorption Experiments*	390
V. SHALLOW AND DEEP ACCEPTORS IN GaN	391
VI. DEFECTS INDUCED BY PARTICLE IRRADIATION IN GaN AND AlN	394
VII. DEVICE-RELATED MAGNETIC RESONANCE STUDIES	396
VIII. TRANSITION METAL IMPURITIES	399
IX. OUTLOOK	403
REFERENCES	404

I. Introduction

The properties of most semiconductors and semiconductor devices are controlled by defects. Therefore, a great number of different and defect-specific spectroscopic techniques have been developed to assess defect properties. The aim is to determine the electronic parameters (binding energies and concentrations), but it is equally important to identify the defects chemically and give conclusive answers about the local surroundings: that is, is the defect on cation or anion sites, or is it on a substitutional or interstitial site? Often two spectroscopic techniques have to be combined, for example, magnetic resonance and optical spectroscopy to optically detected magnetic resonance, to reach the goal (Fig. 1).

In a newly and rapidly developing materials system such as the group III-nitrides (Strite and Morkoc, 1992) the defect identification problem is by no means a simple task. Deep centers and shallow impurities have their own individual importance and relevance for device performance. To give two GaN specific examples:

(i) the source of n-type conductivity of undoped epitaxial films was for a long time far from firmly established. Intrinsic defects such as the nitrogen vacancy V_N and the gallium interstitial Ga_i could be involved

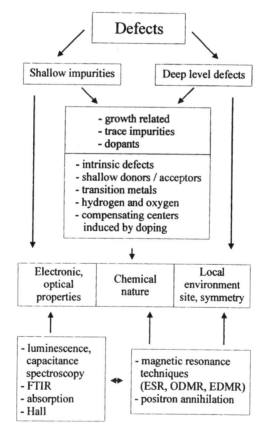

FIG. 1. Defects controlling semiconductors and spectroscopic techniques to identify them.

(Boguslawski et al., 1994; Neugebauer and Van de Walle, 1994). Here GaN plays a unique role among other III-V semiconductors according to theoretical calculations. The V_N and Ga_i are deep-level defects with resonance states within the conduction band. By autoionization of the singly occupied resonance states they can create effective mass-type shallow states. Hence in their electrical activity they would be indistinguishable from shallow *n*-type dopants. However, trace impurities such as Si and oxygen — Si is an efficient *n*-type dopant in GaN — also are present in rather high concentrations and can possibly account for the *n*-type conductivity (Götz et al., 1996; Wickenden et al., 1996).

(ii) The radiative lifetimes of free excitons in GaN are unusually short (Eckey et al., 1996) (around 100 ps at 4.2 K) and deep centers acting as lifetime killers must have a severe impact on lifetime. Prime candidates are transition metal elements (Maier et al., 1994; Pressel et al., 1996) (iron, chromium, and titanium are the top ranking impurities), but intrinsic defects and defect complexes having energy levels deeply in the forbidden gap also have to be considered. In that context, the omnipresent yellow luminescence band in GaN films and its relation to defect properties has to be clarified.

This chapter will review the identification of intrinsic and extrinsic paramagnetic defects in AlGaN by electron spin resonance (ESR) and the related techniques, optically detected magnetic resonance (ODMR), and electrically detected magnetic resonance (EDMR).

II. Magnetic Resonance — The Basis of Identification

1. ELECTRON PARAMAGNETIC RESONANCE (EPR)

Electron spin resonance is one of the most efficient techniques to provide detailed information on the electronic and atomistic structure of paramagnetic defects in semiconductors. One could imagine that a limitation of the EPR method is that the defect under consideration has to carry an unpaired spin in order to give rise to paramagnetism. This is very often the case. Additionally, in semiconductors the position of the Fermi level plays a key role because the occupancy of defects can be influenced by tuning the Fermi level. Shifting the Fermi level from the conduction band in an *n*-type sample by introducing compensating centers (as a result of co-doping or by the creation of defects by particle irradiation) changes the shallow effective mass type donors from the occupied (paramagnetic) to the unoccupied (nonparamagnetic) charge state. Similar arguments hold for deep centers, which are capable of binding two or even more carriers. Examples are the vacancies in elemental and compound semiconductors (the classical example is the isolated vacancy in Si) and the antisite defects in III-V semiconductors. High-spin and low-spin configurations are realized, that is, a three electron center might have $S = 3/2$ or $S = 1/2$, depending on the strength of electron-phonon interactions with respect to electron-electron interactions. Another class of defects is transition metal ions, which in III-V and II-VI semiconductors generally follow Hund's rule and favor high-spin configurations [e.g., Fe^{3+} in GaN has a $S = 5/2$ spin state (see Chapter 8)]. The change in occupancy of a deep or shallow center can also be achieved by light illumination with suitable photon energies. The technique is called photo-EPR or light-induced EPR. Depending on the Fermi level position it is related to carrier capture or carrier emission (photoionization) and, in certain cases, the energy level position and cross section of the defect under investigation can be deduced. In conventional EPR only the ground state of a defect is accessible. This is evident when comparing the energies of the microwaves (10^{-5} eV) with the energy separation between ground and excited states, which can vary between some 100 meV up to some electronvolts. Hence, the occupation of the excited states can neither be realized by the microwave quanta nor by the thermal energy, that is, 25 meV at room temperature.

In all magnetic resonance experiments (EPR, ODMR, EDMR), magnetic dipole transitions are induced between the ground state spin manifold split in the presence of a static magnetic field. At low temperatures there is a population difference between the spin-up and spin-down Zeeman levels. This population difference can be diminished by absorbing microwave radiation if the resonance condition is fulfilled, that is, the microwave energy equals the level splitting energy (Zeeman-energy). Under resonance conditions a fraction of the microwave power is absorbed and results in a detuning of the microwave circuit. To obtain the necessary oscillatory microwave field a microwave resonator is essential. Most EPR investigations make use of a commercial spectrometer. It employs a microwave system operating at around 9 GHz. This frequency determines the size of the resonator and also to some extent the sample sizes. The last point is of prime importance because EPR is a volume sensitive method. The typical sensitivity is 10^{11} spin/gauss linewidth and this fact immediately tells us that the sensitivity in EPR depends on the halfwidth of the EPR signal. The number of spins is given by the area under the resonance line, making EPR a quantitative technique but also setting a limit to the detection of defects in most III-V semiconductors at around mid 10^{15} cm^{-3}.

2. OPTICALLY DETECTED MAGNETIC RESONANCE (ODMR)

The original idea behind ODMR was to combine the advantages of EPR with respect to the defect identification (symmetry, chemical identity) with the high sensitivity and selectivity of optical spectroscopy. Optical spectroscopy is certainly the standard characterization technique of the electronic properties of semiconductors. In this paper we concentrate exclusively on photoluminescence (PL) combined with EPR. In luminescence spectroscopy the specific identification of impurity related recombinations is obtained by studying bound excitons in high-purity films or donor-acceptor pair recombinations by back-doping of high-purity layers. In order to obtain the actual symmetry of a center giving rise to PL, the application of external perturbations such as magnetic and electric fields and uniaxial stress is searched. Also there might be isotope splittings as well as isotope shifts that are observable in the optical spectra. For broadband emissions PL is in a rather poor situation and ODMR has mostly contributed to the defect identification involved in those recombinations. In ODMR the microwave-induced changes within the Zeeman sublevels are detected indirectly by a change of the emission properties. It could be changes in the total emission intensity and/or polarization properties. The essential features of ODMR will be discussed for a donor-acceptor type recombination (Fig. 2). Electron and hole pairs are generated by above bandgap excitation. Depending on the Fermi level position shallow donors might by ionized and shallow acceptors have accepted electrons (hence they are in the singly negative charge state). Thus the following reaction takes place

$$D^+ + A^- \xrightarrow{h\nu} D^+ + e^- + A^- + h^+ \to D^0 + A^0 \to D^+ + A^- + h\nu$$

the recombination by electrons and holes is radiative—the donor-acceptor pair recombination. In the following it is essential whether the acceptor is deep or shallow, that is, whether for the acceptor the effective mass-type description should be used with $J = S = 3/2$ or if for the deep acceptor the orbital momentum is quenched and $J = S = 1/2$ is the appropriate spin state. If donor and acceptor both have $S = 1/2$ the spins of the electron and the hole will arrange in the presence of an external magnetic field into singlet and triplet states (see level scheme in Fig. 2). The splitting is determined by the g-values of the recombining partners (g_D of the donor, g_A of the acceptor). The ground state after recombination of electron and hole is a singlet state. The selection rules for optical transitions within such a system show that for small spin orbit interactions only singlet states can recombine because of spin conservation. The emission from the triplet state is forbid-

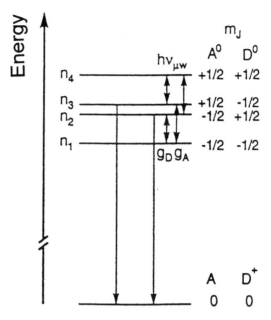

FIG. 2. Level scheme to explain the detection of magnetic resonance in a donor-acceptor pair recombination.

den. Resonant microwave transitions can couple the radiative singlet states with the nonradiative triplet states. Hence both the donor as well as the acceptor resonances will be detectable as changes in total emission intensity. Whether one sees the resonances as a luminescence increase or decrease depends on the occupation of the Zeeman sublevels. For a completely thermalized system ($n_1 > n_2 > n_3 > n_4$) one expects the donor resonance to increase and the acceptor resonance to decrease in the PL intensity. If the system is not thermalized, triplet states have a higher occupation than singlet states ($n_1 > n_3, n_4 > n_2$) and one observes both resonances as a luminescence increase. The recombination might involve shallow donor to deep donor, deep donor to shallow acceptor, but of more interest for GaN is spin-dependent transfer from the shallow donor to a deep donor (Glaser et al., 1995). This transfer can be nonradiative; only the final step, the deep-donor to shallow-acceptor recombination, is radiative. Under certain circumstances both the shallow-donor and the deep-donor resonances can be observed simultaneously (Glaser et al., 1995). The why and how is common for ODMR and another modification of EPR, the electrical detection of magnetic resonance, EDMR.

3. Electrically Detected Magnetic Resonance (EDMR)

In EDMR one searches for resonant changes in electronic transport properties. This could be the recombination or diffusion current in a diode-like structure or dark-/photoconductivity. In a diode structure a constant forward- or reverse-biased voltage is applied and microwave-induced changes in the current are measured. How does magnetic resonance influence the recombination current (forward-biased diode)? We already mentioned the spin-dependent transfer of an electron from the shallow donor to the deep center. Imagine that the deep center is in a paramagnetic charge state, that is, singly occupied and has spin $S = 1/2$. The recombination step would be the capture of the electron and the deep center is now occupied by two electrons. According to the Pauli principle, the two-electron state is in the diamagnetic $S = 0$ singlet state. However, the spins of the two initial states can form both a triplet and a singlet state. As in the case for ODMR due to spin conservation in the nonradiative recombination only singlet states can recombine; the transition is forbidden for triplet states. Inducing spin-flip transitions that act either on the spin of the electron or the defect will transform triplet states into singlet states and enhance the recombination rate. Typical values for the resonantly induced changes in the current are 10^{-4}–10^{-6}. The 4 K characteristic of a NICHIA blue light emitting diode (LED) has at a forward-biased voltage of 4 V a current of 10^{-6} A. With modern current measurement units and using sophisticated lock-in techniques, currents in the femto ampere range can be measured easily and the requirements for EDMR can be met. An important aspect is that EDMR is a device-adopted characterization technique and defects appearing after the degradation of a device can be studied.

III. Shallow Donors in Cubic and Hexagonal GaN (EPR Results)

1. The Conduction Band/Shallow Donor Spin Resonance

Historically the shallow donors in semiconductors were the first defects studied with magnetic resonance techniques. Shortly after their first observation it turned out that from magnetic resonance parameters details of the band structure of a semiconductor could be obtained. Based on a three-band $\mathbf{k}\cdot\mathbf{p}$ calculation, Roth et al. (1959) showed that the g-value of a shallow donor in zinc-blende semiconductors, for example, InSb, is given by

$$g^* - 2 \neq -2/3\,(\Delta_0/E_0)(m_0/m^* - 1) \tag{1}$$

where Δ_0 is the separation between the heavy/light hole band being degenerate at the Γ-point and the spin-orbit split valence band (Γ_{8v}-Γ_{7v} in zinc blende), E_0 is the energy gap and m^* is the electron effective mass at the conduction bandedge in units of m_0. This equation explains why for InSb $g^* = -50$ or for GaAs $g^* = -0.44$ is measured. In this approach it is assumed that the g-factors of shallow donors and of free electrons in the conduction band do not differ. This turned out to be the case. But the g-value is not necessarily isotropic. In semiconductors with wurtzite crystal structure (CdS, CdSe) the degeneracy of the heavy and light hole valence band is lifted and they are split apart by the action of the wurtzite crystal field. The symmetry is hence less than tetrahedral and this leads to an anisotropy in the g-values.

GaN crystallizes in two polytypes, zinc blende and wurtzite. The polytypes can be stabilized by growing on appropriate substrates, for example, for zinc blende on cubic SiC, MgO, Si(100) or GaAs(100), for wurtzite on sapphire, 6H-SiC, etc. Fanciulli et al. (1993) made the first EPR observations of shallow donors in cubic GaN, followed by Carlos et al. (1993) on wurtzite GaN. The results are shown in Figs. 3 and 4. The EPR lines all had a sharp Lorentzian line shape due to motional narrowing.

The Hamiltonian used in the analysis of the EPR results is

$$H = \mu_B \boldsymbol{B} \boldsymbol{g} \boldsymbol{S} \qquad (2)$$

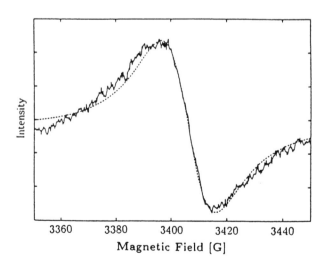

FIG. 3. Electron paramagnetic resonance spectrum of shallow donors in cubic GaN. The dashed line is a fit assuming a Lorentzian lineshape (from Fanciulli et al., 1993).

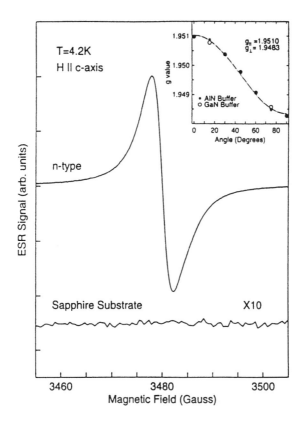

FIG. 4. Electron paramagnetic resonance spectrum of shallow donors in hexagonal GaN. The inset shows the angular dependence (from Carlos et al., 1993).

where μ_B is Bohr's magneton, $\mathbf{g}$ is the g-tensor, $\mathbf{B}$ and $\mathbf{S}$ are the vectors of the magnetic field and spin, respectively. For cubic GaN $\mathbf{g}$ reduces to a scalar and was 1.9533. In wurtzite GaN the axial crystal field leads to the appearance of two components in the g-tensor, $g_\parallel$ and $g_\perp$. The Hamiltonian thus reads

$$H = \mu_B(g_{xx}B_xS_x + g_{yy}B_yS_y + g_{zz}B_zS_z) \qquad (3)$$

The Zeeman energy is given by

$$E = g\mu_B B m_s \qquad (4)$$

where $g = \sqrt{g_\parallel^2 \cos^2(\theta) + g_\perp^2 \sin^2(\theta)}$ and θ is the angle between the c-axis

TABLE I

COMPILATION OF MAGNETIC RESONANCE DATA OBSERVED WITH VARIOUS MAGNETIC RESONANCE TECHNIQUES IN UNDOPED, n-TYPE, SEMI-INSULATING (SI), AND p-TYPE (Mg, Zn-DOPING) GaN AND AlGaN EPITAXIAL FILMS, AND IN DOUBLE HETEROSTRUCTURE (DH) AND SINGLE QUANTUM WELL (SQW)-BASED LIGHT EMITTING DIODES (LED)

Technique	Material	Assignment	$g_\parallel$	$g_\perp$	Linewidth (mT)	References
ESR	n-GaN	EM**	1.9510	1.9483	0.2–2	Carlos et al. (1993)*
ESR	n-GaN	EM	1.9503	1.9483	0.5	Denninger et al. (1996)
ESR	n-GaN	EM	1.9515	1.9488	1	Baur (1998)
ESR	n-GaN	EM	1.9518	1.9489	2	Baur (1998)
ODMR	n,p-GaN	EM	1.9515	1.9485	1–17	Glaser et al. (1995); Kunzer et al. (1994); Kaufmann et al. (1996b); Glaser et al. (1997)
ODMR	GaN:Mg	Donor-like	1.960	—	23	Koschnick et al. (1997)
ODMR	n-GaN	DD	1.988	1.992	13	Glaser et al. (1995)
ODMR	SI-GaN	DD	1.978	1.978	10	Glaser et al. (1997)
ODMR	GaN:Mg	Mg	2.08	2.00	26	Glaser et al. (1995)
			2.067–2.084	1.990–2.022	20	Kunzer et al. (1994); Kaufmann et al. (1996)
ODMR	n-GaN	Shallow trap	1.958	1.958	25	Koschnick et al. (1997)
ODMR	GaN:Mg	Mg-related	2.07	2.03	45	Koschnick et al. (1997)
			2.02	2.00	35	Koschnick et al. (1997)
			2.057	2.045	45	Koschnick et al. (1997)
ODMR	GaN:Zn	Zn	1.997	1.992	7	Kunzer et al. (1994); Kaufmann et al. (1996)
EDMR						
EDMR	n-GaN	EM	—	2.06	8.5	Reinacher et al. (1995)
EDMR	n-GaN	Deep trap	—	1.96	15	Reinacher et al. (1995)
EDMR	DH & SQW	DD**	—	2.00	12	Reinacher et al. (1995)
ELDMR	LEDs			2.00	5	Carlos et al. (1995a, 1995b, 1996); Carlos and Nakamura (1997a); Carlos et al. (1997b)
EDMR	DH LED	Zn	—	2.00	18	Carlos et al. (1995a, 1995b, 1996); Carlos and Nakamura (1997a); Carlos et al. (1997b)
ELDMR						
EDMR	SQW LED	Deep trap	—	2.02	13	Carlos et al. (1995a, 1995b, 1996); Carlos and Nakamura (1997a); Carlos et al. (1997b)
ELDMR						

*For cubic GaN the isotropic g-value of 1.9533 is found (Fanciulli et al., 1993).
**EM means effective mass type defect and DD means deep donor.

and the static magnetic field and the following abbreviations have been used: $g_{xx} = g_{yy} = g_\perp$ and $g_{zz} = g_\parallel$.

For wurtzite GaN $g_\parallel = 1.951$ and $g_\perp = 1.9483$ was obtained for the shallow donors (Table I).

Hermann and Weisbuch (1984) modified the three-band $\mathbf{k} \cdot \mathbf{p}$ calculation of Roth et al. (1959) taking into account higher bands in a five-band $\mathbf{k} \cdot \mathbf{p}$ approach, which is actually better suited for wide bandgap semiconductors ($E_0 > 1.5$ eV) to describe conduction-band spin-resonance experiments on zinc-blende III-V as well as wurtzite II-VI semiconductors. Franciulli et al. (1993) and Carlos et al. (1993) followed the Weisbuch approach and we will do the same.

For the cubic GaN the following equation relates g^* to the band-structure parameters (for details see Fig. 5) and the interband matrix element P^2:

$$g^*/g_e - 1 = -\frac{P^2}{3}\left[\Delta_0/E_0^2 + \lambda^2(\Delta_0'/(E_0' - E_0)^2)\right] \quad (5)$$

where g_e is the free electron g-value, Δ_0' is the spin-orbit splitting in the

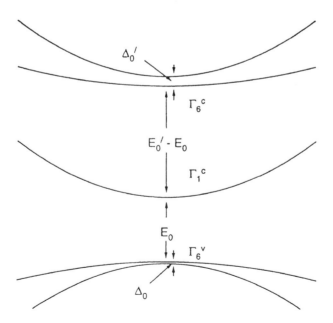

FIG. 5. Energy bands involved in a five-band $\mathbf{k} \cdot \mathbf{p}$ calculation (from Carlos et al., 1993).

conduction band, E'_0 is defined as shown in Fig. 5. In the Weisbuch approach (Hermann and Weisbuch, 1984) λ^2 is the ratio of the interband matrix elements P'^2/P^2. The less-known parameter is the spin-orbit splitting in the higher conduction band and it gives ambiguity in the value for P^2. For instance, using $E_0 = 3.3$ eV, $\Delta_0 = 0.012$ eV, $\Delta'_0 = 0.1$ eV, $\lambda^2 = 0.4$, and $g^* = 1.9533$ one obtains $P^2 = 28$ eV as in Fanciulli et al. (1993, 1995) but for $\Delta'_0 = 0.3$ eV, P^2 is 12.6 eV. Carlos et al. (1993) discuss these uncertainties and estimate that P^2 should range between 17–23 eV.

2. The Conduction Band/Shallow Donor Resonance in AlGaN

The conduction band spin resonance investigations are meanwhile extended towards AlGaN. Carlos (1996) reported on a series of AlGaN samples grown on 6H-SiC substrates where the Al mole fraction was varied from 0 to 0.26. Reinacher et al. (1997) employed AlGaN samples grown by plasma-induced MBE on sapphire substrates. The results are summarized in Fig. 6. The data are arranged in pairs where the lower (higher) g-values correspond to the magnetic field perpendicular (parallel) to the c-axis. Up to approximately 40% molar fraction of Al there seems to be a linear relationship of the g-value upon Al content. Reinacher (1997) reported results on $Al_{0.64}Ga_{0.36}N$ (see open circle in Fig. 6) where a g-value of 1.974 is found, which in terms of magnetic resonance parameters might still arise

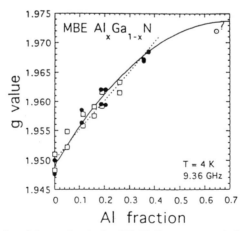

FIG. 6. Variation of the g-values in the AlGaN alloys, open symbols are from Carlos (1996) on MOCVD-grown films, filled and open circles are from Reinacher et al. (1997) on films grown by MBE.

from an effective mass type donor. If true, this would indicate a significant nonlinear contribution to the g-value shift, which in terms of the five-band $\mathbf{k} \cdot \mathbf{p}$ approximation is *a priori* not expectable because the dependence of the bandgap energy in AlGaN on the Al molar fraction has a small bowing parameter. Carlos (1996) analyzed the data in AlGaN with the $\mathbf{k} \cdot \mathbf{p}$ approach and used $P^2 = 17$ eV and $\lambda^2 = 0.4$ (dotted line in Fig. 6) together with the dependence of E_0 and Δ'_0 on the Al molar fraction. He concludes that the decrease of the shift of the g-value from the free electron g-value is primarily due to the decrease in the spin-orbit splitting of the cation (Ga vs Al)-dominated band Γ_{6c}, rather than due to the increase in bandgap with increasing Al mole fraction.

3. Overhauser Shift and Nuclear Double Resonance

Nuclear hyperfine interactions were not resolvable in the shallow donor resonance. They manifest themselves either by a shift of the resonance line position on temperature or on microwave power (Overhauser shift). Carlos et al. (1993) studied the temperature dependence of the ESR line position as well as the resonance linewidth in MOVPE GaN films. Denninger et al.

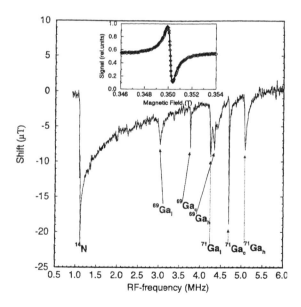

FIG. 7. Overhauser shift in GaN resolving the quadrupole interactions with ^{69}Ga and ^{71}Ga and ^{14}N. In the inset the electron spin resonance is shown (from Denninger et al., 1996).

(1996, 1997) made a detailed investigation combining EPR and the Overhauser shift double resonance technique on a 220-μm thick GaN film grown by hydride vapor phase epitaxy (HVPE). Consistent with the results of Carlos *et al.* (1993), considerable motional or exchange narrowing is present. From the linewidth an estimate for the longitudinal and transversal spin relaxation times was made. In their Overhauser shift experiment (Denninger *et al.*, 1996) the EPR line position was recorded while the radiofrequency (nuclear magnetic resonance, NMR) was swept (see Fig. 7). The shift of all isotopes (^{14}N, ^{69}Ga, ^{71}Ga) is resolved. Due to the nuclear quadrupole splittings one observes three peak shifts for ^{69}Ga (I = 3/2) and ^{71}Ga (I = 3/2). From the quadrupole splittings the electric field gradients V_{zz} can be estimated. The values are: $V_{zz} = (0.594 \pm 0.003) \cdot 10^{20} V/m^2$ (^{14}N); $V_{zz} = (6.43 \pm 0.03) \cdot 10^{20} V/m^2$ (^{69}Ga); $V_{zz} = (6.47 \pm 0.03) \cdot 10^{20} V/m^2$ (^{71}Ga). For the gallium site these results are in good agreement with NMR magic angle spinning (MAS) results (Han *et al.*, 1988) in GaN powder ($V_{zz} = 6.75 \cdot 10 V/m^2$) and recent optically detected electron nuclear double resonance experiments (ODENDOR) (Koschnick *et al.*, 1996; Glaser *et al.*, 1998) (see Section IV.1). The Overhauser effect is obviously different in the Al containing alloys. In Fig. 8 the experimental findings in $Al_{0.37}Ga_{0.63}N$ (Reinacher *et al.*, 1997) and GaN (Carlos *et al.*, 1993) are compared and the data for GaN have been shifted and scaled in magnetic field to account for the different resonance positions. The stronger effect in AlGaN compared to

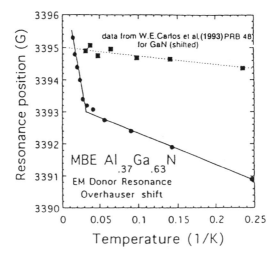

FIG. 8. Overhauser shift in AlGaN and GaN for comparison (data from Reinacher *et al.*, 1997).

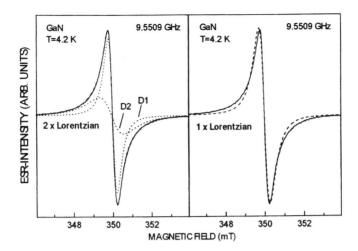

FIG. 9. Simulation of the EPR lineshape of shallow donors in GaN with one (right side) and two Lorentzian (from Baur, 1998).

GaN is probably due to the larger nuclear spin of Al (I = 5/2, compared to 3/2 for Ga) and a larger electronegativity. The slope changes significantly for temperatures above 35 K for reasons to be established.

Baur (1998) in a recent publication could convincingly demonstrate on the basis of Overhauser experiments that indeed two shallow donors with slightly different g-values can be observed in HVPE films (Fig. 9). One coincides within experimental error with the results from Carlos et al. (1993), Denninger et al. (1996), and Glaser et al. (1995) (using ODMR) (see Table I).

The first observation of the EPR of shallow donors was immediately supported by ODMR experiments, which we will discuss in the following.

IV. Shallow and Deep Donors in GaN (ODMR Results)

1. ODMR AND NUCLEAR DOUBLE RESONANCE INVESTIGATIONS

In the first reports on ODMR (Glaser et al., 1995; Glaser et al., 1993) in undoped GaN films, Glaser et al. (1995) used the technique to unravel the source of the n-type behavior of undoped GaN films. At that time the nitrogen vacancy was seriously considered a responsible defect. An essential feature of ODMR not mentioned in the preceding is that the spin flip transition has to occur within the radiative lifetime of a certain recombination. It sets lower limits to the lifetime, which for standard ODMR set-ups should be at least longer than 1 μs. In the films studied the near bandgap luminescence of donor-bound excitons dominated (Fig. 10), but the radiative lifetime is orders too short to allow for ODMR experiments. The notorious yellow luminescence band centered at 2.2 eV has a lifetime distribution extending into the millisecond range and was therefore used as detection monitor (see Fig. 10). The basic findings have been reproduced by several other groups (Koschnick et al., 1996; Hofmann et al., 1995; Hofmann et al., 1996; Kaufmann, 1996). Two resonances appeared as luminescence-increasing signals. One has characteristics of the effective mass-type donor resonance seen also in EPR, the second signal was labeled deep donor (DD). Its full width at half maximum (FWHM) is 13 mT with $g_\parallel = 1.989$ and $g_\perp = 1.992$. The anisotropy is opposite to that for the shallow donor. Depending on the sample the resonantly induced changes were between 0.01

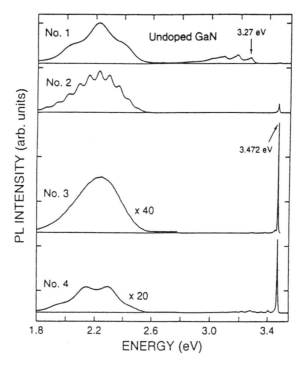

FIG. 10. Photoluminescence in undoped n-type GaN films showing neutral donor-bound exciton recombination at 3.472 eV, donor-acceptor pair recombination at 3.27 eV and the yellow emission at 2.2 eV (from Glaser et al., 1995).

and 0.5% of the emission intensity (Glaser *et al.*, 1995). The spectral dependencies of the ODMR signals match with the yellow luminescence band. A model has been constructed (Fig. 11, model B) to explain the experimental findings. It includes a nonradiative spin dependent electron capture process from the shallow donor to the deep donor located approximately 0.7 eV below conduction band. The deep donor must indeed be a double donor and some of the states must be singly occupied in order to capture an electron. The radiative process is from the deep donor—now occupied by two electrons and hence ODMR inactive—to the shallow acceptor (the residual acceptor in GaN with a binding energy of 220 meV). This part of the cycle should give rise to the yellow emission.

Arguments have been put forth to interpret the luminescent mechanisms in a different way (Hofmann *et al.*, 1995; Ogino and Aoki, 1980; Suski *et al.*, 1995; Perlin *et al.*, 1995). Based on hydrostatic pressure experiments, Suski *et al.* (1995) concluded that it should be of shallow to deep transition (Fig. 11, model A), the deep state is approximately 0.8 eV above valence band. Time-resolved luminescence experiments yielded the same conclusion (Hofmann *et al.*, 1995). In recent photo-EPR investigations, Reinacher *et al.* (1997) studied the light sensitivity of the deep donor resonance. The light-induced EPR signal intensity shows a step-like behavior with a threshold energy of 2.65 eV. The authors interpret the results as an excitation from a level approximately 1 eV above valence band to the conduction band. As to the microscopic origin of the deep donor (or acceptor ?) no information can be given at present.

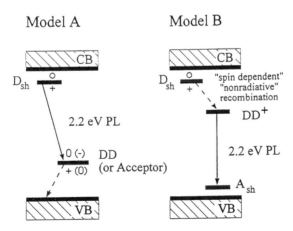

FIG. 11. Two models to account for the ODMR observation of shallow and deep donors within the yellow luminescence band (from Hofmann *et al.*, 1995).

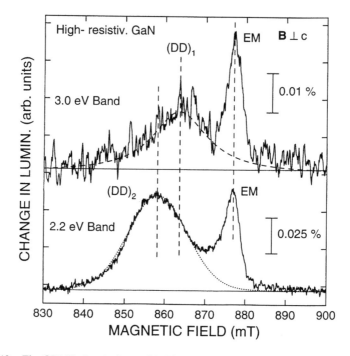

FIG. 12. The ODMR signals detected in high-resistive GaN films on the 3.0-eV band and on the yellow emission. EM denotes effective mass donor, DD stands for deep donor (from Glaser et al., 1997).

Glaser et al. (1997) concentrated on the 3.0-eV emission band found in their high resistive GaN films. Properties of this Gaussian band with partially resolved structure on its high-energy wing have been reported by Kaufmann et al. (1996). The ODMR spectra obtained on the 2.2 eV and 3.0 eV bands from the high-resistive films are shown in Fig. 12. Again two luminescence increasing signals are found, one coinciding in its magnetic resonance parameters with the EM donors. The second resonance has an FWHM of 18 mT, a Lorentzian line shape in contrast to the Gaussian line shape of the deeper donor DD_2, and an isotropic g-value of 1.977. It is concluded that this deep-donor DD_1 is partially effective mass-like and might be associated with the deep donor states 54–57 meV below conduction band. The recombination partner has not been seen in the ODMR experiments.

In addition to the NMR results already mentioned here, two groups (Koschnick et al., 1996; Glaser et al., 1998) employed optically detected electron nuclear double resonance (ODENDOR) to explore defect-related

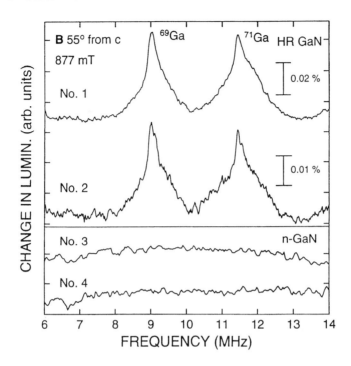

FIG. 13. Optically detected electron nuclear double resonance (ODENDOR) spectra obtained on the shallow donor resonance in four different samples (Nos. 1, 2 are high-resistive GaN; Nos. 3, 4 are n-type GaN) showing the nuclear magnetic resonance of ^{69}Ga and ^{71}Ga (from Glaser et al., 1998).

hyperfine and quadrupolar interactions in GaN. Koschnick et al. (1996) using ODENDOR on a single, undoped film detected rather weak ^{69,71}Ga quadrupole interactions on the EM donor resonance found on the 2.2 eV emission band. Glaser et al. (1998) performed ODENDOR experiments on a set of GaN epitaxial layers. They found strong ^{69,71}Ga lines between 7 and 14 MHz (Fig. 13) on the EM donor resonance from two high-resistive films ($n < 1 \times 10^{16}$ cm^{-3}). The signals could not be observed on as-grown n-type and Si-doped GaN films with $n > 3 \times 10^{16}$ cm^{-3}. This reveals that the observation of ENDOR depends on the carrier concentration. Glaser et al. (1998) could demonstrate that the quadrupole splittings reflect the wurtzite crystal structure and degree of local strain. The ^{69}Ga quadrupole splittings of 2.4 and 2.26 MHz are 15–25% smaller than the values reported by Koschnick et al. (1996) (2.64 MHz), Denninger et al. (1996) (2.85 MHz on a 220-μm thick GaN platelet, which is probably strain-free) and by Han et al. (1988) using the MAS NMR (2.75 MHz on a bulk powder). There are two contributions to the local strain, one arises from the lattice constant mismatch, the other from the difference in thermal expansion coefficients between the GaN film and the sapphire substrate. From x-ray measurements Glaser et al. (1998) could show that the two high-resistive films had different in-plane lattice constants and, hence, could account for the small differences in the quadrupole splittings between the films. Near-neighbor hyperfine interactions, which could contribute to the understanding as to lattice site location of the EM donor, were not observed, instead they were estimated to be smaller than 1 MHz and would contribute only to the line width of the ODENDOR signals. Neither hyperfine nor quadrupole splittings could be seen when monitoring the deep donor resonance.

2. MAGNETO-OPTICAL AND INFRARED ABSORPTION EXPERIMENTS

The binding energy of the EM donors E_D seen in ODMR and EPR can be estimated by a simple effective mass approach using the dielectric constant of 9.7 and the electron-effective mass determined by cyclotron resonance experiments (Drechsler et al., 1995). Neglecting anisotropies in the mass (which are probably small) and the dielectric constant one calculates a binding energy of 31.7 meV for a mass of $m^* = 0.22 m_0$. Recent experiments reported on the Fourier transform infrared absorption experiments on thick HVPE grown GaN films (Meyer et al., 1995). They have been extended meanwhile Meyer (1997) and allowed for the observation of electronic transitions in the range between 200–300 cm^{-1} on a 400-μm substrate-free GaN film. The strongest transition occurs at 209 cm^{-1} and, assuming it to be caused by the $1s - 2p^+$ transition (at 3/4 of E_D), a donor binding energy of 34.5 meV is calculated. Nothing could be stated about the chemical origin of the donor.

From magneto-optical data for Si_{Ga} a value of 29 meV has been obtained (Wang et al., 1996). It was based on the observation of the 1s to $2p^+$ transition in magnetic fields from 15–27 T. From a linear fit Wang et al. (1996) deduced a transition energy at zero magnetic field of 21.7 meV (3/4 of E_D). This linear fit, however, underestimates the binding energy slightly because the behavior of the transition is nonlinear at low magnetic fields. Taking this into account, one obtains for the Si donor a value of 30.4 meV (Meyer, 1997). Moore et al. (1997) have performed similar experiments on HVPE films and found two donors with binding energies of 31.1 and 33.8 meV, respectively. It thus appears that the binding energies measured on different samples and with different approaches agree with each other within a fraction of a meV. In the absence of a chemical fingerprint a positive identification still remains ambiguous, with strong evidence that Si on a gallium site is the shallow donor with E_D of 31 meV. The HVPE films are known to have a rather high concentration of oxygen in connection with high free-carrier concentrations, especially if grown with a ZnO buffer layer, as was the case for the samples used in the experiments of Meyer et al. (1995) and Meyer (1997). It is, therefore, very much likely that oxygen on a nitrogen site accounts for the 34 meV donor but this remains to be confirmed by controlled oxygen doping. Whether the small differences in absolute g-values and anisotropy seen by Baur (1998) could be matched with the two shallow donor binding energies (31, 34 meV) also remains open.

V. Shallow and Deep Acceptors in GaN

Magnesium introduces a shallow acceptor state in GaN with a binding energy of 260 meV (Amano et al., 1989). At low doping levels it gives rise to a characteristic shallow-donor-to-shallow-acceptor pair recombination. The electron-phonon coupling is weak and the zero phonon line is repeated by longitudinal optical phonon replicas of 92 meV. With increasing Mg concentration (the carrier concentration of holes at room temperature is then above 10^{17} cm^{-3}) the donor-acceptor pair band broadens, losing structure and at the highest Mg concentrations ($>10^{19}$ cm^{-3}) eventually shows a red shift (Fig. 14). The deeper acceptor zinc ($E_A = 340$ meV) (Strite, 1994) gives rise to the characteristic blue band and was used in the first light emitting diode structures.

There is an additional shallow donor-to-acceptor recombination involving a residual unidentified acceptor with $E_A = 220$ meV, perhaps involving carbon (Fischer et al., 1995). The ODMR investigations on this particular recombination only revealed the shallow donor resonance and that its amplitude was rather small ($\Delta I/I = 0.015\%$) (Glaser et al., 1996). Nonobservation of the shallow acceptor resonance was explained by the weak spin dependence of the recombination mechanism involving spin thermalized shallow acceptors. Due to random strain fields the acceptor resonance might be broadened beyond detection in analogy to InP:Zn (Viohl et al., 1991). In principle it can also be explained by the crystal structure of GaN on sapphire or 6H-SiC. In the hexagonal modification the heavy hole band is on top of the valence band. Due to the wurtzite crystal field and the spin-orbit interaction the light hole band is split apart by 6–9 meV, depending on the amount of strain in the films. Within the pure $J = 3/2$, $J_z = \pm 3/2$ heavy hole band spin resonance transitions are not allowed; only by

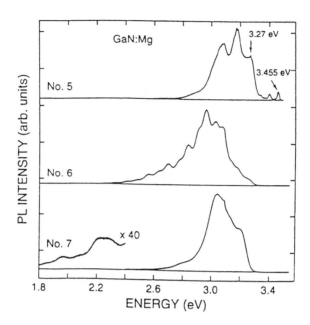

FIG. 14. Photoluminescence spectra of Mg-doped films (No. 5 is lightly doped; Nos. 6 and 7 are highly doped) (from Glaser et al., 1995).

admixture of states due to external perturbations (magnetic field, etc.) do they become weakly allowed. The resonance would be very anisotropic, as the perpendicular component of the g-value is close to zero. Examples are the shallow acceptors in 6H-SiC (Romanov et al., 1986) and CdS (Patel et al., 1981). In a recent magneto-optical study on the neutral donor-bound exciton for the heavy hole band (which might or might not coincide with the g-value of an effective mass-type acceptor), g-values of $g_\| = 1.95$ and $g_\perp = 0$ have been found.

The luminescence band of highly doped GaN:Mg extends from 3.2 eV down to 1.8 eV. Within this spectral window Kunzer et al. (1994; Kaufmann et al., 1996b) followed by Glaser et al. (1995) observed the ODMR signals as shown in Fig. 15. The Mg-related signal has axial symmetry with the g-values $g_\| = 2.080$ and $g_\perp = 2.000$ close to 2 and far away from what would be expected for a shallow state (see the preceding). From the spectral dependence of the Mg ODMR it is concluded that the violet recombination band is of D-A pair character and that the 260 meV acceptor level is involved rather than a deeper (0.5 eV) perturbed Mg-related acceptor. Similar arguments have been given for the blue Zn specific recombination

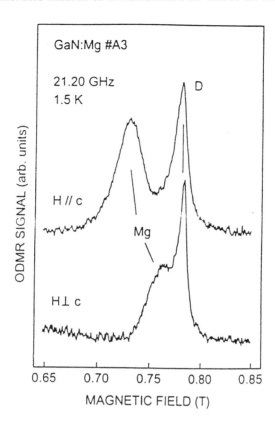

FIG. 15. The ODMR of a Mg-doped GaN film for two orientations of the magnetic field. D is the shallow donor resonance, the Mg acceptor related resonance is also indicated (from Kaufmann et al., 1996).

(Kunzer et al., 1994; Kaufmann et al., 1996b) and the magnetic resonance properties of GaN:Zn (see Table I).

Kaufmann et al. (1996b) present arguments that the Mg is not an EM acceptor and that the hole is distributed mainly over the nearest-neighbor nitrogen ligands surrounding the negatively charged Mg core by analogy to the high-temperature spectrum of the 300 meV deep boron acceptor in cubic SiC (Bratus et al., 1994). However, the shallow acceptors Li and Na in CdS (with similar Bohr radii as Mg in GaN) behave effectively mass-like (Patel et al., 1981). Furthermore, as already outlined, the weak electron-phonon coupling of the Mg D-A pair recombination at least indicates that Mg behaves as a shallow state. For low Mg doping and as long as the D-A pair recombination is well resolvable, to the best of our knowledge no ODMR signal of the Mg acceptor has been observed. The magnetic resonances of Mg and Zn in GaN could also arise from Mg-related centers, if due to the high doping level compensation occurs, and recent ODMR works (Koschnick et al., 1997) at 76 GHz detected two new Mg related defects (see Table I) that were not influenced by annealing to activate the acceptors.

VI. Defects Induced by Particle Irradiation in GaN and AlN

There are two reports (Linde et al., 1997; Watkins et al., 1998) on defect creation in n-type GaN by 2.5-MeV electron irradiation at room temperature with doses of $0.5–2 \times 10^{18}$ cm^{-2}. After irradiation to 1×10^{18} cm^{-2} the bandedge luminescence (the donor-bound exciton and donor-acceptor pair band at 3.28 eV) vanishes completely and the yellow band decreases significantly in intensity. Two new luminescence bands appear in the infrared at 0.85 eV (with a sharp phonon structure) and at 0.95 eV. They anneal partially already at 400 °C and disappear after an anneal at 600 °C, which also leads to a partial recovery of the yellow band. Four new ODMR signals were observed labeled LE1-4, whose g-values are shown in Table II. Three ODMR signals are shown in Fig. 16 in an orientation dependence. The LE 3 shows well-resolved structure and Linde et al. (1997) suggested it to be caused by a gallium interstitial paired off with some other defect. Based on annealing studies (Watkins et al., 1998) this interpretation was modified and the origin of the defects is not clear at present.

Watkins et al. (1998) studied AlN before and after electron irradiation by ODMR. By below bandgap excitation two luminescence bands in the visible spectrum appear and six distinct, well-resolved ODMR spectra are observed

TABLE II

SPIN HAMILTONIAN PARAMETERS FOR RADIATION-INDUCED DEFECTS IN GaN
(from Watkins et al., 1998)

Defect	$g \parallel c$	$g \perp c$	A^{69}/h (MHz)	PL Band
LE1	2.004 ± 0.001	2.008 ± 0.001		both
LE2	1.960 ± 0.002	~2.03		both
LE3		2.002 ± 0.005	1580 ± 50	>0.88 eV
LE4	2.050 ± 0.002	~1.97		<0.83 eV

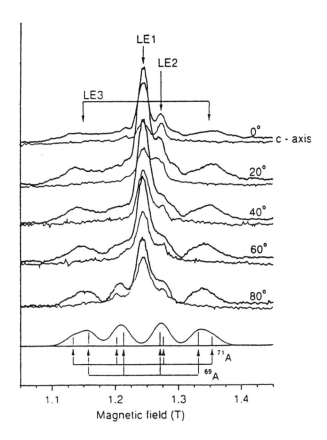

FIG. 16. The ODMR spectra detected in the infrared emission induced by electron irradiation of GaN for different orientations of the sample. The spectrum at 80° is compared with a simulation for hyperfine interaction with a single Ga nucleus and its two isotopes. Light lines show the spectra after anneal at 400 °C (from Watkins et al., 1998).

(Fig. 17). The D5 could be produced by electron irradiation although it was present in the as-grown state of most of the samples studied. It is an $S = 1/2$ center (D1–D4 are $S = 1$ centers) with $g_\parallel = 2.0011$ and $g_\perp = 2.0065$. Its line width varies with the orientation of the crystal with respect to the static magnetic field, and the authors attribute this fact to perhaps unresolved hyperfine interaction with a single nucleus. The D5 spectrum has some similarity with a defect observed by EPR in neutron and electron irradiated polycrystalline AlN and attributed to a nitrogen vacancy (Honda et al., 1990).

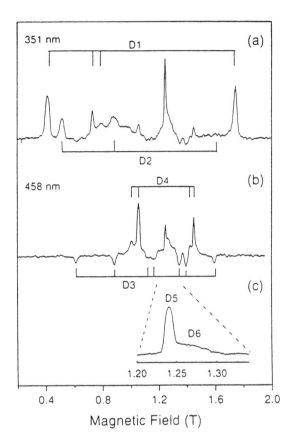

FIG. 17. The ODMR spectra detected in AlN under different below bandgap excitations. D1–D4 are S = 1 triplet centers, D5 is also created by electron irradiation (from Watkins et al., 1998).

VII. Device-Related Magnetic Resonance Studies

Electrical (EDMR) and electroluminescence detection of magnetic resonance (ELDMR) has been used to study defects and their role in the recombination processes in GaN light emitting diodes: GaN $m/i/n/n^+$ diodes, InGaN/GaN double heterostructure devices and blue (In mole fraction 20% in the active region) and green (In mole fraction 43% in the active region) single quantum well diodes (Carlos et al., 1995a, 1995b, 1996; Carlos and Nakamura, 1997; Carlos et al., 1997; Reinacher et al., 1995; Brandt et al., 1996). The room temperature and 4 K current-voltage (I-V) characteristics of a blue InGaN/AlGaN double heterostructure LED together with the size of the resonance-induced change in current is shown in Fig. 18. The dominant defect found by either technique has a g-value of 2.01 and a halfwidth of 13–15 mT (Figs. 19, 20). It is enhanced by high current stressing. Brandt *et al.* (1996) pointed out that, after the low temperature EDMR measurement using low current injection, degradation will appear; this is noticed in Fig. 18 by the difference in the I-V curves at room temperature from the as-shipped state. They also studied the microwave power and temperature dependence of the EDMR signal intensity. The EDMR measurements were restricted to low temperatures; the resonance could not be seen at room temperature. From the rather weak temperature dependence following a $T^{-0.65}$ law they conclude that spin-pair formation

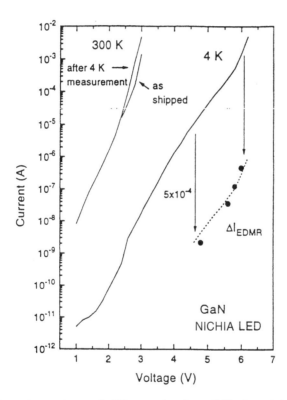

FIG. 18. Room temperature and 4 K current voltage (I-V) characteristics of a blue NICHIA light-emitting diode. Note the difference between the as-shipped state and after 4 K measurements. The full circles show the size of current change at resonance at 4 K (from Brandt et al., 1996).

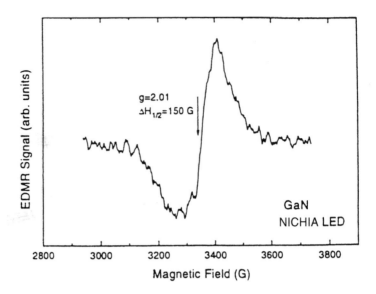

FIG. 19. Electrically detected magnetic resonance (EDMR) signal of the NICHIA LED at 4 K using magnetic field modulation. The resonance parameters are indicated (from Brandt et al., 1996).

is the underlying process causing the spin-dependence. Carlos et al. (1995a, 1995b) compared EDMR and ELDMR in stressed and unstressed LEDs. In EDMR independent on current from 1 μA to 1 mA the $g = 2.01$ resonance is associated with a current increase before stressing and a current decrease after stressing (see Fig. 20, left). This is in contrast to the ELDMR where, depending on bias, the defect is observed in an increase or decrease of the electroluminescence at resonance (see Fig. 20, right). Carlos et al. (1995a, 1995b) pointed out that the defect is present in two locations: (1) a nonradiative recombination path in parallel with the radiative recombination path; and (2) a recombination channel in the depletion region of one of the contacts. At higher bias currents (see Fig. 20, curves labeled b) a second and sharper resonance is seen. It has a value of 1.99 and a line width of 7 mT and shares many similarities with the deep donor resonance seen in ODMR. Reinacher et al. (1997) were able to see the very same defect in light-induced EPR (photo-EPR). It has a characteristic threshold in the excitation spectrum and appears for photon energies above 2.6 eV, already saturates at 2.7 eV, but can be excited with photon energies up to the bandgap energy. This can be explained within model A of Fig. 11. In addition to the signatures of shallow and deep donors, acceptor and

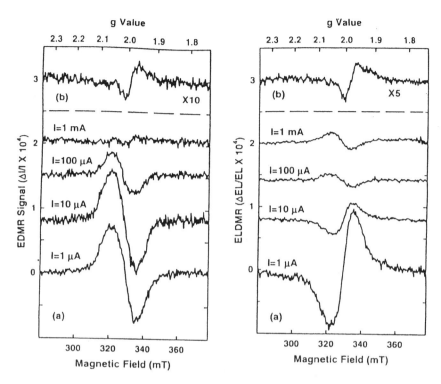

FIG. 20. Electrically detected (EDMR, left) and electroluminescence detected (ELDMR, right) magnetic resonance spectra of a NICHIA InGaN/AlGaN single quantum well light emitting diode. The spectra were taken at 4.2 K using magnetic field modulation ($f = 100\,\text{Hz}$) and show the defect with $g = 2.01$. The defect with $g = 1.99$ is shown in the spectra (b); it was observed under the conditions: $I = 1\,\text{mA}$, 10 kHz for EDMR and $I = 20\,\text{mA}$, 1 kHz for ELDMR (from Carlos et al., 1995a).

deep-trap related resonances are found (see Table I); however, no decisive microscopic defect models have been established to date.

VIII. Transition Metal Impurities

In the near infrared spectral range (0.9–1.3 eV) characteristic luminescence bands (Fig. 21) are observed in GaN films grown by metallorganic vapor phase epitaxy, vapor phase epitaxy, hydride chloride vapor phase epitaxy and the sublimation sandwich technique (SSM). Three zerophonon lines at 1.3 eV, 1.19 eV, 1.047 eV followed by rich phonon sidebands are

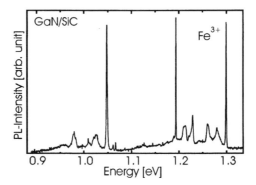

FIG. 21. Photoluminescence spectrum of a GaN on SiC film grown by the sublimation sandwich technique showing internal transitions of transition metal elements present as residual contaminants.

typically observed (Maier et al., 1994; Baur et al., 1994; Wetzel et al., 1994a, 1994b; Pressel et al., 1996). The transition at 1.3 eV is caused by the internal transition $^4T_1(G) \rightarrow {}^6A_1(S)$ as shown by temperature-dependent luminescence and Zeeman experiments (Heitz et al., 1995a). The ODMR (Maier et al., 1994) detected on the internal transition showed the clear fingerprint of a $^6A_1(S)$ ground state of a d^5 $S = 5/2$ configuration, and doping as well as EPR experiments (Baur et al., 1994) presented strong evidence for Fe^{3+} as the responsible defect (Fig. 22). The nature of the other lines was discussed in detail without reaching a conclusive picture as to the chemical nature of

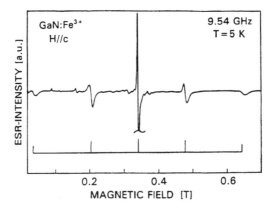

FIG. 22. EPR spectrum of Fe^{3+} in a 40-μm thick GaN layer on sapphire substrate for the magnetic field parallel to the c-axis and at 5 K. The position of the expected five finestructure transitions are indicated in a stick diagram. Additional signals arise from trace impurities in the substrate and from forbidden transitions (from Maier et al., 1994).

the impurities. The experimental results on the 1.047 eV emission fit to a $^4T_2(F) \rightarrow {}^4A_2(F)$ internal electronic transition of a transition metal with a $3d^7$ ($S = 3/2$) electronic configuration (Pressel et al., 1996). It was suggested that the best candidate was Co^{2+}, but Ni^{3+} is equally possible. It should be noted that the emission at 1.047 eV has not been observed as a natural contaminant in GaN samples grown by techniques other than the sublimation sandwich technique. The luminescence band with a zerophonon line at 1.19 eV is attributed to the spin forbidden internal d-d transition $^1E(D) \rightarrow {}^3A_2(F)$ with a d^2 ($S = 1$) configuration. The assignment is based on the observed Zeeman splittings, the small FWHM of the ZPL, the weak phonon sideband and the weak temperature dependence of the luminescence band (Heitz et al., 1995b). Candidates could be Cr^{4+}, V^{3+}, and Ti^{2+}. There is some evidence that the 0.93 eV luminescence (Baur, 1995) arises from V^{3+}. So far, in EPR no $S = 1$ state of a transition metal impurity in GaN could be detected. Figure 23 shows a part of the EPR spectrum of a GaN epitaxial layer on a 6H-SiC substrate (Baranov et al., 1996), recorded at $T = 4$ K. In Fig. 23 a set of six equally intense lines can be seen on both sides of the

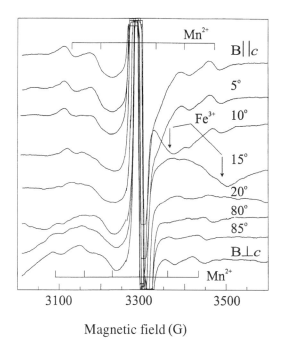

FIG. 23. The EPR spectrum Mn^{2+} in GaN, its orientation dependence, and a stick diagram explaining the hyperfine interaction with $I = 5/2$ of Mn. Additional lines due to Fe^{3+} are indicated (from Baranov et al., 1996).

central line. The strong signal near 3300 G contains, besides the central fine structure line of Fe^{3+}, a signal from the nitrogen donor impurity in the 6H-SiC substrate. The latter was nearly suppressed upon removing substrate. The separations between the lines are about 70 G and the observed splitting corresponds to that of Mn^{2+} in $3d^5$ state (Baranov et al., 1996). The group in Fig. 23 belongs to the Mn^{2+} central fine structure transition ($M_s = 1/2$ to $M_s = -1/2$), split by the hf interaction with nuclear spin $I = 5/2$.

The EPR of Ni^{3+} has the characteristic anisotropy of an $S = 3/2$ system in a strong axial crystalline field (Baranov et al., 1998; Baranov et al., 1997) and positive g shift, which is consistent with the electronic configuration d^7. The ^4F ground term of the free ion is split by a cubic tetrahedral part of crystalline field into two orbital triplets, $^4T_2$ and $^4T_1$, and a groundstate orbital singlet, $^4A_2$. The $^4A_2$ ground state is split into two Kramers doublets by the combined action of the trigonal field and the spin orbit interaction. Omitting the hyperfine interactions, the EPR spectrum (Fig. 24) can be described by a spin Hamiltonian of the form

$$H = g_\parallel \mu_B B_z S_z + g_\perp \mu_B (B_x S_x + B_y S_y) + D[S_z^2 - 1/3S(S+1)]$$

with $S = 3/2$, z denotes the c-axis of the crystal, which is also the principal axis of the center; D is the axial fine structure parameter. Nickel has only one stable odd isotope, ^{61}Ni (natural abundance 1.13%), having nuclear spin $I = 3/2$. No hf structure could be observed. The spin Hamiltonian parameters are listed in Table III. Iron, manganese, and nickel occupy gallium sites in the GaN lattice. It is hence likely that the zerophonon luminescence line at 1.047 eV (Pressel et al., 1996) belongs to the transition $^4T_2(F)-^4A_2(F)$ within 3d levels of a Ni^{3+} ion with a $3d^7$ electronic configu-

TABLE III

SPIN HAMILTONIAN PARAMETERS FOR TRANSITION METALS IN GaN (from Baranov et al., 1998)

			GaN		
	$g_\parallel$	$g_\perp$	$\|D\|$ $10^{-4}\,cm^{-1}$	A $10^{-4}\,cm^{-1}$	References
Fe^{3+}	1.990	1.997	713		Maier et al. (1994)
($3d^5$)	1.995	1.995	715		Baranov et al. (1998)
Mn^{2+}	1.999	1.999	240	70	Baranov et al. (1996)
($3d^5$)					
Ni^{3+}	2.10	$\cong 2.1$	$\geqslant 1.5 \cdot 10^{-4}$		Baranov et al. (1997, 1998)
($3d^7$)					

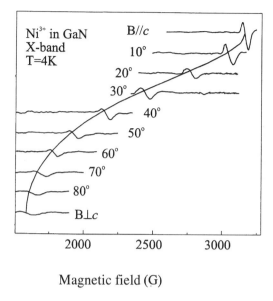

FIG. 24. Orientation dependence of the Ni^{3+} resonance in GaN (from Baranov et al., 1998).

ration, because this luminescence line has been observed only in GaN samples grown by the sandwich technique. On the basis of EPR investigations by Baranov et al. (1996, 1997, 1998), the impurity concentrations in the GaN films could be estimated to be 10^{17}–10^{18} cm^{-3} for Fe, and 10^{16}–10^{17} cm^{-3} for Mn and Ni in the SSM films, around 10^{16} for Fe^{3+} in HVPE and certainly a factor of ten less in MOVPE films.

IX. Outlook

Based on the broad experience with magnetic resonance on intrinsic and extrinsic defects in the compound semiconductors such as GaAs, InP, and GaP, progress in identifying defects in the nitride semiconductors was rather rapid but without being complete. For the transition metal elements EPR and ODMR investigations were followed by detailed magneto-optical experiments. In cases where no EPR results were available (e.g., the 0.93 eV and 1.19 eV luminescence bands in GaN) Zeeman experiments could successfully be performed and allowed conclusions on the spin and charge states of the defects. However, problems remain with the chemical identity (as is the case for the defects monitored in electrically detected magnetic resonance). The status on the shallow effective mass type donors in GaN and AlGaN is best, and there is a convergent picture from other techniques regarding the nature of those shallow impurities. As to the deeper donors and the acceptors, Zn and Mg, work remains to be done if we are to understand their role in the radiative recombination processes. The radiative lifetime of the free and bound excitons is extremely short and, judging from this observation, we have just scratched the surface with defect identification. We are obviously missing or not taking into consideration the extended defects, such as dislocations or inversion domain boundaries. Although yellow luminescence is a prominent defect-related transition, its rather long lifetime renders it unimportant to practical device operation.

REFERENCES

Amano, H., Kito, M., Hiramatsu, K., and Akasaki, I. (1989). *Jpn. J. Appl. Phys.*, **28**, L2112.
Baranov, P. G., Ilyin, I. V., Mokhov, E. N., and Roenkov, A. D. (1996). *Semicond. Sci. Technol.*, **11**, 1843.
Baranov, P. G., Ilyin, I. V., and Mokhov, E. N. (1998). *Mat. Sci. Forum*, **258–263**, 1167.
Baranov, P. G., Ilyin, I. V., and Mokhov, E. N. (1997). *Solid State Comm.*, **101**, 8.
Baur, J., Maier, K., Kunzer, M., Kaufmann, U., and Schneider, J. (1994). *Appl. Phys. Lett.*, **65**, 2211.
Baur, J. (1998). Ph.D. thesis, Freiburg.
Baur, J. Kaufmann, U., Kunzer, M., Schneider, J., Amano, H., Akasaki, I., Detchprohm, T., and Hiramatsu, K. (1995). *Mat. Sci. Forum*, **196–201**, 55.
Boguslawski, P., Briggs, E., White, T. A., Wensell, M. G., and Bernholc, J. (1994). *Mat. Res. Soc. Symp. Proc.*, **339**, 693.
Boguslawski, P., Briggs, E., and Bernholc, J. (1995). *Phys. Rev. B.*, **51**, 17255.
Brandt, M. S., Reinacher, N. M., Ambacher, O., and Stutzmann, M. (1996). *Mat. Res. Soc. Symp. Proc.*, **395**, 657.
Bratus, V., Baran, N., Maksimenko, V., Petrenko, T., Romamenko, V. (1994). *Mat. Sci. Forum*, **143–147**, 81.
Carlos, W. E., Freitas, Jr., J. A., Asif Khan, M., Olson, D. T., and Kuznia, J. N. (1993). *Phys. Rev. B.*, **48**, 17878.
Carlos, W. E. (1996). *Proc. of the Seventh Int. Conf. on Shallow Level Centers in Semiconductors* (1996). C.A.J. Ammerlaan and B. Pajot, eds., Singapore.
Carlos, W. E., Glaser, E. R., Kennedy, T. A., and Nakamura, S. (1995a). *Appl. Phys. Lett.*, **67**, 2376.
Carlos, W. E., Glaser, E. R., Kennedy, T. A., and Nakamura, S. (1995b). *Mat. Sci. Forum*, **196–201**.
Carlos, W. E., Glaser, E. R., Kennedy, T. A., and Nakamura, S. (1996). *J. Electron. Mater.*, **25**, 851.
Carlos, W. E. and Nakamura, S. (1997a). *Appl. Phys. Lett.*, **70**, 2019.
Carlos, W. E., Glaser, E. R., Kennedy, T. A., and Nakamura, S. (1997b). *Mat. Res. Soc. Proc.*, **449**, 757.
Denninger, G., Beerhalter, R., Reiser, D., Maier, K., Schneider, J., Detchprohm, T., and Hiramatsu, K. (1996). *Solid State Comm.*, **99**, 347.

Denninger, G. and Reiser, D. (1997). *Phys. Rev. B.*, **5073**.
Drechsler, M., Meyer, B. K., Detchprohm, D., Amano, H., and Akasaki, I. (1995). *Jap. J. Appl. Phys.*, L1296.
Eckey, L., Podlowski, I., Göldner, A., Hoffmann, A., Broser, I., Meyer, B. K., Volm, D., Streibl, T., Hiramatsu, K., Detchprohm, T., Amano, H., and Akasaki, I. (1996). *Inst. Phys. Conf. Series*, **142**, 943.
Fanciulli, M., Lei, T., and Moustakas, T. (1993). *Phys. Rev. B.*, **48**, 15144.
Fanciulli, M. (1995). *Physica B.*, **205**, 87.
Fischer, S., Wetzel, C., Haller, E. E., and Meyer, B. K. (1995). *Appl. Phys. Lett.*, **67**, 1298.
Glaser, E. R., Kennedy, T. A., Doverspike, K., Rowland, L. B., Gaskill, D. K., Freitas, J. A., Jr., Asif Khan, M., Olson, D. T., Kuznia, J. N., and Wickenden, D. K. (1995). *Phys. Rev. B.*, **51**, 13326.
Glaser, E. R., Kennedy, T. A., Carlos, W. E., Freitas, Jr., J. A., Wickenden, A. E., and Koleske, D. D. (1998). *Phys. Rev. B.*, **57**.
Glaser, E. R., Kennedy, T. A., Crookham, H. C., Freitas, Jr., J. A., Asif Khan, M., Olson, D. T., and Kuznia, J. N. (1993). *Appl. Phys. Lett.*, **63**, 2673.
Glaser, E. R., Kennedy, T. A., Wickenden, A. E., Koleske, D. D., and Freitas, Jr., J. A. (1997). *Mat. Res. Soc. Proc.*, **449**, 543.
Glaser, E. R., Kennedy, T. A., Brown, S. W., Freitas, Jr., J. A., Perry, W. G., Bremser, M. D., Weeks, T. W., and Davis, R. F. (1996). *Mat. Res. Soc. Proc.*, **395**, 667.
Götz, W., Johnson, N. M., Chen, C., Kuo, C., and Imler, W. (1996). *Appl. Phys. Lett.*, **68**, 3144.
Han, O. H., Timken, H. K. C., and Oldfield, E. (1988). *J. Chem. Phys.*, **89**, 6046.
Heitz, R., Thurian, P., Loa, I., Eckey, L., Hoffmann, A., Broser, I., Pressel, K., Meyer, B. K., and Mokhov, E. N. (1995a). *Appl. Phys. Lett.*, **67**, 2822.
Heitz, R., Thurian, P., Pressel, K., Loa, I., Eckey, L., Hoffmann, A., Broser, I., Meyer, B. K., and Mokhov, E. N. (1995b). *Phys. Rev. B.*, **52**, 16508.
Hermann, C. and Weisbuch, C. (1984). In *Optical Orientation*, F. Meier and B. P. Zakharchenya, eds., Amsterdam: North-Holland, p. 463.
Hofmann, D. H., Kovalev, D., Steude, G., Meyer, B. K., Hofmann, A., Eckey, L., Heitz, R., Detchprohm, T., Amano, H., and Akasaki, I. (1995). *Phys. Rev. B.*, **52**, 16702.
Hofmann, D. M., Kovalev, D., Steude, G., Volm, D., Meyer, B. K., Xavier, C., Monteiro, T., Pereira, E., Mokov, E. N., Amano, H., and Akasaki, I. (1996). *Mat. Res. Soc. Proc.*, **395**, 619.
Honda, M., Atobe, K., Fukuoka, N., Okada, M., and Nakagawa, M. (1990). *Jpn. J. Appl. Phys.*, **29**, L652.
Kaufmann, U., Kunzer, M., Merz, C., Akasaki, I., and Amano, H. (1996). *Mat. Res. Soc. Proc.*, **395**, 633.
Koschnick, F. K., Michael, K., Spaeth, J.-M., Beaumont, B., Cibart, P., Off, J., Sohmer, A., and Scholz, F. (1997). *Proc. of the 2nd Int. Conf. on Nitride Semiconductors*, Tokushima, Japan.
Koschnick, F. K., Michael, K., Spaeth, J.-M., Beaumont, B., and Gibart, P. (1996). *Phys. Rev. B.*, **54**, R11042.
Kunzer, M., Kaufmann, U., Maier, K., Schneider, J., Herres, N., Akasaki, I., and Amano, H. (1994). *Mat. Sci. Forum*, **143–147**, 87.
Linde, M., Uftring, S. J., Watkins, G. D., Härle, V., and Scholz, F. (1997). *Phys. Rev. B.*, **55**, R10177.
Maier, K., Kunzer, M., Kaufmann, U., Schneider, J., Monemar, B., Akasaki, I., and Amano, H. (1994). *Mat. Sci. Forum*, **143–147**, 93.
Meyer, B. K., Volm, D., Graber, A., Alt, H. C., Detchprohm, T., Amano, H., and Akasaki, I. (1995). *Solid State Commun.*, **95**, 597.
Meyer, B. K. (1997). *Mat. Res. Soc. Proc.*, **449**, 497.

Moore, W. J., Freitas, Jr., J. A., and Molnar, R. J. (1997). *Phys. Rev. B.*, **56**, 12073.
Neugebauer, J. and Van de Walle, C. G. (1994). *Phys. Rev. B.*, **50**, 8067.
Ogino, T. and Aoki, M. (1980). *Jpn. Appl. Phys.*, **19**, 2395.
Patel, J. L., Nicholls, J. E., and Davies, J. J. (1981). *J. Phys. C.*, **14**, 1339.
Perlin, P., Suski, T., Teysseier, H., Leszczynski, M., Grzegory, I., Jun, J., Porowski, S., Boguslawski, P., Bernholc, J., Chervin, J. C., Polian, A., and Moustakas, T. D. (1995). *Phys. Rev. Lett.*, **75**, 296.
Pressel, K., Nilsson, S., Heitz, R., Hoffmann, A., and Meyer, B. K. (1996). *J. Appl. Phys.*, **79**, 3214.
Reinacher, N. M., Angerer, H., Ambacher, O., Brandt, M. S., and Stutzmann, M. (1997). *Mat. Res. Soc. Proc.*, **449**, 579.
Reinacher, N. M., Brandt, M. S., and Stutzmann, M. (1995). *Mat. Sci. Forum*, **196-201**, 1915.
Romanov, N. G., Vetrov, V. V., and Baranov, P. G. (1986). *Sov. Phys. Semicond.*, **20**, 96.
Roth, L. M., Lax, B., and Zwerdling, S. (1959). *Phys. Rev.*, **114**, 90.
Strite, S. and Morkoc, H. (1992). *J. Vac. Sc. Technol. B.*, **10**, 1237.
Strite, S. (1994). *Jpn. J. Appl. Phys.*, **33**, 699.
Suski, T., Perlin, P., Teysseire, H., Leszczynski, M., Grzegory, I., Jun, J., Bockowski, M., Porowski, S., and Moustakas, T. D. (1995). *Appl. Phys. Lett.*, **67**, 2188.
Viohl, I., Olson, W. D., and Taylor, P. C. (1991). *Phys. Rev. B.*, **44**, 7975.
Wang, Y. J., Ng, H. K., Doverspike, K., Gaskill, D. K., Ikedo, T., Akasaki, I., and Amano, H. (1996). *J. Appl. Phys.*, **79**, 8007.
Watkins, G. D., Linde, M., Mason, P. W., Przybylinska, H., Bozdog, C., Uftring, S. J., Härle, V., Scholz, F., Choyke, W. J., and Slack, G. A. (1998). *Mat. Sci. Forum*, **258-263**, 1087.
Wetzel, C., Volm, D., Meyer, B. K., Pressel, K., Nilsson, S., Mokhov, E. N., and Baranov, P. G. (1994a). *Mat. Res. Soc. Symp. Proc.*, **393**, 453.
Wetzel, C., Volm, D., Meyer, B. K., Pressel, K., Nilsson, S., Mokhov, E. N., and Baranov, P. G. (1994b). *Appl. Phys. Lett.*, **65**, 1033.
Wickenden, A. E., Gaskill, D. K., Koleske, D. D., Doverspike, K., Simons, D. S., and Chi, P. H. (1996). *Mat. Res. Proc.*, **395**, 679.

CHAPTER 10

GaN and AlGaN Ultraviolet Detectors

M. S. Shur

CENTER FOR INTEGRATED ELECTRONICS AND ELECTRONICS MANUFACTURING
AND DEPARTMENT OF ELECTRICAL, COMPUTER, AND SYSTEMS ENGINEERING
RENSSELAER POLYTECHNIC INSTITUTE
TROY, NEW YORK

M. Asif Khan

DEPARTMENT OF ELECTRICAL ENGINEERING
UNIVERSITY OF SOUTH CAROLINA
COLUMBIA, SOUTH CAROLINA

I. INTRODUCTION	407
II. PRINCIPLE OF OPERATION	409
1. *Photovoltaic Detectors*	411
2. *Photoconductive Detectors*	412
III. DETECTIVITY AND NOISE EQUIVALENT POWER (NEP)	415
IV. GaN PHOTODETECTOR FABRICATION	417
V. GaN-BASED PHOTOCONDUCTIVE DETECTORS	419
VI. GaN-BASED PHOTOVOLTAIC DETECTORS	424
1. *GaN p-n Junction Photodetectors*	424
2. *GaN Schottky Barrier Photodetectors*	425
3. *GaN Metal-Semiconductor-Metal Photodetectors*	430
VII. OPTOELECTRONIC AlGaN/GaN FIELD EFFECT TRANSISTORS	430
VIII. CONCLUSIONS AND FUTURE CHALLENGES	436
REFERENCES	437

I. Introduction

Semiconducting materials, whose resistivity may be changed by many orders of magnitude by exposing them to radiation, are a natural choice for photodetectors. They are particularly sensitive to electromagnetic radiation with photon energies exceeding their energy gaps and producing electron-hole pairs. Even silicon with an energy gap of 1.12 eV can be used for the

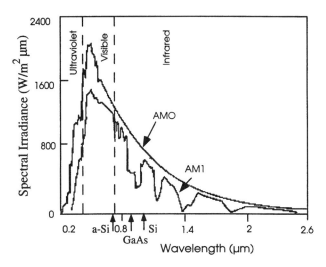

FIG. 1. AM1 and AM0 solar spectra. [From Shur (1990), reproduced by permission of Prentice-Hall, Inc., Englewood Cliffs, NJ.] Also shown are the bandgaps of amorphous Si, GaAs, and Si, which are popular semiconductor materials.

detection of ultraviolet (UV) radiation, corresponding to photon energies higher than 3.1 eV.

However, silicon is not the most appropriate material for this application. This becomes clearer if we analyze the temperature dependence of blackbody radiation as well as the solar spectrum (see Fig. 1).

Such an analysis suggests that under typical conditions, ambient illumination will subject a semiconductor detector to many photons in the visible or infrared (IR) range, and, therefore, it is very desirable to have the UV detector insensitive to the visible and IR radiation (called solar-blind or visible-blind). This carves a unique niche for wide bandgap semiconductors, which are mostly transparent to visible and IR radiation.

Figure 2 shows the bandgaps and lattice constants of semiconductor materials with energy gaps of 3 eV or more.

As GaN is a direct bandgap material with excellent transport and optical properties that has been grown on sapphire, spinel, silicon, and silicon carbide substrates, this semiconductor and related materials, such as AlN, AlGaN, and InGaN, stand out as prime candidates for visible-blind detectors. SiC, which is even further ahead in material quality, has also been used for UV photodetectors. SiC is a better material for UV detectors than silicon but SiC is inferior to GaN for this application, since SiC is an indirect gap semiconductor with a smaller bandgap than GaN.

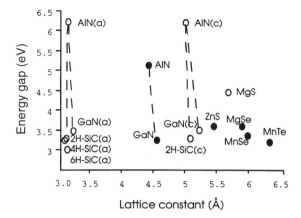

FIG. 2. Bandgaps and lattice constants of semiconductor materials suitable for visible-blind photodetectors. Open circles: wurtzite crystal structure. Solid circles: zinc blende structure. Dashed lines represent ternary compounds (e.g., AlGaN). *a* and *c* correspond to two different lattice constants for wurtzite crystal structure.

II. Principle of Operation

Photons absorbed in a semiconductor may cause different types of transitions. For UV detectors, three types of transitions are the most relevant (see Fig. 3):

1. Intrinsic band-to-band transitions.
2. Valence band to exciton level transitions.
3. Acceptor level to conduction band transitions.

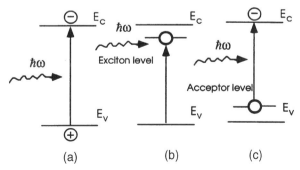

FIG. 3. Intrinsic (valence band to conduction band) transition (a), valence band to exciton level transition (b), and deep acceptor-to-conduction band transition (c).

An exciton can be visualized as an electron-hole pair formation similar to a hydrogen atom. The exciton binding energy in GaN can be estimated as

$$E_B = \frac{q^2}{8\pi\varepsilon_o a_B} = \frac{mq^4}{32\pi^2\varepsilon_o^2\hbar^2} \qquad (1)$$

where ε_o is the dielectric permittivity of GaN ($\varepsilon_o \sim 7.88 \times 10^{-11}$ F/m),

$$m = \frac{m_e m_h}{m_e + m_h} \qquad (2)$$

is reduced effective mass, and

$$a_B = \frac{4\pi\varepsilon_o \hbar^2}{mq^2} \qquad (3)$$

is the effective exciton radius.

For GaN, $m_e \sim 0.23\, m_o$, $m_h \gg m_e$, $m \sim m_e$, where m_o is the free electron mass, and Eqs. (1) and (3) yield $a_B \sim 20.5$ Å and $E_B \sim 40$ meV. These estimates are quite crude because the hydrogen-like model is only valid when $a_B \gg c, a$ where c and a are lattice constants of GaN (see Fig. 2).

The excitons can be dissociated through thermal excitation, or electron-hole pairs forming the excitons can be pulled apart in a high electric field.

The light-generated carriers can produce an electric signal in two ways. They can increase the sample conductivity. If the sample is connected to a voltage source, this will result in an additional photoconductive current. If the sample connected to a current source (which is a voltage source in series with a resistance much greater than the device resistance), the light-generated carriers will cause the voltage across the sample to decrease (see Fig. 4). Such photodetectors are called photoconductors.

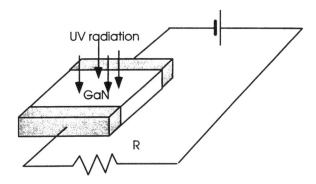

FIG. 4. Circuit diagram for GaN photoconductor.

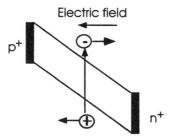

FIG. 5. Electron-hole separation by electric field in a *p-i-n* diode.

Another possibility is to employ a semiconductor device, which has an internal (built-in) electric field of whatever origin. It may be a *p-n* junction diode, a *p-i-n* diode, or a Schottky barrier diode. In either case, electrons and holes may be separated by the built-in electric field or, more typically, by the electric field, which is a combination of the external and built-in electric fields. This charge separation leads to an additional light-generated current. The principle of operation of such a detector is very similar to a solar cell, even though the bias conditions and the device designs are quite different. Such a photodetector is called a photovoltaic detector (see Fig. 5).

1. PHOTOVOLTAIC DETECTORS

The maximum current that can be collected by a photovoltaic detector is given by

$$I_L = qS \int_0^L G_L(x)dx \qquad (4)$$

where q is the electronic charge, S is the device area, L is the dimension of the active region where the light is absorbed, and

$$G_L = \frac{P_l}{\hbar\omega} \alpha \exp(-\alpha x) \qquad (5)$$

is the electron-hole pair generation rate (assuming that each absorbed photon produces an electron-hole pair). Here P_l is the incident radiation intensity (in W/m^2), and $P_l/\hbar\omega$ is photon flux (per unit area per second). We also assume that each light-generated electron and each light-generated hole exit the device contributing to the light-generated current. Substituting Eq. (5) into Eq. (4) and performing the integration we obtain

$$I_L = \frac{qSP_l[1 - \exp(-\alpha L)]}{\hbar\omega} \qquad (6)$$

Equation (6) can be rewritten as

$$I_L = \frac{qSP_l Q_c}{\hbar\omega} \qquad (7)$$

where

$$Q_c = 1 - \exp(-\alpha L) \qquad (8)$$

is the maximum collection efficiency of the photodetector, which should be corrected to account for the light reflection from the surface of the photodetector, that is,

$$Q_c = (1 - R)[1 - \exp(-\alpha L)] \qquad (9)$$

where R is the reflection coefficient for the incident light.

From Eq. (7), we find that the maximum detector responsivity is

$$R_{\max}(A/W) \sim \lambda(nm)/1240 \qquad (10)$$

The ratio

$$Q_e = R_{ph}/R_{\max} \qquad (11)$$

where R_{ph} is the detector responsivity, is called the external quantum efficiency. The upper estimate for R_{ph} is obtained by assuming (quite unrealistically) that $Q_e = 1$. Figure 6 compares the measured values of the responsivity for GaN-based UV photodiodes with the theoretical responsivities for different values of the external quantum efficiency.

2. PHOTOCONDUCTIVE DETECTORS

GaN or AlGaN materials used for UV photodetectors are primarily n-type, because the electron mobility in these compounds is much higher than the hole mobility. This makes the fate of photogenerated electrons and holes quite different. In each of the cases depicted in Fig. 3, electrons wind

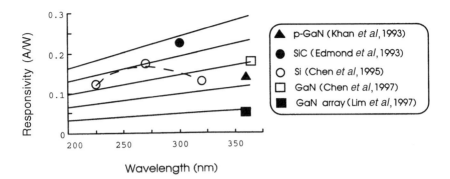

FIG. 6. Responsivities of UV photodiodes. Solid lines show the theoretical limits for the responsivity for different external quantum efficiencies (top line for $Q_e = 1$, step $- 0.2$).

up as free carriers in the conduction band. In an *n*-type material, we expect that electrons supplied by donors fill the deep acceptor traps, which are always present in this partially compensated material. As a consequence, these negatively charged deep acceptor type traps quickly trap the minority carriers (holes). The extra majority carriers can be trapped by neutral acceptor traps, which have already trapped holes. Because the capture cross section for a neutral trap should be much smaller than that for a charged trap, the electron lifetime can be quite long.

Let us now discuss what happens when UV radiation illuminates a piece of GaN with two ohmic contacts and generates electron-hole pairs (see Fig. 4). As discussed previously, holes are trapped but electrons exit at the positively biased ohmic contact and are supplied by a negatively biased ohmic contact.

For a uniform illumination, the generation rate of electron-hole pairs is given by

$$G_L(\omega) = \frac{Q_c P_l}{\hbar \omega t} \qquad (12)$$

where P_l is the absorbed optical power per unit area (in W/m^2), t is the active thickness of the photoconductive film, and Q_c is the quantum efficiency, that is, the number of electron-hole pairs created per absorbed photon.

The recombination rate is given by

$$R_L = \frac{\Delta n}{\tau_l} \qquad (13)$$

where τ_l is the effective electron lifetime. In the steady state, $R_L = G_L(\omega)$. Hence $\Delta n = G_L(\omega)\tau_l$, and the resulting photocurrent is given by

$$I_{ph} = \frac{q\mu_n \Delta n V S}{L} = \frac{q\mu_n G_L \tau_l V S}{L} = \frac{q\mu_n Q_c P_l \tau_l V S}{\hbar\omega L t} \tag{14}$$

where μ_n is the electron mobility, V is the applied voltage, and S is the illuminated device area [see, for example, Shur (1996), p. 467]. It is instructive to compare this current with the maximum light-generated current I_L, produced in a photovoltaic detector with the same quantum efficiency, Q_c, by the same radiation [see Eq. (7)]. The photoconductor gain, A_{ph}, is defined as

$$A_{ph} = \frac{I_{ph}}{I_L} \tag{15}$$

Using Eqs. (7), (13), and (14) we find

$$A_{ph} = \frac{\tau_l}{t_{tr}} \tag{16}$$

where

$$t_{tr} = \frac{Lt}{\mu_n V} \tag{17}$$

is the effective carrier transit time. In large electric fields, the electron velocity saturates, and the transit time $t_{tr} \sim L/v_s$, where v_s is the electron saturation velocity. In short, photoconductors made from materials with long lifetimes and high values of low field mobility and/or saturation velocity, the photoconductor gain may be as high as a million. However, this large gain usually quickly drops with an increase in the signal intensity as the hole traps get filled and additional holes get swept out of the device instead of being trapped.

The transit time t_{tr} of the fastest carriers (usually electrons) determines the detection time of a photodetector. However, the device remains conductive until all photogenerated carriers recombine. Hence, the response time of a photoconductor is proportional to the lifetime of the photogenerated carriers τ_l, which, typically, varies between 10^{-3} and 10^{-10}. In many applications, a detector has to respond to an optical signal with intensity modulated at a certain frequency, ω. To first order, the time dependence of the concentration of the light-generated electrons under such conditions can be found from the simplified continuity equation

$$\frac{d\Delta n}{dt} = G_o \cos(\omega t) - \frac{\Delta n}{\tau_l} \tag{18}$$

The solution of this equation for times much greater than τ_l is given by

$$\Delta n = \frac{G_o \tau_l \cos(\omega t + \varphi)}{\sqrt{1 + \omega^2 \tau_l^2}} \tag{19}$$

Therefore, the responsivity

$$R_{ph}(\omega) = \frac{R_{ph}(0)}{\sqrt{1 + \omega^2 \tau_l^2}} \tag{20}$$

and the maximum frequency at which the detector can follow the variations in the intensity (detector bandwidth) is determined by $1/\tau_l$. As the gain is proportional to the carrier lifetime, there is a clear trade-off between the gain and the speed of a photoconductor. The detector gain-bandwidth product scales approximately with $1/t_{tr}$.

III. Detectivity and Noise Equivalent Power (NEP)

The quality of a photodetector is determined by how little radiation intensity it can detect. This smallest detectable intensity depends on the background radiation and/or on the signal noise level. The quantitative measure is the noise equivalent power (NEP), which is defined as the incident rms optical power required in order to obtain signal-to-noise ratio (S/N) of 1 in a bandwidth of 1 Hz. The NEP can be interpreted as the minimum power that the photodetector can detect and can be presented as

$$\text{NEP} = \langle i_n^2 \rangle / R \tag{21}$$

where $\langle i_n^2 \rangle$ is the noise current, and R is the detector responsivity. The related figure of merit is detectivity. Detectivity is defined as

$$D^* = \frac{\sqrt{SB}}{\text{NEP}} \tag{22}$$

where B is the bandwidth and S is the detector area. For IR detectors, the maximum detectivity is limited by the background radiation. Such detectors are called background limited infrared photoconductor (BLIP) detectors. In the visible-light or UV range, background radiation does not produce that many photons, and the maximum detectivity of detectors operating in these spectral ranges is limited by the signal fluctuation limit (SFL). The maximum detectivity of a BLIP detector is given by

$$D^*_{BLIP}(T, \lambda, B = 1 \text{ Hz}) = \frac{\lambda^2 \exp(\zeta)}{2c[\pi h k_B T(1 + 2\zeta + 2\zeta^2)]} \text{ (cm Hz}^{1/2}/\text{W)} \quad (23)$$

see Kruise et al. (1962). Here c is the speed of light, T is temperature, h denotes Planck constant, λ is the wavelength, and $\zeta = hc/(\lambda k_B T)$.

In the SFL limit,

$$D^*_{SFL}(\lambda, B = 1 \text{ Hz}) = \frac{10^{-4}\lambda}{9.22 hc A^{1/2}} \text{ (cm Hz}^{1/2}/\text{W)} \quad (24)$$

(see Kruise et al., 1962 and Kruise, 1977). Figure 7 shows the computed spectral dependencies of D_{SFL} (straight line) and D^*_{BLIP} for the temperatures 77 K, 300 K, and 1000 K.

As can be seen from the figure, the limiting detectivity value is much higher (and far less achievable) for UV detectors (the SFL limit). Only photomultipliers have approached this limit. The detectivity of semiconductor UV photodetectors is usually limited by noise.

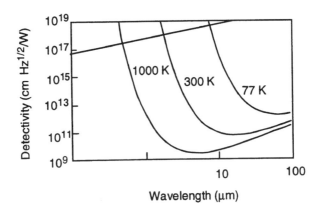

FIG. 7. Computed spectral dependencies of D^*_{SFL} (straight line) and D^*_{BLIP} for temperatures of 77 K, 300 K, and 1000 K.

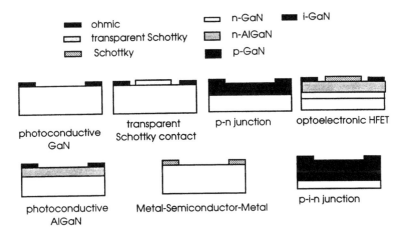

FIG. 8. Different kinds of photodetectors realized using the GaN materials system. (Reprinted with permission from Shur and Khan, 1997.)

Different types of noise include the recombination-generation noise due to the randomness in the recombination and generation processes, the noise caused by the fluctuations in the carrier velocity, shot (or injection) noise, and thermal (Johnson-Nyquist) noise. The detector noise can come from a semiconductor region or originate in the contact regions. At low frequencies (below 1 kHz or so), all these contributions to noise are often exceeded by a $1/f$ noise (also called a flicker noise), the nature of which is still mysterious (van der Ziel, 1986).

Figure 8 schematically shows different kinds of photodetectors realized using the GaN materials system. Recently, a GaN avalanche photo diode (APD) photodetector has been demonstrated (see Osinsky *et al.*, 1998). However, a very high bulk breakdown field in GaN makes the realization of practical GaN and AlGaN APDs a real challenge.

IV. GaN Photodetector Fabrication

GaN may be grown epitaxially on different substrates — sapphire, 6H-SiC, 2H-SiC, Si, and other materials. A typical technique of GaN growth is metalorganic vapor phase epitaxy (MOVPE); see the review paper by Morkoç *et al.* (1994), which contains a detailed discussion of different substrates and different techniques of epitaxial growth. Many GaN-based photodetectors and field effect transistors (FETs) use the epitaxial layers deposited using low pressure metallorganic chemical vapor deposition (MOCVD) over basal plane sapphire substrates. Khan et al. (1991) used triethylgallium, triethylaluminum and ammonia as the precursors for Ga, Al, and N. Typical growth pressure and temperature were 76 torr and 1000 °C, respectively. Typical flows for the two metallorganics were in the range of 1–10 and 1.5–0.6 mmoles/min, respectively. As deposited, the GaN layers are highly resistive with carrier densities well below 10^{15} cm^{-3}. The insulating GaN layers can be doped *n*-type using disilane (Si) as the dopant.

Electron cyclotron resonance (ECR) etching techniques have been used for GaN device fabrication (Pearton et al., 1995). Reactive ion etching and Cl-based etches have been explored as well (Morkoç et al., 1994). Proton or helium bombardment has been used to isolate devices.

The *n*-type and *p*-type regions have been produced in GaN using Si$^+$ and Mg$^+$/p^+ implantation and subsequent anneal at temperatures up to 1100 °C (Pearton et al., 1995). Semi-insulating, *n*-type, and *p*-type GaN and AlGaN have been grown by MOCVD and MBE with electron concentrations from 7×10^{16} cm^{-3} to 10^{19} cm^{-3} and hole concentrations up to 5×10^{18} cm^{-3} (Morkoç et al., 1994).

Just as for the GaAs based material system, good ohmic contacts to GaN-based devices are crucial for obtaining good performance and for realizing reliable devices operating in a wide temperature range. Al and Au ohmic contacts to GaN yielded specific contact resistances of 10^{-4} and 10^{-3} Ω-cm^2, respectively (Foresi and Moustakas, 1993). Khan et al. (1996a) reported on a systematic study of the contact resistance to doped layers of GaN. They demonstrated the direct proportionality between the sheet resistance of the doped GaN layer and specific contact resistance.

The University of Illinois group has developed a new metallization scheme for obtaining very low ohmic contact to *n*-GaN. They deposited a composite metal layer Ti/Al/Ni/Au (150 Å/2200 Å/400 Å/500 Å) on *n*-GaN preceded by a reactive ion etching (RIE) process and annealed at 900 °C for 30 s. The resulting contacts had specific resistivity values of 8.9×10^{-8} Ω-cm^2 or lower for the doping level of 4×10^{17} cm^{-3} (Fan et al., 1996).

Many groups have demonstrated good quality Schottky contacts to GaN. Mohammad et al. (1996) reported on the fabrication and characterization of the Pt/*n*-GaN Schottky barrier. The *n*-GaN was grown by the reactive molecular beam epitaxy (MBE) method for the fabrication of these diodes. The capacitance-voltage (C-V) characteristics indicated a small trap density in the semiconductor, and the current-voltage (I-V) characteristics had an ideality factor very close to unity. The barrier height deduced both from I-V and C-V measurements was about 1.10 eV (Mohammad et al., 1996).

V. GaN-Based Photoconductive Detectors

The energy gap of GaN-based materials changes from 1.89 eV for InN to 3.39 eV for GaN and 6.2 eV for AlN, and a wide range of GaN-AlN and InN-GaN solid state solutions has been demonstrated. Hence, in principle, intrinsic photodetectors based on this materials system may span the wavelength range from 656 nm–200 nm. This includes practically the entire visible range (400–770 nm) as well as much of the ultraviolet range (200–400 nm). Extrinsic GaN-based photodetectors (using donor and acceptor levels) are also possible.

Pankove and Berkeyheiser (1974) reported on photoconductivity of GaN films doped with zinc. Razeghi and Rogalski (1996) and Shur and Khan (1997) reviewed recent work on GaN and related photoconductors.

Khan et al. (1992a) reported on a photoconductive detector using insulating GaN epitaxial layers grown on sapphire substrates. These devices were grown on (0001) sapphire substrates and consisted of 0.8-μm insulating GaN layer grown over a 0.1-μm AlN buffer. Figure 9 shows the detector photoluminescence and optical transmission. As seen from the figure, both the optical absorption and the photoluminescence signal peaked around 365 nm. The detectors had an interdigitated structure (see Fig. 10). The peak detector responsivity was over 1000 A/W (see Fig. 11). The detectors had a very large dynamic range (over five orders of magnitude for the excitation wavelength of 254 nm, see Fig. 12). The characteristic response time was on the order of 1 ms.

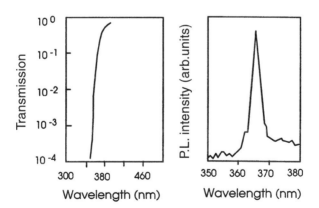

FIG. 9. GaN detector photoluminescence and optical transmission (after Khan et al., 1992a).

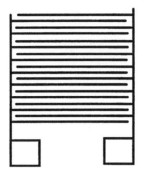

FIG. 10. GaN detector interdigitated finger design (after Khan et al., 1992a).

Stevens et al. (1995) described a photoconductive ultraviolet sensor using Mg-doped MBE-grown GaN on Si(111). Kung et al. (1995) studied the kinetics of photoconductivity in an n-type GaN photodetector. Qiu et al. (1995) presented the results of their study of defect states in GaN films using photoconductivity measurements. More recently, both the Northwestern University and APA Optics, Inc. groups demonstrated AlGaN photoconductive detectors (see Walker, 1996, 1996a; Khan, 1996a; Lim, 1996).

Stevens et al. (1995) reported responsivities on the order of 12 A/W at 4 V bias with a cutoff at 3.3 eV. (At 5 V, the responsivity was 14 A/W for $E = 3.4$ eV, 7 A/W for $E = 3.3$ eV, and 0.1 A/W at 2.9 eV). The characteristic response time was 0.67 ms, which is orders of magnitude higher than the longest measured lifetime for GaAs.

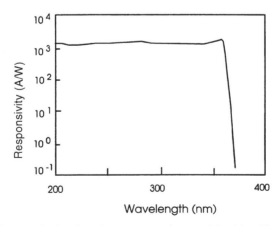

FIG. 11. Photoconductive GaN detector spectral responsivity (after Khan et al., 1992a).

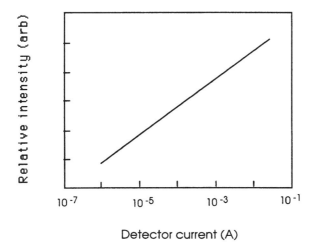

FIG. 12. GaN detector responsivity as a function of incident power at 254 nm (after Khan et al., 1992a).

Qiu et al. (1995) showed that in their MOCVD-grown samples, the photoconductivity response decreased by more than four orders of magnitude with Mg doping. Such a doping seems to generate defect states in the upper half of the energy gap and reduce the density of defect states in the lower half of the energy gap. Figure 13 shows a qualitative distribution of

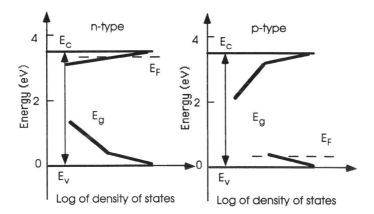

FIG. 13. Qualitative distribution of localized states deduced by Qiu et al. (1995) from the photoconductivity measurements of defect states of n-type, undoped, semi-insulating p-type, and conducting p-type samples (after Qiu et al., 1995).

localized states deduced by Qiu *et al.* from the photoconductivity measurements of defect states of *n*-type, undoped, semi-insulating *p*-type and conducting *p*-type samples. These defect states are much more pronounced than for other crystalline semiconductors.

Johnson *et al.* (1996) studied persistent photoconductivity in Mg-doped *p*-type GaN samples. They showed that at both 240 K and 200 K, the photoconductivity in *p*-type GaN persists for several hours. The time constant of the persistent photoconductivity is thermally activated:

$$\tau_{ppc} = \tau_o \exp\left(\frac{E_{ppc}}{k_B T}\right) \quad (25)$$

where $E_{ppc} \sim 55$ MeV (see Fig. 14).

AlGaN visible photoconductors may use $Al_x Ga_{1-x} N$ epitaxial layers with *x*-values ranging from 0–1. Lim *et al.* (1996) reported on AlGaN photoconductive detectors, which consisted of a 0.5 μm thick AlGaN layer deposited over a basal plane sapphire substrate (with a usual buffer layer). Standard photolithography and lift-off procedures were used to pattern the devices. The detector structure consisted of an interlaced Ti/Au electrode pattern with a 10-μm interelectrode spacing and an approximate area of 1 × 1 mm. The responsivity of the detectors was then measured as a function of wavelength using a UV-visible spectrophotometer as the light source. The wavelength-dependent power output of the spectrophotometer was calibrated using a UV-enhanced Si *p-n* detector. Figure 15 shows a typical spectral response of an AlGaN photoconductive detector. Depending on the molar fraction of Al, the detectors with the Al molar fraction ranging from 0–0.61 had the cutoff wavelengths from 365–250 nm. The transmission spectra for three samples with *x*-values of 0.46, 0.55, and 0.61 are shown in

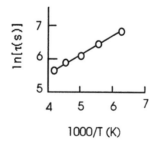

FIG. 14. Time constant of persistent photoconductivity vs temperature (after Johnson *et al.*, 1996).

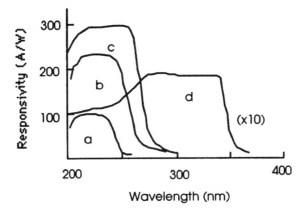

FIG. 15. Typical spectral response of an AlGaN photoconductive detector (after Lim et al., 1996). Curves a, b, c, and d correspond to $Al_xGa_{1-x}As$ with $x = 0.61, 0.55, 0.46$, and 0.05, respectively.

Fig. 16. As seen, the transmission cutoff is very abrupt, indicating a high crystalline and optical quality. Fabry-Perot fringes are also clearly visible in the transmission region. Using the fringe separation and layer refraction indices, Lim et al. (1996) estimated the active layer thicknesses to be between 0.4 and 0.5 μm. These calculated values agreed well with expected thicknesses estimated from layer growth rates. The dc current-voltage characteristic measurements indicated the as-deposited AlGaN layers to be highly

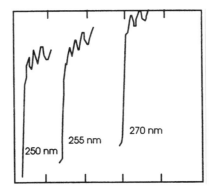

FIG. 16. Transmission spectra for three samples with x-values of 0.46, 0.55, and 0.61 (after Lim et al., 1996). The curves for 250 nm, 255 nm, and 270 nm correspond to $Al_xGa_{1-x}N$ with $x = 0.61, 0.55$, and 0.46, respectively.

resistive. This makes the material well suited for photoconductive detectors operating in the dc mode.

The comparison of Figs. 15 and 16 shows the spectral responsivity to be sharply peaked at the band-edge cutoff wavelength, establishing the intrinsic nature of the photoconductivity. The standard silicon *p-n* detector had a responsivity of about 0.1 A/W in the wavelength region used in these measurements. Assuming that the Si photodetector has a gain of 1, Lim *et al.* (1996) deduced a significant photoconductive gain for these AlGaN sensors. The gain is the result of trapping of minority carriers (holes) at acceptor impurities. This also suggests compensation as one of the causes of the high resistivity of the active layer. This argument is further supported by the measured values of bandwidth (1 kHz) which are lower than expected for direct gap semiconductors.

In addition to speed, noise and detectivity are important characteristics of a photodetector. Preliminary noise studies of noise in GaN photoconductors have shown that $1/f$ noise is the dominant source of noise in these devices (Levinshtein, 1996).

VI. GaN-Based Photovoltaic Detectors

Schottky barrier detectors, transparent Schottky barrier detectors, Schottky barrier detector arrays, metal-semiconductor-metal (MSM) photodetectors, *p-n* junction photodetectors, and *p-i-n* photodetectors have been demonstrated in GaN.

1. GaN p-n Junction Photodetectors

Figure 17 shows a schematic structure of a GaN *p-n* junction detector (Chen *et al.*, 1995; Khan *et al.*, 1995a). The doping level of the *p*- and the *n*-type layers in the junction region was around 5×10^{16} cm^{-3}. The current-

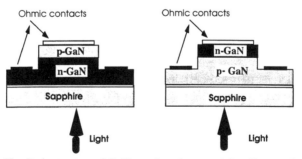

FIG. 17. Basic structure of GaN *p-n* photodetectors (after Chen *et al.*, 1995).

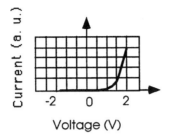

FIG. 18. I-V curve of *p-n* junction GaN UV detector (after Chen *et al.*, 1995).

voltage characteristic for this detector is shown in Fig. 18. A 1.5 V turn-on and a reverse breakdown in excess of 10 V was observed.

Figure 19 shows the responsivity of this detector as a function of wavelength. As seen, no photoresponse is observed for wavelengths larger than the bandgap (365 nm). Below 365 nm, the responsivity is nearly independent of the wavelength, indicating a very low surface recombination.

2. GaN Schottky Barrier Photodetectors

The first photovoltaic detectors realized in GaN were GaN Schottky barrier photodetectors on *p*-type GaN (Khan *et al.*, 1993). Figure 20 shows the basic structure of these GaN Schottky barrier photodetectors. The

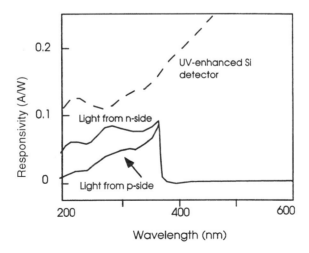

FIG. 19. Spectral response of GaN *p-n* junction photodetector (after Chen *et al.*, 1995).

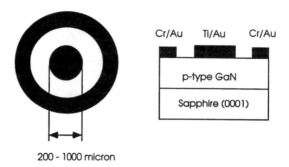

FIG. 20. Basic structure of GaN Schottky barrier photodetectors (after Khan *et al.*, 1993).

p-type doping level was around 5×10^{17} cm^{-3}. The thickness of GaN films was $2\,\mu$m.

The Cr/Au ohmic contacts were relatively poor. The Schottky contact had a turn-on voltage of around 1.5 V and a reverse breakdown voltage of 3 V. The detector was illuminated from the bottom (i.e., from the sapphire side). The measured responsivity was approximately 0.13 A/W at 320 nm and rapidly decreased with an increase in the wavelength beyond 365 nm (see Fig. 21). The detector response time was around 1 μs and was limited by the RC constant of the measurement setup.

More recently transparent Schottky barrier detectors using n-type GaN have been reported (Chen *et al.*, 1997). A transparent Schottky contact allowed for front illumination. The vertical geometry (as opposed to the

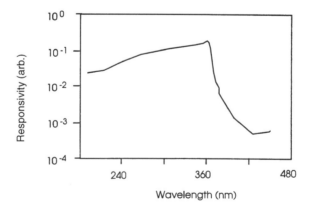

FIG. 21. Spectral dependence of the responsivity of p-type Schottky barrier photodiodes (after Khan *et al.*, 1993).

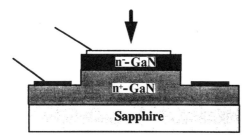

FIG. 22. Schematic structure of transparent Schottky barrier detectors using n-type GaN (after Chen et al., 1997).

mesa structures shown in Fig. 19) leads to a dramatic reduction in the series resistance reducing the RC limited response time down to 118 ns for a 50-Ω load. Figure 22 shows a schematic structure of such a detector.

The epilayer structure used for these devices consisted of a 40 Å AlN buffer layer deposited at 600 °C on a basal plane sapphire substrate followed by a 1 μm thick n^+ GaN layer ($n = 3 \times 10^{18}$ cm^{-3}) and a top n^- 0.4 μm thick GaN layer ($n = 3 \times 10^{16}$ cm^{-3} to 1×10^{17} cm^{-3}) deposited at 1000 °C and 75 torr using a low-pressure MOCVD system. The bottom ohmic contacts were fabricated using e-beam evaporation of 15 nm of Ti and 200 nm of Al followed by annealing for 30 s at 800 °C. The Schottky contact was made using a 5-nm layer of e-beam evaporated Pd. The measured Schottky barrier height was close to 0.9 eV, similar to that for thick Pd Schottky contacts on GaN (Schmitz et al., 1996). The reverse breakdown voltage was approximately 20 V. The reverse saturation current was close to 6×10^{-12} A/cm^2 (estimated for the detector with the area of 4 mm × 4 mm). The leakage current was much larger than the saturation current and is described by the following expression:

$$I_{leak} = \frac{|V|}{R_{leak}} \qquad (26)$$

where $R_{leak} \sim 62.5$ MΩ. Chen et al. (1997) suggested that threading dislocations with an estimated density of up to 10^9 cm^{-2} (Qian et al., 1995) may cause this parallel leakage path.

Figure 23 (Chen et al., 1997) shows the spectral response of the detector. The figure also shows the upper bounds for the responsivity for different values of the overall quantum efficiency. As can be seen from the figure, the responsivity is quite high (up to 0.18 A/W) and, at least at shortwave

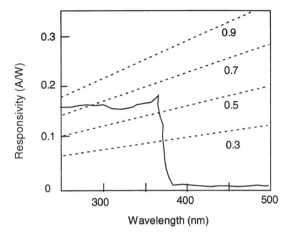

FIG. 23. Spectral response of vertical transparent Schottky GaN detector at −5 V bias. The dashed lines show the maximum theoretical response for the different values of the quantum efficiencies (indicated by numbers near the lines) (after Chen *et al.*, 1997).

lengths, seems to be entirely limited by the Schottky contact transparency.

Figure 24 shows the decay time of photoresponse of GaN Schottky diode detectors vs load resistance (Chen *et al.*, 1997). The decay time constant is proportional to the load resistance and is approximately 118 ns at 50-Ω load. This linear dependence shows that the response is RC limited. These

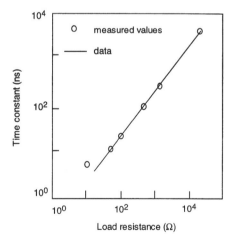

FIG. 24. Decay time of photoresponse vs load resistance (after Chen *et al.*, 1997).

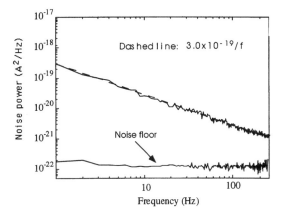

FIG. 25. Measured noise power as a function of frequency for vertical transparent GaN photodetectors (after Chen *et al.*, 1997).

detectors are considerably faster than the *p*-type lateral geometry GaN Schottky detectors described in the preceding text.

Figure 25 (Chen *et al.*, 1997) shows the results of the noise measurements. As can be seen, the $1/f$ noise is dominant. These measurements yielded noise equivalent power (NEP) less than 4×10^{-9} W (which is comparable to commercial Si photodetectors).

Lim *et al.* (1997) fabricated Schottky barrier photodetectors on $Al_{0.26}Ga_{0.74}N$ in order to shift the cut-off wavelength below 300 nm (see Fig. 26). These devices had an electrode spacing shorter than the expected minority diffusion length in AlGaN (about $1-2\,\mu m$). The epitaxial structure for these photovoltaic detectors consisted of a 0.7-μm thick n^+-$Al_{0.26}Ga_{0.74}N$ layer followed by a 0.3-μm thick active layer of $Al_{0.26}Ga_{0.74}N$. These devices had a slower photoresponse, which was RC limited down to about $1\,k\Omega$ and

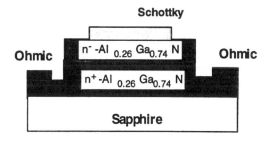

FIG. 26. Schematic device structure of vertical AlGaN photodetectors.

device limited at smaller load resistances. At 50 Ω, the decay time was approximately 1.6 µs, which corresponds to the bandwidth of 100 kHz. The cutoff wavelength for the responsivity was 290 nm. The detector responsivity reached 0.07 A/W, which corresponded to a 30% quantum efficiency.

Lim *et al.* (1997) also demonstrated an 8 × 8 GaN Schottky barrier photodiode array for visible-blind imaging based on the devices already described here.

3. GaN METAL-SEMICONDUCTOR-METAL PHOTODETECTORS

Metal-semiconductor-metal (MSM) photodetectors (see Fig. 27) operate with the same principles as *p-n* or *p-i-n* photodetectors. Their response is related to the current caused by the electron-hole pairs separated by the electric field in the depletion region of two Schottky diodes. The interdigitated design minimizes parasitic resistances. Sometimes, these contacts are made circular to suit circular light beams. Their speed is determined by the transit time, and submicron MSMs have bandwidths of up to 350 GHz.

Several groups studied GaN MSMs (e.g., Carrano *et al.*, 1997; Carrano *et al.*, 1998; Binet *et al.*, 1997). Figure 28 (Carrano *et al.*, 1998) shows measured and modeled dark current-voltage characteristics of such detectors with different interelectrode spacing. These devices exhibited a very low dark current ($\sim$ 57 pA at 10 V reverse bias) and had a typical sharp bandedge cutoff and a high responsivity (possibly enhanced by a photoconductive gain).

VII. Optoelectronic AlGaN/GaN Field Effect Transistors

Khan *et al.* (1995a, 1995b) reported on a photodetector based on a 0.2-µ gate AlGaN/GaN heterostructure field effect transistor (HFET). The epilayer structure and processing details for the gated photodetectors are

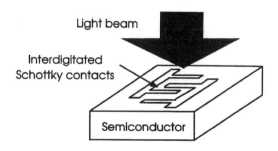

FIG. 27. Schematic diagram of an MSM photodetector (after Shur, 1996).

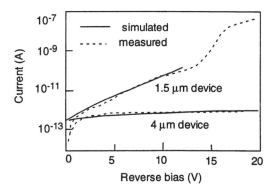

FIG. 28. Experimental (dashed) and modeled (solid) voltage curves for typical 1.5 and 4.0 mm devices (after Carrano et al., 1998).

identical to those for the short gate AlGaN/GaN HFETs (see Fig. 29 from Khan et al., 1994). Figure 30 shows the HFET drain current (at the drain-to-source voltage $V_{ds} = 10$ V) as a function of the gate bias with and without illumination from a 2 mW He-Cd laser from the sapphire substrate side. Figure 31 (solid line) shows the responsivity vs the gate bias for the gated photodetector.

The response time of the HFET photodetector was measured using a mechanical chopper and the synchronous detection of the photocurrent. These transient waveforms indicated an approximate response time of 2×10^{-4} s.

Figure 32 shows the spectral dependence of the responsivity in relative units. As can be seen from the figure, the responsivity falls sharply by several orders of magnitude for wavelengths greater than 365 nm (which corre-

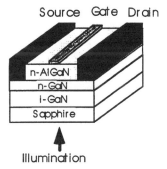

FIG. 29. Schematic diagram of optoelectronic of AlGaN/GaN HFET (from Khan et al., 1995b).

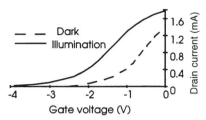

FIG. 30. I-V characteristics of AlGaN/GaN HFETs in the dark and under light (from Khan et al., 1995b).

sponds to the i-GaN bandgap). This clearly demonstrates the visible blind operation of this gated photodetector.

Based on the experimental results, Khan et al. (1995a) proposed the following mechanism of the detector operation. The laser light generates electron-hole pairs in the semi-insulating i-GaN buffer layer and in the

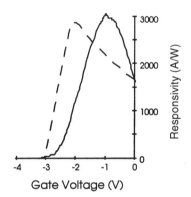

FIG. 31. Measured (solid line) and calculated (dashed line) responsivity of 0.2-μm gate AlGaN/GaN HFET photodetector illuminated with a He-Cd laser from the bottom sapphire-GaN interface. Drain-to-source voltage 10 V. Source-to-drain spacing $L_{sd} = 2\ \mu$m. Gate width $W = 35\ \mu$m. Laser power is $P_{beam} = 2$ mW. Laser beam radius $r_L = 1$ mm. The power of light P is estimated as $P \sim P_{beam} L_{sd} W / \pi r_L^2$. Parameters used in the calculation: electron saturation velocity $v_s = 1.2 \times 10^5$ m/s; GaN doped layer thickness $d = 300$ Å ($\Delta d \sim d/2$); AlGaN barrier thickness $d_i = 300$ Å; wavelength $\lambda = 325$ nm; quantum efficiency (including reflection losses) $Q_c = 0.09$; electron mobility $\mu = 600$ cm^2/Vs; source series resistance $R_{sm} = 30\ \Omega$ mm; gate length $L = 0.2\ \mu$m; i-GaN buffer layer thickness $d_b = 2\ \mu$m; absorption coefficient $\alpha = 3 \times 10^4$ cm^{-1}; effective recombination time $\tau = 0.1$ ms; AlGaN dielectric constant $\kappa_s = 8.9$; threshold voltage $V_T = -2.2$ V; temperature $T = 300$ K; maximum concentration of the 2D electron gas $n_{smax} = 7 \times 10^{11}$ cm^{-2}; subthreshold ideality factor $\eta = 2$; HFET interpolation parameters (defined by Lee et al., 1993, pp. 661–662), $\delta = 2$, and $\gamma = 4$. (Reprinted with permission from Khan et al., 1995b.)

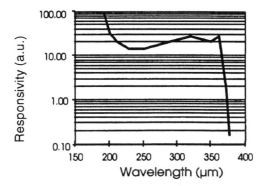

FIG. 32. Measured spectral responsivity of the AlGaN/GaN optoelectronic HFET. (Reprinted with permission from Khan et al., 1995b.)

active n-GaN region. The generation rate G of the electron hole pairs per unit area is given by

$$G = \frac{Q_c P_l \lambda}{2\hbar \pi c} \quad (27)$$

where Q_c is the quantum efficiency, λ is the laser wavelength, $\hbar$ is the reduced Planck constant, and c is the velocity of light in vacuum,

$$P_l = P_o \alpha \int_0^{d_o} \exp(-\alpha y) dy = P_o[1 - \exp(-\alpha d_o)] \quad (28)$$

is the absorbed power of light per unit area, $P_o \sim (1 - R)P_{\text{beam}}/(\pi r^2)$ is the power of incoming radiation at the sapphire-GaN interface per unit area, R is the effective reflection coefficient, P_{beam} and r are the laser beam power and laser beam radius, respectively, α is the absorption coefficient, and d_o is the effective absorption length. d_o may be greater than the combined thickness of the GaN buffer layer and active layer because the light reflections from the top and bottom surfaces may help capture some additional power of light.

For the devices described by Khan et al. (1995a), $\alpha = 3 \times 10^4 \text{ cm}^{-1}$ at 325 nm, and, therefore, $P_l \sim P_o$ because $d_o > 2 \mu$m. As shown in Fig. 33, the photogenerated holes move toward the i-GaN/sapphire interface, and the photogenerated electrons move toward the channel where they are quickly driven into the drain by the high channel field.

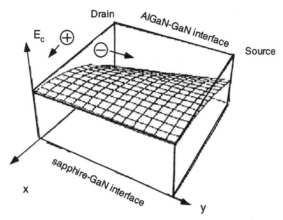

FIG. 33. Qualitative band diagram of gated AlGaN/GaN HFET photodetector. y is the direction from drain to source, x is the direction perpendicular to the AlGaN/GaN interface. (Reprinted with permission from Khan et al., 1995b.)

The total surface concentration of the photoinduced holes is given by

$$p_s = G\tau \tag{29}$$

where τ is the effective hole lifetime. The hole charge causes the shift ΔV_T in the device threshold voltage under illumination (compare with Fig. 30)

$$\Delta V_T \approx -\frac{qp_s(d_i + \Delta d)}{\varepsilon_s} \tag{30}$$

where q is the electronic charge, d_i is the thickness of the AlGaN barrier layer, ε_s is the AlGaN dielectric permittivity, and Δd is the effective thickness of the 2D electron gas.

The sheet electron concentration Δn_s induced by light into the HFET active layer can be estimated as follows. The light-induced device current $\Delta I = q\Delta n_s v_n W$, where v_n is the average electron velocity in the channel and W is the device width. For $v_n \sim 5 \times 10^4$ m/s, $W = 35\ \mu$m, and the measured value of $\Delta I_s \sim 0.5$ mA, we find $\Delta n_s \sim 1.8 \times 10^{11}$ cm^{-2}, $\Delta n = \Delta n_s/d \sim 0.6 \times 10^{17}$ cm^{-3} where $d \sim 300$ Å is the thickness of the GaN active layer.

Khan et al. (1995a) calculated the HFET current-voltage characteristics in the dark and under illumination using the HFET model described by Lee et al. (1993). The effect of illumination was accounted for by evaluating the threshold voltage shift caused by the positive charge of the photoinduced holes [see Eqs. (27)–(30)]. Using this model, we estimated the responsivity

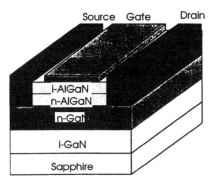

FIG. 34. Schematic diagram of AlGaN/GaN HIGFET. (Reprinted with permission from Khan et al., 1994.)

values for the gated photodetector as a function of the gate voltage. These values are plotted in Fig. 31 (dashed line) along with the measured data (solid line). The figure caption lists the parameters used in the calculation. A good agreement between the measured data and the calculated curves confirms the validity of the model.

The response of AlGaN/GaN heterostructure insulated gate field effect transistors (HIGFETs) to light is different from that of AlGaN/GaN HFETs (see Fig. 34, Khan et al., 1994). The HIGFET exhibits a low resistance state and a persisting high resistance state, before and after the application of a high drain voltage, respectively (see Fig. 35). The device can be returned into the low resistance state by exposing it to optical radiation with sensitivity

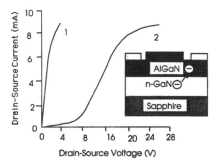

FIG. 35. Current-voltage characteristic collapse. Curve 1 shows the on-state (before the application of a high drain bias). Curve 2 line shows the off-state (after the application of a high drain bias). The insert illustrates the suggested collapse mechanism, which is the electron trapping in the barrier layer at the drain side of the gate. (Reprinted with permission from Khan et al., 1994.)

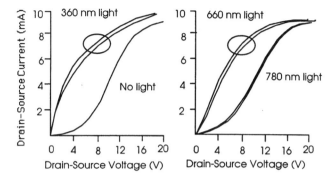

FIG. 36. AlGaN/GaN HIGFET current-voltage characteristics in the dark and under illumination. (Reprinted with permission from Khan et al., 1994.)

peaks at certain wavelengths (see Fig. 36). Electron trapping in the gate insulator near the drain edge of the gate is a possible mechanism for this effect (see the insert to Fig. 35).

Figure 37 shows the dependence of the drain current after the illumination of the device in the collapse state on the wavelength of the optical radiation.

VIII. Conclusions and Future Challenges

GaN-based materials have demonstrated excellent performance potential for applications in visible-blind detectors, which, potentially, may operate in harsh conditions and/or at elevated temperatures. The research and devel-

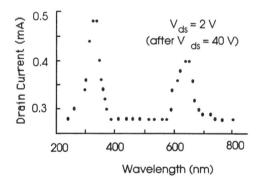

FIG. 37. Dependence of the drain current on the wavelength of the optical radiation. (Reprinted with permission from Khan et al., 1994.)

opment of both GaN-based photodetectors is still in the initial stages. Further improvements in technology and designs, in material quality, contacts, surface passivation, etc. should lead to the development of much better devices and image-processing arrays. New exciting developments may also result from optoelectronic integration of GaN-based photodetectors with GaN-based electronic devices, such as AlGaN/GaN HFETs. The full potential of GaN-based photodetectors is still to be discovered. A major first challenge to be addressed is the reduction of $1/f$ noise, which may be done by improving the material quality.

ACKNOWLEDGMENTS

The work performed at Rensselaer Polytechnic Institute has been partially supported by the Office of Naval Research under contract No. N00014-92-C-0090 (Project Monitor Colin Wood). The work at APA Optics has been supported by ONR under contract No. N00014-95-C-0077 (Project Monitor Colin Wood). It was also partially supported by SDIO and monitored by Dr. Steve Hammonds.

We are grateful to our colleagues and collaborators Dr. J. N. Kuznia, Dr. J. M. Van Hove, Dr. N. Pan, Dr. J. Carter, Dr. M. Blasingame, Dr. A. R. Bhattarai, Dr. Q. Chen, Dr. J. W. Yang, Dr. A. Osinsky, Dr. S. Gangopadhyay, Dr. B. W. Lim, Dr. M. Z. Anwar, Dr. M. A. Khan, Dr. D. Kuksenkov, Dr. H. Temkin, Dr. B. T. Dermott, Dr. J. A. Higgins, Dr. J. Burm, Dr. W. Schaff, Dr. L. F. Eastman, Dr. C. J. Sun, Dr. M. F. Macmillan, Dr. R. P. Devaty, and Dr. J. Choyke, Dr. I. Adesida, and Dr. A. T. Ping, who contributed to the investigations of GaN-based photodetectors described in this chapter. We also thank Professor M. E. Levinshtein and Dr. Stephen O'Leary for critical reading of our manuscript.

REFERENCES

Binet, F., Duboz, J. Y., Laurent, N., Rosencher, E., Briot, O., and Aulombard, R. L. (1997). *J. Appl. Phys.*, **81**, 6449.

Carrano, J. C., Grudowski, P. A., Eiting, C. J., Dupuis, R. D., and Campbell, J. C. (1997). *Appl. Phys. Lett.*, **70**, 1992.

Carrano, J. C., Grudowski, P. A., Eiting, C. J., Dupuis, R. D., and Campbell, J. C. (1998). *Appl. Phys. Lett.*, **72**, No. 5, pp. 542–544.

Chen, Q., Khan, M. A., Sun, C. G., and Yang, J. W. (1995). *Electronics Letters*, **31**, p. 1781.

Chen, Q., Yang, J. W., Osinsky, A., Gangopadhyay, S., Lim, B., Anwar, M. Z., and Khan, M. A. (1997). Schottky barrier detectors on GaN for visible-blind ultraviolet detection, *Appl. Phys. Lett.*, **70**(17), pp. 2277–2279.

Fan, Z-F., Mohammad, S. N., Kim, W., Aktas, Ö., Botchkarev, A. E., and Morkoç, H. (1996). *Appl. Phys. Lett.*, **68**, 1672.

Foresi, S. and Moustakos, T. D. (1993). *Appl. Phys. Lett.*, **62**, 2859.

Edmond, J. A., Kong, H. S., and Carter, Jr., C. H. (1993). *Phys. B*, **185**, p. 453.

Johnson, C., Lin, J. Y., Jiang, H. X., Khan, M. A., and Sun, C. J. (1996). Metastability and persistent photoconductivity in Mg-doped p-type GaN, *Appl. Phys. Lett.*, **68**(13), pp. 1808–1810.

Khan, M. A., Van Hove, J. M., Kuznia, J. N., and Olson, D. T. (1991). *Appl. Phys. Lett.*, **58**, 2408.

Khan, M. A., Kuznia, J. N., Olson, D. T., Van Hove, J. M., Blasingame, M., and Reitz, L. F. (1992a). High-responsivity photoconductive ultraviolet sensors based on insulating single-crystal GaN epilayers, *Appl. Phys. Lett.*, **60**(23), 2917.

Khan, M. A., Kuznia, J. N., Van Hove, J. M., Pan, N., and Carter, J. (1992b). *Appl. Phys. Lett.*, **60**, 3027.

Khan, M. A., Kuznia, J. N., Olson, D. T., Blasingame, M., and Bhattarai, A. R. (1993). Schottky barrier photodetector based on Mg-doped p-type GaN films, *Appl. Phys. Lett.*, **63**(18), 2455.

Khan, M. A., Shur, M. S., Chen, Q., and Kuznia, J. N. (1994). Current-voltage characteristic collapse in AlGaN/GaN heterostructure insulated gate field effect transistors at high drain bias, *Electronics Letters*, **30**, No. 25, p. 2175.

Khan, M. A., Chen, Q., Sun, C. J., Shur, M. S., Macmillan, M. F., Devaty, R. P., and Choyke, J. (1995a). Optoelectronic devices based on GaN, AlGaN, InGaN homoheterojunctions and superlattices. In *Optoelectronic Materials, Devices, and Integrated Circuits, Proc. SPIE*, **2397**, 283.

Khan, M. A., Shur, M. S., Chen, Q., Kuznia, J. N., and Sun, C. J. (1995b). Gated photodetector based on GaN/AlGaN heterostructure field effect transistor, *Electronics Letters*, **31**, No. 5, pp. 398–400.

Khan, M. A., Shur, M. S., and Chen, Q. (1996a). Hall measurements and contact resistance in doped GaN/AlGaN heterostructures, *Appl. Phys. Lett.*, **68**(21), pp. 3022–3024.

Khan, M. A., Chen, Q., Yang, J., Sun, C. J., Lam, B., Anwar, M. Z., Shur, M. S., Temkin, H., Dermott, B. T., Higgins, J. A., Burm, J., Schaff, W., and Eastman, L. F. (1996b). Visible Light Emitters, Ultraviolet Detectors, and High-Frequency Transistors Based on III-N Alloys, *The Physics of Semiconductors*. M. Scheffler and R. Zimmermann, eds., Singapore: World Scientific, pp. 3171–3178.

Kruise, P. W., McGlauchlin, L. D., and McQuistan, R. B. (1962). *Elements of Infrared Technology*. New York: John Wiley and Sons.

Kruise, P. W. (1977). Optical and Infrared Detectors, R. J. Keyes, ed., Berlin: Springer.

Kung, P., Zhang, X., Walker, D., Saxler, A., Piotrowski, J., Rogalski, A., Razeghi, M. (1995). Kinetics of photoconductivity in n-type GaN photodetector, *Appl. Phys. Lett.*, **67**(25), 3792.

Lee, K., Shur, M., Fjeldly, T., and Ytterdal, T. (1993). *Semiconductor Device Modeling for VLSI*, New Jersey: Prentice-Hall.

Levinshtein, M. E. (1996). Private communication.

Lim, B. W., Chen, Q., Khan, M. A., Sun, C. J., and Yang, J. (1996). High responsivity intrinsic photoconductors based on $Al_xGa_{1-x}N$, *Appl. Phys. Lett.*, **68**(26), pp. 3761–3762.

Lim, B. W., Gangopadhyay, S., Yang, J. W., Osinsky, A., Chen, Q., Anwar, M. Z., and Khan, M. A. (1997). 8 × 8 GaN Schottky barrier photodiode array for visible-blind imaging, *Electronics Letters*, **33**, No. 7, pp. 633–634.

Mohammad, S. N., Fan, Z.-F., Kim, W., Aktas, O., Botchkarev, A. E., Salvador, A., and Morkoç, H. (1996). *Electron. Lett.*, **32**, 598.

Morkoç, H., Strite, S., Gao, G. B., Lin, M. E., Sverdlov, B., and Burns, M. (1994). *J. Appl. Phys.*, **76**, 1363.

Osinsky, A., Gangopadhyay, S., Lim, B. W., Anwar, M. Z., Khan, M. A., Kuksenkov, D. V., and Temkin, H. (1998a). Schottky barrier photodetectors based on AlGaN, *Appl. Phys. Lett.* **72**(6), 742–744, Feb.

Osinsky, A., Shur, M. S., Gaska, R., Chen, Q. (1998b). Avalanche breakdown and breakdown luminescence in p-π-n GaN diodes, *Electron. Lett.* **34**(7), 691–692, April 2.

Pankove, J. I. and Berkeyheiser, J. E. (1974). *J. Appl. Phys.*, **45**, 3892.

Pearton, S. J., Vartuli, C. B., Zolper, J. C., Yuan, C., and Stall, R. A. (1995). *Appl. Phys. Lett.*, **67**(10), pp. 1435–1437.

Qian, W., Skowronski, M., de Graef, M., Doverspike, K., Rowland, L. B., and Gaskill, D. K. (1995). *Appl. Phys. Lett.*, **66**, 1252.

Qiu, C. H., Hogatt, C., Melton, W., Leksono, M. W., and Pankove, J. I. (1995). Study of defect states in GaN films by photoconductivity measurement, *Appl. Phys. Lett.*, **66**, 20, pp. 2712–2714.

Razeghi, M. and Rogalski, A. (1996). Semiconductor ultraviolet detectors, *J. Appl. Phys.*, **79**(10), 7433.

Schmitz, A. C., Ping, A. T., Khan, M. A., Chen, Q., Yang, J. W., and Idesida, I. (1996). *Semiconductor Sci. Technol.*, **11**, 1464.

Shur, M. S. (1990). *Physics of Semiconductor Devices*. Englewood Cliffs, NJ: Prentice-Hall, Inc.

Shur, M. S. (1996). *Introduction to Electronic Devices*. New York: John Wiley and Sons.

Shur, M. S. and Khan, M. A. (1997). GaN/AlGaN heterostructure devices: Photodetectors and field effect transistors, *MRS Bulletin*, **22**, No. 2, pp. 44–50.

Stevens, K. S., Kinniburgh, M., and Beresford, R. (1995). Photoconductive ultraviolet sensor using Mg-doped GaN on Si(111), *Appl. Phys. Lett.*, **66**(25), 3518.

van der Ziel, (1986). *Noise in Solid State Devices and Circuits*. New York: John Wiley and Sons.

Walker, D., Kung, P., Saxler, A., Zhang, X. L., Razeghi, M., and Jurgensen, H. (1996a). GaN based semiconductors for future optoelectronics, *Institute of Physics Conference Series*, **145**, pp. 1133–1138.

Walker, D., Zhang, X. L., Kung, P., Saxler, A., Javadpour, S., Xu, J., and Razeghi, M. (1996b). AlGaN ultraviolet photoconductors grown on sapphire, *Appl. Phys. Lett.*, **68**(15), pp. 2100–2101.

CHAPTER 11

III–V Nitride-Based X-ray Detectors

C. H. Qiu and J. I. Pankove

ASTRALUX INC.
BOULDER, COLORADO

C. Rossington

LAWRENCE BERKELEY NATIONAL LABORATORY
BERKELEY, CALIFORNIA

I. INTRODUCTION	441
II. MATERIALS REQUIREMENTS AND CURRENT STATUS	443
III. THE PHOTOCONDUCTIVITY RESPONSE OF NITRIDES	448
1. $\eta\mu\tau$ Product	448
2. Photocurrent Decay	453
3. Mg-Doped GaN	457
IV. EFFECTS OF IRRADIATION ON GaN	458
V. X-RAY RESPONSE OF PROTOTYPE DIODES	460
VI. SUMMARY AND FUTURE WORK	461
1. X-ray and γ-ray Astronomy	462
2. Radiography	463
3. Crystallography	463
4. Environmental Monitoring	463
REFERENCES	463

I. Introduction

The interaction of X-rays with condensed matter materials occurs primarily through interaction with electrons. This interaction produces a wealth of fascinating phenomena, such as X-ray absorption, elastic and inelastic X-ray scattering, X-ray diffraction, and so on. In this chapter, we shall focus the discussion on the detection of X-rays through X-ray absorption in semiconductors.

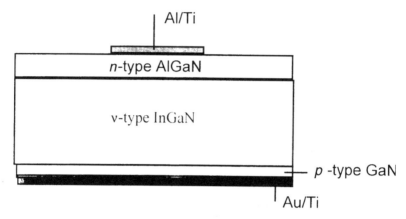

FIG. 1. Schematic diagram of a p-v-n⁺ detector based on III–V nitride semiconductors.

There are many detector structures based on semiconductors, such as a simple photoresistor and a *p-n* junction. A schematic of a p-v-n⁺ structure is shown in Fig. 1. Electron–hole pairs are generated as a consequence of the absorption of X-rays by the semiconductor material. The X-ray-generated electron–hole pairs result in a photocurrent signal that provides the detection of X-rays. In imaging applications, the intensity of X-rays is mapped directly by the photocurrent in detector pixels. An array of semiconductor detectors can function as an electronic "film" substitute, which has not only a broader dynamic range than normal radiographic film, but also offers real-time digital capability. In applications requiring energy resolution, one uses the fact that the energy of the X-ray photon is proportional to the charge collected at the electrodes. This is a consequence of the observation that it takes a specific energy to generate an electron–hole pair in a given semiconductor (Pankove, 1971).

The ability to detect X-rays with good spatial and energy resolution is of great importance because it makes possible a variety of analytical and imaging techniques. However, the popular use of these techniques is often limited by the need to operate the radiation detectors at cryogenic temperatures. The detector is often bulky and maintenance is time consuming. Over the years, many semiconductor materials with bandgaps wider than that of Ge and Si, such as CdZnTe, HgI_2, and diamond have been developed for uncooled X-ray detection (Schlesinger and James, 1995; James et al., 1993). Both CdZnTe and HgI_2 are very promising materials but there are problems associated with their intrinsic softness and chemical activity. The chemical activities of the elements Cd, Te, Hg, and I, for example, cause environmental and human safety concerns. In comparison, GaN and InGaN materials are unique in that they exhibit superior hardness and chemical inertness compared to CdZnTe and HgI_2, and a bandgap energy more favorable for X-ray detection than that of diamond. In addition, it seems that many defects in III–V nitrides (such as dislocations) may not contribute to undesirable recombination of carriers (Lester et al., 1995; Qiu et al., 1996a). The $\eta\mu\tau$ product of GaN and InGaN might be much larger than that of other III–V semiconductors, thus also providing an advantage for X-ray detection.

In this chapter, we shall first discuss briefly what material attributes of nitrides are desired for X-ray detection and what is the current state of the art. Then, we shall discuss the photoconductive properties of nitrides, with a focus on the $\eta\mu\tau$ product of various materials. This is followed by a brief summary of radiation effects on nitrides and the X-ray response of prototype detectors. We shall conclude the chapter with a glimpse of the future.

II. Materials Requirements and Current Status

One key objective for X-ray detection is to achieve high–purity GaN and InGaN with as few defects within their bandgaps as possible. To illustrate this point, some basics of detector physics are discussed briefly here. For a reverse–biased p-v-n structure, for example, deep levels in the v-layer act as carrier generation centers in the presence of a high electric field, and thus lead to a high leakage current and, therefore, noise. Furthermore, charge centers will generate an electric field gradient in the v-layer, which may contribute to premature electric breakdown. In addition, charge centers will also limit the depletion–layer thickness required for radiation detection.

The required thickness of the v-layer in the schematic of Fig. 1, for example, is dictated by the X-ray absorption coefficient of the semiconductor and the depletion–layer thickness for the effective collection of the electron–hole pairs generated by the X-rays. At a reverse bias voltage, V the depletion layer width of the v-p junction is x_i in the v-layer, and x_p in the p-type layer. For an abrupt junction, x_i and x_p can be estimated by (Sze, 1981)

$$x_i = \sqrt{\frac{2\varepsilon_s V}{e(N_D)_i[1+(N_D)_i/N_{A^-}]}} \quad (1a)$$

$$x_p = x_i \frac{(N_D)_i}{N_{A^-}} \quad (1b)$$

TABLE I

DEPLETION LAYER WIDTH IN THE v-LAYER x_i FOR THE v-p JUNCTION AT SEVERAL DONOR AND ACCEPTOR CONCENTRATIONS

$(N_D)_i$ (cm^{-3})	V (volts)	N_{A^-} (cm^{-3})	x_i
1×10^{11}	300	$10^{17}, 10^{18}$	1.73 mm
1×10^{12}	300	$10^{17}, 10^{18}$	0.55 mm
1×10^{13}	300	$10^{17}, 10^{18}$	173 μm
1×10^{14}	300	$10^{17}, 10^{18}$	55 μm
1×10^{15}	100	10^{17}	10 μm
	100	10^{18}	10 μm
1×10^{16}	30	10^{17}	1.65 μm
		10^{18}	1.72 μm

where ε_s is the dielectric constant $\simeq 9\varepsilon_0 = 7.97 \times 10^{-11}$ F/m for GaN (Strite and Morkoç, 1992), e is the electron charge, $(N_D)_i$ is the net ionized donor concentration in the v-layer, and N_{A^-} is the ionized acceptor density in the p-type region. Table I gives x_i values for $(N_D)_i = 10^{11}$–10^{16} cm^{-3} and $N_{A^-} = 10^{17}$–10^{18} cm^{-3}. Similarly, one can estimate the depletion–layer thickness for the v-n junction. Depending on the wavelength of the X-ray, the absorption depth as determined by α^{-1} ranges from a few micrometers to a few millimeters. In the 400- to 450-eV energy region, for example, the absorption depth is about 0.5 μm (Lambrecht et al., 1996). Table I suggests that $(N_D)_i = 10^{12}$ cm^{-3} or less is desired for the detection of hard X-rays.

The depletion layer thickness in Table I is estimated using applied voltages of up to 300 V no matter what the physical layer thickness is. Alternatively, when the maximum applied voltage is not a concern, the depletion layer thickness may be estimated using the maximum electric field ε_m in the junction by

$$x_i = \frac{\varepsilon_s \varepsilon_m}{e(N_D)_i} \qquad (2)$$

The breakdown electric field of GaN is believed to be higher than 5×10^5 V/cm. Suppose the detector is operated under the condition that the maximum electric field ε_m at the junction is approximately 50% of the breakdown field to reduce the leakage current. Then Eq. (2) gives

$$x_i = \frac{1.25 \ \mu m}{(N_D)_i} \qquad (3)$$

where $(N_D)_i$ is in units of 10^{16} cm^{-3}. In other words, for each order of magnitude decrease in the net background donor concentration in the v-layer, the depletion layer thickness is increased by an order of magnitude. It reaches 1.25 mm when $(N_D)_i$ is reduced to 10^{13} cm^{-3}.

Currently, undoped GaN and InGaN are n-type conducting. The best GaN films, as judged by highest room-temperature electron mobility, exhibited a minimum electron concentration of low 10^{16} cm^{-3} (Nakamura, 1991; Kuznia et al., 1993; Molnar, 1946; Keller et al., 1946). The electron concentration in undoped InGaN is usually much higher on the order of 10^{18} cm^{-3} (Strite and Morkoç, 1992). The origin of such a high electron concentration background is not certain. It may be related to the large lattice mismatch between sapphire and the nitride. It may also be related to nitrogen vacancies in the lattice or the residual impurities in the gas sources. Highly resistive unintentionally doped GaN films with resistivity $\sim 10^{10}$ Ω cm were successfully grown using a switched atomic layer epitaxy procedure in a low-pressure metal organic chemical vapor deposition (MOCVD) reactor (Khan et al., 1992). The exact nature of the mechanism responsible for imparting the insulating character to these films is not known. Acceptor compensation and zero bias depletion were suggested as possible causes for high resistivity.

Alternatively, controlled doping of GaN and InGaN by Mg, Zn, and Cd may provide useful photosensitive materials. It is interesting to explore the growth conditions for this purpose. Because both GaN and InGaN can be made p-type conducting (Yamasaki et al., 1995), it seems that nitride films with any resistivity are within reach.

InGaN is particularly interesting for X-ray detection because of the high Z value of In. Assuming the detector is working in the photoelectric absorption region, for example, In$_{0.16}$Ga$_{0.84}$N exhibits an X-ray absorption comparable to that of Ge. Based on the results of Osamura et al. (1972), the energy bandgap of such an alloy is expected to be 3.16 eV, and the lattice constant a of the alloy is expected to be 3.246 Å, very close to $a = 3.252$ Å of wurtzitic ZnO.

The lattice-matched growth of InGaN on ZnO was studied using molecular beam epitaxy (MBE) (Morkoç, 1995) and MOCVD (Matsuoka et al., 1992; Matsuoka, 1995; Qiu et al., 1996c). The smoother Zn surface was prepared by etching with nitric acid. To minimize the degradation of the ZnO surface in ammonia, the substrate was heated in a nitrogen atmosphere, and a thin InGaN layer approximately 20 nm was deposited at 400–500 °C before heating to higher growth temperatures. Matsuoka et al. (1992) were able to grow InGaN films that were lattice-matched along the a axis with ZnO, and found that the indium content at the lattice-matching condition was approximately 22%. X-ray diffraction linewidth of InGaN grown on ZnO was approximately 20% smaller than that of films grown on

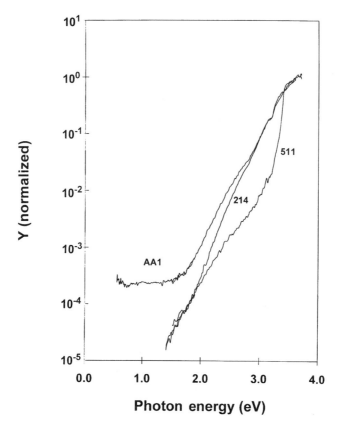

FIG. 2. Normalized photoconductivity spectra of a highly insulating p-type GaN film (sample 214), a p-type conducting GaN film (sample AA1, $p = 6 \times 10^{17} \text{cm}^{-3}$ at 300 K), and an undoped GaN film (sample 511). (From Qiu et al., 1995a.)

sapphire substrates. It is believed that the crystalline quality of InGaN can be improved further as the crystalline quality of ZnO substrates become better and as the surface treatment of ZnO and the growth conditions of InGaN become optimized. The lattice-matched growth may eventually lead to a substantial reduction in the background electron concentration of InGaN because of the better crystallographic structure quality.

Nitride films exhibit high densities of structural and electronic defects. The structural defects, which do not seem to adversely affect the performance of light emitting diodes (LEDs), and the microstructure have been extensively discussed (see, e.g., Ponce, 1997). The electronic defect states within the bandgap of reasonably high-quality nitride films were observed by photoconductivity spectroscopy (Qiu et al., 1995a; Kung et al., 1995; Stevens, Kinniburgh, and Beresford, 1995; Binet et al., 1996a), photothermal deflection spectroscopy (Ambacher et al., 1995) and capacitance and photocapacitance spectroscopy (Götz et al., 1995, 1996; Yi and Wessels, 1996). Figure 2 illustrates normalized photoconductivity spectra for both n-type and p-type GaN (Qiu et al., 1995a). All samples exhibited photoconductivity response at photon energies far below the bandgap energy of GaN, indicating a defect-state distribution within the bandgap. We should emphasize that all the higher-quality GaN films (in terms of higher electron mobility), including the ones we grew by MOCVD at Astralux and the one supplied to us by DenBaars, still exhibited photoconductivity spectra similar to film 511 in Fig. 2, although the optical absorption in the sub-bandgap region appeared lower. The best films, in terms of sharp near bandedge absorption, were Zn-doped GaN films grown using the chloride transport method (Pankove and Berkeyheiser, 1974) and highly compensated insulating GaN films that were grown by switched atomic layer epitaxy (ALE) (Khan et al., 1992). The Zn-doped GaN films exhibited a decrease of two orders of magnitude in normalized photoconductivity from 3.5–3.3 eV and another *five* orders of magnitude decrease from 3.3–1.5 eV. The insulating films of Khan et al. (1992) exhibited a decrease of four orders of magnitude in normalized photoconductivity from 3.5 eV to 3.3 eV. It is worthwhile to point out that the normalized photoconductivity data in Fig. 2 could not be interpreted by two defect levels within the bandgap as done by Balagurov and Chong (1996) for GaN films grown by the chloride transport method. In the latter case, the optical absorption coefficient in the region of 1.5–3.3 eV only increased by a factor of 3, while the increase in the normalized photoconductivity was three or more orders of magnitude. The data in Fig. 2 indicated defect states in n-type GaN extending from the valence bandedge to approximately 0.7 eV below the conduction bandedge in n-type conducting GaN. Such a distribution of electronic defect states supports the electronic compensation as evidenced by the Hall mobility measurements. As pointed out earlier, deep levels in the gap would contribute to leakage current and premature breakdown, and limit depletion layer thickness. The Zn-doped GaN films (Pankove and Berkeyheiser, 1974) and the compensated insulating GaN films (Khan et al., 1992) offer promise for superior detectors.

The current growth rate of nitrides by MOCVD and MBE methods is <4 μm/h. In other words, more than 10 days will be needed in order to grow a 1 mm thick layer. Depending on the wavelength of the X-ray, the desirable thickness of the nitride detector ranges from a few micrometers to a few millimeters. The current growth rate may be adequate for fabricating detectors for soft X-rays. The major issue then is to reduce the background electron concentration. For fabrication of thick detectors for hard X-rays, the growth rate needs to be at least 10 times higher, while also improving the electronic quality. The growth rate for MOCVD may be enhanced by the following methods:

1. Using a tungsten filament to thermally precrack ammonia. The H will help extract carbon from the metalorganics (Yates, Jr., 1994), while the NH and NH_2 radicals are more reactive than NH_3. Such an approach was recently demonstrated for AlN growth (Dupuis and Gulari, 1991). This method seems especially useful for the growth of InGaN, because of the relatively low growth temperature.
2. Using N_2H_4 instead of NH_3. The problem is that the N_2H_4 available now is less pure than NH_3 (Fujida et al., 1987).

Other growth techniques, especially the chloride transport method (Maruska and Tietjen, 1969; Molnar, 1998), should be explored in the future for fabrication of X-ray detectors. The first high-quality thick GaN crystal was grown in Akasaki's group utilizing the chloride transport method and ZnO buffer layer (Detchprohm et al., 1992, 1993). Transparent crystals up to 400 μm thick were produced. For the crystals, room-temperature electron concentration of $(0.9-4) \times 10^{16} \, cm^{-3}$ and electron Hall mobility over 500 cm^2/V s were observed. Several bulk crystal growth methods, such as high-pressure growth from metallic Ga and nitrogen (Porowski et al., 1997), and growth from metallic Ga and ammonia (Sakai et al., 1997; Balkas et al., 1997), have advanced rapidly in the last few years. The crystals, like thin films grown by MOCVD, remain n-type conducting. However, there has been no work reported yet on the growth or doping of thick InGaN crystals by either the chloride transport or the bulk crystal growth methods.

III. The Photoconductivity Response of Nitrides

1. $\eta\mu\tau$ Product

The $\eta\mu\tau$ products for electrons and holes are important parameters determining detector performance. We measured the photoconductivity of a number of undoped n-type GaN and Mg-doped p-type GaN samples, as well as a Zn-doped GaN sample. The photoconductivity was measured using a lock-in technique. A GE 1493 halogen lamp was used as the light source and radiation was filtered by a Perkin–Elmer model 12 monochromator equipped with a quartz prism. The photon flux vs photon energy at the specimen was separately measured by replacing the specimen with an RCA 31025 photomultiplier, a PbS detector, and a calibrated Si detector. Compared to a Xe arc lamp, the GE 1493 supplies much less UV light. The light intensity at 3.4 eV was $0.2\ \mu\text{W/cm}^2$. The photoconductivity was measured in a coplanar geometry with two indium contacts soldered to the GaN surface. For *p*-type and insulating films, a thin layer of gold film was sometimes deposited before indium soldering to improve the I–V characteristics. The I–V curve is linear for *n*-type samples, whereas for *p*-type films the I–V curve is linear for voltage greater than 0.3 V. A dc voltage (10 V for most of the measurements) was applied to the sample through a series resistor. The voltage output from across either the sample or the load resistor, the one with the smaller resistance, was detected by the lock-in technique. For successful measurement, it is important to reduce contact noise.

The photoconductivity response can be represented by $Y = \sigma_{\text{ph}} d / eF$, where σ_{ph} is the photoconductivity corresponding to the incident flux F, d is the film thickness, and e is the electron charge. Accordingly (Qiu *et al.*, 1995a),

$$Y = \frac{\sigma_{\text{ph}} d}{eF} = (1-R)(1-e^{-\alpha d})[(\eta\mu\tau)_e + (\eta\mu\tau)_h] \qquad (4)$$

where R and α are, respectively, the reflection coefficient and the absorption coefficient for the incident light. The subscripts e and h represent, respectively, the contribution of photoexcited electrons and holes to the photoconductivity. η, μ, and τ are, respectively, the quantum efficiency for the photogeneration of mobile electrons and holes, the mobility, and the lifetime of the photoexcited carriers. The $\eta\mu\tau$ product in this work was defined as $(\eta\mu\tau)_e + (\eta\mu\tau)_h$ to include the contribution of both electrons and holes to photoconductivity. The $\eta\mu\tau$ product was calculated from the Y value in the forementioned bandgap ultraviolet (UV) wavelength region using Eq. (4) by assuming $R = 0$, that is, $Y = \sigma_{\text{ph}} d/eF = \eta\mu\tau$. If it is assumed that a voltage V was applied through a gap of separation w, it is easily shown that the responsivity, defined as photocurrent/power (A/W), is

$$\text{Responsivity} = Y\left(\frac{eV}{hv}\right)\bigg/ w^2 \qquad (5)$$

where hv is the photon energy.

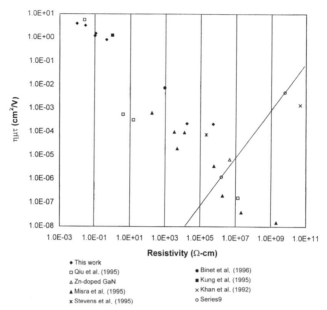

FIG. 3. The $\eta\mu\tau$ product as derived from room temperature photoconductivity measurements, vs the electrical resistivity of GaN films.

Because the electron or hole concentration of the resistive samples cannot be measured by the Hall effect, the $\eta\mu\tau$ data are presented in Fig. 3 as a function of film resistivity. In addition to our recent data, our published data and that of other groups were also included in the plot. Data reported as responsivity were converted to $\eta\mu\tau$ using Eqs. (4) and (5) and the relevant experimental parameters. The films with electrical resistivity of less than 100 Ω cm shown in Fig. 3 were undoped n-type GaN films, and Mg-doped p-type conducting GaN films, which exhibited three orders of magnitude lower $\eta\mu\tau$. The films with electrical resistivity greater than 100 Ω cm were either compensated insulating GaN films (Khan et al., 1992; Misra et al., 1995; Binet et al., 1996a) or intentionally doped insulating GaN films (this work; Stevens, Kinniburgh, and Beresford, 1995). Figure 3 indicates that $\eta\mu\tau$ data decrease as the film becomes more resistive. The $\eta\mu\tau$ data from MBE samples (Misra et al., 1995) seem to be about one order of magnitude less than data of other groups' samples with comparable resistivity. The insulating GaN film of Khan et al. (1992) exhibit a much higher $\eta\mu\tau$ than extrapolated from data of all other groups.

In order to separate the contributions of electrons and holes to photoconductivity, photoconductivity excited by infrared light at 1.0 eV was compared to photoconductivity excited by UV light at 3.6 eV, at several temperatures from 6 K to 400 K (Qiu et al., 1995b). For infrared excitation at 1.0 eV, only one kind of carrier is excited to contribute to the observed photoconductivity. It was suggested that the $\eta\mu\tau$ product of n-type samples was dominated by $(\eta\mu\tau)_e$ for electrons and that the $\eta\mu\tau$ product of p-type samples was dominated by $(\eta\mu\tau)_h$ for holes. With this understanding, Fig. 3 indicates that $(\eta\mu\tau)_e$ in undoped n-type GaN was approximately three orders of magnitude greater than $(\eta\mu\tau)_h$ of p-type conducting GaN. Such a difference between $(\eta\mu\tau)_e$ and $(\eta\mu\tau)_h$ will result in very different collection efficiencies for electrons and holes in X-ray detectors. The single-polarity charge-sensing scheme proposed by Luke (1994) may prove useful for energy resolution applications. Note further that the electron mobility is approximately 20 to 100 times higher than the hole mobility for the conducting samples. It seems that $\eta\tau$ of p-type GaN is approximately 10 times less than $\eta\tau$ of n-type GaN.

For detector applications, it is desirable that the X-ray–generated electrons are not lost in the detector through recombination before they can induce a complete change on the signal electrode. In other words, the electron lifetime is desired to be longer than the time needed to sweep out the electrons, that is,

$$\tau_e > \frac{x_i}{\mu_e \varepsilon_m} \tag{6a}$$

where x_i is the depletion-layer thickness and ε_m is the maximum electric field in the junction as defined in Eq. (2). By using Eq. (2), Eq. (6a) can be rewritten as

$$\tau_e > \frac{\varepsilon_s}{e\mu_e(N_D)_i} = \varepsilon_s \rho \tag{6b}$$

or

$$(\eta\mu\tau)_e > \frac{\eta\varepsilon_s}{e(N_D)_i} = \frac{4.97\eta \times 10^6}{(N_D)_i} \quad \text{cm}^2/\text{V} \tag{6c}$$

where $(N_D)_i$ is in units of cm^{-3}. Note that $\varepsilon_s/e\mu_e(N_D)_i = \varepsilon_s\rho$ is just the RC time constant of the sandwich structure. By assuming $\eta = 1$ and $\mu = 1$ cm^2/V s for the resistive GaN films, the minimum required $\eta\mu\tau$ from Eq. (6c) as a function of resistivity ρ is plotted as a straight line in Fig. 3. Thus, current materials good for detector applications defined by short sweep-out times exhibit a resistivity less than approximately 10^7 Ω cm, but such materials are not quite useful because of high leakage current. In other words, the $\eta\mu\tau$ product for the more resistive films should be improved in the future from below the minimum $\eta\mu\tau$ line to above it to be useful. In this regard, it is interesting to note that the insulating films of Khan et al. (1992) would exhibit a lifetime of ~ 1 ms (assuming a quantum efficiency of 1 and an electron mobility of 1 cm^2/V s) with a sweep-out time and RC time of ~ 36 ms in a sandwiched detector structure. Most of the generated carriers would be recombined and lost in the structure before being collected. However, as there is no doubt that less resistive materials can easily be made while maintaining or improving the $\eta\mu\tau$ product, we feel that the work of Khan et al. (1992) did prove the possibility of achieving useful insulating materials with $\eta\mu\tau > 10^{-3}$ cm^2/V and a reasonable resistivity.

The ratio of photoconductivity to dark conductivity provides a measure of sensitivity of the detector materials. This ratio is proportional to $\eta\mu\tau\rho$, and is independent of mobility as long as the same type of carriers dominate the contributions to both dark and photoconductivity. In Fig. 4, data from Fig. 3 are replotted as $\eta\mu\tau\rho$ vs the resistivity ρ. The data indicate that GaN becomes more sensitive to photons as it becomes more resistive.

For most samples, the $\eta\mu\tau$ data were generally not independent of the

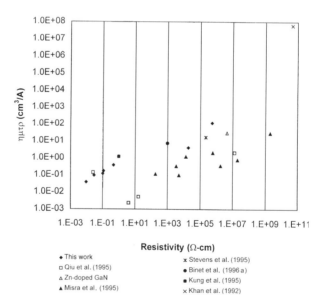

FIG. 4. The sensitivity $\eta\mu\tau\rho$ derived from room-temperature photoconductivity measurements, vs the electrical resistivity of GaN films.

light intensity used during measurement. Our experience was in agreement with the observation of Stevens, Kinniburgh, and Beresford (1995) and Binet et al. (1996a). Typically, when the light intensity was less than a few μW/cm^2, $\eta\mu\tau$ was independent of light intensity, while $\eta\mu\tau$ decreased at higher light intensities. However, an $\eta\mu\tau$ product that was independent of light intensity and varied by several orders of magnitude was reported by Khan et al. (1992).

2. PHOTOCURRENT DECAY

The reported photoconductivity response time as measured from photocurrent decay ranged from a few tens of nanoseconds to a few minutes (Misra et al., 1995; Kung et al., 1995; Binet et al., 1996a). Misra et al. (1995) measured the photocurrent decay in the nanosecond domain. They observed a fast component with a response time of 20 ns and a slower component with a sample dependent response time. Kung et al. (1995), for example, measured the decay for 1 min and found a slow response time of ~ 10 s and an undetermined fast component. The response time was normally deduced from the decay data by plotting the log (photocurrent) vs time. A fraction of the data followed a straight line in the plot and a time constant was obtained from the slope of the straight line.

This ten orders of magnitude difference in response time is interesting. We measured the photocurrent decay of undoped n-type GaN thin films in the time span from 50 ns to 50 s (Qiu et al., 1996a). Figure 5 shows the voltage change due to the photocurrent decay at room temperature for several selected samples, plotted in a log-log scale. A straight line in such a plot implies a decay governed by a power law, $\Delta n = At^{-\alpha}$. In the time span of approximately 1 μs to 0.1 s, the exponent α for all the samples we measured was almost the same, with an average of 0.193 and a standard deviation of 0.011. In a semilog plot of log (photocurrent) vs time, the decay data at $t < 0.5\,\mu$s for a few samples were approximately linear with a sample-dependent time constant in the range of 0.2–0.8 μs. Because of the nonexponential nature of the decay, however, a semilog plot of the same set of data in the millisecond domain resulted in approximately straight-line section yielding a time constant of a few milliseconds; a semilog plot of the same set of data in the second domain yielded a time constant of a few seconds. For curiosity, we measured the photocurrent decay of a few samples at room temperature for several hours (Qiu and Pankove, 1997). The data were still nonexponential, and a time constant of ~ 1 h could be deduced from a semilog plot. Thus, the response time from a semilog plot depends totally on the time domain being studied. The ten orders of

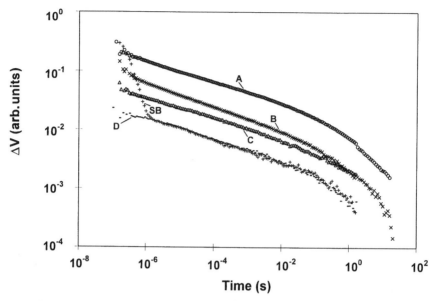

FIG. 5. Photoconductivity decay of n-type GaN measured at room temperature for samples with electron concentration of (A) 9.6×10^{16} cm^{-3}, (B) 5×10^{17} cm^{-3}, (SB) 3.4×10^{17} cm^{-3}, (C) 1.3×10^{18} cm^{-3}, and (D) 1.3×10^{19} cm^{-3}. (From Qiu et al., 1996a.)

magnitude difference in reported response time is merely a reflection of the nonexponential nature of the decay.

The photoconductivity response time can also be determined from the lock-in technique, where the photoconductivity signal depends on chopper frequency (Qiu et al., 1995a; Kung et al., 1995). In the ideal case of a single photoconductivity response time τ_r, the signal amplitude $s(f)$ as a function of frequency f is given by

$$s(f) = \frac{s(0)}{[1 + (2\pi f \tau_r)^2]^{1/2}} \qquad (7)$$

from which the photoconductivity response time τ_r was determined. Figure 6 shows the photoconductivity amplitude as a function of chopper frequency for an n-type sample and a p-type sample at room temperature. In the time domain, the photocurrent decay was nonexponential as illustrated in Fig. 5. The signal amplitude data followed Eq. (7) only approximately, and depended slightly on the light intensity and photon energy. The response

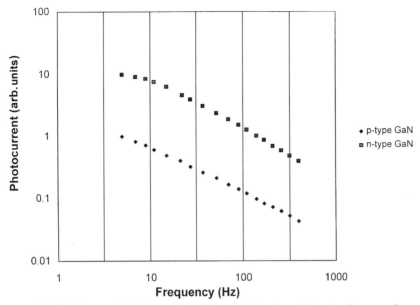

FIG. 6. The photoconductivity signal amplitude as a function of chopping frequency for an n-type sample and a p-type sample.

time could be estimated from the frequency where $s(f) \simeq 50\%\ s(0)$. For both samples, the decrease of the signal amplitude with increasing frequency was almost identical, implying nearly identical τ_r. In fact, a response time of 14 ± 6 ms was obtained for most of the samples measured using the lock-in technique.

Now with such a nonexponential decay picture in mind, the $\eta\mu\tau$ product data in Fig. 3 and the photosensitivity data in Fig. 4 were actually frequency-dependent and time-dependent. This may enable us to understand the data scattering seen among different groups. Our data in Fig. 3 were measured at 7 Hz. The $\eta\mu\tau$ data could be 20 times smaller if a chopping frequency of 400 Hz had been used as shown in Fig. 6. The data could be 100 times greater if the photoconductivity had been measured by waiting a few minutes after the turn-on and turn-off of light source, as shown in Fig. 7. In this case, it was difficult to separate the component of persistent photoconductivity (Johnson et al., 1996; Qiu and Pankove, 1997a) from the observed change in conductivity. Although the data shown in Fig. 7 were collected from a Mg-doped GaN sample, similar differences were observed for undoped n-type samples.

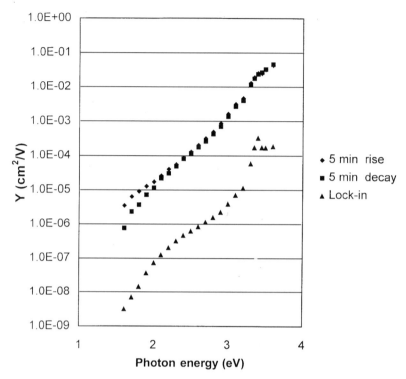

FIG. 7. The photoconductivity spectra at room temperature measured with the lock-in technique at 7 Hz, compared with data measured by waiting for 5 min during the rise and decay of the photocurrent.

It is worthwhile to point out here that one should not confuse the photoconductivity response time as determined from the decay or lock-in technique with the electron and hole lifetimes defined in Eq. (4). Because of trapping, the response time is normally longer than the lifetime (Rose, 1963). In the simple case of one single-energy level of electron traps, the response time is

$$\tau_r = \left(\frac{1 + n_t}{n}\right)\tau_e \qquad (8)$$

where n_t is the concentration of trapped electrons and n is the electron concentration in the conduction band. In the case of a distribution of traps, the factor (n_t/n) is time-dependent because electrons trapped at deeper levels stay there longer before becoming available for recombination. For undoped n-type GaN films, the response time seemed to be determined by hole trapping (Qiu *et al.*, 1996a).

3. Mg-DOPED GaN

Our $\eta\mu\tau$ product data shown in Fig. 3 for samples with electrical resistivity $\rho > 10^3\,\Omega$ cm were collected from samples doped with Mg or Zn. For many Mg-doped samples, the photoconductivity spectra were similar to

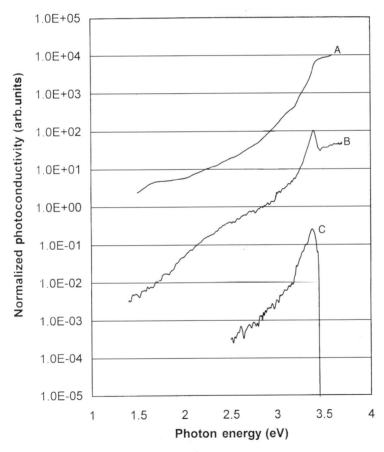

FIG. 8. Typical photoconductivity spectra of Mg-doped GaN films. See test for an explanation of curves A, B, and C.

curve B in Fig. 8, that is, a peak at ~ 3.41 eV was frequently observed. This peak seemed to follow the bandgap energy in temperature dependence, but became more pronounced at lower temperatures and more pronounced under lower electric field across the electrodes, a behavior very different from the exciton peak as observed by Binet et al. (1996b) from undoped n-type GaN. For such Mg-doped GaN samples, the $\eta\mu\tau$ product measured under a high electric field can be 10 or more times greater. In addition, no absorption peak could be observed from transmission measurement. To determine the possible origin of this peak, the photoconductivity spectra of samples grown under different conditions were analyzed. It seemed to us that this peak resulted from a nonuniform Mg distribution. In one extreme case, an undoped GaN layer approximately 0.5 μm thick was grown before the Mg-doped GaN layer. The photoconductivity spectrum of this sample measured in coplanar geometry is shown in Fig. 8, curve C. This sample did not exhibit any photosensitivity at photon energies higher than the bandgap. We then prepared Mg-doped GaN films by introducing the metalorganics into the MOCVD reactor after establishing the cp_2 Mg (a source for p-type dopants) flow for a few minutes through a bypass line of the growth chamber. The photoconductivity spectrum of the resulting sample is shown in Fig. 8, curve A. The photoconductivity peak was no longer observed.

IV. Effects of Irradiation on GaN

The effects of X-ray and γ-ray irradiation on GaN films were studied (Qiu et al., 1997b). Four samples were irradiated with a synchrotron X-ray beam at 4 keV for 48 h. The dose ranged from 1.8 gigarads to 31 gigarads, which was dependent on sample thickness. Three samples were irradiated with a Co-60 source for 17 days, for a dose of 4.5 megarads. Before and after the irradiation, each sample was characterized by photoconductivity and photoluminescence (PL) spectroscopy. The electron concentration of the undoped n-type samples spanned over two orders of magnitudes from low 10^{19} cm^{-3} to high 10^{16} cm^{-3}, while the room temperature electron mobility was in the range of 73–360 cm^2/V s.

After X-ray irradiation, a change of $3 \pm 4\%$ was measured in conductivity and electron mobility. However, this change was not significantly different from zero and was within the limits of sensitivity of our equipment.

To ascertain irradiation induced changes in the $\eta\mu\tau$ product, the same excitation light source, the GE lamp operated at the same voltage and the same chopping frequency was used for photoconductivity measurements

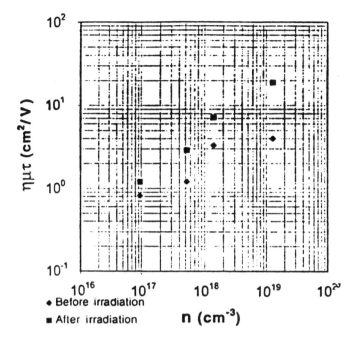

FIG. 9. Irradiation-induced changes in $\eta\mu\tau$ product for samples with different background electron concentrations.

before and after irradiation. Furthermore, an unirradiated GaN sample was used to calibrate the light spectrum at the exit slit of the monochromator. After irradiation, all the samples exhibited a higher $\eta\mu\tau$ product, as illustrated in Fig. 9. Because the irradiation-induced change in electron mobility contributed little to the measured change in $\eta\mu\tau$ product, either the quantum efficiency η or the electron lifetime τ may have been changed by irradiation. The photocurrent decay measurements before and after irradiation seemed to indicate a prolonged electron lifetime after irradiation. The decay data are plotted in Fig. 10. Although no single lifetime could be determined, the slope in the region of $t < 10\,\text{ms}$ became slightly smaller after irradiation, implying a slower decay and a longer response time in the region of $t < 10\,\text{ms}$.

If there had been some increase in the density of gap states that acted as hole traps, the electron lifetime would have become longer. However, the change in the normalized photoconductivity spectra was sample-dependent. One sample exhibited no change in the normalized spectra, while other samples exhibited either an increase or a decrease of the normalized

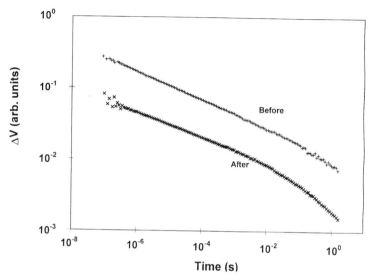

FIG. 10. The photocurrent decay at room temperature of an undoped n-type GaN sample measured before and after irradiation. ΔV is the voltage change across a sampling resistor.

response in the sub-bandgap region. Thus, it is not conclusive whether α, the sub-bandgap optical absorption, is increased or not after the irradiation. The lifetime τ may depend slightly on the photon energy and may exhibit different changes upon irradiation.

After irradiation, the yellow luminescence intensity for undoped n-type samples was slightly changed, and the biggest decrease was observed for a sample with the biggest increase in photoconductivity. The effect of X-ray radiation on Zn-doped GaN (sample GNZN) was similar to the effect of electron beam irradiation, that is, the blue luminescence was enhanced by irradiation as originally observed by Akasaki's group (Amano et al., 1989). Because X-ray detectors are normally made of resistive materials, this effect deserves further study.

V. X-ray Response of Prototype Diodes

There are very few reports on X-ray detectors based on III–V nitrides. Both Misra et al. (1995) and our group have observed X-ray responses from thin film detectors. Misra et al. (1995) observed that the magnitude of the detector response is proportional to the X-ray photon energy in the range of 50–90 kVp. Our measurement results are summarized in the following (Qiu et al., 1996c):

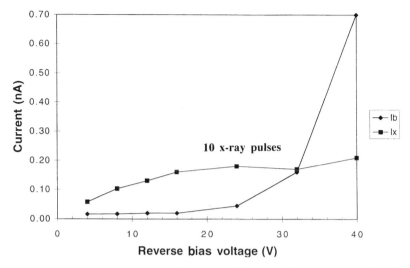

FIG. 11. The detector responses to dental x-rays (I_x) and to light in a normally illuminated room (I_b), as a function of applied reverse bias voltage.

1. A photocurrent due to dental X-rays was observed at room temperature. Under optimal bias conditions (Fig. 11), the magnitude (photocurrent for each single pulse of dental X-rays) was approximately 10 times higher than the dark current and was comparable to the photocurrent due to light in a well-illuminated room.
2. The detector response was rather slow, with a rise time of about 10 s at room temperature.
3. The detector response to a ^{109}Cd source, which provided Ag X-rays of 22 and 25 keV, was observed as a signal peak at around 22 keV. The amplifier peaking time for the measurement was 8 μs. However, the full width at half maximum (FWHM) was approximately 4.3 keV, considerably larger than our theoretical estimate of 1.1 keV.

The slow response and the poor energy resolution are most likely due to carrier trapping and further experiments are under way to determine the exact cause.

VI. Summary and Future Work

We have reviewed in this chapter the photoconductivity properties of GaN thin films. Impressive photosensitivity has been reported by several groups, and the nonexponential photocurrent decay and deep levels are beginning to be understood and to be controlled. As for any other semiconductor material systems, we firmly believe that future crystal growth efforts will lead to not only purer nitride materials but also higher $\eta\mu\tau$ product in the resistive region. Enhanced $\eta\mu\tau$ product in undoped n-type GaN films was observed after irradiation, and the enhancement effect seemed smaller in more resistive materials. The detection of X-rays using thin-film detectors was also demonstrated, while further work is needed to determine the cause of the degraded energy resolution and to fabricate thicker detectors. The growth of thick and insulating GaN and InGaN as well as the photoconductivity properties of InGaN are largely unexplored, and we shall feel satisfied if this chapter engenders interest in further research and development in these areas.

X-ray detectors based on nitrides have some attractive characteristics compared even to cooled Si- and Ge-based detectors. Because of the much wider bandgap of InGaN, the thermal background of carriers should be small, which makes room temperature low noise operation possible. A detector without a cryostat and liquid nitrogen, and that is less responsive to visible light, will be more carefree, less bulky, and more economical to operate.

The most competive technology for X-ray detection is based on coupling CsI(Na) scintillation and a-Si:H (hydrogenated amorphous silicon) photodiodes, where room temperature operation and position sensitive detection are achievable (Perez-Mendez, 1991). However, there are several factors favoring nitride-based X-ray detectors. First, the nitrides are much more stable in air than CsI(Na), and are almost solar blind compared with the strong visible light absorption of a-Si:H. Nitride-based detectors are expected to require less stringent packaging and to last longer. Second, a-Si:H has a stability problem and a substantial thermal background at $T > 130\,°C$. In the high radiation fluxes and higher than room temperature situations that are encountered in nuclear physics research and mining, nitride-based detectors are expected to function better than a-Si:H based detectors. Other areas that may benefit from well-developed nitride-based detectors are listed in the following.

1. X-RAY AND γ-RAY ASTRONOMY

X-ray and γ-ray astronomy relies on detecting radiation using balloons or other outer space vehicles. Here, minimum weight and energy consumption is desirable. The added cost, weight, and energy consumption due to microcoolers can be eliminated.

2. Radiography

Present medical and industrial radiography applications extensively utilize photographic film as the position sensitive detector. Film has excellent spatial resolution, but offers neither realtime nor digital capability, and thus is not very convenient for transferring the film image to computer storage. For realtime X-ray detection in angiography and digital radiology, CsI cathode image intensifiers are used that are bulky and limited in size.

3. Crystallography

X-ray crystallography requires that the position as well as the intensity of the diffraction spots be measured. For protein crystallography, where the crystals are small and can be easily damaged by the radiation, it is advantageous to have a highly efficient, electronic, and position-sensitive detector with a large dynamic range.

4. Environmental Monitoring

It will become easier and less expensive to do environmental monitoring with the use of uncooled X-ray detectors.

Acknowledgments

We thank Dr. M. Leksono and Dr. W. Melton for their help. We are grateful to Dr. H. Amano and Dr. I. Akasaki for supplying two p-type conducting samples, and Dr. S. P. DenBaars for supplying one high-quality n-type GaN sample. We are grateful to Prof. T. Moustaskas for two reprints and Dr. R. Beresford, Dr. K. S. Stevens, and Dr. F. Binet for helpful discussions. This work was supported in part by the U.S. Department of Energy under contract number DE-FG03-95ER86024, and monitored by Dr. Richard Rinkenberger.

References

Amano, H., Akasaki, I., Kozawa, T., Hiramatsu, K., Sawaki, N., Ikeda, K., and Ishii, Y. (1988). *J. Lumines.*, **40-41**, 121.
Amano, H., Kitoh, M., Hiramatsu, K., and Akasaki, I. (1989). *Jpn. J. Appl. Phys.*, **28**, L2112.

Balagurov, L. and Chong, P. J. (1996). *Appl. Phys. Lett.*, **68**, 43.
Balkas, C. M., Sitar, Z., Zheleva, T., Davis, R. F. (1997). *Mat. Res. Soc. Symp. Proc.*, **449**, 41.
Binet, F., Duboz, J. Y., Rosencher, E., Scholz, F., and Harle, V. (1996a). Mechanism of recombination in GaN photodetectors, *Appl. Phys. Lett.*, **69**, 1202.
Binet, F., Duboz, J. Y., Rosencher, E., Scholz, F., and Harle, V. (1996b). Electric field effects on excitons in gallium nitride, *Phys. Rev. B*, **54**, 8116.
Detchprohm, T., Hiramatsu, K., Amano, A., and Akasaki, I. (1992). Hydride vapor phase epitaxial growth of a high quality GaN film using a ZnO buffer layer, *Appl. Phys. Lett.*, **61**, 2688.
Detchprohm, T., Hiramatsu, K., Amano, A., and Akasaki, I. (1993). The growth of thick GaN film on sapphire substrate by using ZnO buffer layer, *J. Cryst. Growth*, **128**, 384.
Dupuis, J. L. and Gulari, E. (1991). *Appl. Phys. Lett.*, **59**, 549.
Fujida, S., Mizuda, M., and Matsumoto, Y. (1987). *Jpn. J. Appl. Phys.*, **26**, 2067.
James, R. B., Schlesinger, T. E., Siffert, P., and Franks, L., eds., (1993). Semiconductors for room temperature radiation detector applications, *Mat. Res. Soc. Proc.* **302**, Pittsburgh, PA.
Johnson, C., Lin, J. Y., Jiang, H. X., Khan, A. M.,. and Sun, C. J. (1996). *Appl. Phys. Lett.*, **68**, 1808.
Khan, Asif, M., Kuznia, J. N., Olson, D. T., Hove, J. M. V., Blasinggame, M., and Reitz, L. F. (1992). High responsivity photoconductive ultraviolet sensors based on insulating single crystal GaN epilayers, *Appl. Phys. Lett.*, **60**, 2917.
Kung, P., Zhang, X., Walker, D., Sasler, A., Piotrowski, J., Rogalski, A., and Razeghi, M. (1995). *Appl. Phys. Lett.*, **67**, 3792.
Lambrecht, W. R. L., Rashkeev, S. N., Segall, B., Lawniczak-Jablonska, K., Suski, T., Gullikson, E. M., Underwood, J. H., and Perera, R. C. C., and Rife, J. C. (1996). X-ray absorption and reflection as probes of the GaN conduction bands: Theory and experiments of the N K-edge and Ga $M_{2,3}$ edges, presented in Fall MRS Symposium Meeting.
Lester, S. D., Ponce, F. A., Craford, M. G., and Steigerwald, D. A. (1995). *Appl. Phys. Lett.*, **66**, 1249.
Luke, P. N. (1994). Single polarity charge sensing in ionization detectors using coplanar electrodes, *Appl. Phys. Lett.*, **65**, 2884.
Maruska, H. P. and Tietjen, J. J. (1969). *Appl. Phys. Lett.*, **15**, 327.
Matsuoka, T., Yoshimoto, N., Sasaki, T., and Katsui, A. (1992). *J. Electron. Mat.*, **21**, 157.
Matsuoka, T. (1995). *Mat. Res. Soc. Symp. Proc.* **395**, 39.
Misra, M., Moustakas, T. D., Vaudo, R. P., Singh, R., and Shah, K. S. (1995). Photoconducting ultraviolet detectors based on GaN grown by electron resonance molecular beam epitaxy, *Proc. SPIE*, **2519**.
Misra, M., Korakakis, D., Singh, R., Sampath, A., and Moustakas, T. D. (1996). Photoconducting Properties of Ultraviolet Detectors Based on GaN and AlGaN Films Grown by ECR MBE, preprint.
Molnar, R. J. (1998). *Semiconductors and Semimetals*, Vol. 57, Boston: Academic Press, p. 1.
Morkoc, H. (Submitted to *Science*.)
Osamura, K., Nakajima, K., Murakami, Y., Shingu, P. H., and Ohtsuki, A. (1972). *Solid State Commun.*, **11**, 617.
Pankove, J. I. (1971). *Optical Processes in Semiconductors*. New York: Dover, p. 256.
Pankove, J. I. and Berkeyheiser, J. E. (1974). Properties of Zn doped GaN. II. Photoconductivity, *J. Appl. Phys.*, **45**, 3892.
Perez-Mendez, V. (1991). In *Amorphous and Microcrystalline Semiconductor Devices: Optoelectronic Devices*, J. Kanicki, ed., Boston: Artech House, p. 297.
Ponce, F. A. (1997). Defects and interfaces in GaN epitaxy, *MRS Bull.*, **22**, 51.

Porowski, S., Boćkowski, M., Lucznik, B., Grzegory, I, Wroblewski, M., Teisseyre, H., Leszczyński, M., Litwin-Straszewska, E., Suski, T., Trautman, P., Pakula, K., and Baranowski, J. (1997). *Acta Physics Polonica A*, **92**, 958.

Qiu, C. H., Hoggatt, C., Melton, W., Leksono, M. W., and Pankove, J. I. (1995a). Study of defect states in GaN films by photoconductivity measurement, *Appl. Phys. Lett.*, **66**, 2712.

Qiu, C. H., Melton, W., and Pankove, J. I. (1995b). *Proc. Mat. Res. Soc.*, **378**, 515.

Qiu, C. H., Melton, W., Leksono, M. W., Pankove, J. I., Keller, B. P., and DenBaars, S. P. (1996a). *Appl. Phys. Lett.*, **69**, 1282.

Qiu, C. H., Leksono, M. W., Pankove, J. I., Rossington, C., and Haller, E. E. (1997b). *MRS Symp. Proc.*, **449**, 585.

Qiu, C. H., Pankove, J. I., Rossington, C., and Haller, E. E. (1996c). Final technical report to Department of Energy (unpublished).

Qiu, C. H. and Pankove, J. I. (1997a). *Appl. Phys. Lett.*, **70**, 1983.

Rose, A. (1963). *Concepts in Photoconductivity and Allied Problems*, New York: John Wiley & Sons.

Sakai, S., Kurai, S., Nishino, K., Wada, K., and Naoi, Y. (1997). *Mat. Res. Soc.*, **449**, 15.

Schlesinger, T. E. and James, R. B., eds. (1995). In Semiconductors for Room Temperature Nuclear Detector Applications, *Semiconductors and Semimetals*, **43**, San Diego: Academic Press.

Stevens, K. S., Kinniburgh, M., and Beresford, R. (1995). Photoconductive ultraviolet sensor using Mg doped GaN on Si(111), *Appl. Phys. Lett.*, **66**, 3518.

Strite, S. and Morkoç, H. (1992). *J. Vac. Sci. Technol. B*, **10**, 1237.

Yamahasi, S., Asami, S., Shibata, N., Koike, M., Manabe, K., Tanaka, T., Amano, H., and Akasaki, I. (1995). *Appl. Phys. Lett.* **66**, 1112.

Yates, Jr., J. T. (1994). Presented at the 2nd workshop on III-nitrides, St. Louis.

Index

Numbers followed by the letter f indicate figures; numbers followed by the letter t indicate tables.

A

AlGaN
  growth by HVPE, 13
  photoconductive detectors made out of, 422–423
  shallow donor resonance in, 382–383
AlGaN films
  devices fabricated on, 110–111
  growth of, 89
  long-range ordering in, 97–100
AlN
  growth by HVPE, 13
  irradiation of, 394–395
  properties of, 276t
Aluminum, in quantum wells, 307
Ammonia, use in MBE, 40–41
Annealing, of GaN crystals, 145–146
Arsenides, growth of, 3

B

Beryllium, covalent radius of, 288t
Binding energy, defined, 159
Bismuth, use as surfactant, 285t, 286

C

Capacitance transient spectroscopy, 195–197
Carbon, covalent radius of, 288t
CdSe/ZnSe, growth mode of, 267
Chloride transport, 34, 448
Chromium, as impurity in GaN, 401
Cobalt, as impurity in GaN, 401
Compensation, defined, 159

D

Deformation potentials, 210–211
Degenerate four-wave mixing (DFWM), 355–361
Density-functional theory, 161
Direct-bandgap II-VI materials, 2
Dislocations, 135, 136, 138, 139f, 150, 283
  threading, 150, 150f, 284f
Dissociation energy, defined, 159
Doping, 10–12
  changing lattice constants, 288–289
  hydrogen, 170–175
  magnesium, 445, 457–458
  n-type, 70–73, 170–171, 188–190
  p-type, 73–80, 157, 171–174, 190–193
  shallow, 186–195
Double-correlation DLTS, 197

E

Electric-field induced second-harmonic generation (EFISH), 346–351
Electrically detected magnetic resonance (EDMR), 377, 396–399
Electro-optic effect (LEO), 338–341
Electroluminescence detection of magnetic resonance (ELDMR), 396–399
Electron cyclotron resonance (ECR) plasma sources, 41–49

Electron paramagnetic resonance (EPR), 373–374
Electron spin resonance (ESR), 373
$\eta\mu\tau$ product, 448–453
Exciton oscillator strength, 234–239

F

Field effect transistors (FETs), optoelectric, 430–436
Frank-Van der Merwe growth mode, 59

G

GaAs
  effects of strain on, 211
  semiconductors, limits of, 2
Gallium, covalent radius of, 288t
γ-ray astronomy, 462
GaN
  bandgap of, 213–217
  breakdown electric field of, 444
  crystal structure of, 320–321
  crystal types of, 379
  crystallography of, 227–234
  cubic, 12–13
  dispersion relations in valence band of, 260–262
  doping of, 10–12, 445
  EDMR and ELDMR studies of, 396–399
  effects of strain on, 217–222
  elastic constants of, 286, 287t, 288–289
  electronic structure of, 222–223
  growth by HVPE, 3–14, 23–25
  growth by OMVPE, 3
  growth kinetics of, 7–8
  infrared absorption of, 390–391
  intentional hydrogenation of, 175–177
  irradiation of, 394–395, 458–467
  lattice mismatch strain with substrate, 244–254
  LEDs made from, 21–23
  magnetic resonance data for, 351t
  magneto-optical studies of, 390
  Mg-doped, 457–458
  nonlinear optical properties of, 322–361
  nuclear double resonance studies of, 388–390
  ODMR studies of, 385–388
  optical waveguiding, 361–364
  photodetectors made out of, 417–430
  properties of, 276t
  shallow and deep acceptors in, 391–395
  shallow donor spin resonance in, 377–382
  temperature stability of, 130
  thermochemistry of, 9–10
  transition metal impurities in, 399–403
GaN crystals
  annealing of, 145–146
  defect distribution in, 135–139
  homoepitaxial layers on, 147–152
  hydrogen effect on growth, 176–177
  hydrogen molecules in, 168
  interstitial hydrogen in, 164–168, 170–175
  large, 146–147
  nanotube defects in, 140–141, 142f
  pinholes on, 152
  PL and point defects in, 142–144
  polarity of, 130–134
  structure of, 320–321
  types of, 379
GaN films
  contacts to, 110
  deep-level defects of, 195–203
  deformation of, 282–293
  deformation potential of, 227–234
  devices fabricated on, 110–111
  electrical properties of, 18–20
  electronic structure of, 80–81
  growth cycle of, 279
  growth of, 51–60, 187, 283, 286
  growth of by HVPE, 14–22
  growth of by OMVPE, 15
  lattice parameters of, 276–277, 297–299
  modes of growth of, 59–60
  morphology of, 14–16
  n-type doping of, 70–73, 170–171, 188–190
  optical properties of, 20, 85–89, 223–225
  p-type doping of, 73–80, 157, 171–174, 190–193
  sapphire as substrate for, 60–66
  screw defect in, 15, 15f
  separation from substrate, 26–27
  shallow dopants of, 187–195
  shear stress on, 281
  silicon as substrate for, 66–70
  strain effects on, 277

stress effects on, 276
structure of, 16–18
substrates for, 51, 129–130
surface of, 285f
thermal strain on, 279–281
thickness dependence of electronic
 properties of, 193–195
transport properties of, 81–82, 83f, 84
See also Wurtzite GaN films; Zinc-blende
 GaN films
GaN/AlGaN quantum wells, 105–110
 effect of strain fields on optical properties
  of, 259–266
 valence band levels and excitons in,
  262–266
GaN/SiC transistors, 117–119
GaP, growth by HVPE, 3
GaSb, effects of strain on, 211
Germanium, covalent radius of, 288t

H

Halide vapor-phase epitaxy (VPE), 3
 raw materials for, 40
Hall-effect measurement of, 186, 187–188
Hartree-Fock-based cluster calculations,
 163
Heterojunction bipolar transistors (HBTs),
 nitride-based, 117–119
Homoepitaxial layers, defects on, 148–152
 growth of, 147–148
HVPE (hydride vapor phase epitaxy)
 by-products of, 5
 commercial potential of, 23–25
 defect reduction for, 25
 description of, 4–5
 epitaxial overgrowth in, 25–26
 film-substrate separation in, 26–27
 growth of arsenides by, 3
 growth of nitrides by, 3
 growth of phosphides by, 3
 history of, 3
 reactor design for, 5, 6, 5f, 6f
 substrates for, 10
Hydrogen
 charge states of, 158–159
 complex with magnesium, 169–170
 as contaminant, 161
 diffusion through lattice, 159, 176
 as dopant, 170–175
 in GaN crystal, 164–168
 on GaN surface, 176–177
 interaction with deep levels, 160
 interaction with dopants, 159, 177–181
 interaction with shallow impurities, 168
 molecules and complexes of, 160
 in nitride growth, 160–161
 studies of, 161–163
Hydrogenation, 161
Hydrostatic pressure
 effect on GaN films, 225–227
 shallow vs. deep behavior under, 257–259

I

InAs/GaAs, growth mode of, 267
Indium, in quantum wells, 304–305, 306f,
 307
InGaN
 growth by HVPE, 13–14
 quality of, 446
 utility for x-ray detection, 445
InGaN films
 growth of, 89
 long-range ordering of, 95–96
 phase separation in, 90–95
InGaN quantum wells, 101–105
InN, 218
 growth by HVPE, 13–14
 properties of, 276t
Iron, as impurity in GaN, 402

K

Knudsen effusion cells, 36

L

LEDs
 GaN, 21–23
 MBE-grown, 111–114
LEO (electro-optic effect), 338–341

M

Magnesium
 complexes with hydrogen, 169t, 177–178,
  180–181
 as dopant, 190–193, 457–458
 in GaN, 169–170

Magnetic resonance
  EDMR, 377
  EPR, 373–374
  ODMR, 375–377
Manganese
  as dopant for GaN films, 73–80
  as impurity in GaN, 402–403
MBE (molecular beam epitaxy)
  advantages of, 34, 38–39
  applications of, 39–40
  described, 34–35
  growth temperature of, 283
  nitrogen sources for, 40–51
  requirements for, 34–38
  speed of deposition of, 447
Metal organic chemical vapor deposition.
    See MOCVD
Metal-semiconductor-metal (MSM)
    photodetectors, 430
Metallorganic molecular beam epitaxy
    (MOMBE), 37, 41
MOCVD (metal organic chemical vapor
    deposition), 157, 267, 445
  advantages of, 34
  growth temperature of, 283
  raw materials for, 40
  speed of deposition of, 447–448
Molecular beam epitaxy. See MBE
Multiquantum wells (MQW)
  GaN/AlGaN, 105–110
  InGaN, 101–105

N

n-type doping, 70–73
  hydrogen and, 170–171
  silicon and, 188–190
Nanotube defects, 140–141, 142f
Neutralization, defined, 159
Nickel, as impurity in GaN, 401, 402
Nitride films
  disadvantages of, 446–447
  growth of, 447–448
Nitrides
  as X-ray detectors, 460–463
  characteristics of, 2
  crystalline structures of, 213
  effects of strain on, 217–219
  growth of, 3

Nitrogen
  ammonia as source of, 40–41
  covalent radius of, 288t
  electron cyclotron resonance (ECR)
      plasma sources of, 41–49
  ionic sources of, 50–51
  radio frequency (RF) plasma sources of,
      49–50
Noise equivalent power (NEP), 415–416
Nonlinear optical phenomena, 319–320
  DFWM, 355–361
  EFISH, 346–351
  electro-optic effect, 338–341
  equations describing, 322–325
  second-harmonic generation, 326–337
  second-order, 325–326
  third-harmonic generation, 342–346
  third-order, 341–342
  two-photon absorption (TPA), 351–355
Nuclear double resonance, 384

O

ODLTS (optical DLTS; photoemission
    capacitance transient spectroscopy),
    197, 200–203
OMVPE (organometallic vapor phase
    epitaxy), growth of GaN by, 3, 187
Optical deep-level transient spectroscopy
    (ODLTS), 197, 200–203
Optical waveguides, 361–364
Optically detected electron nuclear double
    resonance (ODENDOR), 388–390
Optically detected magnetic resonance
    (ODMR), 375–377
Optoelectric field effect transistors (FETs),
    430–436
Organometallic vapor phase epitaxy, 3, 187
Overhauser shift, 383–385
Oxides, partial pressure of oxygen over, 11t
Oxinitrides, 288
Oxygen, covalent radius of, 288t

P

p-n junction photodetectors, 424–425
p-type doping, 73–80, 157, 275
  hydrogen and, 171–174
  magnesium and, 190–193

Passivation, defined, 159
Phosphides, growth of, 3
Photoconductive detectors, 412–415
  AlGaN, 422–423
  GaN, 419–424
Photoconductivity, 449–451
Photocurrent decay, 453–457
Photodetectors
  AlGaN, 115–116
  detectivity of, 415–417
  fabrication of, 417–418
  GaN, 114–115, 417–430
  materials for, 407–408
  MBE-grown, 114–116
  principles of operation, 409–415
Photoemission capacitance transient spectroscopy (ODLTS), 197
Photovoltaic detectors, 411–412
  GaN p-n junction, 424–425
  GaN Schottky barrier, 425–430
  metal-semiconductor-metal, 430
Pinholes, on GaN crystals, 149f, 150, 152
Point defects, 142–144
Precipitate defects, 135
Proton implantation, 161
Pseudopotential-density-functional calculations, 161–163

Q

Quantum boxes, self-organized, 267–268
Quantum wells (QW), 101
  crystal quality and, 305
  effect of strain fields on optical properties of, 259–266
  GaN/AlGaN, 105–110
  InGaN quantum wells, 101–105
  lattice mismatch in, 304–305
  single, 101
  structure of, 259
  valence band levels and excitons in, 262–266

R

Radio frequency (RF) plasma sources, 49–50
Radiography, future of, 463
Rare earth nitrides, growth by HVPE, 14
Rectifying devices, 195–197

Reflection high energy electron diffraction (RHEED), 37–38
Rocksalt structure, 213

S

Sapphire
  properties of, 276t
  as substrate, 60–66
Schottky barrier photodetectors, 425–430
ScN, growth by HVPE, 14
Second-harmonic generation (SHG), 326–337
  electric-field induced (EFISH), 346–351
  enhancement of, 363–366
SiC, 2
  transistors from, 117–119
SiGe, growth mode of, 267
Silicon
  covalent radius of, 288t
  as dopant, 70–73, 188–190
  semiconductors, limits of, 2
Single quantum wells (SQW), 101
6H-SiC, properties of, 276t
Spectroscopy
  capacitance transient, 195–197
  optical deep-level transient (ODLTS), 197, 200–203
  thermal deep-level transient, 197–200
Stacking faults, 135, 136, 137f, 138, 138f
Strain. *See* Stress and strain
Strain fields
  effects on band structure, 217–227
  effects on electronic structure, 213–234
  effects on optical properties, 259–266
  phonons under, 254–257
Stranski-Krastanov growth mode, 59, 212, 267, 28
Stress and strain
  buffer layer and, 292f, 293f, 294–295
  controlling, 296–297
  doping and, 294f, 295
  effects on growth, 283, 286, 308–313
  effects on semiconductors, 210
  effects on solids, 210
  effects on transverse splittings, 239–244
  experimental measurement of, 289–292
  flux ratio and, 293f, 295
  origins of, 277–283
  parameters of, 292–297

Stress and strain (*continued*)
  responses to, 282f
  utility to engineering of, 200–304

T

Thermal deep-level transient spectroscopy, 197–200
Third-harmonic generation (THG), 342–346
Threading dislocations, 150, 150f, 284f
Titanium, as impurity in GaN, 401
Transistors, nitride-based, 117–119
Transition metals, as impurity in GaN, 401
Two-photon absorption (TPA), 351–355

V

Vanadium, as impurity in GaN, 401
Volmer-Weber growth mode, 59, 283

W

Wurtzite GaN films
  deformation potential of, 227–234
  effect of hydrostatic pressure on, 225–227
  electronic structure of, 81
  growth of, 53–56
  optical properties of, 223–225
  substrate for, 53–56
Wurtzite structure, 213
  electronic effects of strain fields on, 213–234

X

X-ray astronomy, 462
X-ray crystallography, 463
X-ray detectors, 441–443
  materials for, 443–461

Y

YN, growth by HVPE, 14

Z

Zinc, as dopant, 186
Zinc-blende GaN films
  electronic structure of, 81
  growth of, 56–60
  substrate for, 56–57
  surface of, 58–59
Zinc-blende structure, 213

Contents of Volumes in This Series

Volume 1   Physics of III–V Compounds

C. Hilsum, Some Key Features of III–V Compounds
Franco Bassani, Methods of Band Calculations Applicable to III–V Compounds
E. O. Kane, The k-p Method
V. L. Bonch-Bruevich, Effect of Heavy Doping on the Semiconductor Band Structure
Donald Long, Energy Band Structures of Mixed Crystals of III–V Compounds
Laura M. Roth and Petros N. Argyres, Magnetic Quantum Effects
S. M. Puri and T. H. Geballe, Thermomagnetic Effects in the Quantum Region
W. M. Becker, Band Characteristics near Principal Minima from Magnetoresistance
E. H. Putley, Freeze-Out Effects, Hot Electron Effects, and Submillimeter Photoconductivity in InSb
H. Weiss, Magnetoresistance
Betsy Ancker-Johnson, Plasma in Semiconductors and Semimetals

Volume 2   Physics of III–V Compounds

M. G. Holland, Thermal Conductivity
S. I. Novkova, Thermal Expansion
U. Piesbergen, Heat Capacity and Debye Temperatures
G. Giesecke, Lattice Constants
J. R. Drabble, Elastic Properties
A. U. Mac Rae and G. W. Gobeli, Low Energy Electron Diffraction Studies
Robert Lee Mieher, Nuclear Magnetic Resonance
Bernard Goldstein, Electron Paramagnetic Resonance
T. S. Moss, Photoconduction in III–V Compounds
E. Antoncik ad J. Tauc, Quantum Efficiency of the Internal Photoelectric Effect in InSb
G. W. Gobeli and I. G. Allen, Photoelectric Threshold and Work Function
P. S. Pershan, Nonlinear Optics in III–V Compounds
M. Gershenzon, Radiative Recombination in the III–V Compounds
Frank Stern, Stimulated Emission in Semiconductors

Volume 3 Optical of Properties III–V Compounds

Marvin Hass, Lattice Reflection
William G. Spitzer, Multiphonon Lattice Absorption
D. L. Stierwalt and R. F. Potter, Emittance Studies
H. R. Philipp and H. Ehrenveich, Ultraviolet Optical Properties
Manuel Cardona, Optical Absorption above the Fundamental Edge
Earnest J. Johnson, Absorption near the Fundamental Edge
John O. Dimmock, Introduction to the Theory of Exciton States in Semiconductors
B. Lax and J. G. Mavroides, Interband Magnetooptical Effects
H. Y. Fan, Effects of Free Carries on Optical Properties
Edward D. Palik and George B. Wright, Free-Carrier Magnetooptical Effects
Richard H. Bube, Photoelectronic Analysis
B. O. Seraphin and H. E. Bennett, Optical Constants

Volume 4 Physics of III–V Compounds

N. A. Goryunova, A. S. Borschevskii, and D. N. Tretiakov, Hardness
N. N. Sirota, Heats of Formation and Temperatures and Heats of Fusion of Compounds $A^{III}B^V$
Don L. Kendall, Diffusion
A. G. Chynoweth, Charge Multiplication Phenomena
Robert W. Keyes, The Effects of Hydrostatic Pressure on the Properties of III–V Semiconductors
L. W. Aukerman, Radiation Effects
N. A. Goryunova, F. P. Kesamanly, and D. N. Nasledov, Phenomena in Solid Solutions
R. T. Bate, Electrical Properties of Nonuniform Crystals

Volume 5 Infrared Detectors

Henry Levinstein, Characterization of Infrared Detectors
Paul W. Kruse, Indium Antimonide Photoconductive and Photoelectromagnetic Detectors
M. B. Prince, Narrowband Self-Filtering Detectors
Ivars Melngalis and T. C. Harman, Single-Crystal Lead-Tin Chalcogenides
Donald Long and Joseph L. Schmidt, Mercury-Cadmium Telluride and Closely Related Alloys
E. H. Putley, The Pyroelectric Detector
Norman B. Stevens, Radiation Thermopiles
R. J. Keyes and T. M. Quist, Low Level Coherent and Incoherent Detection in the Infrared
M. C. Teich, Coherent Detection in the Infrared
F. R. Arams, E. W. Sard, B. J. Peyton, and F. P. Pace, Infrared Heterodyne Detection with Gigahertz IF Response
H. S. Sommers, Jr., Macrowave-Based Photoconductive Detector
Robert Sehr and Rainer Zuleeg, Imaging and Display

Volume 6 Injection Phenomena

Murray A. Lampert and Ronald B. Schilling, Current Injection in Solids: The Regional Approximation Method
Richard Williams, Injection by Internal Photoemission
Allen M. Barnett, Current Filament Formation

R. Baron and J. W. Mayer, Double Injection in Semiconductors
W. Ruppel, The Photoconductor-Metal Contact

Volume 7 Application and Devices
Part A

John A. Copeland and Stephen Knight, Applications Utilizing Bulk Negative Resistance
F. A. Padovani, The Voltage-Current Characteristics of Metal-Semiconductor Contacts
P. L. Hower, W. W. Hooper, B. R. Cairns, R. D. Fairman, and D. A. Tremere, The GaAs Field-Effect Transistor
Marvin H. White, MOS Transistors
G. R. Antell, Gallium Arsenide Transistors
T. L. Tansley, Heterojunction Properties

Part B

T. Misawa, IMPATT Diodes
H. C. Okean, Tunnel Diodes
Robert B. Campbell and Hung-Chi Chang, Silicon Junction Carbide Devices
R. E. Enstrom, H. Kressel, and L. Krassner, High-Temperature Power Rectifiers of $GaAs_{1-x}P_x$

Volume 8 Transport and Optical Phenomena

Richard J. Stirn, Band Structure and Galvanomagnetic Effects in III–V Compounds with Indirect Band Gaps
Roland W. Ure, Jr., Thermoelectric Effects in III–V Compounds
Herbert Piller, Faraday Rotation
H. Barry Bebb and E. W. Williams, Photoluminescence I: Theory
E. W. Williams and H. Barry Bebb, Photoluminescence II: Gallium Arsenide

Volume 9 Modulation Techniques

B. O. Seraphin, Electroreflectance
R. L. Aggarwal, Modulated Interband Magnetooptics
Daniel F. Blossey and Paul Handler, Electroabsorption
Bruno Batz, Thermal and Wavelength Modulation Spectroscopy
Ivar Balslev, Piezopptical Effects
D. E. Aspnes and N. Bottka, Electric-Field Effects on the Dielectric Function of Semiconductors and Insulators

Volume 10 Transport Phenomena

R. L. Rhode, Low-Field Electron Transport
J. D. Wiley, Mobility of Holes in III–V Compounds
C. M. Wolfe and G. E. Stillman, Apparent Mobility Enhancement in Inhomogeneous Crystals
Robert L. Petersen, The Magnetophonon Effect

Volume 11 Solar Cells

Harold J. Hovel, Introduction; Carrier Collection, Spectral Response, and Photocurrent; Solar Cell Electrical Characteristics; Efficiency; Thickness; Other Solar Cell Devices; Radiation Effects; Temperature and Intensity; Solar Cell Technology

Volume 12 Infrared Detectors (II)

W. L. Eiseman, J. D. Merriam, and R. F. Potter, Operational Characteristics of Infrared Photodetectors
Peter R. Bratt, Impurity Germanium and Silicon Infrared Detectors
E. H. Putley, InSb Submillimeter Photoconductive Detectors
G. E. Stillman, C. M. Wolfe, and J. O. Dimmock, Far-Infrared Photoconductivity in High Purity GaAs
G. E. Stillman and C. M. Wolfe, Avalanche Photodiodes
P. L. Richards, The Josephson Junction as a Detector of Microwave and Far-Infrared Radiation
E. H. Putley, The Pyroelectric Detector–An Update

Volume 13 Cadmium Telluride

Kenneth Zanio, Materials Preparations; Physics; Defects; Applications

Volume 14 Lasers, Junctions, Transport

N. Holonyak, Jr. and M. H. Lee, Photopumped III–V Semiconductor Lasers
Henry Kressel and Jerome K. Butler, Heterojunction Laser Diodes
A Van der Ziel, Space-Charge-Limited Solid-State Diodes
Peter J. Price, Monte Carlo Calculation of Electron Transport in Solids

Volume 15 Contacts, Junctions, Emitters

B. L. Sharma, Ohmic Contacts to III–V Compounds Semiconductors
Allen Nussbaum, The Theory of Semiconducting Junctions
John S. Escher, NEA Semiconductor Photoemitters

Volume 16 Defects, (HgCd)Se, (HgCd)Te

Henry Kressel, The Effect of Crystal Defects on Optoelectronic Devices
C. R. Whitsett, J. G. Broerman, and C. J. Summers, Crystal Growth and Properties of $Hg_{1-x}Cd_x$Se alloys
M. H. Weiler, Magnetooptical Properties of $Hg_{1-x}Cd_x$Te Alloys
Paul W. Kruse and John G. Ready, Nonlinear Optical Effects in $Hg_{1-x}Cd_x$Te

Volume 17 CW Processing of Silicon and Other Semiconductors

James F. Gibbons, Beam Processing of Silicon
Arto Lietoila, Richard B. Gold, James F. Gibbons, and Lee A. Christel, Temperature Distribu-

tions and Solid Phase Reaction Rates Produced by Scanning CW Beams
Arto Leitoila and James F. Gibbons, Applications of CW Beam Processing to Ion Implanted Crystalline Silicon
N. M. Johnson, Electronic Defects in CW Transient Thermal Processed Silicon
K. F. Lee, T. J. Stultz, and James F. Gibbons, Beam Recrystallized Polycrystalline Silicon: Properties, Applications, and Techniques
T. Shibata, A. Wakita, T. W. Sigmon, and James F. Gibbons, Metal-Silicon Reactions and Silicide
Yves I. Nissim and James F. Gibbons, CW Beam Processing of Gallium Arsenide

Volume 18 Mercury Cadmium Telluride

Paul W. Kruse, The Emergence of $(Hg_{1-x}Cd_x)Te$ as a Modern Infrared Sensitive Material
H. E. Hirsch, S. C. Liang, and A. G. White, Preparation of High-Purity Cadmium, Mercury, and Tellurium
W. F. H. Micklethwaite, The Crystal Growth of Cadmium Mercury Telluride
Paul E. Petersen, Auger Recombination in Mercury Cadmium Telluride
R. M. Broudy and V. J. Mazurczyck, (HgCd)Te Photoconductive Detectors
M. B. Reine, A. K. Soad, and T. J. Tredwell, Photovoltaic Infrared Detectors
M. A. Kinch, Metal-Insulator-Semiconductor Infrared Detectors

Volume 19 Deep Levels, GaAs, Alloys, Photochemistry

G. F. Neumark and K. Kosai, Deep Levels in Wide Band-Gap III-V Semiconductors
David C. Look, The Electrical and Photoelectronic Properties of Semi-Insulating GaAs
R. F. Brebrick, Ching-Hua Su, and Pok-Kai Liao, Associated Solution Model for Ga-In-Sb and Hg-Cd-Te
Yu. Ya. Gurevich and Yu. V. Pleskon, Photoelectrochemistry of Semiconductors

Volume 20 Semi-Insulating GaAs

R. N. Thomas, H. M. Hobgood, G. W. Eldridge, D. L. Barrett, T. T. Braggins, L. B. Ta, and S. K. Wang, High-Purity LEC Growth and Direct Implantation of GaAs for Monolithic Microwave Circuits
C. A. Stolte, Ion Implantation and Materials for GaAs Integrated Circuits
C. G. Kirkpatrick, R. T. Chen, D. E. Holmes, P. M. Asbeck, K. R. Elliott, R. D. Fairman, and J. R. Oliver, LEC GaAs for Integrated Circuit Applications
J. S. Blakemore and S. Rahimi, Models for Mid-Gap Centers in Gallium Arsenide

Volume 21 Hydrogenated Amorphous Silicon
Part A

Jacques I. Pankove, Introduction
Masataka Hirose, Glow Discharge; Chemical Vapor Deposition
Yoshiyuki Uchida, di Glow Discharge
T. D. Moustakas, Sputtering
Isao Yamada, Ionized-Cluster Beam Deposition
Bruce A. Scott, Homogeneous Chemical Vapor Deposition

Frank J. Kampas, Chemical Reactions in Plasma Deposition
Paul A. Longeway, Plasma Kinetics
Herbert A. Weakliem, Diagnostics of Silane Glow Discharges Using Probes and Mass Spectroscopy
Lester Gluttman, Relation between the Atomic and the Electronic Structures
A. Chenevas-Paule, Experiment Determination of Structure
S. Minomura, Pressure Effects on the Local Atomic Structure
David Adler, Defects and Density of Localized States

Part B

Jacques I. Pankove, Introduction
G. D. Cody, The Optical Absorption Edge of a-Si:H
Nabil M. Amer and Warren B. Jackson, Optical Properties of Defect States in a-Si:H
P. J. Zanzucchi, The Vibrational Spectra of a-Si:H
Yoshihiro Hamakawa, Electroreflectance and Electroabsorption
Jeffrey S. Lannin, Raman Scattering of Amorphous Si, Ge, and Their Alloys
R. A. Street, Luminescence in a-Si:H
Richard S. Crandall, Photoconductivity
J. Tauc, Time-Resolved Spectroscopy of Electronic Relaxation Processes
P. E. Vanier, IR-Induced Quenching and Enhancement of Photoconductivity and Photoluminescence
H. Schade, Irradiation-Induced Metastable Effects
L. Ley, Photoelectron Emission Studies

Part C

Jacques I. Pankove, Introduction
J. David Cohen, Density of States from Junction Measurements in Hydrogenated Amorphous Silicon
P. C. Taylor, Magnetic Resonance Measurements in a-Si:H
K. Morigaki, Optically Detected Magnetic Resonance
J. Dresner, Carrier Mobility in a-Si:H
T. Tiedje, Information about band-Tail States from Time-of-Flight Experiments
Arnold R. Moore, Diffusion Length in Undoped a-Si:H
W. Beyer and J. Overhof, Doping Effects in a-Si:H
H. Fritzche, Electronic Properties of Surfaces in a-Si:H
C. R. Wronski, The Staebler-Wronski Effect
R. J. Nemanich, Schottky Barriers on a-Si:H
B. Abeles and T. Tiedje, Amorphous Semiconductor Superlattices

Part D

Jacques I. Pankove, Introduction
D. E. Carlson, Solar Cells
G. A. Swartz, Closed-Form Solution of I–V Characteristic for a a-Si:H Solar Cells
Isamu Shimizu, Electrophotography
Sachio Ishioka, Image Pickup Tubes

P. G. LeComber and W. E. Spear, The Development of the a-Si:H Field-Effect Transistor and Its Possible Applications
D. G. Ast, a-Si:H FET-Addressed LCD Panel
S. Kaneko, Solid-State Image Sensor
Masakiyo Matsumura, Charge-Coupled Devices
M. A. Bosch, Optical Recording
A. D'Amico and G. Fortunato, Ambient Sensors
Hiroshi Kukimoto, Amorphous Light-Emitting Devices
Robert J. Phelan, Jr., Fast Detectors and Modulators
Jacques I. Pankove, Hybrid Structures
P. G. LeComber, A. E. Owen, W. E. Spear, J. Hajto, and W. K. Choi, Electronic Switching in Amorphous Silicon Junction Devices

Volume 22 Lightwave Communications Technology
Part A

Kazuo Nakajima, The Liquid-Phase Epitaxial Growth of IngaAsp
W. T. Tsang, Molecular Beam Epitaxy for III–V Compound Semiconductors
G. B. Stringfellow, Organometallic Vapor-Phase Epitaxial Growth of III–V Semiconductors
G. Beuchet, Halide and Chloride Transport Vapor-Phase Deposition of InGaAsP and GaAs
Manijeh Razeghi, Low-Pressure Metallo-Organic Chemical Vapor Deposition of $Ga_x in_{1-x} As P_{1-y}$ Alloys
P. M. Petroff, Defects in III–V Compound Semiconductors

Part B

J. P. van der Ziel, Mode Locking of Semiconductor Lasers
Kam Y. Lau and Ammon Yariv, High-Frequency Current Modulation of Semiconductor Injection Lasers
Charles H. Henry, Special Properties of Semiconductor Lasers
Yasuharu Suematsu, Katsumi Kishino, Shigehisa Arai, and Fumio Koyama. Dynamic Single-Mode Semiconductor Lasers with a Distributed Reflector
W. T. Tsang, The Cleaved-Coupled-Cavity (C^3) Laser

Part C

R. J. Nelson and N. K. Dutta, Review of InGaAsP InP Laser Structures and Comparison of Their Performance
N. Chinone and M. Nakamura, Mode-Stabilized Semiconductor Lasers for 0.7–0.8- and 1.1–1.6-μm Regions
Yoshiji Horikoshi, Semiconductor Lasers with Wavelengths Exceeding 2 μm
B. A. Dean and M. Dixon, The Functional Reliability of Semiconductor Lasers as Optical Transmitters
R. H. Saul, T. P. Lee, and C. A. Burus, Light-Emitting Device Design
C. L. Zipfel, Light-Emitting Diode-Reliability
Tien Pei Lee and Tingye Li, LED-Based Multimode Lightwave Systems
Kinichiro Ogawa, Semiconductor Noise-Mode Partition Noise

Part D

Federico Capasso, The Physics of Avalanche Photodiodes
T. P. Pearsall and M. A. Pollack, Compound Semiconductor Photodiodes
Takao Kaneda, Silicon and Germanium Avalanche Photodiodes
S. R. Forrest, Sensitivity of Avalanche Photodetector Receivers for High-Bit-Rate Long-Wavelength Optical Communication Systems
J. C. Campbell, Phototransistors for Lightwave Communications

Part E

Shyh Wang, Principles and Characteristics of Integrable Active and Passive Optical Devices
Shlomo Margalit and Amnon Yariv, Integrated Electronic and Photonic Devices
Takaoki Mukai, Yoshihisa Yamamoto, and Tatsuya Kimura, Optical Amplification by Semiconductor Lasers

Volume 23 Pulsed Laser Processing of Semiconductors

R. F. Wood, C. W. White, and R. T. Young, Laser Processing of Semiconductors: An Overview
C. W. White, Segregation, Solute Trapping, and Supersaturated Alloys
G. E. Jellison, Jr., Optical and Electrical Properties of Pulsed Laser-Annealed Silicon
R. F. Wood and G. E. Jellison, Jr., Melting Model of Pulsed Laser Processing
R. F. Wood and F. W. Young, Jr., Nonequilibrium Solidification Following Pulsed Laser Melting
D. H. Lowndes and G. E. Jellison, Jr., Time-Resolved Measurement During Pulsed Laser Irradiation of Silicon
D. M. Zebner, Surface Studies of Pulsed Laser Irradiated Semiconductors
D. H. Lowndes, Pulsed Beam Processing of Gallium Arsenide
R. B. James, Pulsed CO_2 Laser Annealing of Semiconductors
R. T. Young and R. F. Wood, Applications of Pulsed Laser Processing

Volume 24 Applications of Multiquantum Wells, Selective Doping, and Superlattices

C. Weisbuch, Fundamental Properties of III–V Semiconductor Two-Dimensional Quantized Structures: The Basis for Optical and Electronic Device Applications
H. Morkoc and H. Unlu, Factors Affecting the Performance of (Al, Ga)As/GaAs and (Al, Ga)As/InGaAs Modulation-Doped Field-Effect Transistors: Microwave and Digital Applications
N. T. Linh, Two-Dimensional Electron Gas FETs: Microwave Applications
M. Abe et al., Ultra-High-Speed HEMT Integrated Circuits
D. S. Chemla, D. A. B. Miller, and P. W. Smith, Nonlinear Optical Properties of Multiple Quantum Well Structures for Optical Signal Processing
F. Capasso, Graded-Gap and Superlattice Devices by Band-Gap Engineering
W. T. Tsang, Quantum Confinement Heterostructure Semiconductor Lasers
G. C. Osbourn et al., Principles and Applications of Semiconductor Strained-Layer Superlattices

Volume 25 Diluted Magnetic Semiconductors

W. Giriat and J. K. Furdyna, Crystal Structure, Composition, and Materials Preparation of Diluted Magnetic Semiconductors

W. M. Becker, Band Structure and Optical Properties of Wide-Gap $A^{II}_{1-x}Mn_xB^{IV}$ Alloys at Zero Magnetic Field

Saul Oseroff and Pieter H. Keesom, Magnetic Properties: Macroscopic Studies

Giebultowicz and T. M. Holden, Neutron Scattering Studies of the Magnetic Structure and Dynamics of Diluted Magnetic Semiconductors

J. Kossut, Band Structure and Quantum Transport Phenomena in Narrow-Gap Diluted Magnetic Semiconductors

C. Riquaux, Magnetooptical Properties of Large-Gap Diluted Magnetic Semiconductors

J. A. Gaj, Magnetooptical Properties of Large-Gap Diluted Magnetic Semiconductors

J. Mycielski, Shallow Acceptors in Diluted Magnetic Semiconductors: Splitting, Boil-off, Giant Negative Magnetoresistance

A. K. Ramadas and R. Rodriquez, Raman Scattering in Diluted Magnetic Semiconductors

P. A. Wolff, Theory of Bound Magnetic Polarons in Semimagnetic Semiconductors

Volume 26 III–V Compound Semiconductors and Semiconductor Properties of Superionic Materials

Zou Yuanxi, III–V Compounds

H. V. Winston, A. T. Hunter, H. Kimura, and R. E. Lee, InAs-Alloyed GaAs Substrates for Direct Implantation

P. K. Bhattachary and S. Dhar, Deep Levels in III–V Compound Semiconductors Grown by MBE

Yu. Yu. Gurevich and A. K. Ivanov-Shits, Semiconductor Properties of Supersonic Materials

Volume 27 High Conducting Quasi-One-Dimensional Organic Crystals

E. M. Conwell, Introduction to Highly Conducting Quasi-One-Dimensional Organic Crystals

I. A. Howard, A Reference Guide to the Conducting Quasi-One-Dimensional Organic Molecular Crystals

J. P. Pouquet, Structural Instabilities

E. M. Conwell, Transport Properties

C. S. Jacobsen, Optical Properties

J. C. Scott, Magnetic Properties

L. Zuppiroli, Irradiation Effects: Perfect Crystals and Real Crystals

Volume 28 Measurement of High-Speed Signals in Solid State Devices

J. Frey and D. Ioannou, Materials and Devices for High-Speed and Optoelectronic Applications

H. Schumacher and E. Strid, Electronic Wafer Probing Techniques

D. H. Auston, Picosecond Photoconductivity: High-Speed Measurements of Devices and Materials

J. A. Valdmanis, Electro-Optic Measurement Techniques for Picosecond Materials, Devices, and Integrated Circuits.

J. M. Wiesenfeld and R. K. Jain, Direct Optical Probing of Integrated Circuits and High-Speed Devices

G. Plows, Electron-Beam Probing

A. M. Weiner and R. B. Marcus, Photoemissive Probing

Volume 29 Very High Speed Integrated Circuits: Gallium Arsenide LSI

M. Kuzuhara and T. Nazaki, Active Layer Formation by Ion Implantation
H. Hasimoto, Focused Ion Beam Implantation Technology
T. Nozaki and A. Higashisaka, Device Fabrication Process Technology
M. Ino and T. Takada, GaAs LSI Circuit Design
M. Hirayama, M. Ohmori, and K. Yamasaki, GaAs LSI Fabrication and Performance

Volume 30 Very High Speed Integrated Circuits: Heterostructure

H. Watanabe, T. Mizutani, and A. Usui, Fundamentals of Epitaxial Growth and Atomic Layer Epitaxy
S. Hiyamizu, Characteristics of Two-Dimensional Electron Gas in III–V Compound Heterostructures Grown by MBE
T. Nakanisi, Metalorganic Vapor Phase Epitaxy for High-Quality Active Layers
T. Nimura, High Electron Mobility Transistor and LSI Applications
T. Sugeta and T. Ishibashi, Hetero-Bipolar Transistor and LSI Application
H. Matsueda, T. Tanaka, and M. Nakamura, Optoelectronic Integrated Circuits

Volume 31 Indium Phosphide: Crystal Growth and Characterization

J. P. Farges, Growth of Discoloration-free InP
M. J. McCollum and G. E. Stillman, High Purity InP Grown by Hydride Vapor Phase Epitaxy
T. Inada and T. Fukuda, Direct Synthesis and Growth of Indium Phosphide by the Liquid Phosphorous Encapsulated Czochralski Method
O. Oda, K. Katagiri, K. Shinohara, S. Katsura, Y. Takahashi, K. Kainosho, K. Kohiro, and R. Hirano, InP Crystal Growth, Substrate Preparation and Evaluation
K. Tada, M. Tatsumi, M. Morioka, T. Araki, and T. Kawase, InP Substrates: Production and Quality Control
M. Razeghi, LP-MOCVD Growth, Characterization, and Application of InP Material
T. A. Kennedy and P. J. Lin-Chung, Stoichiometric Defects in InP

Volme 32 Strained-Layer Superlattices: Physics

T. P. Pearsall, Strained-Layer Superlattices
Fred H. Pollack, Effects of Homogeneous Strain on the Electronic and Vibrational Levels in Semiconductors
J. Y. Marzin, J. M. Gerárd, P. Voisin, and J. A. Brum, Optical Studies of Strained III–V Heterolayers
R. People and S. A. Jackson, Structurally Induced States from Strain and Confinement
M. Jaros, Microscopic Phenomena in Ordered Suprlattices

Volume 33 Strained-Layer Superlattices: Materials Science and Technology

R. Hull and J. C. Bean, Principles and Concepts of Strained-Layer Epitaxy
William J. Schaff, Paul J. Tasker, Marc C. Foisy, and Lester F. Eastman, Device Applications of Strained-Layer Epitaxy

S. T. Picraux, B. L. Doyle, and J. Y. Tsao, Structure and Characterization of Strained-Layer Superlattices
E. Kasper and F. Schäffer, Group IV Compounds
Dale L. Martin, Molecular Beam Epitaxy of IV–VI Compounds Heterojunction
Robert L. Gunshor, Leslie A. Kolodziejski, Arto V. Nurmikko, and Nobuo Otsuka, Molecular Beam Epitaxy of II–VI Semiconductor Microstructures

Volume 34 Hydrogen in Semiconductors

J. I. Pankove and N. M. Johnson, Introduction to Hydrogen in Semiconductors
C. H. Seager, Hydrogenation Methods
J. I. Pankove, Hydrogenation of Defects in Crystalline Silicon
J. W. Corbett, P. Deák, U. V. Desnica, and S. J. Pearton, Hydrogen Passivation of Damage Centers in Semiconductors
S. J. Pearton, Neutralization of Deep Levels in Silicon
J. I. Pankove, Neutralization of Shallow Acceptors in Silicon
N. M. Johnson, Neutralization of Donor Dopants and Formation of Hydrogen-Induced Defects in n-Type Silicon
M. Stavola and S. J. Pearton, Vibrational Spectroscopy of Hydrogen-Related Defects in Silicon
A. D. Marwick, Hydrogen in Semiconductors: Ion Beam Techniques
C. Herring and N. M. Johnson, Hydrogen Migration and Solubility in Silicon
E. E. Haller, Hydrogen-Related Phenomena in Crystalline Germanium
J. Kakalios, Hydrogen Diffusion in Amorphous Silicon
J. Chevalier, B. Clerjaud, and B. Pajot, Neutralization of Defects and Dopants in III–V Semiconductors
G. G. DeLeo and W. B. Fowler, Computational Studies of Hydrogen-Containing Complexes in Semiconductors
R. F. Kiefl and T. L. Estle, Muonium in Semiconductors
C. G. Van de Walle, Theory of Isolated Interstitial Hydrogen and Muonium in Crystalline Semiconductors

Volume 35 Nanostructured Systems

Mark Reed, Introduction
H. van Houten, C. W. J. Beenakker, and B. J. van Wees, Quantum Point Contacts
G. Timp, When Does a Wire Become an Electron Waveguide?
M. Büttiker, The Quantum Hall Effects in Open Conductors
W. Hansen, J. P. Kotthaus, and U. Merkt, Electrons in Laterally Periodic Nanostructures

Volume 36 The Spectroscopy of Semiconductors

D. Heiman, Spectroscopy of Semiconductors at Low Temperatures and High Magnetic Fields
Arto V. Nurmikko, Transient Spectroscopy by Ultrashort Laser Pulse Techniques
A. K. Ramdas and S. Rodriguez, Piezospectroscopy of Semiconductors
Orest J. Glembocki and Benjamin V. Shanabrook, Photoreflectance Spectroscopy of Microstructures
David G. Seiler, Christopher L. Littler, and Margaret H. Wiler, One- and Two-Photon Magneto-Optical Spectroscopy of InSb and $Hg_{1-x}Cd_xTe$

Volume 37 The Mechanical Properties of Semiconductors

A.-B. Chen, Arden Sher and W. T. Yost, Elastic Constants and Related Properties of Semiconductor Compounds and Their Alloys
David R. Clarke, Fracture of Silicon and Other Semiconductors
Hans Siethoff, The Plasticity of Elemental and Compound Semiconductors
Sivaraman Guruswamy, Katherine T. Faber and John P. Hirth, Mechanical Behavior of Compound Semiconductors
Subhanh Mahajan, Deformation Behavior of Compound Semiconductors
John P. Hirth, Injection of Dislocations into Strained Multilayer Structures
Don Kendall, Charles B. Fleddermann, and Kevin J. Malloy, Critical Technologies for the Micromachining of Silicon
Ikuo Matsuba and Kinji Mokuya, Processing and Semiconductor Thermoelastic Behavior

Volume 38 Imperfections in III/V Materials

Udo Scherz and Matthias Scheffler, Density-Functional Theory of sp-Bonded Defects in III/V Semiconductors
Maria Kaminska and Eicke R. Weber, El2 Defect in GaAs
David C. Look, Defects Relevant for Compensation in Semi-Insulating GaAs
R. C. Newman, Local Vibrational Mode Spectroscopy of Defects in III/V Compounds
Andrzej M. Hennel, Transition Metals in III/V Compounds
Kevin J. Malloy and Ken Khachaturyan, DX and Related Defects in Semiconductors
V. Swaminathan and Andrew S. Jordan, Dislocations in III/V Compounds
Krzysztof W. Nauka, Deep Level Defects in the Epitaxial III/V Materials

Volume 39 Minority Carriers in III–V Semiconductors: Physics and Applications

Niloy K. Dutta, Radiative Transitions in GaAs and Other III–V Compounds
Richard K. Ahrenkiel, Minority-Carrier Lifetime in III–V Semiconductors
Tomofumi Furuta, High Field Minority Electron Transport in p-GaAs
Mark S. Lundstrom, Minority-Carrier Transport in III–V Semiconductors
Richard A. Abram, Effects of Heavy Doping and High Excitation on the Band Structure of GaAs
David Yevick and Witold Bardyszewski, An Introduction to Non-Equilibrium Many-Body Analyses of Optical Processes in III–V Semiconductors

Volume 40 Epitaxial Microstructures

E. F. Schubert, Delta-Doping of Semiconductors: Electronic, Optical, and Structural Properties of Materials and Devices
A. Gossard, M. Sundaram, and P. Hopkins, Wide Graded Potential Wells
P. Petroff, Direct Growth of Nanometer-Size Quantum Wire Superlattices
E. Kapon, Lateral Patterning of Quantum Well Heterostructures by Growth of Nonplanar Substrates
H. Temkin, D. Gershoni, and M. Panish, Optical Properties of Ga1-$_x$In$_x$As/InP Quantum Wells

Volume 41 High Speed Heterostructure Devices

F. Capasso, F. Beltram, S. Sen, A. Pahlevi, and A. Y. Cho, Quantum Electron Devices: Physics and Applications
P. Solomon, D. J. Frank, S. L. Wright, and F. Canora, GaAs-Gate Semiconductor–Insulator–Semiconductor FET
M. H. Hashemi and U. K. Mishra, Unipolar InP-Based Transistors
R. Kiehl, Complementary Heterostructure FET Integrated Circuits
T. Ishibashi, GaAs-Based and InP-Based Heterostructure Bipolar Transistors
H. C. Liu and T. C. L. G. Sollner, High-Frequency-Tunneling Devices
H. Ohnishi, T. More, M. Takatsu, K. Imamura, and N. Yokoyama, Resonant-Tunneling Hot-Electron Transistors and Circuits

Volume 42 Oxygen in Silicon

F. Shimura, Introduction to Oxygen in Silicon
W. Lin, The Incorporation of Oxygen into Silicon Crystals
T. J. Schaffner and D. K. Schroder, Characterization Techniques for Oxygen in Silicon
W. M. Bullis, Oxygen Concentration Measurement
S. M. Hu, Intrinsic Point Defects in Silicon
B. Pajot, Some Atomic Configurations of Oxygen
J. Michel and L. C. Kimerling, Electical Properties of Oxygen in Silicon
R. C. Newman and R. Jones, Diffusion of Oxygen in Silicon
T. Y. Tan and W. J. Taylor, Mechanisms of Oxygen Precipitation: Some Quantitative Aspects
M. Schrems, Simulation of Oxygen Precipitation
K. Simino and I. Yonenaga, Oxygen Effect on Mechanical Properties
W. Bergholz, Grown-in and Process-Induced Effects
F. Shimura, Intrinsic/Internal Gettering
H. Tsuya, Oxygen Effect on Electronic Device Performance

Volume 43 Semiconductors for Room Temperature Nuclear Detector Applications

R. B. James and T. E. Schlesinger, Introduction and Overview
L. S. Darken and C. E. Cox, High-Purity Germanium Detectors
A. Burger, D. Nason, L. Van den Berg, and M. Schieber, Growth of Mercuric Iodide
X. J. Bao, T. E. Schlesinger, and R. B. James, Electrical Properties of Mercuric Iodide
X. J. Bao, R. B. James, and T. E. Schlesinger, Optical Properties of Red Mercuric Iodide
M. Hage-Ali and P. Siffert, Growth Methods of CdTe Nuclear Detector Materials
M. Hage-Ali and P Siffert, Characterization of CdTe Nuclear Detector Materials
M. Hage-Ali and P. Siffert, CdTe Nuclear Detectors and Applications
R. B. James, T. E. Schlesinger, J. Lund, and M. Schieber, $Cd_{1-x}Zn_xTe$ Spectrometers for Gamma and X-Ray Applications
D. S. McGregor, J. E. Kammeraad, Gallium Arsenide Radiation Detectors and Spectrometers
J. C. Lund, F. Olschner, and A. Burger, Lead Iodide
M. R. Squillante, and K. S. Shah, Other Materials: Status and Prospects
V. M. Gerrish, Characterization and Quantification of Detector Performance
J. S. Iwanczyk and B. E. Patt, Electronics for X-ray and Gamma Ray Spectrometers
M. Schieber, R. B. James, and T. E. Schlesinger, Summary and Remaining Issues for Room Temperature Radiation Spectrometers

Volume 44 II–IV Blue/Green Light Emitters: Device Physics and Epitaxial Growth

J. Han and R. L. Gunshor, MBE Growth and Electrical Properties of Wide Bandgap ZnSe-based II–VI Semiconductors

Shizuo Fujita and Shigeo Fujita, Growth and Characterization of ZnSe-based II–VI Semiconductors by MOVPE

Easen Ho and Leslie A. Kolodziejski, Gaseous Source UHV Epitaxy Technologies for Wide Bandgap II–VI Semiconductors

Chris G. Van de Walle, Doping of Wide-Band-Gap II–VI Compounds — Theory

Roberto Cingolani, Optical Properties of Excitons in ZnSe-Based Quantum Well Heterostructures

A. Ishibashi and A. V. Nurmikko, II–VI Diode Lasers: A Current View of Device Performance and Issues

Supratik Guha and John Petruzello, Defects and Degradation in Wide-Gap II–VI-based Structures and Light Emitting Devices

Volume 45 Effect of Disorder and Defects in Ion-Implanted Semiconductors: Electrical and Physiochemical Characterization

Heiner Ryssel, Ion Implantation into Semiconductors: Historical Perspectives

You-Nian Wang and Teng-Cai Ma, Electronic Stopping Power for Energetic Ions in Solids

Sachiko T. Nakagawa, Solid Effect on the Electronic Stopping of Crystalline Target and Application to Range Estimation

G. Müller, S. Kalbitzer and G. N. Greaves, Ion Beams in Amorphous Semiconductor Research

Jumana Boussey-Said, Sheet and Spreading Resistance Analysis of Ion Implanted and Annealed Semiconductors

M. L. Polignano and G. Queirolo, Studies of the Stripping Hall Effect in Ion-Implanted Silicon

J. Stoemenos, Transmission Electron Microscopy Analyses

Roberta Nipoti and Marco Servidori, Rutherford Backscattering Studies of Ion Implanted Semiconductors

P. Zaumseil, X-ray Diffraction Techniques

Volume 46 Effect of Disorder and Defects in Ion-Implanted Semiconductors: Optical and Photothermal Characterization

M. Fried, T. Lohner and J. Gyulai, Ellipsometric Analysis

Antonios Seas and Constantinos Christofides, Transmission and Reflection Spectroscopy on Ion Implanted Semiconductors

Andreas Othonos and Constantinos Christofides, Photoluminescence and Raman Scattering of Ion Implanted Semiconductors. Influence of Annealing

Constantinos Christofides, Photomodulated Thermoreflectance Investigation of Implanted Wafers. Annealing Kinetics of Defects

U. Zammit, Photothermal Deflection Spectroscopy Characterization of Ion-Implanted and Annealed Silicon Films

Andreas Mandelis, Arief Budiman and Miguel Vargas, Photothermal Deep-Level Transient Spectroscopy of Impurities and Defects in Semiconductors

R. Kalish and S. Charbonneau, Ion Implantation into Quantum-Well Structures

Alexandre M. Myasnikov and Nikolay N. Gerasimenko, Ion Implantation and Thermal Annealing of III-V Compound Semiconducting Systems: Some Problems of III-V Narrow Gap Semiconductors

Volume 47 Uncooled Infrared Imaging Arrays and Systems

R. G. Buser and M. P. Tompsett, Historical Overview
P. W. Kruse, Principles of Uncooled Infrared Focal Plane Arrays
R. A. Wood, Monolithic Silicon Microbolometer Arrays
C. M. Hanson, Hybrid Pyroelectric-Ferroelectric Bolometer Arrays
D. L. Polla and J. R. Choi, Monolithic Pyroelectric Bolometer Arrays
N. Teranishi, Thermoelectric Uncooled Infrared Focal Plane Arrays
M. F. Tompsett, Pyroelectric Vidicon
T. W. Kenny, Tunneling Infrared Sensors
J. R. Vig, R. L. Filler and Y. Kim, Application of Quartz Microresonators to Uncooled Infrared Imaging Arrays
P. W. Kruse, Application of Uncooled Monolithic Thermoelectric Linear Arrays to Imaging Radiometers

Volume 48 High Brightness Light Emitting Diodes

G. B. Stringfellow, Materials Issues in High-Brightness Light-Emitting Diodes
M. G. Craford, Overview of Device issues in High-Brightness Light-Emitting Diodes
F. M. Steranka, AlGaAs Red Light Emitting Diodes
C. H. Chen, S. A. Stockman, M. J. Peanasky, and C. P. Kuo, OMVPE Growth of AlGaInP for High Efficiency Visible Light-Emitting Diodes
F. A. Kish and R. M. Fletcher, AlGaInP Light-Emitting Diodes
M. W. Hodapp, Applications for High Brightness Light-Emitting Diodes
I. Akasaki and H. Amano, Organometallic Vapor Epitaxy of GaN for High Brightness Blue Light Emitting Diodes
S. Nakamura, Group III-V Nitride Based Ultraviolet-Blue-Green-Yellow Light-Emitting Diodes and Laser Diodes

Volume 49 Light Emission in Silicon: from Physics to Devices

David J. Lockwood, Light Emission in Silicon
Gerhard Abstreiter, Band Gaps and Light Emission in Si/SiGe Atomic Layer Structures
Thomas G. Brown and Dennis G. Hall, Radiative Isoelectronic Impurities in Silicon and Silicon-Germanium Alloys and Superlattices
J. Michel, L. V. C. Assali, M. T. Morse, and L. C. Kimerling, Erbium in Silicon
Yoshihiko Kanemitsu, Silicon and Germanium Nanoparticles
Philippe M. Fauchet, Porous Silicon: Photoluminescence and Electroluminescent Devices
C. Delerue, G. Allan, and M. Lannoo, Theory of Radiative and Nonradiative Processes in Silicon Nanocrystallites
Louis Brus, Silicon Polymers and Nanocrystals

Volume 50 Gallium Nitride (GaN)

J. I. Pankove and T. D. Moustakas, Introduction
S. P. DenBaars and S. Keller, Metalorganic Chemical Vapor Deposition (MOCVD) of Group III Nitrides
W. A. Bryden and T. J. Kistenmacher, Growth of Group III-A Nitrides by Reactive Sputtering
N. Newman, Thermochemistry of III-N Semiconductors
S. J. Pearton and R. J. Shul, Etching of III Nitrides
S. M. Bedair, Indium-based Nitride Compounds
A. Trampert, O. Brandt, and K. H. Ploog, Crystal Structure of Group III Nitrides

H. Morkoc, F. Hamdani, and A. Salvador, Electronic and Optical Properties of III–V Nitride based Quantum Wells and Superlattices
K. Doverspike and J. I. Pankove, Doping in the III-Nitrides
T. Suski and P. Perlin, High Pressure Studies of Defects and Impurities in Gallium Nitride
B. Monemar, Optical Properties of GaN
W. R. L. Lambrecht, Band Structure of the Group III Nitrides
N. E. Christensen and P. Perlin, Phonons and Phase Transitions in GaN
S. Nakamura, Applications of LEDs and LDs
I. Akasaki and H. Amano, Lasers
J. A. Cooper, Jr., Nonvolatile Random Access Memories in Wide Bandgap Semiconductors

Volume 51A Identification of Defects in Semiconductors

George D. Watkins, EPR and ENDOR Studies of Defects in Semiconductors
J.-M. Spaeth, Magneto-Optical and Electrical Detection of Paramagnetic Resonance in Semiconductors
T. A. Kennedy and E. R. Glaser, Magnetic Resonance of Epitaxial Layers Detected by Photoluminescence
K. H. Chow, B. Hitti, and R. F. Kiefl, μSR on Muonium in Semiconductors and Its Relation to Hydrogen
Kimmo Saarinen, Pekka Hautojärvi, and Catherine Corbel, Positron Annihilation Spectroscopy of Defects in Semiconductors
R. Jones and P. R. Briddon, The Ab Initio Cluster Method and the Dynamics of Defects in Semiconductors

Volume 51B Identification of Defects in Semiconductors

Gordon Davies, Optical Measurements of Point Defects
P. M. Mooney, Defect Identification Using Capacitance Spectroscopy
Michael Stavola, Vibrational Spectroscopy of Light Element Impurities in Semiconductors
P. Schwander, W. D. Rau, C. Kisielowski, M. Gribelyuk, and A. Ourmazd, Defect Processes in Semiconductors Studied at the Atomic Level by Transmission Electron Microscopy
Nikos D. Jager and Eicke R. Weber, Scanning Tunneling Microscopy of Defects in Semiconductors

Volume 52 SiC Materials and Devices

Kenneth Järrendahl and Robert F. Davis, Materials Properties and Characterization of SiC
V. A. Dmitriev and M. G. Spencer, SiC Fabrication Technology: Growth and Doping
V. Saxena and A. J. Steckl, Building Blocks for SiC Devices: Ohmic Contacts, Schottky Contacts, and p-n Junctions
Michael S. Shur, SiC Transistors
C. D. Brandt, R. C. Clarke, R. R. Siergiej, J. B. Casady, A. W. Morse, S. Sriram, and A. K. Agarwal, SiC for Applications in High-Power Electronics
R. J. Trew, SiC Microwave Devices
J. Edmond, H. Kong, G. Negley, M. Leonard, K. Doverspike, W. Weeks, A. Suvorov, D. Waltz, and C. Carter, Jr., SiC-Based UV Photodiodes and Light Emitting Diodes
Hadis Morkoç, Beyond Silicon Carbide! III–V Nitride-Based Heterostructures and Devices

Volume 53 Cumulative Subject and Author Index Including Tables of Contents for Volume 1–50

Volume 54 High Pressure in Semiconductor Physics I

William Paul, High Pressure in Semiconductor Physics: A Historical Overview
N. E. Christensen, Electronic Structure Calculations for Semiconductors under Pressure
R. J. Neimes and M. I. McMahon, Structural Transitions in the Group IV, III-V and II-VI Semiconductors Under Pressure
A. R. Goni and K. Syassen, Optical Properties of Semiconductors Under Pressure
Pawel Trautman, Michal Baj, and Jcek M. Baranowski, Hydrostatic Pressure and Uniaxial Stress in Investigations of the EL2 Defect in GaAs
Ming-fu Li and Peter Y. Yu, High-Pressure Study of DX Centers Using Capacitance Techniques
Tadeusz Suski, Spatial Correlations of Impurity Charges in Doped Semiconductors
Noritaka Kuroda, Pressure Effects on the Electronic Properties of Diluted Magnetic Semiconductors

Volume 55 High Pressure in Semiconductor Physics II

D. K. Maude and J. C. Portal, Parallel Transport in Low-Dimensional Semiconductor Structures
P. C. Klipstein, Tunneling Under Pressure: High-Pressure Studies of Vertical Transport in Semiconductor Heterostructures
Evangelos Anastassakis and Manuel Cardona, Phonons, Strains, and Pressure in Semiconductors
Fred H. Pollak, Effects of External Uniaxial Stress on the Optical Properties of Semiconductors and Semiconductor Microstructures
A. R. Adams, M. Silver, and J. Allam, Semiconductor Optoelectronic Devices
S. Porowski and I. Grzegory, The Application of High Nitrogen Pressure in the Physics and Technology of III-N Compounds
Mohammad Yousuf, Diamond Anvil Cells in High Pressure Studies of Semiconductors

Volume 56 Germanium Silicon

J. C. Bean, Growth Techniques and Procedures
D. E. Savage, F. Liu, V. Zielasek, and M. G. Lagally, Fundamental Crystal Growth Mechanisms
R. Hull, Misfit Strain Accommodation in SiGe Heterostructures
M. J. Shaw and M. Jaros, Fundamental Physics of Strained Layer GeSi: Quo Vadis?
F. Cerdeira, Optical Properties
S. A. Ringel and P. N. Grillot, Electronic Properties and Deep Levels in Germanium-Silicon
J. C. Campbell, Optoelectronics in Silicon and Germanium Silicon
K. Eberl, K. Brunner, and O. G. Schmidt, $Si_{1-y}C_y$ and $Si_{1-x-y}Ge_xC_y$ Alloy Layers

ISBN 0-12-752166-6